# Theoretical Chemistry and Computational Modelling

Modern Chemistry is unthinkable without the achievements of Theoretical and Computational Chemistry. As a matter of fact, these disciplines are now a mandatory tool for the molecular sciences and they will undoubtedly mark the new era that lies ahead of us. To this end, in 2005, experts from several European universities joined forces under the coordination of the Universidad Autónoma de Madrid, to launch the European Masters Course on Theoretical Chemistry and Computational Modeling (TCCM). This course is recognized by the ECTNA (European Chemistry Thematic Network Association) as a Euromaster; it has been part of the Erasmus Mundus Masters Program of the EACEA (European Education, Audiovisual and Culture Agency) since 2010. The aim of this course is to develop scientists who are able to address a wide range of problems in modern chemical, physical, and biological sciences via a combination of theoretical and computational tools. The book series, Theoretical Chemistry and Computational Modeling, has been designed by the editorial board to further facilitate the training and formation of new generations of computational and theoretical chemists.

Ángel Martín Pendás · Julia Contreras-García

# Topological Approaches to the Chemical Bond

Springer

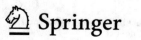

Ángel Martín Pendás
Departamento de Química Física y Analítica
Universidad de Oviedo
Oviedo, Spain

Julia Contreras-García
Laboratoire de Chimie Théorique
Sorbonne Université
Paris, France

ISSN 2214-4714 ISSN 2214-4722 (electronic)
Theoretical Chemistry and Computational Modelling
ISBN 978-3-031-13668-9 ISBN 978-3-031-13666-5 (eBook)
https://doi.org/10.1007/978-3-031-13666-5



This Springer imprint is published by the registered company Springer Nature Switzerland AG
The registered company address is: Gewerbestrasse 11, 6330 Cham, Switzerland

*Cuando no entiendas un problema,*
*disminuye su dimensión.*

*El trabajo en los tiempos del COVID19*

# Foreword

Within this series dedicated to Theoretical Chemistry and Computational Modelling (TCCM), the subject developed in this book is of special relevance with regards precisely to computational modeling, since the simulation of the real world through the use of quantum-chemical methods is unavoidably based on the mathematical item that describes that reality, that is, on the wave function. When the Schrödinger equation for any chemical system, $\widehat{H}\psi = E\psi$, is solved, the energy of the system and the wave function that describes it, are obtained simultaneously. The problem, however, is that the Schrödinger equation can only be solved exactly in very few situations, which implies that for almost the totally of the systems we wish to describe, both energy and wave function are approximate entities This means that, although in most cases when the Hamiltonian, $\widehat{H}$, in the Schrödinger equation is the exact one, there is no way of obtaining the exact wave function of the system to be described, and therefore neither the exact energy can be evaluated. But it is precisely the wave function the entity that allows us to "see", understand and interpret the properties of the system, simply because according to one of the postulates of quantum mechanics, although no physical meaning can be associated with the wave function, such a physical meaning can be ascribed to its square, admitting that the square of the wave function has a probabilistic interpretation. Thus, if one wishes to understand the properties of the particles, nuclei and electrons ensemble, that make up any chemical system, one must analyze the properties of a mathematical function of many particles, which is the square of the wave function. The consequences of this assertion are obvious, so that in quantum modeling the different ways to withdraw useful information from the square of the wave function become crucial to obtain a precise view, or at least as precise as possible, of the particles distribution and, consequently, of the properties of the system that we are trying to model, and this is precisely the objective of the book you have in your hands, to provide a timely and complete overview of the different formalisms able to get useful information from the wavefunction obtained by solving the Schrödinger equation.

Let me add that the methodological compilation that is the contents of this book it is not easily found, at present, in other publications in this area. Here, you will be able to follow rigorous deductions and explanations of the different mathematical and physical concepts behind a significantly large variety of the methodologies nowadays available, to analyze a wavefunction, specifically adapted for postgraduate students.

One can consider that the story began in the fifties of the past century, when profiting that in the framework of the Born–Oppenheimer approximation, the square of the wave function provides a probabilistic description of the distribution of the electrons, the first methods able to calculate the probability of finding electrons associated with the different nuclei in the physical space were developed. These formalisms, usually known as Population Analyses, were the first attempt to understand the rules that govern the polarity within the different chemical entities, and to formulate rules able to understand the values of multipoles (dipoles, quadrupoles, octuples…) within the realm of molecular chemistry.

However, we cannot forget that, after all, the information to be obtained is associated with a mathematical function that, as mentioned in the Introduction dwells in Hilbert, rather than in the physical space. The consequence is that the methodology to be used to obtain the information speaks a language very different from that used by chemists in their everyday activity, simply because molecules live in the physical space, but as said above the function that provides that information dwells in a Hilbert space. This is one of the virtues of this book, to lead us by the hand, without losing the necessary mathematical rigor in the description of the methodology to be used, to a deep understanding and clear comprehension of the properties of atoms, molecules, bonds, non-covalent interactions and other endless "real" world facts. From my personal viewpoint, a paradigmatic example of this virtue would be, for instance, the deduction in Chap. 2 of Morse's equality that establishes the relationship between the number of cage points, ring points, bond points and the number of nuclei for any chemical system; but there are many more all along the following chapters.

The organization of the book, on the other hand, is perfect to achieve this goal, starting with the Introduction in which, it is made clear to the reader that the analysis of the electron density allows to obtain a very rich information contained in the wavefunction. An introductory chapter on topological spaces, including scalar fields and some very interesting advanced material, such as the section devoted to topological invariants, is indeed essential to understand correctly the analysis of the electron density, which is precisely a scalar field whose topological characteristics are of a paramount importance to achieve a sound and complete view of a concept central to chemistry: the chemical bond. In this sense the section dedicated to topological invariants constitutes a novelty that is difficult to find completely developed in previous publications, which through the analysis of the critical points of a dynamical systems provides a perfect background to understand the basis of the QTAIM (Quantum Theory of Atoms in Molecules) developed in the following chapter, and that culminates with the deduction of Morse's equality I mentioned above.

Very interesting is the presentation of chemical reality as a fascinating game in which both, localization and delocalization of the electron density, play fundamental roles, and how they are connected with other pioneering ideas as, for instance, the theory of loges of R. Daudel, the link between localization criteria and the formation of electron pairs, the equivalence between hole-localization and electron-localization, the usefulness of the localization and delocalization indexes, or how these effects characterize well-known chemical concepts, among them the inductive and the mesomeric effects, or why rather old models, such as the Berlin's regions, are recovered when many of these new formalisms are applied.

Many people, among them many chemists, are compelled to think that ionic and covalent bonding are the fundamental necessary concepts to rationalize the whole chemical world, and the analysis of chemical bonding in this book is really a nice and complete piece of work, but in this book emphasis is made on the fact that nowadays is more and more evident that it is impossible to understand chemical reality ignoring the non-covalent interactions. However, non-covalent interactions always represent a challenge because, by nature, they are weak and therefore very difficult to be adequately and accurately described by theory. They are for sure essential to understand many biochemical phenomena, among other reasons because they are behind chemical recognition and molecular assembly; but more recently they were also found to play important roles in material science and crystals, as well as in atmospheric chemistry and astrochemistry. Another richness of the book you have in your hands is that it explores advanced methodology for the analysis of non-covalent interactions which cannot be found in other books on theoretical and computational chemistry. Indeed, some of the approaches described here are very recent contributions in the field. To this group belong methods such as the Localized Orbital Locator (LOL) formulated in the year 2000, the Electron Localization Indicator (ELI) proposed four years later, the Electron Pair Localization Function (EPLF) first defined in 2004, but modified in 2011 to make it analytically computable with standard wavefunctions, obtained with both DFT or ab initio methods, the Localized Electron Detector (LED) initially proposed in 2010, or the Density Overlap Regions Indicator (DORI) formulated in 2014; but the book also presents the first successful attempts to well describe and classify intermolecular interactions through the use of the Quantum Theory of Atoms in Molecules (QTAIM), the Electron Localization Function (ELF) and the Non-Covalent Interactions (NCI) formalism, in whose development, one of the authors of this book (J. C.-G.) had an important role. To complete this perspective, the authors were able to describe in detail how these formalisms change the perspective of modern chemistry using representative cases of weak interactions such as hydrogen bonds, the coordination of biomolecules around metal centers, steric repulsions or van der Waals interactions.

Although QTAIM has been, I would say, heavily used by theoretical and computational chemists along the last two decades of the last century and along the first two decades of the present one and, therefore it could be considered a "well known" tool in this domain, the presentation of such a formalism in this book is original indeed and very rigorous. Many readers perhaps are aware that QTAIM theory is related with the topological analysis of the electron density, $\rho$, of a molecular system, but perhaps not many people is conscious that $\rho$ is not a truly differentiable field, due to the cusps of this function at the position of the nuclei, which in principle should be a serious problem when trying to analyze its topology. Details like this, which are currently omitted in other analyses, make this work particularly attractive and rigorous. Beautiful are also the illustrations used to explain the link between the chemical concept of bond and the (3,−1) critical points of the density, and the way in which the molecular graphs, which are very useful components of the topological analysis of $\rho$ are built up. Very importantly, however, the authors also warn the reader about the cases in which the molecular graph does not necessarily coincide with the chemical graph, indicating also that situation may easily appear when dealing with molecular structures outside equilibrium. Along this line, it is also very interesting the distinction, outlined by the authors following the suggestions of Prof. Paul Popelier, an authority in this field, between bond path, bond interaction line and the plain chemical concept of bond. It is also a charming novelty the parallelism established, in one of the sections of the same chapter, of the flux of the electron density with the flux of an incompressible fluid. No less interesting is the discussion of relativistic effects in this formalism, or its extension to the description of the time evolution of open quantum subsystems. All this background will allow in subsequent sections to discuss concepts like momentum density an force density that, only in this book, are analyzed in deep showing that they constitute the fundamental ideas of what is nowadays known as quantum field dynamics.

Something similar can be said as far as the chapter devoted to the ELF (Electron Localization Function) is concerned, in the sense that the authors provide the reader beautiful arguments justifying the model, that are not easy to find in other presentations of the theory, playing with the intrinsic interactions between fermions and the symmetry requirements the wavefunction that represent them must fulfill, or addressing questions like 'if electrons are delocalized, why do we observe Lewis pairs?' to build a bridge between this theoretical scheme and the classical ideas of Lewis and Linnet, and the fundamental role of the spin concept. As a matter of fact, to be honest I must say that the historical introduction to this methodology is a wonderful piece of work.

Another important feature of this book concerning the topic of non-covalent interactions is the section dedicated to very recent developments aiming at providing a quantification of the formalisms normally used to analyze them. In general the different methods mentioned above, in this preface, provide different ways to look at the electron redistributions associated with the appearance of non-covalent interactions, but still it is a challenge to establish a quantitative relationship between them and the energetics of the interaction. A section of the book addresses this important question, even if this quantification remains still limited. It is also remarkable the section devoted to the use of NCI formalisms to analyze non-covalent interactions by revealing electron localization through the relationship between NCI and the kinetic energy density. The examples provided to illustrate this point are really beautiful.

The last section of the book dedicated to applications was carefully designed, trying to present a wide scope of fields in which the use of the methodology described in the previous six chapters may become an almost mandatory tool to well understand many features of the systems under scrutiny. One important added value of Chap. 7, the first chapter of this section, was the ability of the authors to include, among the different examples chosen to illustrate the potentiality of these theoretical tools to help in the understanding of the structure, bonding, reactivity and a great variety of properties of molecules, a good deal of controversial cases where other molecular descriptors are not able to provide a clear view of the aforementioned properties. On top of that the last part of the seventh chapter is not only dedicated to explain a great variety of characteristics of molecules and clusters, but also to predict the behavior of these systems, mainly in the domain of the chemical reactivity. For this purpose, again, the illustrations used were carefully and competently selected.

Chapter 8 that closes the second part of this book is like a final gift, because it addresses a topic not easily found well compiled in the literature: how to extend the topological methods described in the previous chapters to the solid state and in particular to periodic crystals, which requires a careful analysis of the relationship between topology and periodicity. This opens the door for the understanding of many of the properties of molecular crystals through the analysis of the intermolecular behavior in these crystals. The new viewpoints provided by these topological approaches imply to extend, as a novelty, many of the ideas used in the chemical domain to the understanding of properties and bonding in the solid state, but not restricted only to periodic domains but also extended to the changes that take place along phase transitions.

The Part III of the book was to me a wonderful surprise. In general, in most of textbooks in science, the section devoted to exercises is just a compilation of well selected numerical problems or conceptual questions related with the topic developed in the book. In this particular case, Chap. 10 is an extremely important part of this book, where the reader can find a description not usually found in text books, the way of connecting the scientific concepts of the different formalisms to the numerical results that allow to interpret them. This requires to built up efficient algorithms which are summarized along this chapter, paying particular attention to the location of critical points and to basin integration, including implicit, explicit, grid and adaptive algorithms, even though basin integration has not yet reached an optimal handling. The last section of this excellent chapter provides an extremely useful overview of most of the codes, freely available, to analyze the topological features of a variety of scalar fields.

This work closes with a last chapter devoted to Exercises; but again this is not a conventional chapter in which different questions or numerical problems are proposed to the reader, as in most textbooks. Here the authors guide the reader along the different steps of the process, starting with how to obtain the wavefunction, and continuing with the route to be followed to efficiently use the different codes, always wonderfully illustrated, in the last step with exercises aiming at getting the reader familiar with the characteristics of each formalism and code.

My final words will be to assure the reader that the excursion along the pages of this book is going to be a wonderful experience.

June 2020

Manuel Yáñez
Emeritus Professor, Department
of Chemistry
Universidad Autónoma de Madrid
Madrid, Spain

# Preface

When I started my Ph.D., I was confronted to a topic that had no "academic" book to start. I was given many specialized "papers" and had to find my way through.

This book is intended to provide a thorough (even if not complete) background on topological analysis in real space of electronic structure for students at the master level.

This book expects to make their life easier. This said, it is based on the TCCM lectures we have been giving for several years now, so it does not intend to be complete, and we apologize for all the things that have been left aside !

Paris, France
Oviedo, Spain
Mayo 2020

Julia Contreras-García
Ángel Martín Pendás

# Acknowledgements

(This scene happens in a Parisian appartment during confinement)

**Julia Contreras-García:** Many students have directly or indirectly contributed to this book. More specifically, a good part of the artwork present in this monograph (in most cases, the most visually appealing one) has been possible thanks to them. Thank you so much for improving my natural (in)capacity to create nice figures. Julen Munarriz and Roberto A Boto have had a great contribution, but I also want to thank Ruben Laplaza and Francesca Peccati for their contribution to Chap. 5.

I want to acknowledge them for their work, and for enlightening my daily life. They have really given me a great working environment.

This book started long time ago, as part of my Ph.D. thesis. Hence, I have to thank my family.Whoever has written a thesis, knows that your dearest ones have to put up with a thesis, but in this case, my father was the most insisting person on this book being written. Bernard, for insisting that I should write this book, and Manuel for crystallizing it. Ángel for making it so easy to work together. And again my father who kept putting pressure on me all the way through! And of course, thank so much to Eduard for his thorough proof-reading! And since this book was mainly written during confinement, I also want to recall the Sopeñas for the time spent together in confinement. But above all, to my family who are a proof of unconditional support, confinement being just any other example.

(Simultaneously in the North of Spain...)

**Ángel Martín Pendás:** Co-writing a book seems a hard enterprise at first. In this case it has been a piece of cake, for Julia and I had already discussed many times about how the main topic of this monograph should be taught to students. A previous set of handouts written years ago in Spanish to help Master students were also an invaluable help. I thank all those students who convinced me that those notes were helpful. Without them, and without Manuel's insistence, this initiative would have never come to light. I also want to thank Miguel Gallegos for his help with figures.

Finally, we want to thank Bruno Landeros and Fernando Jiménez Grávalos for making sure that the exercises where doable!

# Contents

# Acronyms

| | |
|---|---|
| 3D | Three Dimensional |
| AB | Attraction Basin |
| AELF | Approximate Electron Localization Function |
| a.u. | atomic units |
| BCC | Bonded Charge Concentration |
| BCM | Bond Charge Model |
| BCP | Bond Critical Point |
| bip | Bond Interaction Point |
| CAS | Complete Active Space |
| CCP | Cage Critical Point |
| CD | Cumulant Density |
| CP | Critical Point |
| CS | Charge-Shift |
| CT | Charge Transfer |
| CVB | Core-Valence Bifurcation |
| DAFH | Domain Averaged Fermi Hole |
| DFT | Density Functional Theory |
| DI | Delocalization Index |
| DNA | Deoxyribonucleic Acid |
| DNO | Domain Natural Orbital |
| DORI | Density Overlap Regions Indicator |
| DS | Dynamical System |
| *e* | electron(s) |
| EDA | Energy Decomposition Analysis |
| ELF | Electron Localization Function |
| ELI | Electron Localizability Indicator |
| ELI-D | Fixed-pair Electron Localizability Indicator |
| ELI-q | Fixed-charge Electron Localizability Indicator |
| EMLF | Electron Momentum Localization Function |
| EPLF | Electron Pair Localization Function |
| GGA | Generalized Gradient Approximation |

| | |
|---|---|
| HB | Hydrogen Bond |
| HF | Hartree–Fock |
| IAM | Independent Atom Model |
| IAS | Interatomic Surface |
| IP | (1st) Ionization Potential |
| IQA | Interacting Quantum Atoms |
| IRC | Intrinsic Reaction Coordinate |
| KS | Kohn–Sham |
| LED | Localized Electron Detector |
| LI | Localization Index |
| LOCCC | Ligand Opposed Core Charge Concentration |
| LOL | Localized Orbital Locator |
| LS | Linearized System |
| LSDA | Local Spin Density Approximation |
| NCC | Non-bonded Charge Concentration |
| nCD | $n$-th order Cumulant Density |
| NCI | Non-Covalent Interaction |
| NR | Newton–Raphson |
| nRD | $n$-th order Reduced Density |
| nRDM | $n$-th order Reduced Density Matrix |
| OCM | Outer Core Maxima |
| ODE | Ordinary Differential Equation |
| QM | Quantum Mechanics |
| QT | Quantum Theory |
| QTAIM | Quantum Theory of Atoms in Molecules |
| RB | Repulsion Basin |
| RCP | Ring Critical Point |
| RD | Reduced Density |
| RDG | Reduced Density Gradient |
| RDM | Reduced Density Matrix |
| RE | Reference Electron |
| RGB | Red, Green, Blue |
| SAM | Self-assembled Monolayer |
| SAPT | Symmetry Adapted Perturbation Theory |
| SCF | Self-consistent Field |
| SOAO | Single Occupied Atomic Orbital |
| STO | Slater Type Orbital |
| TDELF | Time Dependent Localization Function |
| vdW | van der Waals |
| VSCC | Valence Shell Charge Concentration |
| VSEPR | Valence Shell Electron Pair Repulsion |
| ZPVE | Zero-point Vibrational Energy |

# Chapter 1
# Introduction

**Abstract** Chemical bonding, at the core of Chemistry as a Science, did only acquire a physically sound status after the development of Quantum Mechanics. In the hundred years passed since the first application of Quantum Theory to chemical problems, the molecular orbital paradigm has become a de facto standard to analyze the chemical bond, although it is manifestly reference dependent. Thanks to Hohenberg and Kohn, exploring the behavior of the electron density offers an independent approach to chemical bonding. This road has led to the topological theories of the chemical bond.

## 1.1 Quantum Theory and the Chemical Bond

In these first two decades of the new century we are celebrating a large number of exceptional anniversaries in the physical and chemical sciences. Just to name a few, two hundred years separate us from Dalton's atomic theory (1805–1806) [3], and 2016 marks the centennial of the cubical atom and the electron pair as proposed by Gilbert Newton Lewis [6]. After immersing ourselves in the zeitgeist of this new world, a large part of the program imagined by Paul Dirac, that was actually started by Heitler, London, Slater and Mulliken, among many others, may be considered complete. The new scenario opened by the availability of automatic computing methods after the seminal advances of von Neumann and Turing on the one hand, and the development of solid state transistors by Bardeen, Shockley and Brattain on the other, allows us to obtain basically exact solutions of Schrödinger's equation for systems with a number of electrons close to the first power of ten, $N \simeq 10$. Interestingly, already in 1936, J. H. Van Vleck [10], co-awarded the 1977 Nobel prize in physics, foretold the severe problems envisaged in the nearbies of $N \simeq 1000$ in what he called the wavefunction information problem, or Van Vleck's catastrophe. His disciple Walter Kohn recalled his words in the 1998 Nobel prize in physics lecture, a must for every practicing computational chemist, declaring: "*In general the many-electron wave function for a system of N electrons is not a legitimate scientific concept, when $N > N_0$, where $N_0 \simeq 10^3$*. I will use two criteria for defining "legitimacy": (a). That $\Psi$ can be calculated with sufficient accuracy and (b) that $\Psi$



Á. Martín Pendás and J. Contreras-García, *Topological Approaches to the Chemical Bond*, Theoretical Chemistry and Computational Modelling,
https://doi.org/10.1007/978-3-031-13666-5_1

can be recorded with sufficient accuracy". His arguments are easily understood by examining the vast amount of information contained in the wavefunction of a system about that threshold size. For instance, if we only needed to store 1 bit per variable, the size of the Hilbert space for $n \simeq 1000$ would approach $10^{1200}$ bits, a number to be gauged upon the estimated number of barions in the accessible universe ($\simeq 10^{80}$).

Fortunately, entanglement phenomena, where the full size of Hilbert space is indeed explored, do not play a dominant role for almost all the relevant applications of Quantum Mechanics to Chemistry. This is the reason why most of the vast amount of information that lies in the state vector of a system is irrelevant at the chemical level. Even more provocatively, we may say that Quantum Theory (QT) does not speak the language of Chemistry. Actually, the basic concepts used by chemists—atoms, functional groups, bonds—lie still today much closer to Dalton's paradigm than to Quantum Physics. However much we dress the language of chemistry with orbital brushstrokes or with rules based on the conservation of several symmetries that only exist in QT, chemists keep on visualizing reactive processes through entities embodied with particular properties, entities that are transferred from one place to another, that accept or donate electrons or more complex atomic groups, all of them inhabiting the physical space. This primordial hypothesis, the existence of separable and transferable objects in $\mathbb{R}^3$ that make up chemistry, has turned out to be very difficult to extract in a rigorous manner from the properties of quantum state vectors, that dwell in Hilbert, not in the physical space. As Paul Popelier pointed out with the concept of farsightedness [9], we face one of the various epistemological problems that shocked science in the twentieth century. Many of these problems appear when trying to marry the emergent phenomena observed at higher levels of complexity with the basic theories that govern them at lower levels. In this sense, Chemistry possesses a personal structure and language, emerged from the systematization of a large number of observations recorded at a time when Physics was a discipline clearly outside its scope. Attempts to put both paradigms closer together have been made from both the Chemistry and the Physics ends.

On the one hand, Chemistry has immersed itself in QT, and the latter theory has completely imbued the standard chemical reasoning. At a cost. As more and more precise techniques to solve Schrödinger's equation were developed, extracting chemical information from the resulting state vectors became more and more dependent on simplified models. We arrive at the paradoxical conclusion that *the more accurate the calculations become, the more concepts tend to vanish into thin air*. Notice that this assertion was already made by R. S. Mulliken in 1965 [8], not merely a couple of years ago. In the historical evolution of chemistry we have come to an interesting situation as the basic concept of modern Chemistry, the chemical bond, is regarded: being, in principle, fully contained in the frame of QT, it is *interpreted* using a model to solve the exact quantum mechanical solution (the orbital model) together with a usually poor approximation (that of considering orbitals as linear combination of a small number of fixed basis functions or atomic orbitals) within the model.

At the other end, Theoretical Chemistry has embarked in a long-term project centered on creating a general framework, compatible with QT, in which the standard chemical concepts fit naturally. To that end it is necessary to compact the information stored in the wave vector in the physical space. These tow ideas guide us immediately towards the study of reduced densities (RDs) or reduced density matrices (RDMs). Actually, a key in this search process comes from the foundational theorems of Pierre Hohenberg and Walter Kohn [4], together with the implementation of a new paradigm in quantum physics which is now known as Density Functional Theory (DFT), after the work by Kohn and Sham [5]. Hohenberg and Kohn first theorem assures that the diagonal elements of the first order density matrix, the electron density ($\rho$), determine the molecular Hamiltonian, the position of the nuclei, and the state vector of the system. In this way, the study of $\rho$, be it local or non-local, should allow us to recover all the information contained in $\Psi$. More recently, Paul Mezey has demonstrated an even stronger version of this result. His holographic theorem [7] makes it clear that it suffices to know the density at a given point and its close neighborhood to reconstruct the full density and, with it, $\Psi$.

During the course of history, and while the orbital paradigm started to become chemists' favorite, probably thanks to the outstanding agreement between computed and experimentally determined molecular properties, some voices started to remind the community that the orbitals and monoelectronic functions that were pervading chemical thinking were actually arbitrary. Those voices fought to recover the true quantum mechanical observables as the basic objects that should be used to build a truly objective theory of the chemical bond. During the last 1960s, Bader, Henneker and Cade [1] started a program of analysis of the properties of the electron density in molecular systems, convinced that it was the distribution of electrons in space, and not in orbital space, the key to incorporate QT to the normal chemical language.

## 1.2 Analyzing Scalar Fields

Obtaining chemically relevant information from the $\rho$ scalar field was for a long time a problem without a unanimously accepted solution. A first tip that paves the way out comes from the immediate relationship that exists between the electron density and the forces exerted or felt by the nuclei in a molecule. The Hellmann–Feynman theorem, in its electrostatic version, establishes that

$$\mathbf{F}_\alpha = -\nabla_{\mathbf{R}_\alpha}\left\{-Z_\alpha \int d\mathbf{r}\frac{\rho(\mathbf{r})}{|\mathbf{r}-\mathbf{R}_\alpha|} + \sum_{\beta\neq\alpha}\frac{Z_\alpha Z_\beta}{|\mathbf{R}_\alpha-\mathbf{R}_\beta|}\right\}, \tag{1.1}$$

where $\boldsymbol{F}_\alpha$ and $\boldsymbol{R}_\alpha$ stand for the force felt by nucleus $\alpha$ and its corresponding position vector, respectively. In diatomic molecules, $AB$, in cylindrically symmetric states, it is easy to prove that the projection on the internuclear axis ($z$) of the force that a unit test-charge exerts over both nuclei is equal to

$$f_z = Z_A\frac{\cos\theta_a}{r_a^2} - Z_B\frac{\cos\theta_b}{r_b^2}, \tag{1.2}$$

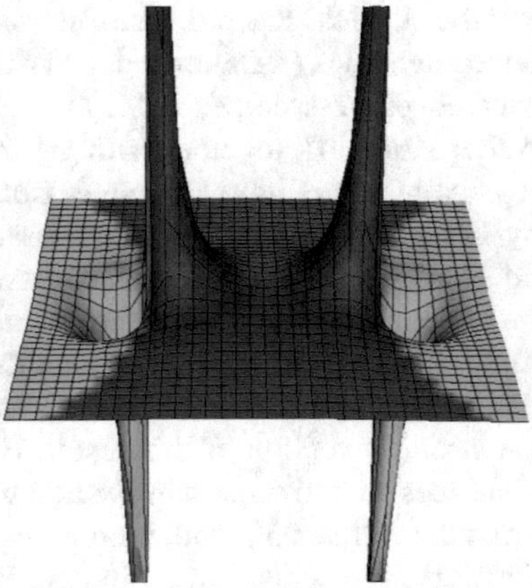

**Fig. 1.1** Projection onto the $xz$ plane of the $f_z$ function for a homodiatomic molecule. The two colors refer to the bonding ($f_z > 0$, red) and antibonding ($f_z < 0$, green) regions, respectively

where $r_a$ and $r_b$ are the module of the vectors that join the test charge with the nuclei, and $\theta_a$ and $\theta_b$ measure the angles that those vectors form with $z$. The geometrical locus of the points where this force vanishes is a hyperbolic revolution surface passing through the nuclei. This surface divides the space into *bonding* regions, in which an electron contributes to an attractive force upon the nuclei, and *antibonding* regions, in which the contrary occurs. Figure 1.1 offers a projection of $f_z$ onto a plane containing the two nuclei of a homodiatomic molecule. The bonding region lies between the nuclei (in red), while the two antibonding ones appear on their rear part (in green). This type of analysis, originally due to Berlin [2], correctly explains the coarser aspects of the charge redistribution that accompanies the formation of chemical bonds: electrons are pumped from the rear nuclear regions into the internuclear ones. Notice that it is in the vicinity of the nuclei, and not exactly in the midpoint along the internuclear axis, where the charge density accumulation has a more intense bonding character.

These early insights show that a chemical bonding based analysis of the electron density should be based on the identification of the electron redistribution following chemical processes. The simplest, most direct way to measure those deviations is based on the construction of density difference maps. Namely, if we wish to compare two initial and final distributions, $\rho_i$ and $\rho_f$, we construct $\Delta\rho = \rho_f - \rho_i$ and we isolate the regions where charge is accumulated and/or depleted. These maps were very popular at the end of the 1960s and during the 1970s, and are still used today (e.g. they are very common in crystallography). However, they soon faced a basic difficulty: the appropriate choice of a reference density. Let us imagine the extremely simple case of an homodiatomic molecule. What is the appropriate initial or reference density, $\rho_i$ with respect to which we compute the difference map? A straightforward answer is to choose the promolecular density, the sum of the *in vacuo* atomic densities. Figure 1.2 contains the difference maps for the $H_2$ and $N_2$ molecules, computed under the HF/6-311G(p,d) approximation, at the respective theoretical equilibrium

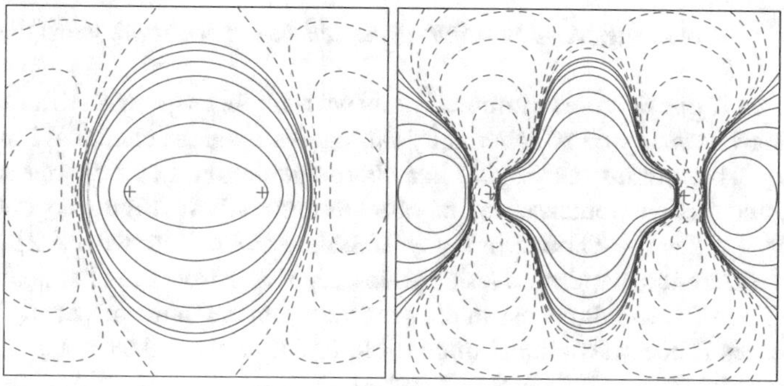

**Fig. 1.2** Isolines map of $\Delta\rho$ for the $H_2$ (left) and $N_2$ (right) molecules in a plane that contains the nuclei. HF/6-311G(p,d) calculations at the theoretical equilibrium distances. Isolines in blue show charge accumulation regions, while those in red correspond to charge depletion zones. The position of the nuclei is marked with red crosses

distances. The reference density is the sum of the HF/6-311G densities for the H and N atoms in their $^2$S and $^4$S states, respectively.

As we can see, in both cases the regions where charge is accumulated or depleted correspond, grossly speaking, with Berlin ones. Important difficulties arise, however, if we try to generalize this procedure. The first stands out when the atomic densities are not spherically symmetric. If this is the case, how do we orient the atoms? There is no single answer to this question. A well-established possibility consists in using spherically averaged atomic densities (see Sect. 3.1). Another procedure insists on generating appropriate atomic *valence states* from them. For example, Fig. 1.3 shows HF/6-311G(p,d) $\Delta\rho$ maps for the $F_2$ molecule obtained using two different reference atomic densities. On the left, the spherically average density of the $^2$P state of the F atom was used, whereas two fluorine atoms in their $^2$P states oriented such that their

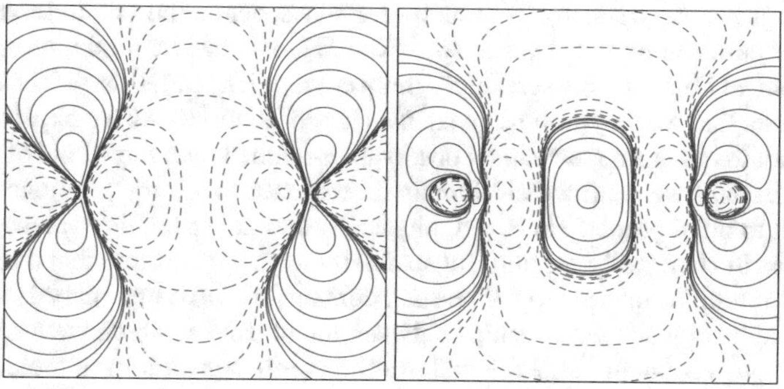

**Fig. 1.3** $\Delta\rho$ isolines map for $F_2$ in a plane containing the nuclei from HF/6-311G(p,d) calculations at the theoretical equilibrium distance. The atomic reference is made up of spherically averaged $^2$P fluorines (left) or oriented F atoms with $2p_z$ SOAO'S (right)

$2p_z$ orbitals become the SOAO (*single occupied atomic orbital*) were used on the right.

In contrast to our previous examples, now two references give rise to almost opposite bonding views. If we use spherically averaged atoms, the chemical image is not satisfactory. The internuclear region is a charge depletion, not a charge accumulation zone, and this phenomenon can be observed not only in difluorine, but in many other cases, e.g. the O—O bond in $H_2O_2$. If, on the other hand, we use $2p_z$ oriented atoms, the difference map improves considerably with respect to chemical wisdom, but the feeling of non-uniqueness in the procedure lingers in the air. In slightly more complex cases it becomes almost impossible to define the valence state and/or the appropriate orientation of the atoms involved.

The second difficulty that difference maps face is more subtle and theoretically deep. Promolecular densities violate Pauli's principle, not being $N$-representable under the molecular potential. In other words, the promolecular density is not the solution of any quantum mechanical problem related to the system under scrutiny. This means that the two densities we subtract in a deformation map come from incompatible models. To conclude, we should avoid to make our analyses of molecular electron densities rest on quantities with such weak and subjective theoretical grounds.

## 1.3 Topological Analyses

It is quite clear from the above reasoning that if we aim to avoid the problems posed by using references, any sensible procedure to analyze the electron density should be based on studying the density itself. Let us briefly enumerate the properties we would demand for such a method. On one hand, it should be able to provide quantities measuring the accumulation or depletion of charge autonomously, without recourse to external (reference) densities. If, besides this, we also wish to recover the atoms and functional groups of chemistry from such an analysis, our method should contemplate a means of partitioning the physical space into regions that we might associate with those chemical objects. This seemingly innocent requisite hides one of the strongest requirements that we may impose to such a class of theories. As we expect to show, many standard quantum mechanical observables defined over the whole of $\mathbb{R}^3$ cease to be hermitian if we restrict their domain of existence to arbitrary regions of space. If we want to build properly behaved, physically sound expectation values for these observables in chemically meaningful spatial regions, we will find that there appear stringent limitations on how to define such domains. Finally, our analysis should be able to establish binary relationships between the objects in which we have divided the space: our chemical language endows the atoms constituting a molecule with such kind of relations, that we usually call the *graph* of chemical bonds.

The first of the conditions that we need to impose, the absence of external references, seems to have a unique solution. Paraphrasing Paul Mezey, if we wish to analyze the behaviour of a scalar function, we can only turn to the values of the function itself together with those of its derivatives. Just like in the construction of density functionals, the higher the order of the latter that we include (admitting convergence of the Taylor series of our aim function) the greater the degree of non-locality that we will allow in our quantities. Provided that the behaviour of the derivatives of a scalar field is intimately linked to that of its gradient field, during the last 1970s and early 1980s a large collective effort was put on integrating all these elements into a well-structured method that is known today as the topological approach. It naturally integrates the three conundrums that we have just introduced: it analyzes the density without any external reference, it provides an exhaustive partition of the space into well-defined regions, and it easily establishes binary chemical relationships among them.

The topological approach, whose development has involved a large number of researchers, was pioneered by Richard F. W. Bader, from McMaster University in Canada. It is a greatly intuitive and easy to learn theory, which may be applied to a large number of scalar fields, many of them related to $\rho$ and its derivatives, but not only (the Laplacian of the density, the electron localization function ELF, the molecular electrostatic potential, the momentum density, etc). It has generated an impressive number of images in the 3D space, much closer to chemical intuition than any other coming from the orbital paradigm. It is now considered a part on its own right of many chemistry curricula.

In the coming chapter we will briefly examine some of the mathematical apparatus that will be needed to walk at ease the rest of our journey.

## References


1. Bader, R.F.W., Henneker, W.H., Cade, P.E.: Molecular charge distributions and chemical binding. J. Chem. Phys. **46**, 3341 (1967)
2. Berlin, T.: Binding regions in diatomic molecules. J. Chem. Phys. **19**, 208 (1950)
3. Dalton, J.: New System of Chemical Philosophy, vol. 1. Bickerstaff, London (1808)
4. Hohenberg, P., Kohn, W.: Inhomogeneous electron gas. Phys. Rev. B **136**, 864 (1964)
5. Kohn, W., Sham, L.J.: Self-consistent equations including exchange and correlation effects. Phys. Rev. **140**, A1133 (1965)
6. Lewis, G.N.: The atom and the molecule. J. Am. Chem. Soc. **38**, 762 (1916)
7. Mezey, P.G.: The holographic electron density theorem and quantum similarity measures. Mol. Phys. **96**, 169 (1999)
8. Mulliken, R.S.: Molecular scientists and molecular science: some reminiscences. J. Chem. Phys. **43**, 1849 (1965)
9. Popelier, P.L.A.: Quantum chemical topology: on bonds and potentials. In: Wales, D.J. (ed.) Structure and Bonding, vol. 115, pp. 1–56. Springer, Berlin (2005)
10. Vleck, J.H.V.: Nonorthogonality and ferromagnetism. Phys. Rev. **49**, 232 (1932)

# Chapter 2
# Topological Spaces

**Abstract** We will devote this chapter to present a minimal set of mathematical concepts in Topology and the theory of Dynamical Systems. Many of the ideas that will be introduced are well known from introductory courses in Mathematical Analysis (i.e. critical points of many-variables functions and their characterization by means of the Hessian matrix), while others (i.e. the axiomatic definition of a topological space) will be probably less familiar to the audience. All of them are simple and intuitive, offering a unified vision of some of the concepts that will plague our future discussions.

## 2.1 Topological Spaces

The concept of topological space is built around the need to generalize the notion of continuity (e.g. that of a continuous function) to sets of general objects. It is through this idea that we may further introduce relations among the elements of an abstract space, getting rid of the need of introducing any kind of metric.

A set of objects together with a collection of subsets, is a topological space whenever the following three conditions are satisfied. Let $\mathbf{X}$ be a set and $\mathcal{U}$ a collection of subsets of $\mathbf{X}$ satisfying:

- $\emptyset \in \mathcal{U}, \mathbf{X} \in \mathcal{U}$,
- the intersection of any two members of $\mathcal{U}$ belongs to $\mathcal{U}$,
- The union of any number of members of $\mathcal{U}$ belongs to $\mathcal{U}$.

Any collection $\mathcal{U}$ that satisfies the above conditions is called a *topology* in $\mathbf{X}$. The set $\mathbf{X}$, together with the collection of $\mathcal{U}$ subsets, define a *topological space* denoted as $(\mathbf{X}, \mathcal{U})$. It is customary nonetheless to simplify the above notation and simply use $\mathbf{X}$. The members of $U \in \mathcal{U}$ are called *open sets* of the topological space $\mathbf{X}$. From the third of the above conditions it turns out that if we call $\mathcal{S}(\mathcal{U})$ to the set of all subsets of $\mathbf{X}$, a topology in $\mathbf{X}$ is no more than a choice of $\mathcal{U} \subseteq \mathcal{S}(\mathcal{U})$ that satisfies the previous conditions. Different choices give rise to different topologies in $\mathbf{X}$. Several examples are shown in Fig. 2.1. Note that the examples crossed out lack the empty and the complete set, which means they do not qualify as topological spaces.

 
Á. Martín Pendás and J. Contreras-García, *Topological Approaches to the Chemical Bond*, Theoretical Chemistry and Computational Modelling,
https://doi.org/10.1007/978-3-031-13666-5_2

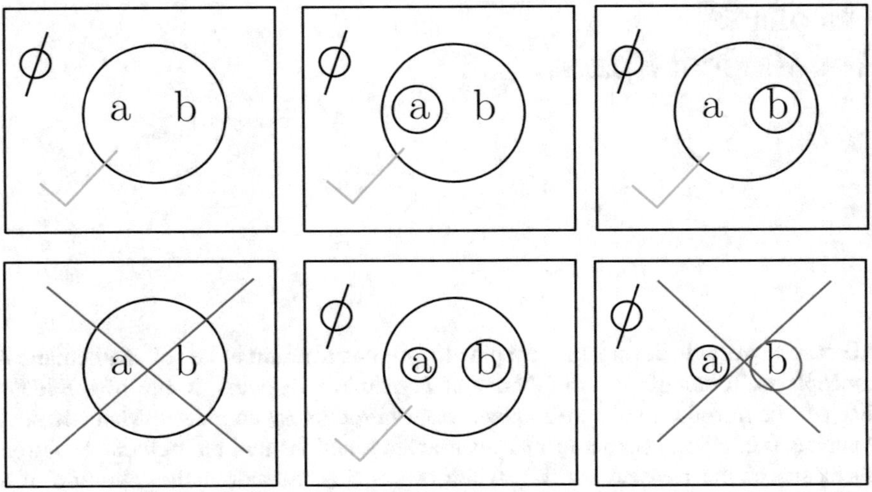


**Fig. 2.1** An example of topological space, as described in the text. The left- and rightmost examples in the bottom row lack the empty and the complete set, respectively, and do not qualify as topological spaces

Let us now consider two extreme cases of classes of subsets of a discrete set **X** complying with the conditions such that **X** is a topological space. The first of them consists of taking $\mathcal{U} = \{\emptyset, \mathbf{X}\}$, i.e. the empty set and the complete set. It is clear that this choice satisfies the axioms of a topological space. The topology we have just introduced, that can be defined for any space **X** is called the *indiscrete* or *coarse* topology of **X**. On the contrary, if we choose $\mathcal{U}$ as the set $\mathcal{S}(\mathcal{U})$ of all the possible subsets of **X**, the topology is now called the *discrete topology* of **X**.

We will illustrate these concepts with a simple example. If **X** contains two objects $\{a, b\}$, we can endow **X** with four different topologies:

$$\mathcal{U}_1 = \{\emptyset, \mathbf{X}\}\,; \qquad \mathcal{U}_2 = \{\emptyset, \{a\}, \mathbf{X}\}\,;$$
$$\mathcal{U}_3 = \{\emptyset, \{b\}, \mathbf{X}\}\,; \; \mathcal{U}_4 = \{\emptyset, \{a\}, \{b\}, \mathbf{X}\}\,.$$

Notice that $\mathcal{U}_1$ y $\mathcal{U}_4$ correspond to the coarse and discrete topologies, respectively. Let us actively show that $\mathcal{U}_2$ is a topology in **X**, and by analogy, that the same applies to $\mathcal{U}_3$: it contains the empty and complete sets $\emptyset$, and **X**, satisfying the first condition. Similarly, the intersection of $\emptyset$ with $\{a\}$ or with **X** gives again the empty set, and the intersection of $\{a\}$ with **X** is $\{a\}$: all possible intersections thus belong to $\mathcal{U}_2$. Last, but not least, we can verify that the union of the empty set with $\{a\}$ is $\{a\}$, and the union of any of these two subsets with **X** is **X** itself, so all unions belong to $\mathcal{U}_2$. In this way the three conditions are satisfied, and we have proven that both $\mathcal{U}_2$ and $\mathcal{U}_3$ are topologies. We can then say for example for $\mathcal{U}_2$ that $\emptyset$ and $\{a\}$ are the open subsets of the topological space of $\mathcal{U}_2$.

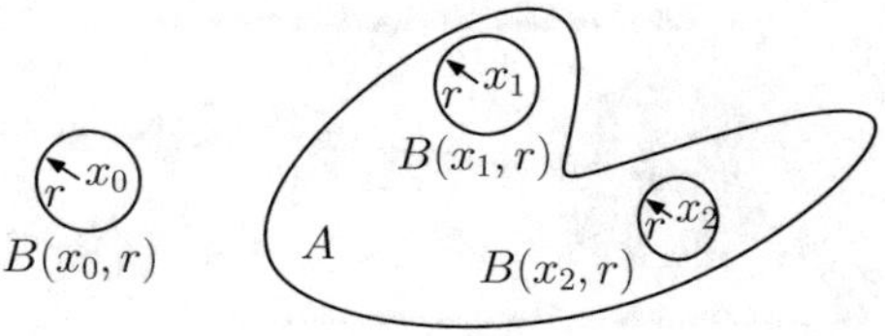


**Fig. 2.2** Metric topology in $\mathbb{R}^2$. $B$ is an open ball, and $A$ a general open set obtained as union of open balls

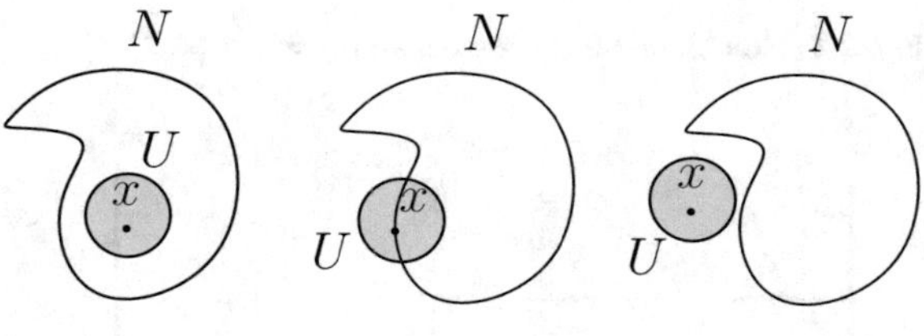


**Fig. 2.3** Neighborhoods. $N$ is a neighborhood of $x$ only in the leftmost case

It is important to note that if a metric is specified for a given space, then the use of distances allows us to induce immediately a topology in that space. That is called the metric topology. The open sets are all the subsets that can be defined as union of open balls, $B(x_0, r) = \{x,\ ||x - x_0|| < r\}$. This is illustrated in Fig. 2.2. On the left we can see an open ball around point $x_0$. On the right we can see how an open set can be easily constructed from several open balls. If this is done with the Euclidean metric to define neighborhoods in $\mathbb{R}^n$, we obtain the so-called standard topology. With our focus set in future discussions, notice that any topology of a set carries a partition of the set into subsets. In this way, a topology of the physical space will partition it into regions.

We can now use the above ideas to generalize the concept of neighborhood. Let **X** be a topological space. We say that a subset N $\subseteq$ X is a *neighborhood* of x $\in$ N if there exists an open set U such that x $\in$ U $\subseteq$ N. In particular, every open set is a neighborhood of its points. In a more general way, any set A with non-empty interior is a neighborhood of each of the points in the interior of A. Overall, this is easily visualized in Fig. 2.3: $N$ is a neighborhood of $x$ in the leftmost case, but not in the other two.

Let $\phi : \mathbf{X} \to \mathbf{Y}$ be a map between two topological spaces. We say that $f$ is *continuous* if the inverse map $\phi^{-1}(U)$ of every open set U of **Y** is an open set in **X**. If we also demand that $\phi$ be one to one and that its inverse be also continuous, we say that such a map is a *homeomorphism* between the topological spaces **X** and **Y**. This is illustrated in Fig. 2.4: it is easy to see how the space on the left (curved) can be deformed into the flat space on the right. All the points on the curved space have their equivalent on the flat plane. Overall, these concepts allow to analyze space and its continuous transformations. For instance, an orange is a body with no holes. We can easily imagine a continuous transformation from an orange shape to a dish shape (just by smashing a sphere of Play-Doh©). This means they are homeomorphic. A donut has as many holes as a pottery mug. Hence, they are homeomorphic: we can imagine a continuous transformation from one into the other. But the orange cannot be transformed into a donut: they belong to different topological classes.

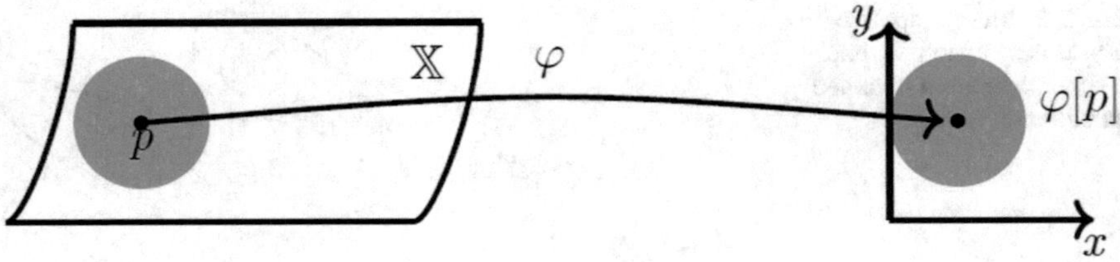


**Fig. 2.4** Homomorphism $\phi$ onto an open subset of $\mathbb{R}^2$

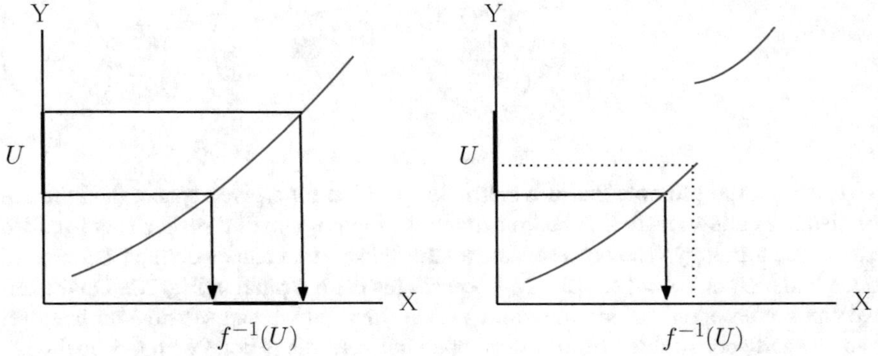


**Fig. 2.5** Continuity. The inverse map of an open set in **Y** is only open in the leftmost case

It is easy to show that the above definitions of neighborhood and continuity coincide with the more traditional ones if we reintroduce a metric. Then we can also show that the Euclidean continuity and the topological continuity are equivalent.

Let us now introduce *closed sets*. A subset C of a topological space **X** is said to be closed if and only if $\mathbf{X} - C$ is open. This is easily understood by analyzing the Euclidean space $\mathbb{R}^3$ endowed with the standard topology. A closed ball is there defined as $\bar{B}(x_0, r) = \{x \quad \text{such that} \quad ||x - x_0|| \leq r\}$, and thus contains also its boundary points. $\mathbb{R}^3 - \bar{B}$ is thus open, not containing the boundary points of $\bar{B}$. The open and closed sets here defined thus coincide with the usual ones. It is also easy to show that a map $f : \mathbf{X} \to \mathbf{Y}$ between two topological spaces is continuous if, and only if, for all closed subset C in **Y**, $f^{-1}(C)$ is closed. It can be verified that this definition of continuity is equivalent to the previous one. Figure 2.5 shows how this definition agrees with the well known concept of continuity for a 1D curve: whereas the set in the closed set $U$ chosen in the y axis leads under the transformation by $f^{-1}$ to a closed set on the left picture, but to an open one on the right panel.

## 2.2 Dynamical Systems (DS)

Out of the large number of topologies that we can define to partition space, there is one that is extremely intuitive. When looking at a mountain range, we involuntarily divide it into regions associated to its peaks. In mathematical terms, this partitioning is based on the gradient of the height field, i.e. it is induced by a gradient dynamical system.

We call a dynamical system DS to any vector field $\mathbf{y}$ defined over a $n$–dimensional manifold $M$. We understand here a manifold as a topological space in which each point has a neighborhood homeomorphic to the $n$-dimensional Euclidean space. If the field is differentiable, then the system of differential equations $d\mathbf{r}/dt = \mathbf{y}$ univocally defines the so-called trajectories $\mathbf{r}(t)$ of the DS. We will recall here just a few features of gradient dynamical systems in $\mathbb{R}^3$.

Let us consider a *scalar function* $\rho$.

$$\begin{aligned} \mathbb{R}^3 &\rightarrow \mathbb{R} \\ \boldsymbol{r} &\rightarrow \rho(\boldsymbol{r}), \end{aligned} \tag{2.1}$$

that we will call *potential function* in the following.

Given $\rho$, we can associate a DS system to it via its gradient field $\rho$ [2], the $\nabla\rho = \boldsymbol{f}$ vector field defined by the action of the gradient operator over the scalar field:

$$\nabla\rho = \boldsymbol{\imath}\,\frac{\partial\rho}{\partial x} + \boldsymbol{\jmath}\,\frac{\partial\rho}{\partial y} + \boldsymbol{k}\,\frac{\partial\rho}{\partial z} = \rho_x\boldsymbol{\imath} + \rho_y\boldsymbol{\jmath} + \rho_z\boldsymbol{k}. \tag{2.2}$$

The trajectories of this DS become defined by the system of ordinary differential equations $\dot{\boldsymbol{r}} = \boldsymbol{f} = \nabla\rho$, whose solution can be written in terms of the following parameterized curves in $\mathbb{R}^3$:

$$\boldsymbol{r}(t) = \boldsymbol{r}(t_0) + \int_{t_0}^{t} \nabla\rho(\boldsymbol{r}(t))dt. \tag{2.3}$$

These curves are also known as flux, force, field, or gradient lines. We will use almost all of these writings indistinctly. Some well-known properties of flux lines:

1. There exists one and only one trajectory of $\nabla\rho$ passing through a given point of the space. Equivalently, field lines do not cross each other. The only exception to this rule occurs at the so-called special points of the field.
2. At each point $\boldsymbol{r}$ the $\nabla\rho(\boldsymbol{r})$ vector is tangent to the field line passing through this point.
3. As the gradient field always points toward the direction where the scalar field $\rho$ increases as fast as possible, the trajectories of $\nabla\rho$ are *orthogonal to the isoscalar lines*.
4. *Each field trajectory must start or end either at a point characterized by* $\nabla\rho(\boldsymbol{r}) = \mathbf{0}$, or at infinity.

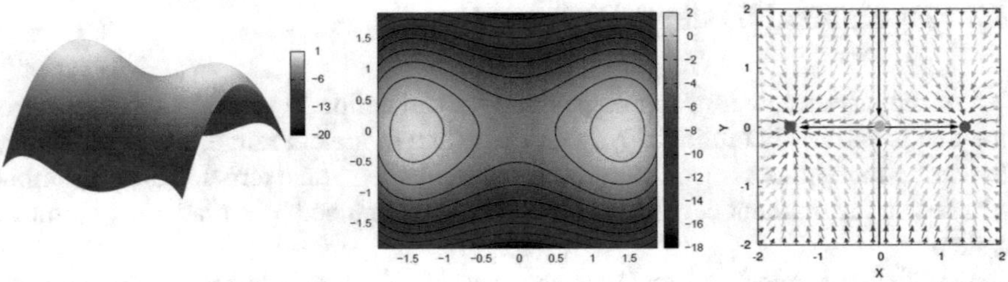


**Fig. 2.6** $f(x, y) = -x^4 + 4(x^2 - y^2) - 3$. Left: $f(x, y)$. Middle: Isolines. Right: vector field, $\nabla f(x, y)$. The thick black lines are symbolic representations of the integral lines connecting the CPs. Topological partition obtained from $\nabla f(x, y)$, $a$ and $b$ stand for the stable manifolds of each of the maxima of $f(x, y)$. Reproduced from R. Boto's Ph.D. thesis with permission [1]

These properties are illustrated in Fig. 2.6: only one vector is associated to non-critical points, these vectors are tangent to the isolines, vectors point toward the maxima, where the vector module is zero.

### 2.2.1 *Critical Points of a Scalar Field*

Those points of $\mathbb{R}^3$ satisfying the $\dot{\boldsymbol{r}} = \mathbf{0}$ condition are called critical or stationary points of the DS vector field. The number, type, and position of the CPs of a field determine its basic features. For instance, the set of asymptotic points of the DS, $\lim_{t\to\pm\infty} \boldsymbol{r}(t)$, necessarily include all of its CPs.

Let us define the $\alpha$– and $\omega$–limit sets of a point $\boldsymbol{p} = \boldsymbol{r}(0)$ of a DS as the limits $\lim_{t\to-\infty} \boldsymbol{r}(t)$ and $\lim_{t\to+\infty} \boldsymbol{r}(t)$, respectively. The manifold defined by all the trajectories sharing the same $\omega$–limit set is called the **stable manifold** of the limit set. If we consider the $\alpha$–limit instead, the manifold is called **unstable manifold**. Imagine that the map in Fig. 2.6 represents some kind of attractive potential. If an object is placed on the different parts of the curve, it will feel the attraction from the two attraction points (in yellow) and will approach one of them as time evolves. For $t \to \infty$ we would find the object either at one or the other maxima. All the trajectories that lead to the maximum on the left constitute its stable manifold, with an equivalent definition for the maximum on the right. With this formalization, we recover the intuitive mountain peak partition sketched at the beginning of the section.

The characterization of the CPs of a field is based on the behavior of the DS in their vicinities. To unveil it, it is useful to study the so-called linearized system (LS), an approximation to the DS truncated to first order in the distance to a given critical point, $\boldsymbol{r}_c$:

$$\begin{aligned}\dot{\boldsymbol{r}} &= \boldsymbol{f}(\boldsymbol{r}) \simeq \boldsymbol{f}(\boldsymbol{r}_c) + \mathbb{J}(\boldsymbol{r} - \boldsymbol{r}_c) \\ &= \mathbb{J}(\boldsymbol{r} - \boldsymbol{r}_c),\end{aligned} \tag{2.4}$$

where $\mathbb{J}$ is the Jacobian matrix of the DS at $\boldsymbol{r}_c$,

$$\mathbb{J} = \frac{\partial f(x, y, z)}{\partial (x, y, z)}. \tag{2.5}$$

An exhaustive enumeration of all the possible types of CPs of the LS in 3D would take too long. Fortunately, gradient DSs are much simpler. In these special cases, the Jacobian matrix at $\boldsymbol{r}_c$ is nothing but the Hessian matrix of $\rho$, the matrix of second derivatives of the field:

$$\mathbb{J}(\nabla\rho) = \mathbb{H}, \tag{2.6}$$
$$\mathbb{H} = \begin{pmatrix} \frac{\partial^2\rho}{\partial x^2} & \frac{\partial^2\rho}{\partial x\partial y} & \frac{\partial^2\rho}{\partial x\partial z} \\ \frac{\partial^2\rho}{\partial y\partial x} & \frac{\partial^2\rho}{\partial y^2} & \frac{\partial^2\rho}{\partial y\partial z} \\ \frac{\partial^2\rho}{\partial z\partial x} & \frac{\partial^2\rho}{\partial z\partial y} & \frac{\partial^2\rho}{\partial z^2} \end{pmatrix}_{\boldsymbol{r}_c}.$$

Translating the origin of coordinates to the CP under scrutiny, the trajectories of the linearized system are given by the following expression:

$$\dot{\boldsymbol{r}} = \mathbb{H}\boldsymbol{r}. \tag{2.7}$$

The classification of the CPs of a gradient field is simpler than in the general case due to the symmetric character of the $\mathbb{H}$ matrix. The latter can always be diagonalized by an orthogonal transformation, and their eigenvalues are necessarily real. Since orthogonal transformations are equivalent to coordinate system rotations, a new change of coordinate system to that of the principal axes of $\mathbb{H}$ will uncouple the system of differential equations, that is now straightforwardly solvable.

Let then $\mathbb{U}$ be the orthogonal matrix that diagonalizes $\mathbb{H}$ at a critical point:

$$\mathbb{U}^t\mathbb{H}\mathbb{U} = \Lambda = \text{diag}(\lambda_i), \tag{2.8}$$

where the three $\lambda_i$'s are the eigenvalues of $\mathbb{H}$. Be $\boldsymbol{\eta}$ also the reference frame defined by

$$\begin{aligned} \boldsymbol{r} &= \mathbb{U}\boldsymbol{\eta}, \\ \dot{\boldsymbol{r}} &= \mathbb{U}\dot{\boldsymbol{\eta}}. \end{aligned} \tag{2.9}$$

If we change to this new frame, the DS equations transform into:

$$\dot{\boldsymbol{\eta}} = \Lambda\boldsymbol{\eta}, \tag{2.10}$$

whose trivial solution is

$$\eta_i(t) = \eta_i(t_0)e^{\lambda_i(t-t_0)}, \; \{i = 1, 3\}. \tag{2.11}$$

The numbers $\lambda_i$ are also known as characteristic or Lyapunov exponents. They are used to determine the DS in the neighborhood of a critical point in more general treatments. In general, these exponents may be complex numbers, giving rise to spiral trajectories around the CP. Spiral points do not exist in gradient DSs. If $\Re(\lambda) \neq 0$ we say the CP is hyperbolic, and degenerate if this condition is not met. When a non-hyperbolic CP is found, we face a situation similar to that of a simple one-variable function that has a stationary point with null second derivative. In order to classify it we must examine further derivatives. In our case, this is equivalent to saying that the linearized system does not have enough information about the original non-linearized one. As we will see, degenerate CPs are rather rare in the topologies we will study. An example will be shown in Chap. 4 for illustrative purposes.

From Eq. 2.11 it follows that a negative(positive) $\lambda$ exponent signals a direction belonging to the stable(unstable) manifold of the CP under scrutiny. In practice, to find the $\lambda_i$'s, we solve the secular determinant

$$\det |\mathbb{H} - \lambda\mathbb{I}| = 0. \tag{2.12}$$

Once this has been done, we order its eigenvalues such that $\lambda_1 \leq \lambda_2 \leq \lambda_3$. The change of basis matrix $\mathbb{U}$ that determines the eigen orthonormal frame in which the DS equations are uncoupled is the matrix constructed from the eigenvectors of $\mathbb{H}$, and may be obtained explicitly by solving the linear system

$$\mathbb{H}\mathbb{U} = \mathbb{U}\Lambda. \tag{2.13}$$

Let's solve Eq. 2.13 for the system in Fig. 2.6 with $f(x, y) = -x^4 + 4(x^2 - y^2) - 3$. We have

$$\begin{aligned} \mathbb{R}^2 &\rightarrow \mathbb{R} \\ f(x, y) &\rightarrow -x^4 + 4(x^2 - y^2) - 3 \end{aligned}$$

The CPs of this function are such points where its $\nabla f(x, y)$ gradient vanishes: $\nabla f(x, y) = (-4x^3 + 8x)\mathbf{i} - 8y\mathbf{j} = 0\mathbf{i}, 0\mathbf{j}$. This equation is easy to solve, leading to three CPs $(+\sqrt{2}, 0)$ , $(-\sqrt{2}, 0)$ and $(0, 0)$. This can be qualitatively verified at Fig. 2.6. We can see that critical points at $(\pm\sqrt{2}, 0)$ are maxima whereas $(0, 0)$ is a saddle point.

In order to verify the nature of these critical points, it suffices to calculate the Hessian matrix:

$$\mathbb{H}(x, y) = \begin{pmatrix} -12x^2 + 8 & 0 \\ 0 & -8 \end{pmatrix} \tag{2.14}$$

$\mathbb{H}(x, y)$ is already diagonal (otherwise we would have to diagonalize it). Its eigenvalues are given by $-12x^2 + 8$ and $-8$. At $(\pm\sqrt{2}, 0)$ both eigenvalues are negative, $(-16, -8)$, which confirms these points are maxima. At $(0, 0)$, the eigenvalues have different sign, $(8, -8)$, indicating therefore the presence of a saddle point.

We usually call the eigen reference frame the principal curvatures frame, and its axes, principal axes of curvature, since they correspond to the directions in which the scalar field varies maximally. Let us notice that if two or three of the eigenvalues are equal, then two or three of the principal axes are not uniquely defined. In these degeneracy cases we can always construct the principal frame by choosing any pair (or trio) of orthogonal axies that belong to the subspace spanned by the degenerate set.

The classification of hyperbolic CPs is done through the signs of their $\lambda_i$ values. In $\mathbb{R}^3$ we may find the following cases (the full list is available in Table 2.1):

1. $\lambda_1, \lambda_2, \lambda_3 < 0$. All the curvatures are negative. Then all the field trajectories converge toward the critical point, giving rise to a *sink* or *attractor* or *maximum.*
2. $\lambda_1, \lambda_2 < 0, \lambda_3 > 0$. In the plane defined by the $\boldsymbol{\eta}_1$ and $\boldsymbol{\eta}_2$ vectors the field lines approach the critical point; however, in the orthogonal $\boldsymbol{\eta}_3$ direction they escape from it. We call this situation *saddle point of the first kind.*
3. $\lambda_2, \lambda_3 > 0$ and $\lambda_1 < 0$. Two field trajectories approach the CP along the $\boldsymbol{\eta}_1$ principal curvature axis. The rest of the field lines emerge from the critical point, and at it are tangent to the two-dimensional plane defined by the $\boldsymbol{\eta}_2$ and $\boldsymbol{\eta}_3$ vectors. We call this type of CP a *saddle point of the second kind.*

**Table 2.1** Classification of the CPs of a scalar field in $\mathbb{R}^3$ according to their rank, $r$, and their signature $s$. AB and RB signal the dimension of the attraction and repulsion basins of each CP, respectively. Notice that if we classify critical points using repulsion basis, saddles of the first kind will behave as those of the second kind, and viceversa

| $(r, s)$ | Curvatures | Description | AB | RB |
|---|---|---|---|---|
| $(3, -3)$ | $\lambda_i < 0 \;\; \forall i$ | Local maximum (attractor) | 3D | 0D |
| $(3, -1)$ | $\lambda_1, \lambda_2 < 0; \lambda_3 > 0;$ | Saddle of the first kind | 2D | 1D |
| | | Maximum along $\boldsymbol{\eta}_1$ and $\boldsymbol{\eta}_2$; minimum along $\boldsymbol{\eta}_3$ | | |
| $(3, +1)$ | $\lambda_1, \lambda_2 > 0; \lambda_3 < 0;$ | Saddle of the second kind | 1D | 2D |
| | | Maximum *along*$\boldsymbol{\eta}_3$ and minimum along $\boldsymbol{\eta}_1$ and $\boldsymbol{\eta}_2$ | | |
| $(3, +3)$ | $\lambda_i > 0 \;\; \forall i$ | Local minimum (source) | 0D | 3D |
| $(2, -2)$ | $\lambda_1, \lambda_2 < 0; \lambda_3 = 0;$ | Local maximum along $\boldsymbol{\eta}_1$ and $\boldsymbol{\eta}_2$ | 2D | 0D |
| $(2, 0)$ | $\lambda_1 < 0; \lambda_2 = 0; \lambda_3 > 0$ | Saddle point. | 1D | 1D |
| | | Maximum along $\boldsymbol{\eta}_1$ and minimum along $\boldsymbol{\eta}_3$ | | |
| $(2, +2)$ | $\lambda_1 = 0; \lambda_2, \lambda_3 > 0$ | Local minimum along $\boldsymbol{\eta}_2$ and $\boldsymbol{\eta}_3$ | 0D | 2D |
| $(1, -1)$ | $\lambda_1 < 0; \lambda_2 = \lambda_3 = 0$ | Local maximum along $\boldsymbol{\eta}_1$ | 1D | 0D |
| $(1, +1)$ | $\lambda_1 = \lambda_2 = 0; \lambda_3 > 0$ | Local minimum along $\boldsymbol{\eta}_3$ | 0D | 1D |
| $(0, 0)$ | $\lambda_1 = \lambda_2 = \lambda_3 = 0$ | | 0D | 0D |

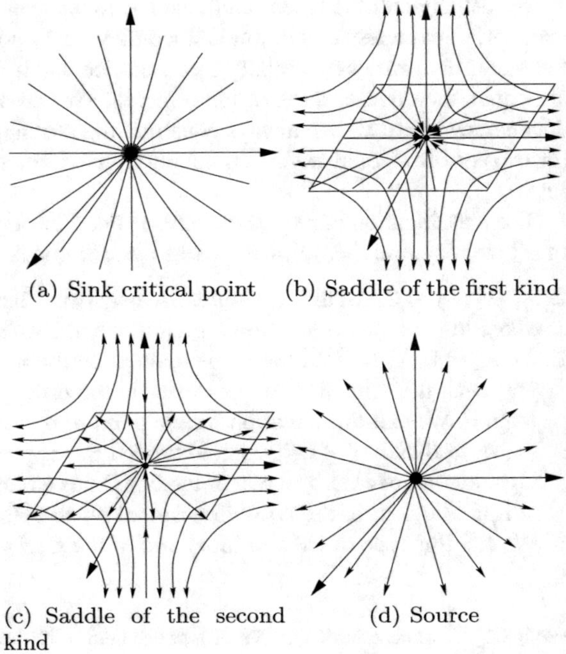


**Fig. 2.7** A representation of the four types of critical points of a scalar field in the three-dimensional space

4. $\lambda_1, \lambda_2, \lambda_3 > 0$. Then all the curvatures are positive. Every field line escapes the CP in three dimensions, along the three curvature axes. Such a critical point is known as a *source* or *repellor* or *minimum* of the field.

A pictorial representation of the general structure of the field trajectories in the vicinity of each of the four types of critical points of a 3D scalar is found in Fig. 2.7.

Before finishing this subsection, we would like to briefly consider some notation issues regarding CPs. A rather well-known way to refer to the CPs of a field consists of using two integer indices, written as $(r, s)$. The *rank*, $r$, is defined as the number of non-zero curvatures of the CP, and the *signature*, $s$, as the difference between the number of positive and negative curvatures. A hyperbolic CP displays a rank always equal to the dimension of the space, $r = n$. The four non-degenerate cases described above thus correspond to $(3, -3)$, $(3, -1)$, $(3, +1)$ and $(3, +3)$ points, respectively.

It is important to note that the number of critical points must fulfill a sum rule, which depends on the type of system. For non-periodic systems in $\mathbb{R}^3$, we have

$$n - b + r - c = 1, \tag{2.15}$$

where $n$ stands for the $(3, -3)$ points, $b$ are the $(3, -1)$, $r$ are the $(3, +1)$ ane $c$ are the $(3, +3)$ points. This is known as the Euler-Poincaré relationship.

Periodic systems follow Morse's equality:

$$n - b + r - c = 0, \tag{2.16}$$

The derivation of these formulae is included in the Advanced Sect. 2.3. There, we can also check Euler-Poincaré's relationship for our example in $\mathbb{R}^2$.

## 2.2.2 Basins, Separatrices and Induced Topologies

As a given DS trajectory approaches a CP, $\nabla\rho$ decreases and the $|t|$ parameter increases. At the CP $\nabla\rho = 0$ and $t \to \pm\infty$. The $t$ parameter for a field line thus varies between the $-\infty$ limit (at the source PC or $\alpha$ point) and $+\infty$ (at the attractor or $\omega$ point). Since to each $\boldsymbol{r} \in \mathbb{R}^3$ non-critical point we can associate a unique trajectory, and each of these has two well-defined limit CPs, it is possible to establish a map among points in $\mathbb{R}^3$ and the different CPs of $\rho$. We will call $\alpha(\omega)$ limit set of a CP the geometrical locus of all the points in $\mathbb{R}^3$ whose trajectories start(end) at the CP.

Let us now focus on a given principal direction of one CP. Its $\alpha$-limit set will be one-dimensional and its $\omega$-limit zero-dimensional if the CP is a minimum of the potential function along that direction, while the contrary will be true if the CP is a maximum of the scalar field. In both of the above situations, the CP is a source or an attractor of the field lines, respectively. Similar considerations may be done for the remaining principal axes. It is interesting to check this for the saddle point (the mountain pass) in the 2D potential of Fig. 2.6. This point is a minimum of the potential along the line joining the mountain peaks ($y = 0$), and a maximum in the direction perpendicular to it ($x = 0$). These two 1D limit sets are shown in Fig. 2.8 with their trajectories: the $\omega$ limit set in between the mountains has a leaving trajectory (1D), whereas the $\alpha$ omega set (1D) has an entering one.

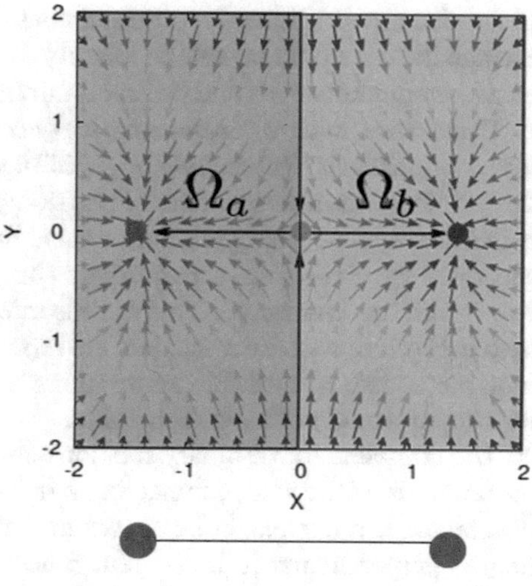


**Fig. 2.8** $f(x, y) = -x^4 + 4(x^2 - y^2) - 3$. Topological partition obtained from $\nabla f(x, y)$, $a$ and $b$ stand for the stable manifolds of each of the maxima of $f(x, y)$. The resulting graph is shown below

In general, and from a formal point of view, each attractor (or source) in $\mathbb{R}^3$ shows a three-dimensional neighborhood which is invariant to the $\nabla\rho$ flux, in such a way that any trajectory originating (or ending) in such neighborhood ends (starts) at the attractor (source). The largest open neighborhood of a CP that fulfills this property is known as its attraction, AB (or repulsion, RB) basin. Going back to Fig. 2.6, each mountain is the AB of its corresponding maximum (peak): the semispace $x < 0$ for the peak at $(-\sqrt{2}, 0)$ and the semispace $x > 0$ for that at $(+\sqrt{2}, 0)$ (Fig. 2.8).

Similarly, we call separatrix the set of points that belong neither to the AB or RB of a given attractor or repulsor. In Figs. 2.6 and 2.8, the separatrix is given by the $y = 0$ line. Notice that this is the attraction basin of the saddle, and that $x = 0$ is its repulsion basin.This result is immediately generalizable to three-dimensional spaces. If the function were a 2D projection of a 3D scalar generated by revolution symmetry around the $x$ axis, it is easy to convince ourselves that the separatrix would now consist in the $y = 0$ plane, and that the full 3D space would have been partitioned into two semi-spaces, separated by a plane. Note nevertheless that after convolution, the repulsion basin of the saddle point would still be a line joining the maxima (the saddle point would be of type $(3, -1)$).

With these basic concepts, it is possible to partition the physical space, $\mathbb{R}^3$, in a finite number of subsets, that correspond to the $\alpha$- and $\omega$-limits of the CPs of the potential function. These are, respectively, either the RBs of the sources, or the ABs of the attractors. With this, we endow $\mathbb{R}^3$ with a structure generated by a collection of open subsets, $\{A_j\}$, that we shall specify by the $(\mathbb{R}^3, A)$ pair. It can be easily proven that this new space satisfies all the criteria of a topological space. We call $A$ the topology induced by $\rho$ in $\mathbb{R}^3$. From a formal point of view, we have transformed an Euclidean space into a non-metric abstract construct. Turning again to our example in Fig. 2.6, this mathematical apparatus has allowed us to divide the space into the two attraction basins associated to each of the maxima, that may be considered as the open sets of a topology of the space. *The function $f$ has induced a topology in* $\mathbb{R}^2$. Note that this leads to two disjoint regions that recover the intuitive partitioning of a mountain range into its different peaks. Would the mountains deform under an earthquake we would be able to identify the changes in terms of the new number of peaks, mountain passes and potential minima (lakes) that would have appeared.

*Whenever a separatrix appears between two attractors, we will necessarily find a pair of gradient lines linking them.* In Fig. 2.8, the separatrix at $x = 0$ is source at the saddle point of two gradient lines going to the maxima along the line $y = 0$. It is in this way that binary relations appear among attractors, whose basins are either separated by a common separatrix or completely disconnected. In our case, these two basins are connected. This is illustrated by the graph below the gradient field map in Fig. 2.8, where the maxima are represented as red circles. However, we could imagine adding a third maximum along the $x = 0$ line, so that the first and third maximum would not have such a line.

The complete set of binary relations among the attractors, allows us to simplify the structure of the field, compacting it in its adjacency graph: the set of related pairs of attractors. It can easily be shown that at each point of a separatrix the gradient field is perpendicular to its tangent space, $\boldsymbol{n}(\boldsymbol{r})$. We then say that separatrices are local zero-flux surfaces:

$$\int_S \nabla f(\boldsymbol{r}) \cdot \boldsymbol{n}(\boldsymbol{r}) ds = 0, \tag{2.17}$$

$s$ being the surface element. This can be intuitively seen in Fig. 2.8, where gradient lines run parallel to the separatrix.

Summarizing, the $\alpha$– and $\omega$–limit sets of the CPs of a scalar field induce two simple topologies in the space:

1. In one, the open sets in $\mathbb{R}^3$ are the $\alpha-$limit sets of the critical points of the scalar field $\rho$.
2. In the other, the open sets of $\mathbb{R}^3$ are the $\omega-$limit sets of the critical points of $\rho$.

These two partitions transform $\mathbb{R}^3$ into a *topological space*. We have seen throghout an example of a partition into $\omega$-limit sets, but you can see an example of $\alpha$ limit set in the Exercises (Chap. 9).

## 2.3 Topological Invariants (Advanced)

The number and type of CPs of a DS is limited by the nature of the mathematical manifold (e.g. dimension, periodicity) in which the field exists. Knowing the relationships that these objects must fulfil is extremely useful to carry out consistency checks on numerically calculated topologies.

Let us introduce the connectivity, or Betti, number, $R_n$, of a domain $\mathcal{D}$ as the number of topologically different $n$–dimensional regions in $\mathcal{D}$ that have no boundaries and are not the boundaries of any $(n+1)$–dimensional region of $\mathcal{D}$. This is, by no means, a formal definition, but will serve our purposes. We will name these regions $C_i^n$, where $n$ is the dimension of the region, and $i$ a numerical label to distinguish the different topologically non-equivalent regions sharing the same dimensionality. As an aid, all $C_i^0$ will be equivalent to a point, $C_i^1$ to a circumference, $C^2$ to a sphere and so on. They have no boundaries, but may be boundaries of higher dimensional regions. For instance, a circumference is a one-dimensional region which is the boundary of a 2D region in $\mathbb{R}^2$, the circle it encloses.

Let's see the example of a 2D torus, $T^2$. We have regions of dimension $i = 0, 1$ and 2: $C^0$ is any point of the torus, $C^2$ is the whole torus surface, and there are two regions in 1D, $C_1^1$ and $C_2^1$, which correspond to the two topologically distinct rings that we can construct on its surface in the directions of its two periods. All the $C_1^1$ lines are topologically equivalent: if we start from the $C_1^1$ line, we can transform it into another one of its category just by rolling/distorting it around the torus. The same procedure can be applied to $C_2^1$. We can nevertheless not continuously deform any of the $C_1^1$ lines into one of the $C_2^1$ sets (see Fig. 2.9). They are not topologically equivalent among themselves. Thus, these three types of regions meet the conditions to define

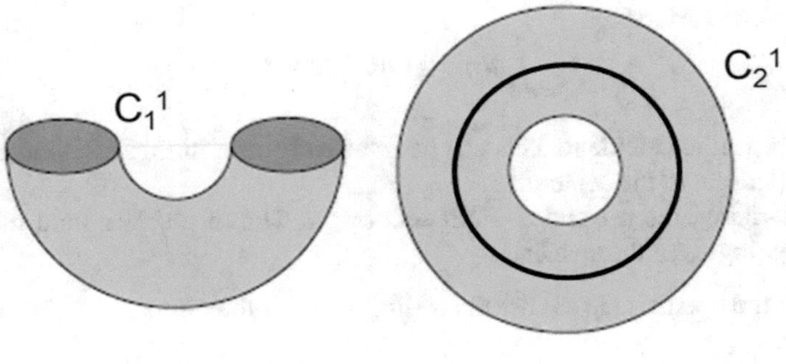


**Fig. 2.9** Examples of the $C_1^1$ and $C_2^1$ lines

our connectivity numbers: they have no boundaries, thanks to periodicity, and are not boundaries of any two-dimensional regions, since in that case we could not deform them into a point. Hence, the different connectivity numbers are $R_0 = R_2 = 1$ and $R_1 = 2$.

We can similarly examine a $T^3$ torus (the Cartesian product of three circles). We assume that $x$, $y$ and $z$ are the three variables in which the $f$ field is periodic. Let us consider a closed line $C^1$ with constant $x$ and $y$, in which $z$ is allowed to run over all possible values: this line is closed thanks to the $z$ periodicity, so that the points $z$ and $z + T_z$, where $T_z$ is the period in $z$, are equivalent. We can define such kind of lines for any possible value of the $x, y$ pairs, and we will label them as $C_{xy}^1$. All the $C_{xy}^1$ lines are topologically equivalent. Similarly, we can define the $C_{xz}^1$ lines, in which now $x$ and $z$ remain fixed for any value of $y$, as well as $C_{yz}^1$ lines. Moreover, since the $T^3$ is triply periodic, it is clear that we cannot find more one-dimensional regions satisfying our criteria, so that the $R_1$ connectivity number of $T^3$ is equal to 3. Similarly, $R_2 = 3$, since we can find build $x, y$ $x, z$ or $y, z$ doubly periodic surfaces at each $z$, $y$, or $z$ values, respectively. Finally, $R_0 = R_3 = 1$.

Let us now propose a topological characterization of the CPs of a gradient DS that, as we will see, lacks any analytical argument. This presentation follows that of Jones and March [3], and classifies critical points in terms of repulsion basins. Let us assume that at a given point $\boldsymbol{x}$, $f(\boldsymbol{x}) = \varepsilon$. We now define the region $\{f < \varepsilon\}$ (that may contain two or more disconnected regions) as the set of points $\boldsymbol{y} \in \mathcal{D}$ fulfilling $f(\boldsymbol{y}) < \varepsilon$. Let us also take an $n$-dimensional region $Z_n$ completely contained in $\{f < \varepsilon\} + \boldsymbol{x}$ and containing also the point $\boldsymbol{x}$, possibly at its boundary. When this region cannot be continuously deformed so as to exclude $\boldsymbol{x}$ such that it fully remains within $\{f < \varepsilon\}$, *and such that its boundary does not cross* $\boldsymbol{x}$ by means of deformations that always lie within $\{f < \varepsilon\} + \boldsymbol{x}$, we then say that $Z_n$ is an $n$–undeformable region. If there exists any $Z_n$ region, $\boldsymbol{x}$ is a critical point. Its degree is the smallest $n$ value for which $Z_n$ regions appear. For our purposes, a point has no boundaries, and the boundary of a line is composed of its two end points, like the a and b points in Fig. 2.10a. Using our Fig. 2.6 example in 2D: if we choose $\boldsymbol{x}$ at one of the maxima, all points around it fulfill $\{f < \varepsilon\}$. We can thus deform $\boldsymbol{x}$ by moving it either way, and we

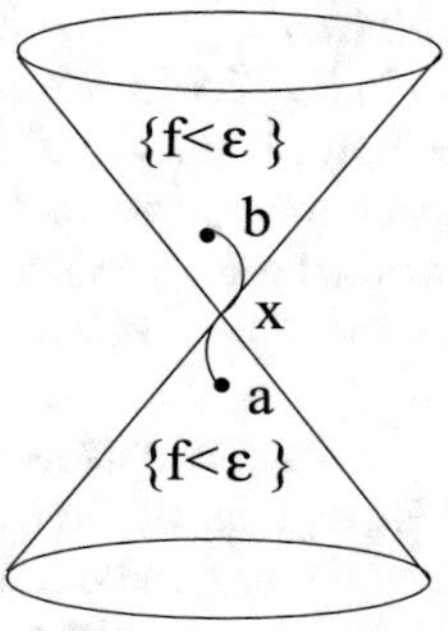


(a) The $(a,b)$ line contains a CP of the first kind, i.e. (3,+1)

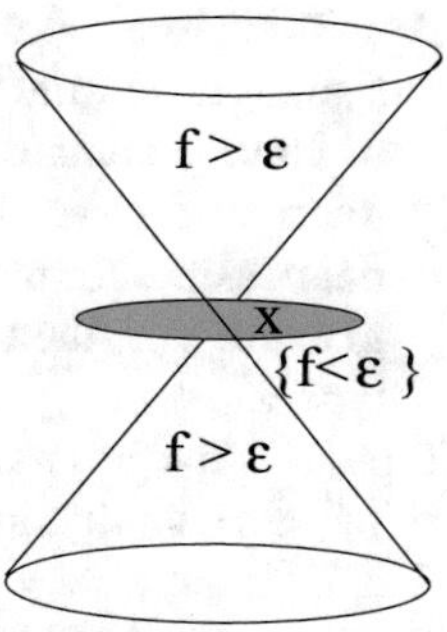


(b) The disk in shadow contains a CP of the second kind, i.e. (3,-1)

**Fig. 2.10** Topological characterization of critical points of degree 1 and 2. In Fig. 2.10a $Z_n = Z_1$ is a line. In Fig. 2.10b, $Z_n = Z_2$ is a two-dimensional surface

can build an infinite number of lines that include $\boldsymbol{x}$ and slide them without leaving $\{f < \varepsilon\}$. $Z_0$ and $Z_1$ do not exist. However, a 2D region will have $\boldsymbol{x}$ in its interior, so that deforming them to exclude that point, will make its boundary cross it at some point. $Z_2$ exists, and $\boldsymbol{x}$ is a maximum. If $\boldsymbol{x}$ is not a critical point, we can place $\boldsymbol{x}$ at the boundary of points, lines, and 2D regions, and exclude the point continuously without leaving $\{f < \varepsilon\}$.

Let us examine the full set of possibilities in the three dimensional case, showing how the analytical definition of CP matches the topological one we are introducing.

Type 0. If $\boldsymbol{x}$ is a minimum, then $Z_0$ does only contain the $\boldsymbol{x}$ point itself, which cannot be continuously deformed in the interior of $\{f < \varepsilon\}$, for there does not exist any such region in the vicinity of $\boldsymbol{x}$.

Type 1. If we focus on the reference frame formed by the principal axes at $\boldsymbol{x}$, we can Taylor expand the field in its neighborhood,

$$f = \varepsilon + \sum_{i=1}^{3} \alpha_i \delta_i^2. \tag{2.18}$$

If $\alpha_1 > 0, \alpha_2 > 0$ and $\alpha_3 < 0$, then the region $\{f < \varepsilon\}$ is a volume located at both sides of $\boldsymbol{x}$ that unfolds about it as a double cone (see Fig. 2.10a). A surface or volume within $\{f < \varepsilon\} + \boldsymbol{x}$ that contains $\boldsymbol{x}$ can be only defined if $\boldsymbol{x}$ lies on its boundary. In this way, such a volume or surface can deform continuously until it is fully contained in $\{f < \varepsilon\}$ (excluding $\boldsymbol{x}$). This is done simply by extracting $\boldsymbol{x}$ from its boundary: the latter needs not pass over $\boldsymbol{x}$, for $\boldsymbol{x}$ was already on it. However, the $(a, b)$ line shown in Fig. 2.10a includes the $\boldsymbol{x}$ point, and if we deform it in order to exclude it, one of its boundaries, either $a$ or $b$, must pass over $\boldsymbol{x}$. The existence of such a line shows that this critical point is of the first kind.

Type 2. For a saddle point in which $\alpha_1 > 0$ and $\alpha_2$ and $\alpha_3 < 0$ in Eq. 2.18, we face a situation similar to that found in Fig. 2.10b. We can observe how any point, line or volume containing $\boldsymbol{x}$ can be deformed so as to exclude $\boldsymbol{x}$ and be placed in the interior of $\{f < \varepsilon\}$ such that its boundary does not cross over $\boldsymbol{x}$. This is, however, not possible for the two-dimensional region shown in Fig. 2.10b: if we would like to exclude $\boldsymbol{x}$, then its boundary must pass over it, so that this region would be of type $Z_2$.

Type 3. If $\boldsymbol{x}$ is a maximum, all the volume in its close surroundings is itself within $\{f < \varepsilon\}$, so that no $Z_0$, $Z_1$ or $Z_2$ regions would exist. There exists, nevertheless, a $Z_3$ region: Any three-dimensional region containing $\boldsymbol{x}$ that starts from region $\{f < \varepsilon\}$ can not be deformed into $\{f < \varepsilon\}$ without its boundary passing over $\boldsymbol{x}$.

Just as we saw in the 2D example, it is important to recognize that if $\boldsymbol{x}$ is a point such that $\nabla f \neq \mathbf{0}$, any region in $\mathcal{D}$ that belongs to $\{f < \varepsilon\} + \boldsymbol{x}$ can only include $\boldsymbol{x}$ if it is on its boundary, such that in order to deform the region into $\{f < \varepsilon\}$ it suffices to slide its boundary excluding $\boldsymbol{x}$, without passing over it necessarily. In this way, a point that is not critical in the analytical sense, is also not critical according to the topological concepts described in this section.

The Morse relationships link the number of $Z_n$ regions with the previously introduced connectivity numbers. Let us consider a parameter $\varepsilon$ ranging from the absolute minimum and maximum values of $f(\boldsymbol{x})$. If we now increase $\varepsilon$ from its minimum, we generate a region $\{f < \varepsilon\}$ with connectivity numbers that are defined in the same way as those used for the domain $\mathcal{D}$ itself. Let us consider a particular point $\boldsymbol{x}$, and examine the variation of the connectivity numbers of $\{f < \varepsilon\}$ as $\varepsilon$ changes so that the boundary of $\{f < \varepsilon\}$ passes over $\boldsymbol{x}$.

We now take $f(\boldsymbol{x}) = \varepsilon$, and we consider, in the first place, that $\boldsymbol{x}$ is not a critical point. The connectivity numbers of $\{f < \varepsilon\} + \boldsymbol{x}$ are then equal to those of $\{f < \varepsilon\}$. Since $\boldsymbol{x}$ is not a CP, any region of $\{f < \varepsilon\} + \boldsymbol{x}$ can be deformed toward region $\{f < \varepsilon\}$. The number of regions that we use to define the connectivity numbers will thus remain constant in this process.

On the other hand, let us suppose that $\boldsymbol{x}$ is a critical point, being associated to a region $Z_n$ of $\{f < \varepsilon\} + \boldsymbol{x}$ that cannot be deformed into $\{f < \varepsilon\}$ without its boundary passing over $\boldsymbol{x}$. We have to consider two cases:

1. The boundary of $Z_n$ encloses a second region, $U_n$, completely contained in $\{f < \varepsilon\}$; that is, the boundary of $Z_n$ is also a boundary of $U_n$. The region $Z_n + U_n$ will thus have no boundary, for $Z_n$ and $U_n$ have a common unique boundary. Moreover, $Z_n + U_n$ cannot enclose an $n + 1$–dimensional region of $\{f < \varepsilon\} + \boldsymbol{x}$. If this were the case, we could continuously deform $U_n$ into $Z_n$ while fully within $\{f < \varepsilon\} + \boldsymbol{x}$ such that the enclosed region would collapse. This would violate the conditions for a CP to exist: a region in $\{f < \varepsilon\}$ would be deformable to include $\boldsymbol{x}$ such that its boundary would not pass over it. Let us suppose that this were indeed possible. Then one might, deforming in the contrary sense, deform $Z_n$ until it would be entirely contained in $\{f < \varepsilon\}$, its boundary never crossing over $\boldsymbol{x}$, again

violating the definition of $Z_n$. We come there to the conclusion that $Z_n + U_n$ is a closed $n$-dimensional region $C^n$ of $\{f < \varepsilon\} + \boldsymbol{x}$, thus fulfilling the criterion to add to the connectivity number $R_n$ (remember we saw several examples with the torus). Since it cannot be continuously deformed into any region of $\{f < \varepsilon\}$, the connectivity number $R_n$ of $\{f < \varepsilon\}$ would increase by one unit when the boundary of $\{f < \varepsilon\}$ passes over $\boldsymbol{x}$. We will thus call such a $\{f < \varepsilon\}$ region for a CP the *increasing region*, and we will also name the number of CPs for which such regions exist as $N_+^n$.

2. Let us suppose now that the boundary of $Z_n$ is not simultaneously the boundary for another region of $\{f < \varepsilon\}$. Then, by definition, it is a $C^{n-1}$ region for $\{f < \varepsilon\}$, but not for $\{f < \varepsilon\} + \boldsymbol{x}$, since in the latter it is a boundary of $Z_n$. Consequently, that boundary contributes to the $R_{n-1}$ connectivity number of $\{f < \varepsilon\}$, but not to its equivalent in $\{f < \varepsilon\} + \boldsymbol{x}$. We will thus call $\{f < \varepsilon\}$ a *decreasing region*, for the connectivity number $R_{n-1}$ of $\{f < \varepsilon\}$ decreases in one unit when its boundary passes over the CP at $\boldsymbol{x}$, and we will also define $N_-^n$ as the number of CPs of degree $n$ displaying this property.

It is then found that when $\varepsilon$ runs over its possible values within the $\mathcal{D}$ domain the connectivity numbers must fulfil the condition

$$R_n = N_+^n - N_-^{n+1}, \tag{2.19}$$

where the total number of CPs of type $n$ is

$$N^n = N_+^n + N_-^n. \tag{2.20}$$

Let us illustrate this with a very minimal example. The function $f(x) = \sin(x)$ defined over the circle $x \bmod 2\pi$, where we impose periodic boundary conditions that make 0 equivalent to $2\pi$. The circumference has a single connected type of region, so $R_0 = 1$, and one single 1D region without boundary giving $R_1 = 1$. The function displays a single maximum, $N^1 = 1$ and a single minimum, $N^0 = 1$. If we start at the minimum value of $f$, $\epsilon = -1$, it is clear that the domain $\{f < \epsilon\}$ does not exist, so that crossing the critical point $\boldsymbol{x}$ increases the $R_0$ Betti number by 1. This means that $N_+^0 = 1$. As we increase $\epsilon$, the domain $\{f < \epsilon\}$ expands. Similarly, on crossing $\epsilon = 1$, which corresponds to the maximum, the domain closes, and $R_1$ increases by one, so $N_+^1 = 1$. Obviously, $R_0 = N_+^0$, $R_1 = N_+^1$, and $R_0 - R_1 = N^0 - N^1 = 0$. In plain words, the number of maxima and minima on a circle must coincide, something which is easy to grasp. Figure 2.11 shows the case $f(x) = \sin(4x)$

Gathering both equations,

$$N^n - R_n = N_-^n + N_-^{n+1}. \tag{2.21}$$

If $\mathcal{D}$ is an $l$-dimensional region, then there are neither descending regions of dimension 0, since crossing minima always increases the number of connected

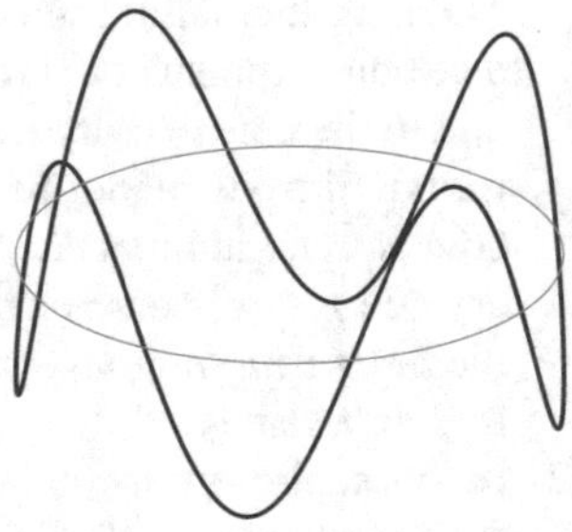

**Fig. 2.11** Plot of $f(x) = \sin(4x)$ over a ring with polar angle $x$. The function is periodic and shows four maxima and four minima

regions. Similarly, there are no regions of dimension $l + 1$, since $l$ is the final dimension. Hence, $N_-^0 = N_-^{l+1} = N_+^{l+1} = 0$. In this way, Eq. 2.21 can be used to generate a series of relations among the connectivity numbers and the number of CPs by recursively eliminating the different $N_-^i$. For $n = 0$ we get

$$N^0 - R_0 = N_-^0 + N_-^1 = N_-^1 \geq 0, \tag{2.22}$$

for $N_-^0 = 0$ and any number of critical points must be non-negative. Substituting $N_-^1$ in Eq. 2.21 for $n = 1$,

$$N^1 - R_1 = N_-^1 + N_-^2 = N^0 - R_0 + N_-^2, \tag{2.23}$$

for $N_-^2 \geq 0$:

$$N^1 - N^0 \geq R_1 - R_0. \tag{2.24}$$

Repeating the process for $n = 2$,

$$N^2 - R_2 = N_-^2 + N_-^3 = N^1 - R_1 - N^0 + R_0 + N_-^3, \tag{2.25}$$

that can be transformed into

$$N^2 - N^1 + N^0 \geq R_2 - R_1 + R_0. \tag{2.26}$$

The recursion may be propagated until $n = l$, where $N_-^{l+1} = 0$, thus closing the inequality chain with another equality.

In its most general form, the Morse relations for an $l$-dimensional domain $\mathcal{D}$ become

$$\begin{aligned} N^0 &\geq R_0 \\ N^1 - N^0 &\geq R_1 - R_0 \\ N^2 - N^1 + N^0 &\geq R_2 - R_1 + R_0 \\ &\vdots \\ \sum_{i=0}^{n}(-1)^{n-i}N^i &\geq \sum_{i=0}^{n}(-1)^{n-i}R_i \\ &\vdots \\ \sum_{i=0}^{l}(-1)^{i}N^i &= \sum_{i=0}^{l}(-1)^{i}R_i. \end{aligned} \tag{2.27}$$

If we now particularize to the $\mathbb{T}^3$ torus analyzed above, $l = 3$ and the connectivity numbers become $R_0 = R_3 = 1$ and $R_1 = R_2 = 3$, so that:

$$\begin{aligned} N^0 &\geq 1 \\ N^1 - N^0 &\geq 2 \\ N^2 - N^1 + N^0 &\geq 1 \\ N^3 - N^2 + N^1 - N^0 &= 0. \end{aligned} \tag{2.28}$$

If we now recall that $N^0$, $N^1$, $N^2$ and $N^3$ stand for the number of cage ($c$), ring ($r$), bond ($b$) and nuclear ($n$) points, respectively, the first Morse inequality states that $c \geq 1$. Substituting this result in the second inequality we get $r \geq 3$. Doing the same with $r - c \geq 2$ in the third we obtain that $b \geq 3$, and from the equality $N^3 = N^2 - N^1 + N^0$, which should be greater or equal than 1 from the third inequality, we get that $n \geq 1$. The last equality is probably the most important Morse relation, that we usually write in the form

$$n - b + r - c = 0. \tag{2.29}$$

This equality will be useful in solids systems, which are just like the $\mathbb{T}^3$ torus, periodic along the three directions (Chap. 7).

It can be found that repeating this process in $\mathbb{R}^3$ Morse's equality reads

$$n - b + r - c = 1, \tag{2.30}$$

which is also known as the Euler-Poincaré relationship. In the case of a scalar field like the electron density, which we will examine in detail in later Chapters, we can reason as follows. The field decays to zero exponentially at large distances, so that the infinity may be imagined as a minimum and then use the torus expression leaving this extra-minimum in the right-hand side of the Morse equality.

Let's check this equality in our well known example of Fig. 2.6. In $\mathbb{R}^2$, Morse's relationship reads $N^2 - N^1 + N^0 = 1$. Our surface has two maxima ($N^2 = 2$), one saddle point ($N^1 = 1$) and zero minima ($N^0 = 0$). We thus have: $2 - 1 + 0 = 1$.

## References

1. Boto, R.A.: Development of the NCI method. Ph.D. thesis, Sorbonne University (2016)
2. Hirsch, M.W., Smale, S., Devaney, R.L.: Differential Equations, Dynamical Systems, and an Introduction to Chaos, 2nd edn. Elsevier Academic Press, San Diego (2004)
3. Jones, W., March, N.H.: Theoretical Solid State Physics, vol. 1, Chap. A1.6. Dover Publications, Inc., New York (1975)

# Part I
# Descriptors

# Chapter 3
# The Electron Density

**Abstract** In this chapter we analyze the basic scalar field in the topological theories of the chemical bond, the electron density, $\rho$. After a succinct revision of the general properties of the $\rho$ field, the topology that it induces in real space through its attractor basins is presented. This gives rise to an atomic partition, which is generally known as the Quantum Theory of Atoms in Molecules (QTAIM). In this Chapter, the types and properties of the different critical points of $\rho$ are reviewed, and the physical roots of the partition are examined. As it will be shown, only when QTAIM basins are used several basin observables, like the atomic kinetic energy, can be defined consistently. In many instances, we also need to know the joint probability of finding not only one electron at a point, but a pair of electrons at a couple of positions, for instance. This leads us to conclude the chapter by presenting the general formalism of reduced densities and density matrices, and to use them to partition the energy in the interacting quantum atoms (IQA) approach or to construct localization and delocalization descriptors.

## 3.1 Introduction

The topological analysis of the electron density uses the ideas of the previous Chapter and applies them to a basic scalar field, the electron density, $\rho$. It will suffice our purposes to use the stationary electron density that is obtained from the state vector $\Psi(\boldsymbol{x}_1, \boldsymbol{x}_2, \ldots \boldsymbol{x}_N, \boldsymbol{R}_1, \ldots \boldsymbol{R}_M)$ as:

$$\rho(\boldsymbol{r}) = N \sum_{s_1} \int \cdots \int d\boldsymbol{x}_2 \ldots d\boldsymbol{x}_N \int \cdots \int d\boldsymbol{R}_1 \ldots d\boldsymbol{R}_M$$
$$\Psi^*(\boldsymbol{x}_1 \ldots \boldsymbol{x}_N, \boldsymbol{R}_1 \ldots \boldsymbol{R}_M)\Psi(\boldsymbol{x}_1 \ldots \boldsymbol{x}_N, \boldsymbol{R}_1 \ldots \boldsymbol{R}_M). \qquad (3.1)$$

In this equation, $s_1$ is electron 1 spin coordinate, $\boldsymbol{x}_i$ refer to spin-spatial coordinates of the various $N$ electrons and $\boldsymbol{R}_i$ are the coordinates of the nuclei of the $M$ atoms. The density function defined by expression Eq. 3.1 is general, since no condition has been imposed on the state vector. It is usual, however, to work under the Born–

 
Á. Martín Pendás and J. Contreras-García, *Topological Approaches to the Chemical Bond*, Theoretical Chemistry and Computational Modelling,
https://doi.org/10.1007/978-3-031-13666-5_3

Oppenheimer approximation, in which the electronic and nuclear motions have been uncoupled from each other. Under these conditions, the stationary electronic wavefunction depends explicitly on the spin-spatial coordinates of the $N$ electrons, $\boldsymbol{x}_i$, and parametrically on the nuclear positions, $\boldsymbol{R}$:

$$\Psi(\boldsymbol{x}_1, \boldsymbol{x}_2, \ldots \boldsymbol{x}_N; \boldsymbol{R}). \tag{3.2}$$

Admitting these approximations, the static density

$$\rho(\boldsymbol{r}_1; \boldsymbol{R}) = N \sum_{s_1} \int \ldots \int d\boldsymbol{x}_2 \ldots d\boldsymbol{x}_N \Psi^*(\boldsymbol{x}_1 \ldots \boldsymbol{x}_N; \boldsymbol{R}) \Psi(\boldsymbol{x}_1 \ldots \boldsymbol{x}_N; \boldsymbol{R}), \tag{3.3}$$

is a scalar field that depends parametrically on the $3M$ nuclear coordinates. It is not difficult to generalize this definition, using statistical mechanics, to introduce a thermally averaged electron density that takes into account thermal effects on the nuclear motion. It is also possible to define, at fixed nuclear positions, the so-called density matrix $\rho(\boldsymbol{r}_1; \boldsymbol{r}'_1)$ if in the above equation the spatial coordinate of the first electron in $\Psi^*$ is distinguished from that in $\Psi$ by calling it $\boldsymbol{r}'_1$. However, we will focus in the following on simple densities as defined by Eq. 3.3 and defer a more general treatment to Sect. 3.5.

Being $\rho$ an observable, it is amenable to experimental determination [24]. In the last two decades, we have lived an extraordinary development of such experimental techniques: X-ray diffraction, Compton scattering, electron diffraction, mixed neutron, X-ray analyses, etc. As a result of these advances, there are at the time of writing a number of techniques able to provide experimental densities of similar quality to those obtained via theoretical methods of medium quality. Our classical, basic introduction will just present the simplest techniques used in single-crystal X-ray diffraction. For a modern, authoritative account, see the book by Macchi and collaborators [35].

In an idealized X-ray diffraction experiment, the total intensity of the diffracted rays, we suppose here elastic and coherent dispersion, can be written as

$$I(\boldsymbol{k}) = \left| \int \rho(\boldsymbol{r}) e^{2\pi i \boldsymbol{k} \cdot \boldsymbol{r}} d\boldsymbol{r} \right|^2, \tag{3.4}$$

where $\boldsymbol{k}$ is the dispersion vector, that bisects the angle between the diffracted ray and the inverse of the incident ray, with module $2 \sin(\theta)/\lambda$. The dispersion amplitude

$$A(\boldsymbol{k}) = \int \rho(\boldsymbol{r}) e^{2\pi i \boldsymbol{k} \cdot \boldsymbol{r}} d\boldsymbol{r} \equiv F[\rho(\boldsymbol{r})], \tag{3.5}$$

coincides with the Fourier transform of the crystal's electron density. The latter is a three-dimensional periodic function, which can be described with the convolution

$$\rho_{\text{crystal}}(\boldsymbol{r}) = \sum_h \sum_k \sum_l \rho_{\text{unit cell}}(\boldsymbol{r})\delta(\boldsymbol{r} - n\boldsymbol{a} - m\boldsymbol{b} - p\boldsymbol{c}), \tag{3.6}$$

where $n$, $m$ y $p$ are integers, $\boldsymbol{a}$, $\boldsymbol{b}$, y $\boldsymbol{c}$ are the primitive cell translation vectors, and $\delta$ is Dirac's delta function. Now, applying Fourier's convolution theorem,

$$A(\boldsymbol{k}) = \hat{F}_{\boldsymbol{k}}[\rho_{\text{unit cell}}(\boldsymbol{r})] \sum_n \sum_m \sum_p \delta(\boldsymbol{k} - h\boldsymbol{a}^* - k\boldsymbol{b}^* - l\boldsymbol{c}^*), \tag{3.7}$$

we find that a perfect crystal does only disperse X-rays in directions $\boldsymbol{k} = h\boldsymbol{a}^* + k\boldsymbol{b}^* + l\boldsymbol{c}^*$, where $\boldsymbol{a}^*, \boldsymbol{b}^*, \boldsymbol{c}^*$ are the reciprocal lattice cell vectors. The dispersion amplitude in such directions is the Fourier transform of the unit cell electron density. We call this quantity **structure factor**,

$$F(h, k, l) = \hat{F}_{\boldsymbol{k}}[\rho_{\text{unit cell}}(\boldsymbol{r})] = \int_{\text{unit cell}} \rho_{\text{unit cell}}(\boldsymbol{r}) e^{2\pi i \boldsymbol{k}\cdot\boldsymbol{r}} d\boldsymbol{r}. \tag{3.8}$$

Whenever diffraction experiments are used with structural purposes, a simplified model in which the density is modelled via isolated spherical atoms is used such that

$$\rho_{\text{unit cell}}(\boldsymbol{r}) \simeq \sum_j \rho_{\text{atom},j}(\boldsymbol{r})\delta(\boldsymbol{r} - \boldsymbol{r}_j), \tag{3.9}$$

where $\rho_{\text{atom},j}(\boldsymbol{r})$ is taken as the electron density of a free atom, which is supposed not to change upon crystal formation. Using this approximation $F$ adopts the following simplified form,

$$F(h, k, l) = \sum_j f_j e^{2\pi i \boldsymbol{k}\cdot\boldsymbol{r}_j}, \tag{3.10}$$

where

$$f_j(h, k, l) = \hat{F}[\rho_{\text{atom},j}] = \int \rho_{\text{atom},j}(\boldsymbol{r}) e^{2\pi i \boldsymbol{k}\cdot\boldsymbol{r}} d\boldsymbol{r} \tag{3.11}$$

is the so-called *atomic form factor*, which coincides with the Fourier transform of the electron density of the atom in vacuum. Equation 3.10 allows to obtain the atomic positions within the unit cell by minimizing the deviation between the $F's$ experimental values and those calculated at trial atomic positions. This independent atom model can obviously not be used to obtain accurate electron densities from X-ray experiments. Several refined techniques bypass this limitation [35]. A fruitful framework, the Hansen-Coppens method [24], still uses atomic densities which are allowed to distort from sphericity up to a given multiple. By means of this and other improvements, experiments yield today crystalline electron densities of great quality.

The rough morphology of the electron density $\rho$ is well-known since calculations provided approximate molecular wavefunctions. It is strongly determined by the position of the system's nuclei, which provide the strongest of all the interactions felt by the electrons. Unfortunately, there are just a few analytical results known about the mathematical properties of $\rho$. The first and probably most important result is Kato's atomic theorem [42] about the nuclear cusp, generalized by Steiner y Bingel [12, 69] to molecular systems. If we define the spherically averaged density, $\bar{\rho(r)}$:

$$\bar{\rho(r)} = \frac{1}{4\pi} \int \rho(\boldsymbol{r}) \mathrm{sen}\theta d\theta d\phi, \tag{3.12}$$

then,

$$-2Z_\alpha = \left.\frac{\partial \ln \bar{\rho}(r)}{\partial |\boldsymbol{r} - \boldsymbol{R}_\alpha|}\right|_{r=R_\alpha}, \tag{3.13}$$

so that molecular densities display exponential cusps at the nuclear positions. Notice that this theorem allows to determine the position and charge of the nuclei from a simple analysis of $\rho$, in agreement with Hohenberg and Kohn's first theorem. Another analytical result regards the asymptotic behavior of the density in isolated molecules, where it is possible to show that

$$\bar{r} \sim e^{2r\sqrt{2\,IP}} \tag{3.14}$$

In this expression, IP is the first ionization energy of the system, and $Z_{\mathrm{total}}$ is the total number of protons in the molecule. Finally a number of inequalities related to the density are also known, such as the one for atoms demonstrated by Hoffmann-Ostenhof and Hoffmann-Ostenhof in 1977 [39]

$$-\frac{1}{2}\nabla^2\rho(\boldsymbol{r}) + \left(\mathrm{IP} - \frac{Z}{r}\right)\rho(\boldsymbol{r}) \leq 0. \tag{3.15}$$

Despite these positive results, some basic postulates, like that on the alleged *monotonicity* of atomic densities, have never been proven. Others, like convexity, have been falsified [2]. Anyhow, the gross features of molecular densities are relatively well known, exhibiting large nuclear cusps and an exponential decay as we get far from the nuclei in all directions (in an overwhelmingly large number of cases).

As an example, Fig. 3.1 shows a projection of the HF/6-311G(p, d) density for the HCN molecule in a plane that contains the three nuclei. The cusps on the carbon and nitrogen atoms have been cut to allow for the density around the H atom to be visible.

Note that $\rho$ is not a truly differentiable field, since there are cusps, not maxima, around the nuclei. We can nevertheless build a homeomorphic field that is equal to $\rho$ at all points save small neighborhoods around the nuclei. In these small regions we substitute the true non-differentiable density by an approximation showing a true local maximum. Under these conditions, the nuclear cusps become sinks or attractors of $\rho$, and the topology induced by the density through the attraction basins of its nuclear critical points becomes a natural partition. Note that this homeomorphic field is also naturally found when we introduce Gaussian basis sets.

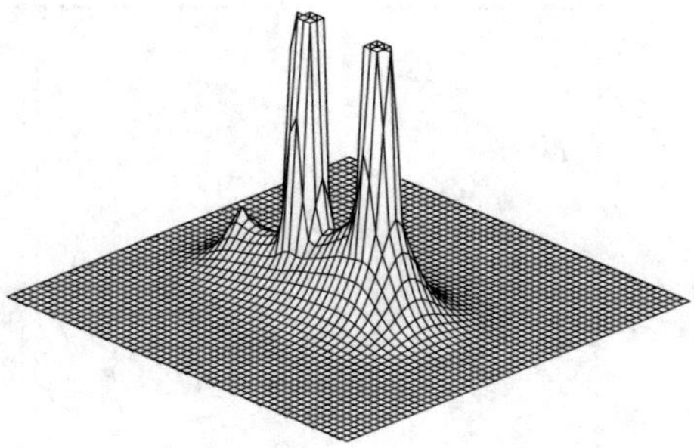

**Fig. 3.1** Relief electron density plot of HCN in a plane containing the nuclei. HF/6-311G(p, d) calculation

## 3.2 Topological Analysis of $\rho$

Let us now turn to the topology induced by $\rho$ through its attraction basins as we saw in Chap. 2. Since, in general, the attractors of this field coincide in number and position with the nuclei of the systems, there will exist as many three-dimension attraction basins of $\rho$ as atoms in the system. This partition is particularly pleasant to the chemist: by defining an atom in a molecule as the set formed by a nucleus and its attraction basin, we divide the physical space into non-overlapping atomic domains; or into functional groups, if we gather together appropriately such regions. The chemical importance of this formal link stems, in the first place, from a relevant empirical fact: the basins of equal atoms that find themselves in similar bonding situations turn out to be very similar in shape, density at their inner points, etc. In other words, the atomic basins of chemically-alike atoms are transferable. This recovers the fundamentals of Chemistry, where atomic properties are collected in order to predict group properties and molecular behavior in general. Figure 3.2 shows, as an example, the isolines of $\rho$ for the LiF and LiCl molecules calculated at the HF/6-311G* level, in which the separatrix between the two atomic basins is shown as a continuous line. Beyond this qualitative argument, there exist powerful theoretical reasons to embrace this topology. They will be postponed to subsequent sections.

Let us continue for now with the chemical characterization of the different types of critical points. To that end we will examine the topology of a set of model molecules. All the mathematically possible CPs have been found in both theoretical and experimental electron densities. And they are all endowed with a clear and intuitive meaning, although we warn against inflexible interpretations.

Our first example will be the lithium hydride molecule, LiH. Its ground state is a $^1\Sigma^+$ singlet with an internuclear distance approximately equal to 1.60 Å. In standard chemistry textbooks, we read that a *bond* exists between the lithium and hydrogen atoms. Figure 3.3a shows a relief representation of its electron density obtained at the Hartree-Fock level with a polarized triple-zeta basis set (HF/TZV* level) in a plane that contains the nuclei. Besides the two clear nuclear cusps we appreciate another critical point on the internuclear line. This is a minimum in the direction that connects the two atoms, and a maximum of the density in the two other orthogonal directions. The last assertion is clear after Fig. 3.3b, in which we represent the charge

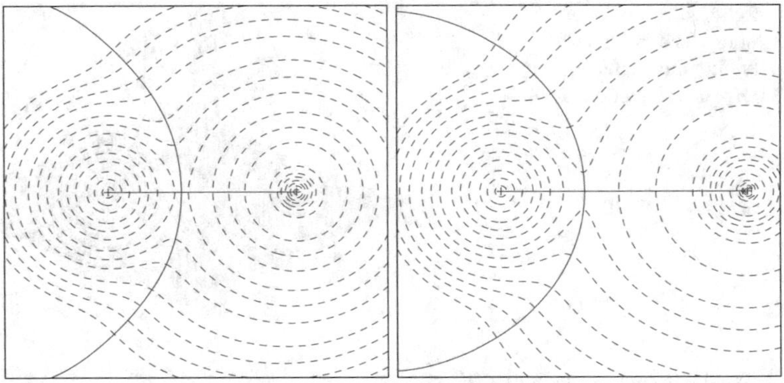

**Fig. 3.2** Isolines of $\rho$ for the LiF and LiCl molecules in a plane that contains the nuclei (Li atom on the left). The density has been obtained from HF/6-311G* functions. The scales are common, and the Li nuclei have been made to coincide in space. Full lines indicate the separatrices separating the two atomic basins. Notice the similarity between both Li atoms

density in a plane perpendicular to the internuclear axis passing through this new critical point. This is then a $(3, -1)$ CP, and due to these properties we call it a **bond critical point**. Its repulsion basin contains a pair of gradient lines that connect both nuclei. We call this line **bond path or bond line**, although it might be better called simply an **interatomic path** to avoid over-interpretations. We can also observe that the bond critical point lies closer to the lithium than to the hydrogen atom, for the density decreases much more quickly in the neighborhood of the former than of the latter.

There are other common ways to represent these density data. Sometimes it is useful to examine the isolines of the field, as in Fig. 3.3c, d. In this type of projections, the bond critical points are easily identified as saddle points. In other cases, it is necessary to look at the gradient field in more detail, displaying a set of field lines on their way to their final attractors, as in Fig. 3.3e, f. These diagrams clearly identify the different attraction and repulsion basins, allowing us to grasp the relative size of the different atomic basins as well as the structure of the separatrices. For instance, Fig. 3.3e shows very neatly the position of the bond critical point, its bond path (which coincides with its repulsion basin) and the separatrix between the two attractors (its 2D attraction basin).

Finally, it is also rather popular to examine sets of isosurfaces of the scalar field (better with an interactive computational package) so that we can vary the isoscalar value in a continuous way. Figure 3.4 shows this variation in our toy molecule. We check how at small enough density values ($\rho = 0.01$ a.u.) the density isosurface encloses both atoms. Actually, it has been shown that the isosurfaces of $\rho \approx 0.002$ a.u. are indeed very similar to the traditional van der Waals molecular envelopes. As we increase the value of the isodensity the isosurfaces adapt themselves to the atomic configuration, decreasing in volume. There exists a critical density value, in this case $\rho = 0.0401$ a.u., at which the isosurface splits and become multiply connected,

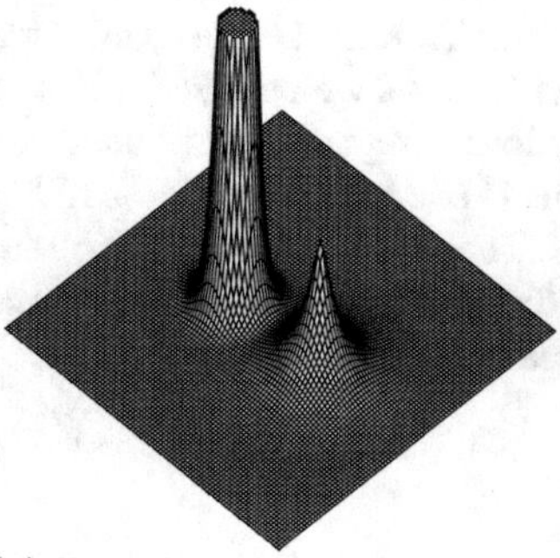

(a) Relief map in a plane containing the nuclei (plane 1).

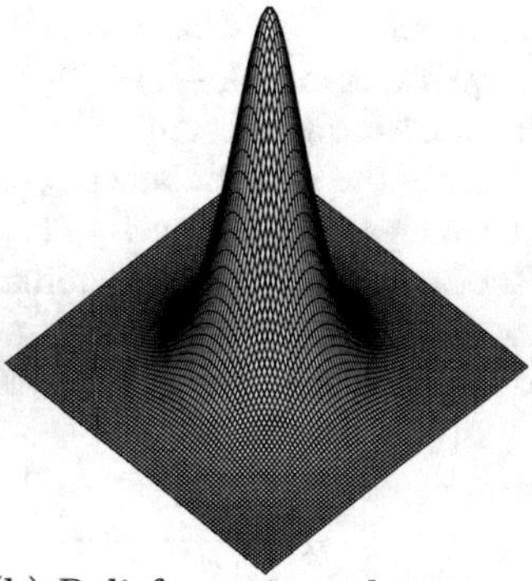

(b) Relief map in a plane perpendicular to the internuclear axis at the bond critical point (plane 2).

(c) Isolines in plane 1.

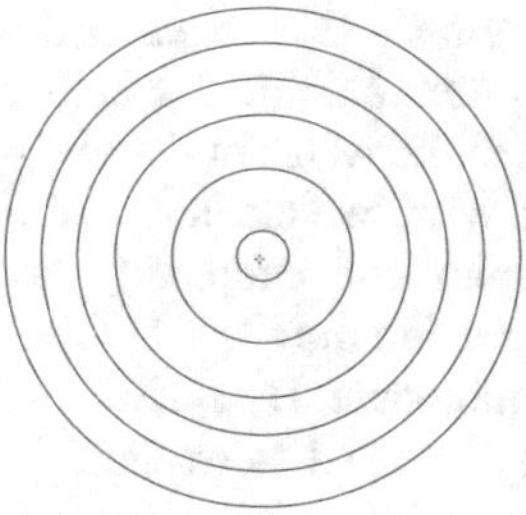

(d) Isolines in plane 2.

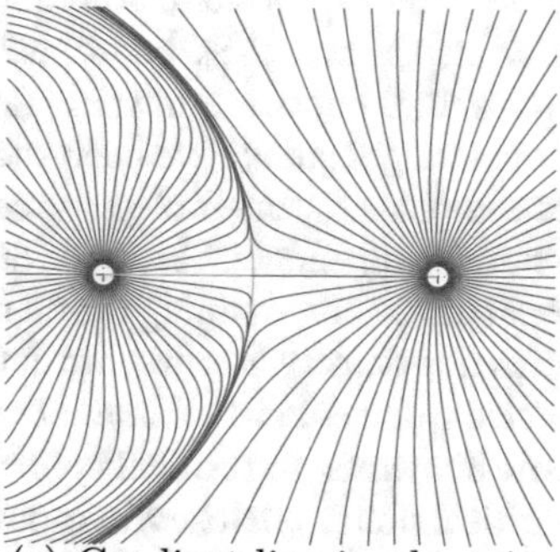

(e) Gradient line in plane 1.

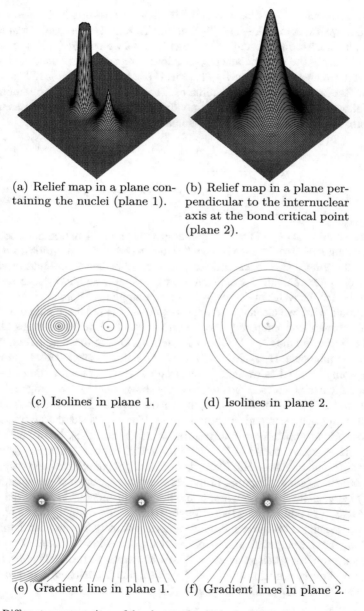

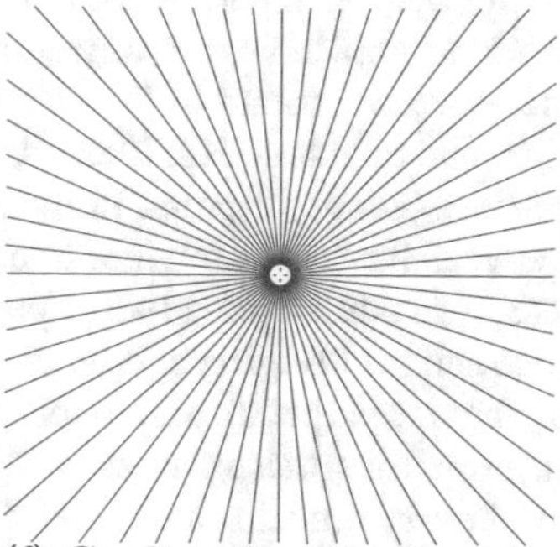

(f) Gradient lines in plane 2.

**Fig. 3.3** Different representations of the electron density in the LiH molecule changing its topology. It is easy to show that these topological changes can only occur when we traverse a $(3, -1)$ critical point. Actually, all the topological changes suffered by the isosurfaces occur on crossing CPs. As we can see in the Figure, if we further increase $\rho$ the two unconnected surfaces compact around their nuclei. If the density at the nuclear point of the lighter atom is overcome (at $\rho = 0.3297$ a.u.), only the isosurface enclosing the lithium remains. The overall process can be summarized by the following bifurcation diagram, in which the evolution of the connectivity of the density isosurfaces is shown as we increase the value of $\rho$:

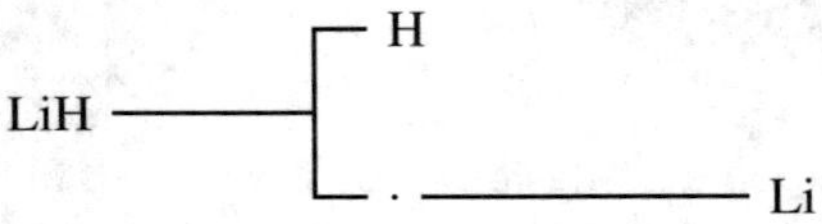


The association of $(3, -1)$ critical points as bond points is, in principle, empirical. It is surprising how an object so far from the standard chemical intuition corresponds so faithfully, via their associated bond paths, to what chemists identify with the dashes with which chemical graphs are drawn. As an example, Fig. 3.5 shows the gradient field of the water molecule in the molecular plane. There are two O-H bond critical points with two clear separatrices, but not a H-H bond critical point.

More complex and interesting cases corroborate the link between the chemical concept of bond and the $(3, -1)$ CPs of the density. One of the classical examples is the diborane molecule, $B_2H_6$, in which two B-H-B three-center two-electron bonds are commonly invoked to explain its electron structure, but not a direct B-B link. The topological analysis of its electron density shows exactly this bridged structure, as shown in Fig. 3.6. In the first place, there exist bond critical points between each of the bridging hydrogen and the boron atoms. These CPs share common features with those found until now, but a peculiarity stands out: their bond paths are not straight lines, as those found in the water molecule, but curve inwards (Fig. 3.6c). The curvature and total length of these far from straight paths have been successfully used as *bond tension* descriptors. In the second place, at the molecular center, just half way between the two boron atoms another CP appears that is not of the $(3, -1)$ type, but a $(3, +1)$ point. Along two orthogonal directions, defining the plane formed by the borons and the bridging hydrogen atoms, the density is minimum at this point, while along the line perpendicular to this plane the density is maximum. Due to these geometrical properties, these classes of CPs are known as ring critical points, and they tend to appear every time that a set of bond critical points enclose a two-dimensional region. The density bifurcation diagram of the molecule is as follows:

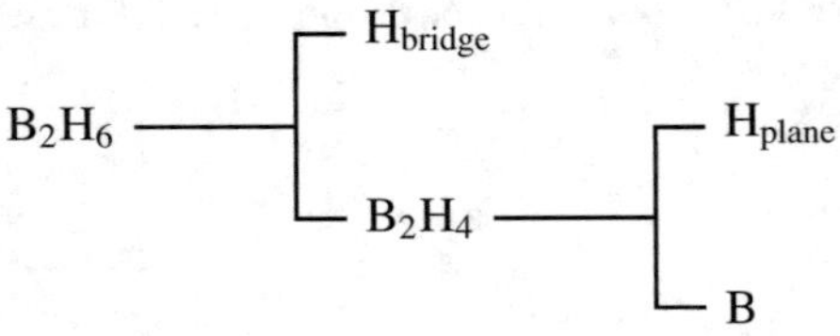

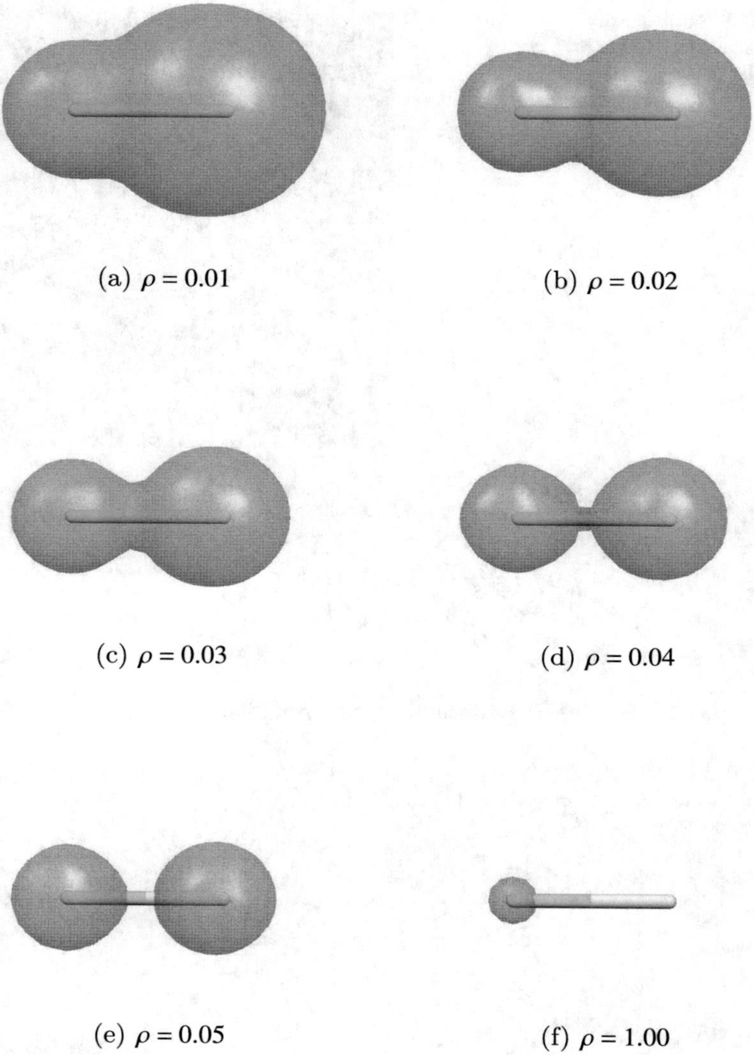


**Fig. 3.4** $\rho$ isosurfaces in LiH. All data in a.u

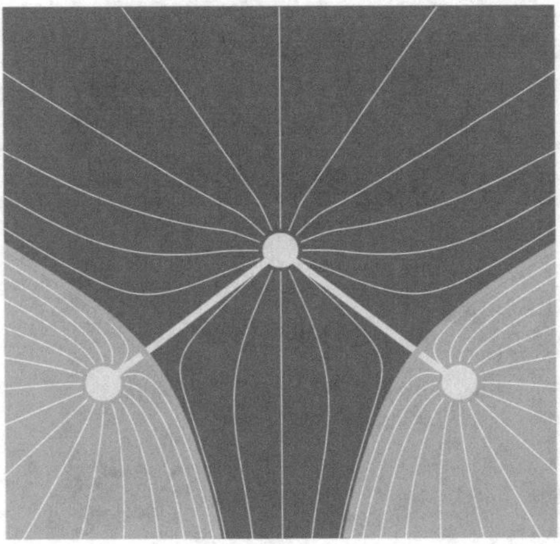

**Fig. 3.5** Electron density gradient field for $H_2O$ in the molecular plane

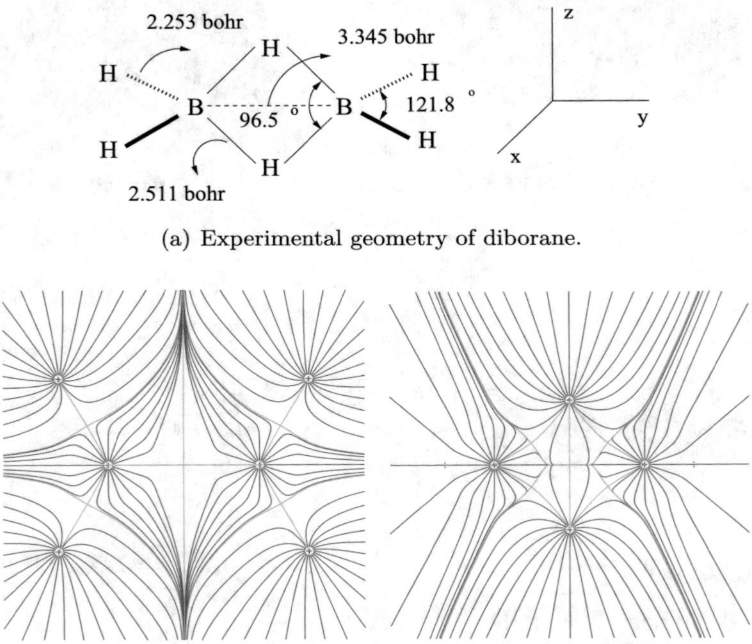


(a) Experimental geometry of diborane.

(b) Gradient field in the *xy* plane. (c) Gradient field in the *yz* plane.

**Fig. 3.6** Topology of $\rho$ in the diborane molecule. HF/6-311G* data

It is now clear that with the set of pairs of atoms linked by bond paths we can build a non-directed graph that we identify with the **molecular graph** (i.e. the "lines-joining atoms" representation of bonds). Through its now long history, the topological analysis we are now describing has convincingly shown that at equilibrium geometries the molecular graphs obtained with this technique are basically isomorphic to the graphs used in Chemistry. Since chemical dashes are not observables, but useful constructs, there will always be territory for discrepancy, and over the years a growing number of examples in which the chemical graph does not exactly coincide with the topological graph have been found. This has given rise to never ending debates which we will not reproduce in this introductory text (see Ref. [47], for instance, and the references therein). Anyhow, it is more than interesting that such a simple and basic object as the electron density encodes that kind of information in such a simple geometrical structure. We warn, however, that whenever we examine molecular geometries outside equilibrium or in the case of weak interactions, the probability of disagreement between the topological and the admitted chemical graphs increases. Do not forget, however, that both in the case of non-equilibrium as well as when weak interactions are involved, the rules for drawing chemical dashes do not really exist, so a direct comparison with the results of the topological method is simply out of place. Let us abound briefly on the first situation by commenting, for instance, on the meaning of the existence of a bond critical point between two helium atoms at any internuclear distance, forced by symmetry, even when the density we use to perform the analysis comes from a Hartree-Fock calculation, which leads to an unbound potential energy curve. There are researchers, like Paul Popelier from the University of Manchester, who consider it necessary to add the condition of vanishing forces over the nuclei before performing a topological analysis, thus introducing a difference between **bond path, bond interaction line, and the plain chemical concept of bond**. For others, the presence of a bond CP is always accompanied by an accumulation of electron density from regions orthogonal towards the bond line, a behavior that has always been related to bond formation. Be that as it may, topological chemical graphs have been successfully used in a wealth of non-standard situations, like excited electronic states or transition states along chemical reactions. Other scalar fields, like the Electron Localization Function (Chap. 4) and the Non-covalent Interaction index (Chap. 5), allow to differentiate between these different cases.

There is only one type of non-degenerate critical point that has not been yet commented, corresponding to a three-dimensional source of the density field. This is a local minimum of the electron density, a $(3, -3)$ point. Reasoning by analogy, it is likely that such a type of CP appears whenever a set of rings enclose a three-dimensional region. A simple example is found in the cubane molecule, $C_8H_8$, that exhibits a minimum of $\rho$ at the center of the cube.

A HF geometry optimization in this molecule leads to C—C and C—H distances of about 1.54 Å and 1.07 Å, respectively. In Fig. 3.7a we find the electron density in a $C_4H_4$ plane that passes through the center of the cube, and that contains all

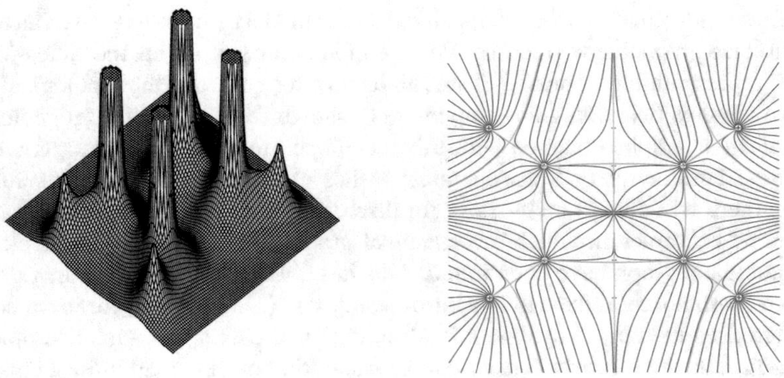

(a) Relief map. (b) Gradient field.

**Fig. 3.7** Electron density and its gradient field in a $C_8H_8$ plane of the cubane molecule that passes through the center of the cube center of the cube and contains a $C_4H_4$ fragment. The calculation is HF/TZV*

the possible types of non-degenerate CPs in three dimensions. The nuclear maxima corresponding to the C and H atoms are clearly visible in the relief map, as well as the $(3, -1)$ C-H bond critical points. The saddle points appearing between each pair of carbons are actually of two different types: those between the closest C atoms are bond critical points, while those between the more distant ones, lying at the center of the faces of the cube, are ring critical points. The geometrical center of the Figure contains a minimum, or cage critical point. There is additional chemical information in Fig. 3.7b, where we notice some bond tension. The vertical C—C bond lines are slightly bent. It has been found that the topological C—C bonding angle (formed by two C—C bond paths emanating from a given C atom) tends to be much closer to the 109° tetrahedral angle than the purely geometrical bond angle that is usually reported. This is a common observation. In tensioned bonds, the bond paths thus tend to recover the tetrahedral angles, curving in the process.

## 3.3 Properties at Critical Points

The existence of distinguished points in $\mathbb{R}^3$ effectively discretizes the physical space, so that instead of examining a complex 3D object itself we may try to correlate several other scalar or vector fields at those points with chemical concepts. All this leads, in a natural way, to linking bonds, rings or cages with the properties of the density, or of related quantities in the proximity of these CPs. For instance, given that it is a basic tenet of our knowledge of the electronic structure of molecules that the electron density accumulates along the internuclear axis upon bond formation, the value of $\rho_b$ itself at the bond critical point (BCP) could be used as a characteristic value, univocally defined for each type of bond, of the charge accumulation. The comparative study of a large set of quantities calculated at the different types of CPs in molecules has provided a rich set of bonding indices or descriptors. Many of them have been empirically found to be strongly correlated with a considerable number of physical and chemical properties: electronegativity, basicity, hardness, electrophilic or nucleophilic character, etc. We stress that, in most cases, these are statistical correlations lacking physical soundness. Notice that properties at other types of critical points can also be related to more specific chemical behaviors.

In classical topological analysis there are three properties computed at BCPs which are widely spread: **the density,** $\rho_b$, the **ellipticity,** $\epsilon$ and the **bond radius,** $r_b$. Similarly, ring critical point properties have been related to ring-specific features, like aromaticity [60], and so on.

### *3.3.1 The Density at a Bond Critical Point*

The quantitative value of the density at a BCP is clearly related to the alleged charge accumulation in bonding regions. It is, however, not simple to compare $\rho_b$ values for links between different pairs of bonded atoms, since $\rho_b$ depends, above all, on the parental atomic densities, which scale as the third power of the nuclear charge, roughly [6]. This precaution should always be taken, and it is only safe to compare $\rho_b$ values for sets of molecules that share the same pair of bonded atoms. In these cases, good correlations have been found between bond critical point densities and standard measures of bond strength.

Table 3.1 contains some numerical results obtained in the second and third period diatomic hydride molecules. We can first check that the values of $\rho_b$ in LiH, BeH, NaH and MgH are small, well separated from those in the rest of the systems. This is usually interpreted in terms of the large ionic character of these links, bound mainly due to electrostatic forces rather than by the accumulation of density in the internuclear axis. We can also observe how the bond density increases over a period and how it can be considered a periodic property.

When $\rho_b$ is compared across the same pair of atoms it can even be used to define a topological bond order measure, $n_b$ which is obviously invariant under orbital transformations. This was first proposed for C—C bonds in ordinary organic molecules. For instance, the values of $\rho_b$ for the carbon-carbon bonds in the series ethane, ethene and ethyne are, approximately, 0.249, 0.356 y 0.426 a.u., respectively. A simple linear fit of these numbers to the classical bond orders 1, 2 and 3, provides $n_b = 1.6$ for the C—C bond in benzene, in reasonable agreement with other estimations based on completely different grounds. For instance, Coulson's Hückel-like bond order for benzene is 1.444. The $n_b$–$\rho_b$ fits are different for each pair of atoms, tending to provide bond order values in line with those usually accepted. More detailed analyses, as well as theoretically based arguments lead to the conclusion that $\rho_b$ is an exponentially decreasing function of the bond distance for a wide window of interatomic distances. Since the relation between bond distance and bond order is one of the

**Table 3.1** Electron density, $\rho_b$, bond radius, $r_b$ and equilibrium distance, $R_e$, for diatomic hydrides of the second and third period. All calculations are reported at the B3LYP/6-311+G(2d,2p)//HF/6-31(d) level, in a.u

| System | $\rho_b$ | $r_b$(H) | $R_e$ |
|---|---|---|---|
| HH | 0.270 | 0.690 | 1.379 |
| LiH | 0.038 | 1.723 | 3.091 |
| BeH | 0.095 | 1.454 | 2.547 |
| BH | 0.192 | 1.279 | 2.316 |
| CH | 0.281 | 0.711 | 2.094 |
| NH | 0.336 | 0.531 | 1.935 |
| OH | 0.372 | 0.380 | 1.811 |
| FH | 0.380 | 0.285 | 1.721 |
| NaH | 0.032 | 1.709 | 3.618 |
| MgH | 0.050 | 1.619 | 3.304 |
| AlH | 0.076 | 1.590 | 3.122 |
| SiH | 0.117 | 1.457 | 2.863 |
| PH | 0.167 | 1.311 | 2.665 |
| SH | 0.217 | 0.904 | 2.513 |
| ClH | 0.249 | 0.707 | 2.393 |

pillars of chemical reasoning, it was soon proposed that the bond order concept is simply a scaled version of $\rho_b$. Today these arguments have been qualified through the use of second order density matrix arguments, which we will explore in Sect. 3.5.1.

### 3.3.2 *Ellipticity*

The density along a bond path is minimum at a BCP. At this point the hessian matrix is characterized by a positive eigenvalue, with eigenvector tangent to the bond path and two negative eigenvalues that generate the tangent plane to the interatomic surface at the BCP. Following standard practice, let us call these three curvatures, $\lambda_1$, $\lambda_2$ and $\lambda_3$, where $\lambda_1 < \lambda_2 < \lambda_3$. In a diatomic molecule in an axially symmetric electronic state, $\lambda_1 = \lambda_2$, and the perpendicular density accumulation toward the internuclear axis is axially isotropic. We then define the ellipticity at a bond critical point as the ratio $\epsilon = \lambda_1/\lambda_2 - 1$. This quantity thus measures the asymmetry in the perpendicular density accumulation, determining the easy and hard directions for this accumulation of charge. The standard example found in most texts is the ellipticiy of the C—C bond in the ethylene molecule, $\epsilon \approx 0.30$, which indicates a clear accumulation of charge in the plane that is orthogonal to the molecular plane, a result that can be interpreted in terms of the $\pi$ cloud above and below the plane. Ellipticity is then related to the $\pi$ character of a bond, and the value $\epsilon = 0.23$ that one finds for the C—C bond in benzene matches a partial double bond nicely.

We find this an appropriate moment to warn about over-interpretations in quantum chemical topology. In the first place, the $\pi$ character of a chemical bond is not an observable quantity, but one based on a given type of orbital description. A rotation from canonical $\sigma - \pi$ to banana-like orbitals destroys the $\pi$-bond concept, for instance. Similarly, $\epsilon$ is actually sensing **anisotropies** in the density distribution, and does not contain any direct information of orbital contributions to such anisotropy. The case of the ethyne molecule, in which the axial symmetric distribution leads to $\epsilon = 0$ makes a particularly strong case. According to the naïve interpration of the ellipticity concept, the C—C bond in acetylene has no $\pi$ character. Surely no-one would get trapped in such a case, but when $\epsilon$ is used in not so clear problems everything blurs. For instance, the B—H link between the boron and the bridging H atoms in diborane has a large ellipticity, with $\epsilon = 0.33$. This case is clearly more related to the curvature of the bond path found in Fig. 3.6 than to any $\pi$ contribution. For some, we should actually turn our arguments upside down, and we should start to think about the possibility that some of the properties we usually assign to $\pi$ bonds are simply reflecting some kind of asymmetry of the density distribution.

### *3.3.3 Bond Radii*

The distance from the bond critical point to each of the two nuclei with whom it binds is called the bond or bonding radius of each of the nuclei. It is interesting to recall that this quantity was already used as a measure of bond radius back in the 1960s when Shannon [68] introduced the crystallographic radius. In simple symmetric crystals like NaCl, in which experimentally accurate electron densities were starting to be determined, the point along the internuclear axis in which the electron density is minimum actually coincides with the topological bond critical point. Shannon proposed it as an experimental definition of ionic radius. The great success of his proposal can be immediately translated to its topological analogue, with the added value that it provides a generalization beyond ionic systems, i.e. an effective size of an atom in a molecule along the bond direction. Figure 3.8 shows how these radii evolve for the second row hydrides. We can check how the size of the neutral atoms decreases as we traverse the period. This fact reflects the increase in effective nuclear charge, $Z^*$. We will see in future sections that the formal charge of the H atom becomes more and more positive as we move from Li to F. Notice also that the curvature of the separatrix at the BCP changes sign in methane, as the difference in electronegativity between the central atom and hydrogen does also change sign. This observation is interesting to keep in mind as a visual qualitative approach to relative electronegativity.

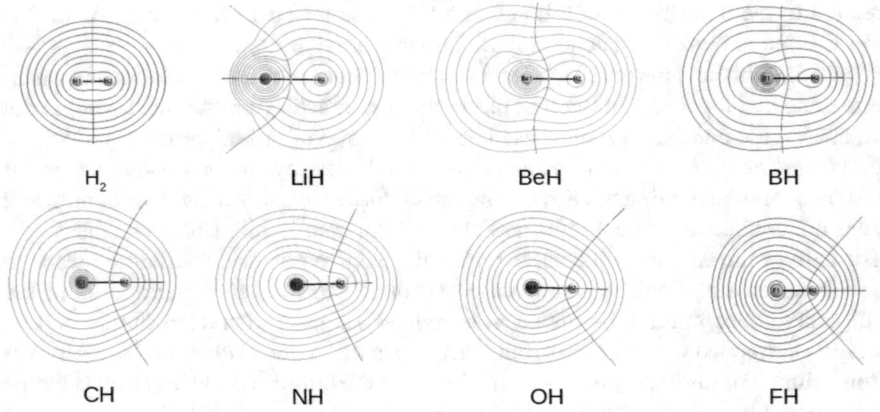


**Fig. 3.8** Isolines of the electron density $\rho$ together with the separatrix and the position of the bond critical point for second period diatomic hydrides computed at the B3LYP/aug-cc-pVTZ level. The outermost isoline corresponds to $\rho = 0.001$ a.u. in each case

## 3.4 The Quantum Theory of Atoms in Molecules

We have shown how the topological analysis of some scalar fields based on the electron density provide an orbital invariant chemical intuitive picture of quantum mechanical calculations. Upon partitioning the physical space into atomic regions, and after inducing pairwise relationships among those regions that can be linked to chemical bonds, we start to understand how a new language to make quantum chemistry go back to chemists might be devised.

There are much deeper reasons that justify (for some researchers only partially) the intimate correlation between the quantities defined in previous sections and the real world. R. F. W. Bader and coworkers tried to show that the topology induced by $\rho$ stems from the generalization of quantum mechanics (QM) to subsystems (or regions) in $\mathbb{R}^3$. The partition of the space by means of the separatrices of the gradient field of the density warrants that QM be still applicable to those open regions [6].

The generalization of QM to subsystems has been a classical recurring theme from the 1930s. Such an enterprise faces important obstacles that are not easy to overcome: among them, the presence of **surface terms** that do not necessarily vanish within the finite limits of the subsystems. Such difficulties strike the very pillars of QM, since the Hermitian character of its basic operators (particularly those depending on the momentum operator, $\hat{\mathbf{p}}$) is based on the vanishing of those surface terms (see the example of the kinetic energy density in Sect. 3.4.2). The quantum theory of atoms in molecules (QTAIM) provides a possible way out to from this problem.

The QTAIM is a theory of open subsystems, in the thermodynamic sense: since the number operator, $\hat{N}$, does not commute with position operators, no partition of the space can assign a definite number of electrons to a region in $\mathbb{R}^3$. Within a subsystem we can only speak about the average number of particles, so that its frontiers are

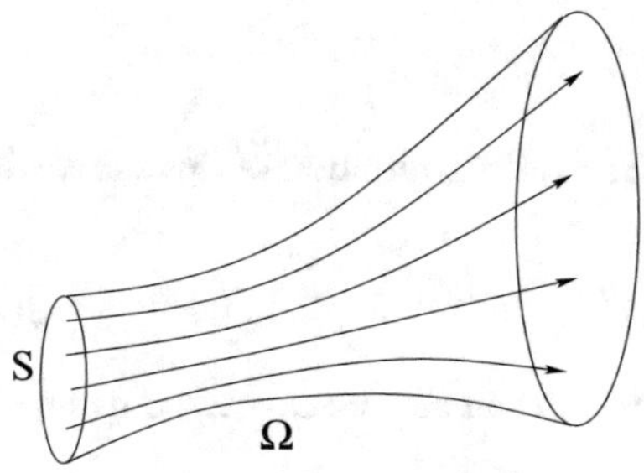


**Fig. 3.9** Flux of an incompressible fluid through a varying section pipe. Matter conservation forces a change in the fluid's velocity between the entry and the exit

necessarily permeable to electrons. The QTAIM provides expectation values for the traditional quantum chemically relevant operators within those subsystems. We will call them **atomic observables**. As we will see, most standard molecular observables can be reconstructed from atomic observables.

The atomic properties provided by the QTAIM satisfy a number of theorems, usually obvious generalizations of those satisfied by the total molecular expectation values. Gathering a set of atoms to form what in chemistry is known as a *functional group* leads to QTAIM group properties that recover the additivity known to the experimentalist. For instance, the QTAIM group dipole moments, energies, polarizabilities, etc, are in extremely good agreement with those experimentally tabulated. It can be said that the QTAIM provided, for the first time, a QM basis for the chemical transferability of functional groups. Summarizing, it is the stability of the shape and inner features of the atomic basins that justifies it. Since it is relatively important to keep track of units in this section, we will maintain the use of $\hbar$ and $m$, the electron mass, which we usually drop when working in atomic units.

### *3.4.1 Surface Terms*

In 1926, Max Born [16] proposed a probabilistic interpretation of the wavefunction. According to this orthodox Copenhaguen interpretation, the complex square of the wavefunction, $\Psi^*\Psi = |\Psi|^2$, must be interpreted as the probability density of finding the system in a given spatial (or spinspatial) configuration. The origin of this interpretation, within a non-relativistic framework, comes from particle number conservation. It is shown that the probability density fulfils a continuity equation, so that no sources or sinks exist for it. Using a hydrodynamic analogy, the QM probability density is incompressible (Fig. 3.9).

Let us consider a fluid of density $\rho$ running with velocity $\boldsymbol{v}$ through a pipe. The mass of the fluid enclosed in a given region, $\Omega$, varies over time due to the flow across the separation surface of $\Omega$, $S$. That mass can be quantified in terms of the density of the fluid (which is a local quantity) and the volume $\Omega$ as follows:

$$m = \int_{\Omega} \rho \, d\boldsymbol{r}. \tag{3.16}$$

The rate of variation of this mass with time will be

$$\frac{dm}{dt} = \int_{\Omega} \frac{\partial \rho}{\partial t} d\boldsymbol{r}, \tag{3.17}$$

which can also be expressed using the mass current density $\boldsymbol{j} = \rho \boldsymbol{v}$,

$$\frac{dm}{dt} = -\oint_{S} \boldsymbol{j} \cdot d\boldsymbol{S}, \tag{3.18}$$

This integral form shows that the change in the total mass enclosed by $\Omega$ corresponds to the flux of $\boldsymbol{j}$ across the frontier surface, $S$.

In the absence of mass sources or sinks we can equate expressions (3.17) y (3.18), so that applying the divergence theorem we obtain

$$\int_{\Omega} \frac{\partial \rho}{\partial t} d\boldsymbol{r} = -\oint_{S} \boldsymbol{j} \cdot d\boldsymbol{S} = -\int_{\Omega} \nabla \cdot \boldsymbol{j} \, d\boldsymbol{r}. \tag{3.19}$$

Reorganizing slightl y,

$$\int_{\Omega} \left( \frac{\partial \rho}{\partial t} + \nabla \cdot \boldsymbol{j} \right) d\boldsymbol{r} = 0, \tag{3.20}$$

an expression valid for any volume $\Omega$, thus also locally, so that

$$\frac{\partial \rho}{\partial t} + \nabla \cdot \boldsymbol{j} = 0, \tag{3.21}$$

which is known as the continuity equation of an incompressible fluid.

Conservation of the number of particles in non-relativistic QM leads to a similar equation, written in terms of a probability current density, $\boldsymbol{j}$, and Born's probability density $\rho = \Psi^* \Psi$. The fulfilment of Eq. 3.21 by this pair of objects is what allows us to interpret $\rho$ as a true, faithful probability density, thus the existence of neither sources nor sinks of particles.

In order to simplify, we will consider in the following a simple system, composed by a spinless particle. If $\rho = |\Psi|^2 = \Psi^* \Psi$, then

$$\frac{d}{dt} \int_{\Omega} \Psi^* \Psi \, d\boldsymbol{r} = \int_{\Omega} \frac{\partial}{\partial t} \left( \Psi^* \Psi \right) d\boldsymbol{r}, \tag{3.22}$$

so that

$$\frac{\partial}{\partial t} \left( \Psi^* \Psi \right) = \dot{\Psi}^* \Psi + \Psi^* \dot{\Psi}. \tag{3.23}$$

This equation allows us to introduce Schrödinger's evolution equation by using $\hat{H}\Psi = i\hbar\dot{\Psi}$. All the expressions we will obtain are simple, intuitive and generalizable to the multiparticle case. The single particle obeys

$$\hat{H}\Psi = -\frac{\hbar^2}{2m}\nabla^2\Psi + V(\boldsymbol{r})\Psi. \tag{3.24}$$

taking its complex conjugate,

$$(\hat{H}\Psi)^* = -\frac{\hbar^2}{2m}\nabla^2\Psi^* + \hat{V}(\boldsymbol{r})\Psi^*, \tag{3.25}$$

and substituting both expressions in Eq. 3.23,

$$\begin{aligned}\frac{\partial\rho}{\partial t} &= \frac{-i}{\hbar}\left[\frac{\hbar^2}{2m}\nabla^2\Psi^*\Psi - \hat{V}(\boldsymbol{r})\Psi^*\Psi - \Psi^*\frac{\hbar^2}{2m}\nabla^2\Psi + \Psi^*\hat{V}(\boldsymbol{r})\Psi\right] \\ &= \frac{i\hbar}{2m}\left[\Psi^*\nabla^2\Psi - \Psi\nabla^2\Psi^*\right].\end{aligned} \tag{3.26}$$

This expression summarizes the time evolution of the charge density, and does not include the potential operator anymore. We can compute the evolution of the density by using the wavefunction and its derivatives only.

Taking into account that $\nabla^2 = \nabla \cdot \nabla$, we can rewrite the above equation in terms of the divergence of a gradient as follows:

$$\frac{\partial\rho}{\partial t} = \frac{i\hbar}{2m}\nabla\cdot\left[\Psi^*\nabla\Psi - \Psi\nabla\Psi^*\right]. \tag{3.27}$$

Now, defining

$$\boldsymbol{j} = -\frac{i\hbar}{2m}\left[\Psi^*\nabla\Psi - \Psi\nabla\Psi^*\right], \tag{3.28}$$

we can write

$$\frac{\partial\rho}{\partial t} + \nabla\cdot\boldsymbol{j} = 0. \tag{3.29}$$

As we see, this expression is equivalent to that in Eq. 3.21, and allows us to identify $\boldsymbol{j}$ as a true probability density current: it defines how the charge density varies due to its time flux.

### 3.4.2 Hermiticity

If, instead of using a specific form for the Hamiltonian operator as we have just done, we use Schrödinger's equation to simplify the time dependence of the state vector

$\Psi$, ($\dot{\Psi} = (-i/\hbar)\hat{H}\Psi$), then Eq. 3.23 can be rewritten in the following way,

$$\frac{\partial}{\partial t}\left(\Psi^*\Psi\right) = \frac{-i}{\hbar}\left[-(\hat{H}\Psi)^*\Psi + \Psi^*(\hat{H}\Psi)\right]. \tag{3.30}$$

Using this expression, it is easy to verify that for a closed systems with frontiers at infinity,

$$\frac{\partial}{\partial t}\int \Psi^*\Psi\, d\boldsymbol{r} = 0, \tag{3.31}$$

if and only if the two members of Eq. 3.30 are equal,

$$\int (\hat{H}\Psi)^*\Psi\, d\boldsymbol{r} = \int \Psi^*(\hat{H}\Psi)\, d\boldsymbol{r}. \tag{3.32}$$

In other words, the conservation of particle number, Eq. 3.31 is equivalent to the fulfilment of condition Eq. 3.32 on the $\hat{H}$ operator. We recognize here the usual notion of Hermitian character in $\mathcal{L}^2(\mathbb{R}^3)$, $\hat{H} = \hat{H}^\dagger \Leftrightarrow \langle\Psi|\hat{H}|\Psi\rangle = (\langle\Psi|\hat{H}^\dagger|\Psi\rangle)^*$. $\hat{H}$ must therefore be Hermitian.

The presence of a surface term in the continuity equation opens a path to generalize the concept of hermiticity to open subsystems. In the case of the Hamiltonian, by repeating the argument that has led to Eq. 3.32, although keeping the surface term, we arrive at

$$\int_\Omega \left[(\hat{H}\Psi)^*\Psi - \Psi^*(\hat{H}\Psi)\right] d\boldsymbol{r} = \oint_S \boldsymbol{j}\cdot d\boldsymbol{S}, \tag{3.33}$$

and the condition that needs to be satisfied so that the hermiticity of $\hat{H}$ is warranted imposes a physical requirement on the system: the current density must have a vanishing flux through the surface $S$ that limits the system.

We come to the following conclusions:

1. Let us suppose that $\Omega \to \mathbb{R}^3$; then

$$\frac{\partial}{\partial t}\int_{\mathbb{R}^3} \Psi^*\Psi\, d\boldsymbol{r} = 0 \Leftrightarrow \oint_{\partial\mathbb{R}^3} \boldsymbol{j}\cdot d\boldsymbol{S} = 0. \tag{3.34}$$

   This condition is fulfilled for well-behaved state vectors, for then the wavefunction and its derivatives (like $\boldsymbol{j}$) vanish at infinity. Actually, it is this property that is used to define properly the concept of well-behaved function. We say that a wavefunction is well-behaved when it fulfils a continuity equation and when the flux of $\boldsymbol{j}$ through a surface at infinity is null.

2. For a bound system, $\Psi$ and its derivatives vanish at infinity, so that the fulfilment of the zero flux condition is assured. This is usually written in terms of the Hermitian character of $\hat{H}$:

$$\hat{H} = \hat{H}^\dagger \Leftrightarrow \oint_{\partial\mathbb{R}^3} \boldsymbol{j}\cdot d\boldsymbol{S} = 0. \tag{3.35}$$

3. If we limit the analysis to a finite subsystem $\Omega$, then

$$\int_{\Omega} \left[ (\hat{H}\Psi)^* \Psi - \Psi^* (\hat{H}\Psi) \right] dr = 0 \Leftrightarrow \oint_{S=\partial\Omega} j \cdot dS = 0, \tag{3.36}$$

and the hermiticity of operators is not assured in the usual sense. If we wish that the structure of QM is kept unaltered in open subsystems, we must demand the fulfilment of new conditions at the subsystem frontiers.[1]

Another important issue that appears when trying to generalize QM to subsystems is related to the definition of operators. Even the founders of the theory, in particular Dirac, understood that expressions that are equivalent in classical mechanics need not be so in the new formalism. For instance, given the incompatibility between the $p$ and $x$ operators, the classical form $p^2x$ admits several possibilities in QM: $xpp$, $pxp$, $ppx$. Dirac proposed to symmetrize all this possibilities in order to define the classical analogue to such an expression.

In our case, and by using the standard correspondence rules, a similar difficulty appears when we find that it is possible to write different expressions for an operator whose expectation values are equal when integrating over $\mathbb{R}^3$, but which differ in surface terms when a subsystem is chosen. All these forms, which are equivalent in $\mathbb{R}^3$, cease to be so in subsystems.

One of the basic operators in quantum chemistry that is affected by this type of indefinition is the kinetic energy operator, $\hat{T}$. In what follows, and with the purpose of avoiding the additional problem that comes from the fact that expectation values in subregions may not be real valued, we will symmetrize all the following results in $\mathbb{C}$, defining

$$\langle \hat{A} \rangle_\Omega = \frac{1}{2} \int_\Omega \Psi^* \hat{A} \Psi \, d\tau + \frac{1}{2} \int_\Omega \Psi \hat{A} \Psi^* \, d\tau. \tag{3.37}$$

In the case of the kinetic energy operator, $\hat{T}$,

$$\hat{T} = \frac{p^2}{2m} = \frac{-\hbar^2}{2m} \nabla^2, \tag{3.38}$$

and upon acting on the probability density,

[1] It is interesting to recall that in the beginnings of quantum mechanics a significant debate arose related to the difficulty to find relativistic wave equations with positive definite density. For instance, the Klein–Gordon equation [40] $(\Box^2 - m^2c^4)\Psi = 0$ (The D'Alambertian operator is defined as $\Box^2 = \nabla^2 - (1/c^2)\partial^2/\partial t^2$), which was the first relativistic equation to be proposed, has solutions with $\rho > 0$ and solutions with $\rho < 0$. The problem was finally solved after a revolution that proposed the existence of particles with proper positive energy as well as of a new set of $E < 0$ counterparts (antiparticles).

$$\frac{-\hbar^2}{2m}\nabla^2(\Psi^*\Psi) = \frac{-\hbar^2}{2m}\nabla\cdot\nabla(\Psi^*\Psi)$$
$$= \frac{-\hbar^2}{2m}\left[\nabla^2\left(\Psi^*\Psi\right) + 2\left(\nabla\Psi^*\right)\left(\nabla\Psi\right) + \Psi^*\nabla^2\Psi\right], \quad (3.39)$$

or, generalizing to a multielectron system,

$$\frac{-\hbar^2}{2m}\sum_i \nabla_i^2(\Psi_i^*\Psi_i) = \frac{-\hbar^2}{2m}\sum_i(\Psi_i^*\nabla_i^2\Psi_i + \Psi_i\nabla_i^2\Psi_i^* + 2\nabla_i\Psi_i^*\nabla_i\Psi_i). \quad (3.40)$$

We recognize in the rightmost part of the previous equality two terms similar to those in Eq. 3.37, together with a third one whose meaning we now comment upon.

Each of the terms can be integrated out over the spin coordinates of the particle under consideration and over all the coordinates (spinspatial) of the rest. Taking into account the indistinguishability of the electrons,

$$N\frac{-\hbar^2}{4m}\int\nabla_1^2(\Psi^*\Psi)ds_1d\boldsymbol{x}_1\ldots d\boldsymbol{x}_N = N\frac{-\hbar^2}{4m}\int(\Psi^*\nabla_1^2\Psi + \Psi\nabla_1^2\Psi^*)ds_1d\boldsymbol{x}_2\ldots d\boldsymbol{x}_N$$
$$+ N\frac{-\hbar^2}{2m}\int\nabla_1\Psi^*\nabla_1\Psi ds_1d\boldsymbol{x}_2\ldots d\boldsymbol{x}_N. \quad (3.41)$$

Since we will be singling out electron 1 in the following, it is useful to tell coordinates in $\Psi^*$ from those in $\Psi$ by labelling the first ones with an apostrophe. This allows us to differentiate them independently before performing any integration step. The primed coordinates are tacitly assumed to be made equal to the unprimed ones when the integration is finally done:

$$\int f(\boldsymbol{r};\boldsymbol{r}')d\boldsymbol{r} \equiv \int f(\boldsymbol{r};\boldsymbol{r}')|_{\boldsymbol{r}'\to\boldsymbol{r}}d\boldsymbol{r}. \quad (3.42)$$

We can recognize the $\rho(\boldsymbol{r}_1;\boldsymbol{r}_1')$ density matrix in Eq. 3.41 (see Sect. 3.5 for details). Relabelling $\boldsymbol{r}_1$ as $\boldsymbol{r}$ and integrating over $\boldsymbol{r}$, we arrive at:

$$\frac{-\hbar^2}{4m}\int_\Omega\nabla^2\rho(\boldsymbol{r})\,d\boldsymbol{r} = \frac{-\hbar^2}{4m}\int_\Omega(\nabla^{2\prime}+\nabla^2)\rho(\boldsymbol{r};\boldsymbol{r}')d\boldsymbol{r} + \frac{-\hbar^2}{2m}\int_\Omega\nabla'\nabla\rho(\boldsymbol{r};\boldsymbol{r}')d\boldsymbol{r}. \quad (3.43)$$

Applying now the divergence theorem to the left hand side,

$$\frac{-\hbar^2}{4m}\int_\Omega(\nabla^{2\prime}+\nabla^2)\rho(\boldsymbol{r};\boldsymbol{r}')d\boldsymbol{r} = \frac{\hbar^2}{2m}\int_\Omega\nabla'\nabla\rho(\boldsymbol{r};\boldsymbol{r}')d\boldsymbol{r} - \frac{\hbar^2}{4m}\oint_S\nabla\rho\cdot d\boldsymbol{S}. \quad (3.44)$$

This last relation allows us to introduce two plausible kinetic energy densities, $K$ and $t$,

$$K = -\frac{\hbar^2}{4m}(\nabla^{2\prime} + \nabla^2)\rho(\boldsymbol{r};\boldsymbol{r}'),$$
$$t = \frac{\hbar^2}{2m}\nabla' \cdot \nabla\rho(\boldsymbol{r};\boldsymbol{r}'), \tag{3.45}$$

as well as an operator directly related to the Laplacian of the electron density,

$$L = -\frac{\hbar^2}{4m}\nabla^2\rho. \tag{3.46}$$

Note that $t$ is also used frequently as $G$ in the context of QTAIM, but we have favored the $t$ notation in order to be coherent with the most common notation used when dealing with electron localization (Chap. 4).

Equation 3.44 is usually rewritten in terms of these quantities,

$$\int_\Omega K(\boldsymbol{r})d\boldsymbol{r} = \int_\Omega t(\boldsymbol{r})d\boldsymbol{r} - \int_\Omega \nabla^2\rho d\boldsymbol{r},$$
$$K_\Omega = t_\Omega + L_\Omega. \tag{3.47}$$

Note that for a closed system, the density decays exponentially to zero at infinity and so does $\nabla^2\rho$, meaning that the right-most term becomes null. Hence, $K$ and $t$ become equivalent in this case.

These simple manipulations are a clear example of some of the difficulties that we have previously introduced. On the one hand, we see how the kinetic energy admits several non-equivalent definitions in subsystems, characterized by different properties. For instance, $t$ is a positive definite density, and it is usually known as the "positive-definite kinetic energy density" (being thus closer to classical concepts), while $K$ is not. This is easily proven, for instance, if we introduce a natural expansion of the density matrix $\rho(\boldsymbol{r};\boldsymbol{r}') = \sum_i n_i\phi_i(\boldsymbol{r})\phi_i^*(\boldsymbol{r}')$ for $r \to r'$. Then,

$$t(\boldsymbol{r}) = \frac{\hbar^2}{2m}\sum_i n_i \nabla\phi_i(\boldsymbol{r}) \cdot \nabla\phi_i^*(\boldsymbol{r}), \tag{3.48}$$

and each term is the norm of a vector, thus positive. $K$, on the contrary, contains the Laplacian of orbitals, which oscillates in sign. This means that if we choose $K$ as the kinetic energy functional, we will very likely find negative values at some points in space. It is interesting to recall that Schrödinger chose $t$ initially in its wavefunction formulation of usually obvious generalizations of those satisfied by the total molecular expectation values. In the second place, the expectation values of these two densities in subsystems differ in a surface term, which obviously vanishes if the surface extends to infinity. Finally, both expectation values will also be equal if the boundary of our integration domain, $S = \partial\Omega$, is a **zero flux surface** of the $\nabla\rho$ field. This is precisely the condition we have introduced based on the electron density to induce a topology in $\mathbb{R}^3$. It is possible to show that a wide family of kinetic energy densities [22] known as Laplacian densities, obtained from Wigner's distribution function, provide identical results when they are integrated over regions limited by zero flux surfaces of $\nabla\rho$. Their general expression is $t(\boldsymbol{r}) + \epsilon\nabla^2\rho(\boldsymbol{r})$, where $\epsilon$ is an arbitrary parameter, so that their equivalence when integrated over a zero-flux bounded region is obvious. Since in the standard non-relativistic Hamiltonians the rest of the energetic contributions do not depend on momentum operators we can conclude that, under this approximation, the regions defined by the QTAIM possess well-defined expectation values for all the energetic contributions. This is one of the main advantages of the electron density over other fields.

### 3.4.3 Time Evolution of Open Systems

In this subsection we will briefly present some results related to the time evolution of open quantum subsystems. All of these results are heirs of several general equations of continuity, and are formalized considerably more easily in Heisenberg's picture. We will consider a spinless one-electron system with coordinates $\boldsymbol{r} \equiv (x, y, z)$ to decrease the algebraic burden, but the results can be written in the case of many particle-systems using the same tricks already shown. $x_i$ will then refer to any of the $x, y, z$ cartesian coordinates of our electron.

Let us thus briefly recall, to fix the notation that will be used in the following, that Heisenberg's picture is characterized by static state operators and time dependent operators:

$$|\Psi_S(t)\rangle = \hat{U}(t, t_0)|\Psi_H\rangle, \tag{3.49}$$

where $\hat{U}$ is the time evolution operator and $|\Psi_S(t)\rangle$ and $|\Psi_H\rangle$ the state kets in Schrödinger's ($S$) and Heisenberg's ($H$) pictures, respectively. Similarly,

$$\hat{A}_H = \hat{U}^\dagger \hat{A}_S \hat{U}. \tag{3.50}$$

Whenever there is no explicit time dependency of an operator $\hat{A}$ in the $S$ picture, it follows that the equation of motion of the operators acquires the following form:

$$i\hbar\frac{d\hat{A}_H}{dt} = i\hbar\frac{\partial\hat{A}_H}{\partial t} + [\hat{A}_H, \hat{H}], \tag{3.51}$$

We also recall that these are exactly the canonical equations in classical mechanics if we change commutators by Poisson's brackets. Moreover, the equation of motion of an operator in the $H$ picture, is kept untouched in the $S$ representation if we change the operators by their expectation values, In the case that the $\hat{A}$'s are the position and momentum operators, we get:

$$\frac{d}{dt}\langle\hat{\boldsymbol{r}}\rangle = \frac{\langle\hat{\boldsymbol{p}}\rangle}{m}, \tag{3.52}$$

$$\frac{d}{dt}\langle\hat{\boldsymbol{p}}\rangle = \langle\hat{\boldsymbol{F}}\rangle, \tag{3.53}$$

the well-known first and second Ehrenfest equations.

We can now try to apply all these relations to subsystems. Some examples are particularly important, so we will develop them further in the next sections.

#### 3.4.3.1 Momentum Density

Ehrenfest's first equation (Eq. 3.52) in the $S$ picture, relates the expectation values of the linear momentum with the temporal variation of the expectation values of the position vector. Differentiating the latter with respect to time leads to:

$$\frac{d\langle\hat{x}_i\rangle_\Omega}{dt} = \int_\Omega x_i \frac{\partial}{\partial t}\left(\Psi^*(\boldsymbol{r},t)\Psi(\boldsymbol{r},t)\right) d\boldsymbol{r}, \tag{3.54}$$

where we have taken into account that the position operator is time independent.

Provided that in a one-electron system, $\rho = \Psi^*\Psi$,[2] we can use Eq. 3.29 to prove that

$$\frac{d\langle\hat{\boldsymbol{r}}\rangle_\Omega}{dt} = \int_\Omega \boldsymbol{r}\frac{\partial\rho(\boldsymbol{r},t)}{\partial t} d\boldsymbol{r} = -\int_\Omega \boldsymbol{r}\nabla\boldsymbol{j}\, d\boldsymbol{r}. \tag{3.55}$$

If we introduce now the tensor (or direct) product of two vectors,

$$\boldsymbol{a}\otimes\boldsymbol{b} = \begin{pmatrix} a_x \\ a_y \\ a_z \end{pmatrix}\begin{pmatrix} b_x & b_y & b_z \end{pmatrix} = \begin{pmatrix} a_xb_x & a_xb_y & a_xb_z \\ a_yb_x & a_yb_y & a_yb_z \\ a_zb_x & a_zb_y & a_zb_z \end{pmatrix}, \tag{3.56}$$

an operation that fulfils the following important equality

$$\boldsymbol{\nabla}\cdot(\boldsymbol{a}\otimes\boldsymbol{b}) = \boldsymbol{a}\cdot(\boldsymbol{\nabla}\otimes\boldsymbol{b}) + \boldsymbol{b}\,(\boldsymbol{\nabla}\cdot\boldsymbol{a}), \tag{3.57}$$

we can develop the expression for $\boldsymbol{r}\otimes\nabla\boldsymbol{j}$ and use it in Eq. 3.54

$$\frac{d\langle\boldsymbol{r}\rangle_\Omega}{dt} = -\int_\Omega \left[\nabla(\boldsymbol{j}\otimes\boldsymbol{r}) - \boldsymbol{j}\,(\nabla\cdot\boldsymbol{r})\right] d\boldsymbol{r}. \tag{3.58}$$

Notice that we will be using also boldface $\boldsymbol{\nabla}$ in these pages to highlight the vector nature of the gradient operator when needed.

---

[2] A similar demonstration for multielectron systems would integrate all but the coordinates that are averaged.

Since $\nabla \otimes \boldsymbol{r}$ is the unit matrix, $(\nabla \otimes \boldsymbol{r})_{ij} = \delta_{ij}$, and applying the Gauss theorem to each of the three components of $\nabla\,(\boldsymbol{j} \otimes \boldsymbol{r})$, we arrive at

$$\frac{d\langle \boldsymbol{r}\rangle_\Omega}{dt} = -\oint_S (\boldsymbol{j} \otimes \boldsymbol{r})\, d\boldsymbol{S} + \int_\Omega \boldsymbol{j}\, d\boldsymbol{r}. \tag{3.59}$$

Upon substitution in Eq. 3.52,

$$\langle \boldsymbol{p}\rangle_\Omega = m \int_\Omega \boldsymbol{j}\, d\boldsymbol{r} - m \oint_S (\boldsymbol{j} \otimes \boldsymbol{r})\, d\boldsymbol{S}. \tag{3.60}$$

Expression Eq. 3.60 demonstrates that $m\boldsymbol{j}$ is a linear momentum density and that $\langle \boldsymbol{p}\rangle$ includes a genuine surface term that is absent otherwise.

For a closed system, with borders at infinity, $\boldsymbol{j}$ vanishes quickly enough at the frontier, since we always deal with well behaved wavefunctions. We then recover the standard Ehrenfest result. For open systems, the term

$$-m \oint_S (\boldsymbol{j} \otimes \boldsymbol{r})\, d\boldsymbol{S}$$

corresponds to the transfer of momentum density across the limiting surface, and only if this flux is null we have

$$m\frac{d\langle \boldsymbol{r}\rangle_\Omega}{dt} = \langle \boldsymbol{p}\rangle_\Omega \Leftrightarrow -m \oint_S (\boldsymbol{j} \otimes \boldsymbol{r})\, d\boldsymbol{S} = \boldsymbol{0}. \tag{3.61}$$

Notice that in all these expressions, all terms are physically sound, well known in fluid mechanics. In other words, they can be applied safely to any subsystem in other contexts.

#### 3.4.3.2 Force Density

Let us now turn to examine the quantum mechanical analogue of a force. To do that we have to recall some basic aspects of the classical mechanics of deformed objects.

In an undeformed object in mechanical equilibrium, the resultant of all the forces exerted upon any volume element must vanish. When deformations exist, **internal stresses** appear that will try to recover the original equilibrium state. Following the line of reasoning followed by Landau and Lifshitz [45], the origin of such forces is microscopic and short-ranged. As a consequence, we can consider that all internal forces act exclusively on the surface separating two parts of the body under consideration.

If we define the **force density**, $\boldsymbol{f}$, acting upon each volume element of a body, and we integrate over a given finite region, $\Omega$, all the internal forces acting on the points interior to that region will cancel by the action and reaction principle, and we will be left only with those exerted by the environment over the surface of $\Omega$, $\partial S$. To transform the volume integral into a surface integral we use the divergence theorem and define a second order tensor, $\sigma$, whose divergence provides the force density:

$$\nabla \cdot \sigma = \boldsymbol{f}. \tag{3.62}$$

In other words, we demand that

$$\int_{\Omega} \boldsymbol{f} d\boldsymbol{r} = \int_{S=\partial\Omega} \sigma \cdot d\boldsymbol{S}. \tag{3.63}$$

Such object is known in mechanics as the **stress tensor**. The product $\sigma \cdot d\boldsymbol{S}$ is the force exerted over the surface element $d\boldsymbol{S}$ (see Fig. 3.10). The diagonal elements of $\sigma$ are *pressures*, while the non-diagonal one are *shears*. One can easily deduce from Fig. 3.10 that the total moment of the forces acting on a volume element only vanishes if, and only if, $\sigma$ is a symmetric tensor, $\sigma = \sigma^t$. When a body is subjected to hydrostatic pressure, $p$, a force $-pd\boldsymbol{S}$ acts on each surface element and, therefore, $\sigma_{ij} = -p\delta_{ij}$. This is usually written in the following way, $p = -1/3\text{Tr}\sigma$.

After this brief revision, let us come back to the Ehrenfest equations. Let us consider Eq. 3.53 with conservative forces ($\hat{\boldsymbol{F}} = -\nabla \hat{V}$):

$$\frac{d\langle \hat{p} \rangle}{dt} = -\langle \nabla \hat{V} \rangle. \tag{3.64}$$

We have written before the expected value of the linear momentum operator in terms of a momentum density (first term in Eq. 3.60). Differentiating with respect to time, we will obtain a force density:

$$\frac{d\langle \hat{p} \rangle_{\Omega}}{dt} = m \int_{\Omega} \frac{\partial \boldsymbol{j}}{\partial t} d\boldsymbol{r}. \tag{3.65}$$

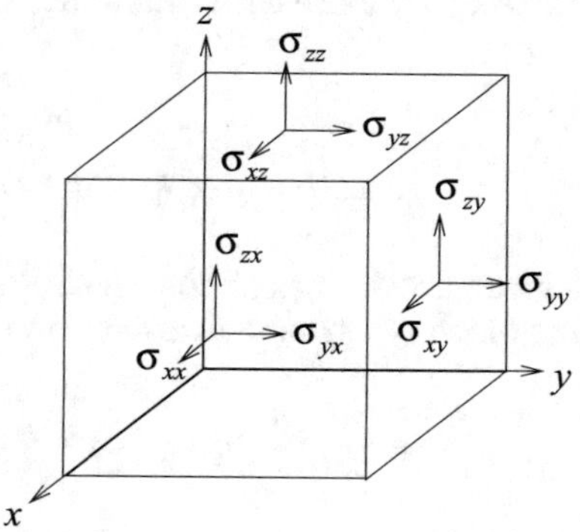


**Fig. 3.10** Physical meaning of the components of the classical stress tensor. The cube represents a differential volume element, and its faces the differential surface elements on which the environment exerts forces

Recalling the definition of $\boldsymbol{j}$ (Eq. 3.28),

$$\boldsymbol{j} = -\frac{i\hbar}{2m}(\Psi^* \nabla \Psi - \Psi \nabla \Psi^*), \tag{3.66}$$

and taking the time derivative,

$$m\frac{\partial \boldsymbol{j}}{\partial t} = -\frac{i\hbar}{2}\{\dot{\Psi}^* \nabla \Psi + \Psi^* \nabla \dot{\Psi} - \dot{\Psi} \nabla \Psi^* - \Psi \nabla \dot{\Psi}^*\}. \tag{3.67}$$

If we now substitute derivatives by using the evolution equation,

$$m\frac{\partial \boldsymbol{j}}{\partial t} = -\frac{1}{2}\{-(\hat{H}\Psi)^* \nabla \Psi + \Psi^* \nabla (\hat{H}\Psi) - (\hat{H}\Psi) \nabla \Psi^* + \Psi \nabla (\hat{H}\Psi^*)\}. \tag{3.68}$$

We can now again try to simplify as much as possible, so we will use a one-particle system with Hamiltonian

$$\hat{H} = -\frac{\hbar^2}{2m}\nabla^2 + \hat{V}(r). \tag{3.69}$$

Substituting into Eq. 3.68, we obtain:

$$\begin{aligned} m\frac{\partial \boldsymbol{j}}{\partial t} = -\frac{1}{2}\Bigg\{&\frac{\hbar^2}{2m}(\nabla^2\Psi)^* \boldsymbol{\nabla}\Psi - \hat{V}\Psi^* \boldsymbol{\nabla}\Psi - \frac{\hbar^2}{2m}\Psi^* \boldsymbol{\nabla}(\nabla^2\Psi) + \Psi^* \boldsymbol{\nabla}(\hat{V}\Psi) \\ &+ \frac{\hbar^2}{2m}\nabla^2\Psi \boldsymbol{\nabla}\Psi^* - \hat{V}\Psi \boldsymbol{\nabla}\Psi^* - \frac{\hbar^2}{2m}\Psi \boldsymbol{\nabla}(\nabla^2\Psi^*) + \Psi \boldsymbol{\nabla}(\hat{V}\Psi^*)\Bigg\}. \end{aligned} \tag{3.70}$$

It is easy to show that all terms multiplied by the potential cancel each other: using the relation $\boldsymbol{\nabla}(fg) = g\boldsymbol{\nabla}f + f\boldsymbol{\nabla}g$ we find, for instance, that

$$-\hat{V}\Psi^* \boldsymbol{\nabla}\Psi + \Psi^* \boldsymbol{\nabla}(\hat{V}\Psi) - \hat{V}\Psi \boldsymbol{\nabla}\Psi^* + \Psi \boldsymbol{\nabla}(\hat{V}\Psi^*) = 2\Psi^*\Psi \boldsymbol{\nabla}\hat{V}. \tag{3.71}$$

Hence, only those terms multiplied by $\boldsymbol{\nabla}\hat{V}$ survive in Eq. 3.70, and we have:

$$\begin{aligned} m\frac{\partial j}{\partial t} &= -\Psi^*\Psi \boldsymbol{\nabla}\hat{V} \\ -\tfrac{\hbar^2}{4m}\big\{&\nabla^2\Psi^* \boldsymbol{\nabla}\Psi - \Psi^* \boldsymbol{\nabla}(\nabla^2\Psi) + \boldsymbol{\nabla}\Psi^* \nabla^2\Psi - \boldsymbol{\nabla}(\nabla^2\Psi^*)\Psi\big\}. \end{aligned} \tag{3.72}$$

This expression can be considerably simplified if we introduce **Pauli's stress tensor**, $\sigma_t$, which is a symmetric second rank tensor:

$$\sigma_t = \frac{\hbar^2}{4m}\left\{\Psi^* \boldsymbol{\nabla} \otimes (\boldsymbol{\nabla}\Psi) + \boldsymbol{\nabla} \otimes (\boldsymbol{\nabla}\Psi^*)\Psi - \boldsymbol{\nabla}\Psi^* \otimes \boldsymbol{\nabla}\Psi - \boldsymbol{\nabla}\Psi \otimes \boldsymbol{\nabla}\Psi^*\right\} \tag{3.73}$$

Pauli's stress tensor is not uniquely defined, it can be subjected to gauge transformations that do not alter its divergence, and this arbitrariness has been the source of some debate in the literature [28]. Now, taking into account that $\boldsymbol{\nabla} \cdot \boldsymbol{\nabla} = \nabla^2$ is a scalar operator, while $\nabla \otimes \nabla$, the hessian, which is a second rank tensor operator, we can use once again $\nabla \cdot (\boldsymbol{a} \otimes \boldsymbol{b}) = \boldsymbol{a} \cdot (\nabla \otimes \boldsymbol{b}) + \boldsymbol{b}\,(\nabla \cdot \boldsymbol{a})$ (Eq. 3.57) we find that

$$\begin{aligned}\frac{4m}{\hbar^2}\nabla\sigma_t &= \nabla\Psi^*\nabla \otimes (\nabla\Psi) + \Psi^*\nabla(\nabla^2\Psi) + \nabla(\nabla^2\Psi^*)\Psi + \nabla \otimes (\nabla\Psi^*)\nabla\Psi - \\ &- \nabla\Psi^*\nabla \otimes \nabla\Psi - \nabla\Psi\nabla^2\Psi^* - \nabla\Psi\nabla \otimes \nabla\Psi^* - \nabla\Psi^*\nabla^2\Psi = \\ &= \Psi^*\nabla(\nabla^2\Psi) + \nabla(\nabla^2\Psi^*)\Psi - \nabla\Psi\nabla^2\Psi^* - \nabla\Psi^*\nabla^2\Psi,\end{aligned}$$

which can be used directly to transform Eq. 3.72 into:

$$m\frac{\partial \boldsymbol{j}}{\partial t} = -\Psi^*\Psi\nabla\hat{V} + \nabla \cdot \sigma_t. \tag{3.74}$$

This is one of the pillars of **quantum fluid dynamics**, a theory that establishes a parallelism between the time evolution of the quantum density and a fluid subjected to two force fields: one of them classical in nature, $-\nabla V$, and another of purely quantum origin, $\nabla \cdot \sigma_t$. The latter is also called the **quantum force field**.

If we now integrate expression Eq. 3.74 over a volume $\Omega$, and we apply the divergence theorem, we obtain

$$\int_\Omega m\frac{\partial \boldsymbol{j}}{\partial t}\,d\boldsymbol{r} = -\langle\nabla\hat{V}\rangle_\Omega + \oint_S \sigma_t \cdot d\boldsymbol{S}. \tag{3.75}$$

We can distinguish several cases:

1. When $\Omega$ is the whole space, then the surface term vanishes for well-behaved wavefunctions, and:
$$\int m\frac{\partial \boldsymbol{j}}{\partial t}\,d\boldsymbol{r} = -\langle\nabla\hat{V}\rangle. \tag{3.76}$$
In this expression, analogous to Newton's law, we identify $m\partial\boldsymbol{j}/\partial t$ with the force density.
2. On the contrary, if $\Omega$ is finite, then the surface term in Eq. 3.75 can be interpreted as that coming from the quantum forces that must be exerted on the surface of the subsystem to keep its equilibrium.
3. For stationary states, Eq. 3.75 can be written as
$$\langle -\nabla\hat{V}\rangle_\Omega = -\int_\Omega \nabla \cdot \sigma_t\,d\boldsymbol{r}. \tag{3.77}$$
The above expression can be generalized to the multiparticle case, if we understand that
$$\langle -\nabla\hat{V}\rangle_\Omega = \int_\Omega d\boldsymbol{r}\int d\boldsymbol{r}'(-\nabla\hat{V})\rho(\boldsymbol{r},\boldsymbol{r}'), \tag{3.78}$$
and that

$$\sigma_t = \frac{\hbar^2}{4m}\left\{\nabla\otimes\nabla + \nabla'\otimes\nabla' - \nabla'\otimes\nabla - \nabla\otimes\nabla'\right\}\rho(\boldsymbol{r},\boldsymbol{r}'). \tag{3.79}$$

The last equation allows us to identify the total force density acting upon the electrons, $\boldsymbol{f}(\boldsymbol{r})$, with minus the divergence of Pauli's tensor,

$$\boldsymbol{f}(\boldsymbol{r}) = -\nabla\cdot\sigma_t(\boldsymbol{r}). \tag{3.80}$$

This force density is known as the **Ehrenfest force**. It contains both electron-electron as well as electron-nuclei interactions.

#### 3.4.3.3 The Virial Theorem

The QTAIM started its now long way in the 1970s, after the seminal results proven by R. F. W. Bader regarding the fulfilment of the virial theorem in finite regions of the 3D space. Using the temporal evolution of an operator in Heisenberg's picture and taking into account that neither the position nor the momentum operators depend explicitly on time in this picture, we use (Eq. 3.51) to write

$$\frac{d}{dt}(\hat{\boldsymbol{r}}\cdot\hat{\boldsymbol{p}})_H = \frac{i}{\hbar}\left[\hat{H},(\hat{\boldsymbol{r}}\cdot\hat{\boldsymbol{p}})_H\right], \tag{3.81}$$

where the product $\hat{\boldsymbol{r}}\cdot\hat{\boldsymbol{p}}$ is known, for reasons that will soon be clear, as the **virial generator**. If we particularize to a stationary system, then

$$\frac{d}{dt}(\hat{\boldsymbol{r}}\cdot\hat{\boldsymbol{p}})_H = 0, \tag{3.82}$$

so that the commutator in Eq. 3.81 must be null. With a little algebra we can show that:

$$\left[\hat{H},(\hat{\boldsymbol{r}}\cdot\hat{\boldsymbol{p}})_H\right] = \hat{\boldsymbol{r}}_H\cdot\left[\hat{H},\hat{\boldsymbol{p}}_H\right] + \left[\hat{H},\hat{\boldsymbol{r}}_H\right]\cdot\hat{\boldsymbol{p}}_H = \hat{\boldsymbol{r}}_H\cdot(i\hbar\nabla\hat{V}_H) - \frac{i\hbar\hat{p}_H^2}{m}, \tag{3.83}$$

where we have used the commutators rule $[A, BC] = B[A, C] + [A, B]C$. [3] Multiplying by $i/\hbar$,

$$-\boldsymbol{r}_H\cdot\nabla\hat{V}_H + \frac{\hat{p}_H^2}{m} = -\boldsymbol{r}_H\cdot\nabla\hat{V}_H + 2\hat{T}_H = 0. \tag{3.85}$$

[3] Recall that for any function $f$, $\left[\hat{f},\hat{\boldsymbol{p}}\right] = i\hbar\nabla f$. Using the commutator rule $[AB, C] = A[B, C] + [A, C]B$ it can also be deduced that:

$$\left[\hat{H},\hat{r}\right] = [\hat{T}+\hat{V}] = \frac{\hat{\boldsymbol{p}}}{m}[\hat{\boldsymbol{p}},\hat{r}] = -\frac{i\hbar\hat{\boldsymbol{p}}}{m} \tag{3.84}$$

since $[\hat{V},\hat{r}] = 0$.

This is known as the quantum virial theorem. If the potential is an $n$th degree homogeneous function in the electron coordinates, then $\boldsymbol{r}_H \cdot \nabla \hat{V}_H = n\hat{V}$, so that the theorem acquires the well known form;

$$2\hat{T}_H = n\hat{V}_H, \tag{3.86}$$

which, turning back to the $S$ picture, reads

$$\langle \boldsymbol{r} \cdot \nabla \hat{V} \rangle = 2\langle \hat{T} \rangle. \tag{3.87}$$

We will now generalize this expression for a partition of space into regions $\Omega$. We start by taking the time derivative of the virial generator, expressed in terms of the momentum density and Pauli's tensor (Eq. 3.74):

$$m\boldsymbol{r} \cdot \frac{\partial \boldsymbol{j}}{\partial t} = \boldsymbol{r} \cdot \nabla \sigma_t - \Psi^* \Psi \boldsymbol{r} \cdot \nabla \hat{V}. \tag{3.88}$$

Taking into account the vector identity

$$\begin{aligned} \nabla \cdot (\boldsymbol{r}\sigma) &= \left( \frac{\partial}{\partial x}\ \frac{\partial}{\partial y}\ \frac{\partial}{\partial z} \right) \left[ \left( x\ y\ z \right) \begin{pmatrix} \sigma_{xx} & \sigma_{xy} & \sigma_{xz} \\ \sigma_{yx} & \sigma_{yy} & \sigma_{yz} \\ \sigma_{zx} & \sigma_{zy} & \sigma_{zz} \end{pmatrix} \right]^t \\ &= \left( \frac{\partial}{\partial x}\ \frac{\partial}{\partial y}\ \frac{\partial}{\partial z} \right) \begin{pmatrix} x\sigma_{xx} + y\sigma_{xy} + z\sigma_{xz} \\ x\sigma_{yx} + y\sigma_{yy} + z\sigma_{yz} \\ x\sigma_{zx} + y\sigma_{zy} + z\sigma_{zz} \end{pmatrix} \\ &= \mathrm{Tr}\sigma_t + \boldsymbol{r} \cdot \nabla \sigma_t, \end{aligned} \tag{3.89}$$

the first term in Eq. 3.88 can be written as

$$\nabla \cdot (\boldsymbol{r}\sigma_t) - \mathrm{Tr}\sigma_t, \tag{3.90}$$

due to the symmetry of $\sigma_t$. In this way, Eq. 3.88 transforms into

$$m\boldsymbol{r} \cdot \frac{\partial \boldsymbol{j}}{\partial t} = -\mathrm{Tr}\sigma_t + \nabla \cdot (\boldsymbol{r}\sigma_t) - \Psi^* \Psi \boldsymbol{r} \cdot \nabla \hat{V}. \tag{3.91}$$

Integrating over the volume $\Omega$ and applying again Gauss's theorem,

$$\int_\Omega m\boldsymbol{r} \cdot \frac{\partial \boldsymbol{j}}{\partial t} d\boldsymbol{r} = -\int_\Omega \mathrm{Tr}\sigma_t d\boldsymbol{r} + \oint_S \boldsymbol{r}\sigma_t d\boldsymbol{S} - \langle \boldsymbol{r} \nabla \hat{V} \rangle_\Omega. \tag{3.92}$$

If we take a detailed look at the trace of the stress tensor (Eq. 3.73), taking into account that $\mathrm{Tr}(\nabla \otimes \nabla) = \nabla^2$,

$$-\mathrm{Tr}\sigma_t = -\frac{\hbar^2}{4m} \sum_i \left[ \Psi^* \nabla_i^2 \Psi + (\nabla_i^2 \Psi^*)\Psi - \nabla_i \Psi^* \nabla_i \Psi - \nabla_i \Psi \nabla_i \Psi^* \right]. \tag{3.93}$$

We can now identify previously defined quantities in Eqs. 3.45 and 3.46 to find that

$$-\,\mathrm{Tr}\sigma_t = K + t = 2K - L, \tag{3.94}$$

(expressed most commonly in QTAIM literature using $G$ instead of $t$: $-\mathrm{Tr}\sigma_t = K + G = 2K - L$). Integrating over $\Omega$ and applying the divergence theorem:

$$-\int_\Omega \mathrm{Tr}\sigma_t d\boldsymbol{r} = 2\langle\hat{T}\rangle_\Omega + \frac{\hbar^2}{4m}\oint_S \nabla\rho \cdot d\boldsymbol{S}. \tag{3.95}$$

Substitution in Eq. 3.92 leads to

$$\int_\Omega m\boldsymbol{r} \cdot \frac{\partial \boldsymbol{j}}{\partial t} d\boldsymbol{r} = 2\langle\hat{T}\rangle_\Omega - \langle\boldsymbol{r}\,\nabla\hat{V}\rangle_\Omega + \oint_S \boldsymbol{r}\sigma_t d\boldsymbol{S} + \frac{\hbar^2}{4m}\oint_S \nabla\rho \cdot d\boldsymbol{S}. \tag{3.96}$$

If the system is in a stationary state,

$$2\langle\hat{T}\rangle_\Omega = \underbrace{\langle\boldsymbol{r}\,\nabla\hat{V}\rangle_\Omega}_{\mathcal{V}^b_\Omega} \underbrace{- \oint_S \boldsymbol{r}\sigma_t d\boldsymbol{S} - \frac{\hbar^2}{4m}\oint_S \nabla\rho \cdot d\boldsymbol{S}}_{\mathcal{V}^s_\Omega}, \tag{3.97}$$

so $2\langle\hat{T}\rangle_\Omega$ equals the sum of the virial of the classical and quantum forces exerted over the surface of the subsystem plus a term that vanishes over atomic basins. It is in this case that we can talk of a generalized version of the virial theorem. The volume and surface terms are known as the **basin virial**, $\mathcal{V}^b$, and the **surface virial**, $\mathcal{V}^s$:

$$2\langle\hat{T}\rangle_\Omega = \mathcal{V}^b_\Omega + \mathcal{V}^s_\Omega. \tag{3.98}$$

Summarizing, we stress that all the observables thus characterized, among which we find all the non-relativistic energy components, the virial and the force, admit generalizations compatible with QM in subsystems whenever the regions in which we divide the space are separated by zero flux surfaces. This is not a demonstration *stricto sensu* that the QTAIM partition is the only one compatible with these facts, but we will not proceed any further. There has been much debate in the literature about the deepness of the link between the QTAIM and QM, and the reader is referred to the more specialized literature for further information [6].

### *3.4.4 Equations of Motion in Open Subsystems*

The brief discussion in the last subsection should have convinced us about the need to (i) introduce surface terms when using open systems, (ii) to study the time evolution of those terms, when present. In this subsection we will show how openness modifies the temporal behavior of a system. In the first place we will consider an open system in a stationary state. Then we will free this constraint in order to consider non-stationary states.

#### 3.4.4.1 Evolution of Stationary States

In order to describe the time evolution of a global observable $A$, we will use again the general equation in the $H$ picture,

$$\frac{d\hat{A}_H}{dt} = \frac{\partial \hat{A}_H}{\partial t} + \frac{i}{\hbar}\left[\hat{H}, \hat{A}_H\right], \tag{3.99}$$

that adopts the following form in the Schrödinger's image:

$$\frac{d}{dt}\langle\Psi|\hat{A}|\Psi\rangle = \langle\Psi|\frac{\partial \hat{A}}{\partial t}|\Psi\rangle + \frac{i}{\hbar}\langle\Psi|\left[\hat{H}, \hat{A}\right]|\Psi\rangle. \tag{3.100}$$

In a stationary state,

$$\frac{\partial \hat{A}}{\partial t} = 0, \tag{3.101}$$

so

$$\frac{\partial \langle\hat{A}\rangle}{\partial t} = 0. \tag{3.102}$$

This means that

$$\frac{i}{\hbar}\langle\Psi|\left[\hat{H}, \hat{A}\right]|\Psi\rangle = 0, \tag{3.103}$$

a generalization of the expression that we have repeatedly used for particular cases in the last subsection. This generalization is known as the **hypervirial theorem** or **generalized virial theorem**. Let us constrain now the integration to the open region under investigation. To simplify, we will restrict ourselves to a one-particle system, although the multiparticle generalization is straightforward. Manipulating Eq. 3.103 and introducing two cancelling ($\mp \int_\Omega (\hat{H}\Psi)^* \hat{A}\Psi d\boldsymbol{r}$) terms we arrive at

$$\begin{aligned}\int_\Omega \Psi^*(\hat{H}\hat{A} - \hat{A}\hat{H})\Psi d\boldsymbol{r} = &\int_\Omega \Psi^* \hat{H}\hat{A}\Psi d\boldsymbol{r} - \int_\Omega (\hat{H}\Psi)^* \hat{A}\Psi d\boldsymbol{r} \\ &+ \int_\Omega (\hat{H}\Psi)^* \hat{A}\Psi d\boldsymbol{r} - \int_\Omega \Psi^*(\hat{A}\hat{H})\Psi d\boldsymbol{r}. \end{aligned} \tag{3.104}$$

Given that the system is in a stationary state, $\hat{H}\Psi = E\Psi$. Substituting in Eq. 3.104, the last two terms cancel out. Taking the other two terms and working out the expression we arrive at

$$\int_{\Omega} \Psi^*(\hat{H}\hat{A} - \hat{A}\hat{H})\Psi d\boldsymbol{r} = \int_{\Omega} \frac{-\hbar^2}{2m}(\Psi^*\nabla^2\hat{A}\Psi - \nabla^2\Psi^*\hat{A}\Psi)\, d\boldsymbol{r}. \tag{3.105}$$

Since our aim is to write a divergence, we can write the Laplacians as the divergences of gradients. After doing it and multiplying by factor $i/\hbar$ we get:

$$\begin{aligned} \frac{i}{\hbar}\int_{\Omega} \Psi^*(\hat{H}\hat{A} - \hat{A}\hat{H})\Psi\, d\boldsymbol{r} &= \frac{i}{\hbar}\frac{-\hbar^2}{2m}\int_{\Omega} \boldsymbol{\nabla}\cdot[(\Psi^*\boldsymbol{\nabla}\hat{A}\Psi) - (\boldsymbol{\nabla}\Psi^*\hat{A}\Psi)]\, d\boldsymbol{r} \\ &= \oint_S \boldsymbol{j}_A \cdot d\boldsymbol{S}, \end{aligned} \tag{3.106}$$

where $\boldsymbol{j}_A$ is the current density of operator $\hat{A}$, defined as

$$\boldsymbol{j}_A = \frac{-i\hbar}{2m}\{(\Psi^*\boldsymbol{\nabla}\hat{A}\Psi) - (\boldsymbol{\nabla}\Psi^*\hat{A}\Psi)\}. \tag{3.107}$$

With this we have just shown that

$$\frac{i}{\hbar}\langle\Psi|\left[\hat{H},\hat{A}\right]|\Psi\rangle_{\Omega} = \oint_S \boldsymbol{j}_A \cdot d\boldsymbol{S}. \tag{3.108}$$

The expression above is shown to recover the hypervirial theorem if we integrate extending the frontier to infinity. It is thus the generalization we were looking for. In open subsystems, the temporal evolution of a quantity requires taking into consideration the flux of its density through the surface(s) of the region(s) of choice. Only when these fluxes vanish we can talk about conserved quantities for subsystems in stationary states.

#### 3.4.4.2 Non-stationary States

We still have to consider the most general case, the one in which the stationary system constraint is released. We use the general equation for temporal evolution in the $S$ picture:

$$\frac{d}{dt}\langle\Psi|\hat{A}|\Psi\rangle = \langle\Psi|\frac{\partial\hat{A}}{\partial t}|\Psi\rangle + \frac{i}{\hbar}\langle\Psi|\left[\hat{H},\hat{A}_H\right]|\Psi\rangle. \tag{3.109}$$

Transforming the commutator in the same way as we did in the last subsection, and also introducing two mutually cancelling terms needed in further manipulations,

$$\begin{aligned} \int_{\Omega} \Psi^*(\hat{H}\hat{A} - \hat{A}\hat{H})\Psi d\boldsymbol{r} &= \int_{\Omega} \Psi^*\hat{H}\hat{A}\Psi d\boldsymbol{r} - \int_{\Omega} (\hat{H}\Psi)^*\hat{A}\Psi d\boldsymbol{r} \\ &+ \int_{\Omega} (\hat{H}\Psi)^*\hat{A}\Psi d\boldsymbol{r} - \int_{\Omega} \Psi^*(\hat{A}\hat{H})\Psi d\boldsymbol{r}. \end{aligned} \tag{3.110}$$

The difference between the first two sums gives rise, after analogous steps to those performed in the stationary case, to the flux of the current density associated with the operator $\hat{A}$ through the surface, $S$,

$$\int_{\Omega} \Psi^* \hat{H}\hat{A}\Psi d\boldsymbol{r} - \int_{\Omega} (\hat{H}\Psi)^* \hat{A}\Psi d\boldsymbol{r} = -i\hbar \oint_{S} \boldsymbol{j}_A \cdot d\boldsymbol{S}. \tag{3.111}$$

Bearing in mind that $\hat{H}\Psi = i\hbar\dot{\Psi}$, and substituting in the last two terms of Eq. 3.110, we obtain:

$$\int_{\Omega} (\hat{H}\Psi)^* \hat{A}\Psi d\boldsymbol{r} - \int_{\Omega} \Psi^* (\hat{A}\hat{H})\Psi d\boldsymbol{r} = \int_{\Omega} -i\hbar(\dot{\Psi}^* \hat{A}\Psi + \Psi^* \hat{A}\dot{\Psi}) d\boldsymbol{r}. \tag{3.112}$$

Using the following identity in Eq. 3.112

$$\frac{\partial}{\partial t}(\Psi^* \hat{A}\Psi) = \dot{\Psi}^* \hat{A}\Psi + \Psi^* \frac{\partial \hat{A}}{\partial t}\Psi + \Psi^* \hat{A}\dot{\Psi}, \tag{3.113}$$

we verify that

$$\begin{aligned} \int_{\Omega} (\hat{H}\Psi)^* \hat{A}\Psi d\boldsymbol{r} - \int_{\Omega} \Psi^* (\hat{A}\hat{H})\Psi\, d\boldsymbol{r} &= -i\hbar \int_{\Omega} \left( \frac{\partial}{\partial t}(\Psi^* \hat{A}\Psi) - \Psi^* \frac{\partial \hat{A}}{\partial t}\Psi \right) d\boldsymbol{r} \\ &= -i\hbar \frac{d}{dt} \langle \Psi | \hat{A} | \Psi \rangle_{\Omega} + i\hbar \langle \Psi | \frac{\partial \hat{A}}{\partial t} | \Psi \rangle_{\Omega}. \end{aligned} \tag{3.114}$$

Gathering the expressions obtained for the four terms in Eq. 3.110 (Eqs. 3.111 and 3.114), and reorganizing terms, we arrive at

$$\frac{i}{\hbar}\langle \left[ \hat{H}, \hat{A} \right] \rangle_{\Omega} + \langle \frac{\partial \hat{A}}{\partial t} \rangle_{\Omega} = \oint_{S} \boldsymbol{j}_A \cdot d\boldsymbol{S} + \frac{d}{dt}\langle \hat{A} \rangle_{\Omega}, \tag{3.115}$$

which is the **hypervirial equation**, or general equation of motion for an open subsystem in a non-stationary state.

As a corollary, whenever an operator that does not depend explicitly on time commutes with the Hamiltonian, its eigenvalues will be constants of motion. In this way, provided time-independent operators, Eq. 3.115 reduces to a continuity equation, analogous to that of an incompressible fluid: the temporal variation of the expectation value of the quantity within the open system is opposite to the flux of its corresponding operator density through the separating surface of the subsystem,

$$\oint_{S} \boldsymbol{j}_A \cdot d\boldsymbol{S} + \frac{d}{dt}\langle \hat{A} \rangle_{\Omega} = 0. \tag{3.116}$$

On the contrary, if the operator $\hat{A}$ does not commute with the Hamiltonian, Eq. 3.115 includes quantum fluctuation terms, whose ultimate origin has to be searched for in Heisenberg's uncertainty principle.

Let us see several particularizations of the general operator $\hat{A}$.

### 3.4.5 *One-Electron Atomic Observables*

Once the validity (or usefulness) of the QTAIM partition is admitted, we will examine some of its atomic observables in specific cases. We will start from the existence, as shown in previous subsections, of well defined **operator densities** that can be integrated over atomic basins. As we will see, the form of some of the operators will depend directly on the structure of the open regions, so that their integration over regions not separated by zero flux surfaces does not necessarily lead to physically sound results. We will start by considering **one-electron operators**.

Let $\hat{\mathcal{O}}$ be the spatial density associated with a one-electron operator $\hat{O}$. Under these circumstances,

$$\langle \hat{O} \rangle = \sum_{\Omega} \int_{\Omega} \hat{\mathcal{O}} \, d\boldsymbol{r} = \sum_{\Omega} \hat{O}_{\Omega}, \tag{3.117}$$

and all global expectation values are additively reconstructed through the **atomic observables**. The **transferability** shown by the latter, as we already noticed, is the origin of the constancy and experimental additivity of the so-called **group properties**. Notice that the above definition may require a complex symmetrization if the integration procedure leads to a non-zero imaginary part.

Out of the many one-electron densities that we can integrate over an atomic region we find the unit density, which determines the volume of the region:

$$V_{\Omega} = \int_{\Omega} d\boldsymbol{r}. \tag{3.118}$$

In isolated molecules it is difficult to find a truly finite atomic region. It is however common use to limit the integration region by the intersection between the atomic region and a density isosurface $l$ with, for instance $\rho = 0.001$ a.u. The atomic volumes of these density-cut atoms $V_{\Omega}$, have been shown to be rather transferable, and their variation successfully related with changes in electron population or chemical environment.

There exist other simple examples that have been widely used. The most important of all, and probably the most misinterpreted atomic observable is the electron population of a basin, obtained from integrating the electron density itself, $\rho$:

$$N_{\Omega} = \int_{\Omega} \rho(\boldsymbol{r}) \, d\boldsymbol{r}. \tag{3.119}$$

These electron populations are usually rewritten as **net atomic charges**, $Q_\Omega = Z - N_\Omega$. Since the atomic populations are not constants of motion, their variances,

$$\sigma^2(\Omega) = \langle N^2 \rangle_\Omega - N_\Omega^2 \tag{3.120}$$

can be used to determine the quantum-mechanical uncertainty associated with the number of electrons contained in a region. We will consider these objects in more detail in Sect. 3.5.3.

Similarly, we can define the moments of the atomic charge distribution, which are of great importance when studying intermolecular interactions. It is useful to that end to introduce **real spherical harmonics**, $S_{lm}$, related with their complex counterparts $Y_{lm}$ through the following unitary transformation:

$$S_{lm}(\theta, \phi) = \begin{cases} \frac{(-1)^m}{\sqrt{2}}(Y_{lm} + Y_{lm}^*) & m > 0, \\ Y_{l0} & m = 0, \\ \frac{(-1)^m}{i\sqrt{2}}(Y_{l|m|} - Y_{l|m|}^*) & m < 0. \end{cases} \tag{3.121}$$

Using this notation, $S_{00} = s$, $S_{1\bar{1}} = p_y$, $S_{10} = p_z$, $S_{11} = p_x$, $S_{2\bar{2}} = d_{xy}$, $S_{2\bar{1}} = d_{yz}$, $S_{20} = d_{z^2}$, $S_{21} = d_{xz}$, y $S_{22} = d_{x^2-y^2}$.

We thus define the **Spherical atomic multipole moments**, $N_\Omega^{lm}$, as

$$N_\Omega^{lm} = \int_\Omega r^l S_{lm}(\theta, \phi)\rho(\boldsymbol{r})d\boldsymbol{r}. \tag{3.122}$$

$N^{00}$ is obviously the atomic population, the trio $(N^{11}, N^{1\bar{1}}, N^{10})$ determines the **atomic dipolar moment** vector, $\mu_\Omega$, and the moments with $l = 2, 3, 4, \ldots$ are identified with the well known quadrupole, octopole, hexadecapole, etc, moments. Atomic moments measure the distortion of the charge distribution with respect to a sphere. Only when all the $N^{lm}$ with $l > 0$ vanish the charge distribution is spherical. Bear in mind that the atomic multipole moments are not invariant under translations. They are strongly dependent on the reference system that is chosen to define them. By default, in the QTAIM we center those reference systems on the nucleus of each basin.

If the $N_\Omega^{lm}$ for all the atomic basins in a system are known we can reconstruct the global molecular moments with respect to a given reference system. This is a non-trivial task in the general case. For $l = 1$, however, the transformation is simple enough:

$$\begin{aligned} \boldsymbol{\mu} &= \int \boldsymbol{r}\rho(\boldsymbol{r})d\boldsymbol{r} = \sum_A \int_{\Omega_A} \boldsymbol{r}\rho(\boldsymbol{r})d\boldsymbol{r} = \sum_A \int_{\Omega_A} (\boldsymbol{r} - \boldsymbol{R}_A)\rho(\boldsymbol{r})d\boldsymbol{r} + \sum_A \int_{\Omega_A} \boldsymbol{R}_A \rho d\boldsymbol{r} \\ &= \sum_A \boldsymbol{\mu}_A + \sum_A N_A \boldsymbol{R}_A = \boldsymbol{\mu}_{\text{pol}} + \boldsymbol{\mu}_{\text{CT}} \end{aligned} \tag{3.123}$$

where we have labeled with index $A$ the different basins and $\boldsymbol{R}_A$ is the position vector of the nucleus in basin $A$. We note that to recover the molecular dipole moment it is necessary to know both the atomic *dipoles* and *monopoles*, and that the term $\boldsymbol{\mu}_{\mathrm{CT}}$ is only origin independent if the molecule is overall neutral, a known condition for the global invariance of the dipole moment. The above expression neatly shows the origin of a usual confusion in structural chemistry. A molecule's dipole moment cannot be reconstructed from atomic charges, i.e. through the charge transfer (CT) term, $\boldsymbol{\mu}_{\mathrm{CT}}$. On the contrary it requires the consideration of the density distortions around each atomic nucleus. This last polarization term, $\boldsymbol{\mu}_{\mathrm{pol}}$ plays an important role in the vast majority of the cases.

We can now consider dynamic observables, starting with several important energetic components. The kinetic energy has already been discussed. In the case of QTAIM basins, $L_\Omega = 0$ in Eq. 3.47, so that the atomic kinetic energies are well defined: $T_\Omega = t_\Omega = K_\Omega$, with $T = \sum_A T_{\Omega_A}$.

The electron-nucleus potential energy can also be partitioned easily. We define $V_{AB}^{en}$ as the attraction energy among the electrons in basin $A$ and nucleus $B$:

$$V_{en}^{AB} = -Z_B \int_{\Omega_A} \frac{\rho(\boldsymbol{r})}{|\boldsymbol{r} - \boldsymbol{R}_B|} d\boldsymbol{r}, \tag{3.124}$$

It is also convenient to introduce the attraction energy between a nucleus and the electrons of an atomic basin. In an explicit notation: $V_{ne}^{AB} = V_{en}^{BA}$. The interaction among the electrons in a basin and its own nucleus is thus determined by the diagonal element $V_{en}^{AA} = V_{AA}^{ne}$. With these new elements,

$$V_{en} = -\sum_A Z_A \sum_B \int_{\Omega_B} \frac{\rho(\boldsymbol{r})}{|\boldsymbol{r} - \boldsymbol{R}_A|} d\boldsymbol{r} = \sum_{A,B} V_{en}^{BA}. \tag{3.125}$$

We have to be careful when considering interelectronic repulsion, since it is a two-electron property. In order to partition it, it is necessary to use the second order density, $\rho_2$. Recalling that the total electron repulsion admits the following expression,

$$V_{ee} = \frac{1}{2} \int\int \rho_2(\boldsymbol{r}_1, \boldsymbol{r}_2) r_{12}^{-1}\, d\mathbf{r}_1\, d\mathbf{r}_2, \tag{3.126}$$

we introduce the following quantities,

$$V_{ee}^{AA} = \frac{1}{2} \int_{\Omega_A} \int_{\Omega_A} \rho_2(\boldsymbol{r}_1, \boldsymbol{r}_2) r_{12}^{-1}\, d\boldsymbol{r}_1, d\boldsymbol{r}_2, \tag{3.127}$$

and

$$V_{ee}^{AB} = \frac{1}{2} \int_{\Omega_A} \int_{\Omega_B} \rho_2(\boldsymbol{r}_1, \boldsymbol{r}_2) r_{12}^{-1}\, d\boldsymbol{r}_1, d\boldsymbol{r}_2, \tag{3.128}$$

whose meanings are as follows. $V_{ee}^{AA}$ is the total electron repulsion among the electrons in basin $A$. Similarly, $V_{ee}^{AB}$ is the repulsion among the electrons in basin $A$ and those in basin $B$. Given the exhaustivity of the QTAIM partitioning,

$$V^{ee} = \frac{1}{2}\sum_A \int_{\Omega_A} \sum_B \int_{\Omega_B} r_{12}^{-1}\rho_2(\boldsymbol{r}_1, \boldsymbol{r}_2)d\mathbf{r}_1\, d\mathbf{r}_2 = \sum_A V_{ee}^{AA} + \frac{1}{2}\sum_A \sum_{B\neq A} V_{ee}^{AB}. \tag{3.129}$$

We finish this subsection by considering the atomic force partition. Using Eqs. 3.78 and 3.80, the atomic Ehrenfest force adopts the following expressions

$$\begin{aligned} \boldsymbol{f}_\Omega = \int_\Omega \boldsymbol{f}(\boldsymbol{r})d\boldsymbol{r} = \langle -\nabla\hat{V}\rangle_\Omega = \int_\Omega d\boldsymbol{r} \int (-\nabla\hat{V})\rho(\boldsymbol{r}, \boldsymbol{r}')d\boldsymbol{r}' \\ = -\int_\Omega \nabla \cdot \sigma_t\, d\boldsymbol{r} = -\oint_S \sigma_t \cdot d\boldsymbol{S}. \end{aligned} \tag{3.130}$$

The gradient of the potential contains electron-electron as well as electron-nuclei terms. Each of these are the expected value of the respective gradient in Eqs. 3.124, 3.127 y 3.128. Namely,

$$\boldsymbol{f}_{en}^{AB} = -Z_B \int_{\Omega_A} \frac{\rho(\boldsymbol{r})}{|\boldsymbol{r} - \boldsymbol{R}_B|^3}(\boldsymbol{r} - \boldsymbol{R}_B)d\boldsymbol{r}, \tag{3.131}$$

$$\boldsymbol{f}_{ee}^{AA} = \frac{1}{2}\int_{\Omega_A}\int_{\Omega_A} r_{12}^{-3}\,(\boldsymbol{r}_1 - \boldsymbol{r}_2)\rho_2(\boldsymbol{r}_1, \boldsymbol{r}_2)\, d\boldsymbol{r}_1\, d\boldsymbol{r}_2 = \boldsymbol{0}, \tag{3.132}$$

$$\boldsymbol{f}_{AB}^{ee} = \int_{\Omega_A}\int_{\Omega_B} r_{12}^{-3}\,(\boldsymbol{r}_1 - \boldsymbol{r}_2)\rho_2(\boldsymbol{r}_1, \boldsymbol{r}_2)\, d\boldsymbol{r}_1\, d\boldsymbol{r}_2. \tag{3.133}$$

Let us notice the very important relation $\boldsymbol{f}_{ee}^{AA} = \boldsymbol{0}$, which is based on Newton's third law. Electrons within each of the subsystems are not subjected to any force, $\boldsymbol{f}_{ee}^{AA} = \boldsymbol{0}$, since $\rho_2(\boldsymbol{r}_1, \boldsymbol{r}_2)$ is symmetric in its coordinates and $(\boldsymbol{r}_1 - \boldsymbol{r}_2) = -(\boldsymbol{r}_2 - \boldsymbol{r}_1)$. This is Newton's third law. In the inter-domain case the two regions are not equivalent, in general, and the $\boldsymbol{f}_{AB}^{ee}$ does not vanish. We come thus to the conclusion that

$$\boldsymbol{f}_A = \boldsymbol{f}_{en}^{AA} + \sum_{B\neq A}(\boldsymbol{f}_{en}^{AB} + \boldsymbol{f}_{ee}^{AB}). \tag{3.134}$$

Similarly, the Hellmann-Feynman force over a given nucleus $A$, $\boldsymbol{F}_A$, has the following expression:

$$\boldsymbol{F}_A = \sum_B \boldsymbol{f}_{ne}^{AB} = -\sum_B \boldsymbol{f}_{en}^{BA}. \tag{3.135}$$

Obviously, both the sum of the Ehrenfest atomic forces as well as that of the Fellmann-Feynman ones, cancel out.

With these observables we can prove a large number of identities, the so-called **atomic theorems**. Some of them may be found in Ref. [9].

### 3.4.6 Atomic Properties

We will devote this section to present what type of chemically relevant information can be obtained from the previously defined atomic observables. We will pay attention to a pair of toy systems, the $H_2$ and LiF molecules, that we will take as archetypal systems of covalent and ionic bonds, respectively. We will examine some other systems with taxonomic aims.

Table 3.2 contains some of the atomic observables for a dihydrogen molecule in a HF/cc-pVQZ calculation at the theoretical equilibrium distance, $R_e = 1.3866$ bohr. we choose a standard reference frame in which the H atoms are located at $(0, 0, \pm R_e/2)$. Save explicit mention, we will refer to the properties of the uppermost hydrogen, that at $(0, 0, R_e/2)$. It is possible to extract a large amount of information from these data. We will start with the charge distribution.

Since the molecule is homodiatomic, the separatrix between both atomic basins is analytic, a plane perpendicular to the internuclear axis passing through the center of mass. The average number of electrons in each basin is determined by symmetry, coinciding with that of each H atom in vacuum. Despite the molecule lacks a total dipole moment, the electron distribution in each basin is not spherical, so each basin has a non-vanishing atomic electronic dipole. For $l = 1$ the only non-zero moment is $N^{10}$, and its centroid is displaced towards positive $z$ values. This is a very common situation that can be understood after Fig. 1.2. Although a clear accumulation of electron charge between the nuclei appears after bonding, the atomic electronic centroid is polarized in the contrary direction, since there is *more room* in the right than in the left part of the nucleus.

The usefulness of spherical moments with respect to cartesian ones is easily appreciated. Only the $(lm) = (l0)$ components are non-zero. It is important to recognize that atomic moments usually diverge with $l$, and that in our case those with even $l$ increase much more quickly than those with odd $l$. The negative value of $N^{20}$ shows, for instance, that the charge distribution is oblate with respect to the internuclear axis.

**Table 3.2** Some atomic properties of the H atom in a $H_2$ HF/cc-pVQZ molecule at the theoretical equilibrium distance. The molecule is aligned on the $z$ axis and the atom under consideration is located at $(0, 0, R_e/2)$. All values in a.u. In the isolated H atom, the only non-vanishing quantities are: $N = 1.0$, $T = 0.5$, $V_{en} = -1.0$, $\mathcal{V}^b = -1.0$ a.u. Data taken from Ref. [6]

| Quantity | Value | Quantity | Value | Quantity | Value |
|---|---|---|---|---|---|
| $N_{\mathrm{H}}$ | 1.0000 | $T_{\mathrm{H}}$ | 0.5664 | $(f_{en}^{\mathrm{HH}})_z$ | 0.1767 |
| $N_{\mathrm{H}}^{10}$ | 0.0986 | $V_{en}^{\mathrm{HH}}$ | −1.2216 | $(f_{en}^{\mathrm{HH}'})_z$ | −0.3406 |
| $N_{\mathrm{H}}^{20}$ | −0.3740 | $V_{en}^{\mathrm{HH}'}$ | −0.6023 | $(f_{ee}^{\mathrm{HH}'})_z$ | 0.1262 |
| $N_{\mathrm{H}}^{30}$ | −0.0731 | $V_{ee}^{\mathrm{HH}}$ | 0.1985 | $(f_{nn}^{\mathrm{HH}'})_z$ | 0.5201 |
| $N_{\mathrm{H}}^{40}$ | 1.0052 | $V_{ee}^{\mathrm{HH}'}$ | 0.2638 | $\mathcal{V}_{\mathrm{H}}^{b}$ | −1.1069 |
| $N_{\mathrm{H}}^{50}$ | 0.2678 | $V_{nn}^{\mathrm{HH}'}$ | 0.7212 | $\mathcal{V}_{\mathrm{H}}^{s}$ | −0.0262 |

As energetic quantities are regarded, the analysis can be performed at several levels. Let us show some of these possibilities. In the first place, we can simply compare the atomic energetic properties with those of the free atoms. Usually this is equivalent to choosing an arbitrary energetic reference, but in a homodiatomic molecule, where no charge transfer exists, the *in vacuo* reference is rather clear. The atomic kinetic energy is larger than the *in vacuo* reference. Since the average number of electrons $N$ has not changed, this implies that the average density has compacted, much along Slater's arguments on the process of formation of chemical bonds. We do not forget that in this case the partition of the kinetic energy is trivial, since the global value is divided equally into two identical atoms, $T_{\mathrm{H}} = 1/2T$.

The partition of the electronic potential energy of the system provides richer insights. On the one hand, the compaction of the density that we have just introduced increases the nucleus-electron attraction within a basin in approximately 0.2 a.u. ($\simeq$140 kcal/mol). The presence of a second nucleus is another essential factor. The interaction of the electrons in a basin with its neighboring nucleus stabilizes the atom in about 400 kcal/mol. On the contrary, the presence of other electrons generates new inter- and intrabasin repulsions. The existence of the latter is shocking (think of the average $N = 1$ electron population) until we realize that the number of electrons fluctuates in a basin. If we add together all the one-center contributions for a given atom we come to a quantity that has been called the atomic **net energy** or **self-energy** (see Sect. 3.5.2 for a formal definition). $E_{self} = -0.4566$ a.u. for a Hartree-Fock H atom, indicating that the atom is less stable in about 30 kcal/mol in-the-molecule than in the free state. It is almost always found that atoms are destabilized upon bonding. If we extend the integration to infinity in all directions, $E_{self}$ would be the expectation value of the atomic Hamiltonian, with a value bounded by below by the true in vacuo atomic energy. The finiteness of basins in certain directions, together with the fluctuation of the electron population alters this zero-order image. What is clear from this image is that the stabilization of the molecule comes, as expected, from the interaction with the neighboring atom. This interaction energy, including the internuclear repulsion, is $E_{int} = -0.2196$ a.u. ($\simeq$140 kcal/mol), thus exceeding by far the one-center destabilization.

The analysis of the components of the interaction energy for a given pair of basins contains valuable information about the nature of the chemical links. The relative share of the attractive and repulsive terms is particularly interesting. The attraction among the electrons in a basin and the nucleus of another, as well as the repulsion among the electrons in those two basins depend on the asphericity of the electron distribution and on the classical interpenetration of the atomic densities. For non-overlapping spherical distributions only the monopole-monopole terms would survive. Thus, the difference between $V_{ee}^{\mathrm{HH}'}$ and $1/R_e = 0.7212$ can be interpreted as due to a considerable delocalization of the electrons, so that a part of the initial interatomic repulsion has been transported to the intra-basin electron repulsion. As a result, the interaction energy is very negative. As we will see, this is one of the cleanest energetic signatures characterizing **covalent** interactions (see Sect. 3.5.3).

There is a growing scientific community in favor of recovering the concept of force in QM. Let us consider briefly the force being exerted on the electrons in dihydrogen.

On one side, the nucleus pulls the electrons in its own basin in a dissociative manner. Although this fact is against the direction of the atomic dipole moment $N^{10}$, it does not contradict any principle, since the Berlin forces depend on the polarization of the density in the vicinity of the nucleus, and this polarization may not coincide with the total one. Similarly, the electrons of a basin also exert a dissociative force against those of the basin considered. Both contributions are **almost** cancelled by the attractions exerted by the neighboring nuclei, leaving a total cohesive Ehrenfest force of −0.0378 a.u. which is now cancelled by the quantum (Pauli) forces.

On the contrary, forces acting on the nuclei are only electrostatic within our theoretical framework. Both the electrons from its own basin as well as those from other basins exert cohesive forces upon the nuclei, not very different in magnitude. We should recall here Berlin's interpretation. It is well known that using gaussian expansions for variational wavefunctions lead to good energies and to the satisfaction of the virial theorem. However, this is not the case with the Hellmann-Feynman theorem, which is not satified unless extremely large basis sets are used. Thus, the force acting on nuclei obtained from the derivatives of potential energy surfaces have much higher quality than those obtained from direct calculations of forces. As seen from Table 3.2, the total force acting on the H nucleus is $\boldsymbol{F}_{\mathrm{H}} = -0.003$ a.u., close to, but not exactly zero. Any calculation with a smaller basis set would have provided a much worse value.

A similar, although not exhaustive, analysis for the LiF molecule is worthwhile. Table 3.3 contains some atomic properties calculated for both basins. In the first place we should emphasize the large net topological charge of both atomic basins.

**Table 3.3** Some atomic properties of the lithium and fluorine atoms in the LiF molecule in a HF/cc-pVTZ calculation at the theoretical equilibrium distance. The molecule is placed along the $z$ with the lithium atom at (0, 0, −2.20784) a.u. and the fluorine atom at (0, 0, 0.73595) a.u. All data in a.u. As a comparison, for the free ions, $E_{\mathrm{Li}^+} = -7.2362$, $E_{\mathrm{F}^-} = -99.4594$, $E_{\mathrm{Li}} = -7.4327$, $E_{\mathrm{F}} = -99.4093$ a.u

| Quantity | Value | Quantity | Value | Quantity | Value |
|---|---|---|---|---|---|
| $N_{\mathrm{Li}}$ | 2.0618 | $T_{\mathrm{Li}}$ | 7.3492 | $(f_{en}^{\mathrm{LiLi}})_z$ | 0.1936 |
| $N_{\mathrm{Li}}^{10}$ | −0.0188 | $V_{en}^{\mathrm{LiLi}}$ | −16.3423 | $(f_{en}^{\mathrm{LiF}})_z$ | 2.1181 |
| $N_{\mathrm{Li}}^{20}$ | −0.0266 | $V_{en}^{\mathrm{LiF}}$ | −6.2734 | $(f_{ee}^{\mathrm{LiF}})_z$ | −2.2175 |
| $N_{\mathrm{Li}}^{30}$ | −0.0437 | $V_{ee}^{\mathrm{LiLi}}$ | 1.7715 | $(f_{nn}^{\mathrm{LiF}})_z$ | 3.1157 |
| $N_{\mathrm{Li}}^{40}$ | 0.0875 | $V_{ee}^{\mathrm{LiF}}$ | 6.8624 | $\mathcal{V}_{\mathrm{Li}}^{b}$ | −14.5571 |
| $N_{\mathrm{Li}}^{50}$ | −0.2770 | $V_{nn}^{\mathrm{LiF}}$ | 9.1785 | $\mathcal{V}_{\mathrm{Li}}^{s}$ | −0.1429 |
| $N_{\mathrm{F}}$ | 9.9383 | $T_{\mathrm{F}}$ | 99.5966 | $(f_{en}^{\mathrm{FLi}})_z$ | 0.8813 |
| $N_{\mathrm{F}}^{10}$ | −0.2416 | $V_{en}^{\mathrm{FF}}$ | −243.7996 | $(f_{en}^{\mathrm{FLi}})_z$ | −3.3775 |
| $N_{\mathrm{F}}^{20}$ | −0.4038 | $V_{en}^{\mathrm{FLi}}$ | −10.0977 | $(f_{ee}^{\mathrm{FLi}})_z$ | 2.2175 |
| $N_{\mathrm{F}}^{30}$ | 1.6586 | $V_{ee}^{\mathrm{FF}}$ | 44.7796 | $(f_{nn}^{\mathrm{FLi}})_z$ | 3.1157 |
| $N_{\mathrm{F}}^{40}$ | −3.1958 | $V_{ee}^{\mathrm{FLi}}$ | 6.8624 | $\mathcal{V}_{\mathrm{F}}^{b}$ | −109.0919 |
| $N_{\mathrm{F}}^{50}$ | 0.5726 | $V_{nn}^{\mathrm{FLi}}$ | 9.1785 | $\mathcal{V}_{\mathrm{F}}^{s}$ | −0.1208 |

We have an almost complete electron transfer, compatible with a lithium cation. The atomic multipole moments of the Li basin (the $Li^+$ cation in the following) are very small, indicating a considerable sphericity in its electron distribution. $N^{10}$ is negative, so that the density of the cation is slightly counter-polarized, displaced by the effect of the almost ten electrons of the neighboring fluoride. In a similar manner, the atomic dipole of the F atom shows a clear polarization toward the $Li^+$ cation. As the energies are regarded, in the case of the Li moiety all of them are compatible with a slightly perturbed lithium cation. For instance, the electron-electron repulsion of a free $Li^+$ is about 1.65 a.u., in contrast with the 2.28 a.u. repulsion of a neutral Li atom. A similar conclusion can be obtained from examining energetic quantities in the F basin. More interesting, however, is the consideration of the interaction energy between the two atoms. The total interaction energy, $E_{int} = -0.3369$ a.u., is to be compared with the electrostatic attraction energy between two point opposite sign unit charges separated by the internuclear distance: $V_{cl} = -0.3397$ a.u. This means that **99.2% of the interaction is classical in nature**. This is the signature of an ionic bond.

The simple ionic model provides a very good approximation to any of the components of the interaction energy in this system. Let us, for instance, take the attraction between the electrons of $Li^+$ and the F nucleus, which is computed to be $-6.2734$ a.u. This value can be compared with $-6.1145$ a.u., if we use a model with complete ionization, or with $-6.3041$ a.u. if we use the computed topological charges.

Not less relevant is the interbasin electron-electron repulsion. Its value, 6.8623 a.u., is very close to the classical one ($2 \times 9/R_e$), 6.7949 a.u. This is to be compared with the drastic decrease of about 0.5 a.u that was found due to electron delocalization in dihydrogen .

The systematization of all these results has given rise to an energetic and dynamic picture of the forces that control the charge distribution in molecules. Eventually, it led to the development of the interacting quantum atoms (IQA) scheme (Sect. 3.5.2). However, out of all the observables that can be defined, atomic populations have played an important role in the success or rejection of the QTAIM by the community, probably due to the emotional component related to the colorful history of population analyses in theoretical chemistry. This debate has two different sides. On the one hand, atomic populations are not experimentally available (notice however that QTAIM populations can be obtained from X-ray diffraction). On the other hand, every community has tried over the years to define atomic charges that would be useful to reproduce the observables under scrutiny: electrostatic potentials, force fields, thermodynamic properties, etc. In general, the net charges provided by the QTAIM are *too large* for many. Larger, let us say, than those coming from Mulliken's analysis. This is a very well known, misinterpreted fact.

As we have shown, it has no physical sense to use QTAIM charges exclusively to reconstruct molecular dipole moments. There is no direct physical relation between atomic monopoles and molecular dipoles. Table 3.4 shows the atomic volumes as well as the monopoles and dipoles of the second and third period hydrides at the B3LYP level. A quick look at these results shows what is to be expected from electronegativity differences. Perhaps surprisingly, LiH shows a very large formal charge, close to 0.9

**Table 3.4** Some atomic properties for the second and third row hydrides *A*H. B3LYP/6-311+G(2d,2p)//HF/6-31(d) calculations, with all data in a.u. $\mu_{\mathrm{mol}}$ is the total molecular dipole, and volumes are obtained from the $\rho = 0.001$ a.u isosurface. Data taken from Ref. [6]

| Molecule | $R_e$ | $V_A$ | $V_{\mathrm{H}}$ | $Q_A$ | $\mu_A$ | $\mu_{\mathrm{H}}$ | $\mu_{\mathrm{CT}}$ | $\mu_{\mathrm{mol}}$ |
|---|---|---|---|---|---|---|---|---|
| HH | 1.3792 | 60.5 | 60.5 | 0.0000 | −0.1140 | +0.1140 | 0.0000 | 0.0000 |
| LiH | 3.0908 | 36.1 | 196.7 | +0.8869 | −0.0092 | −0.4765 | +2.7413 | +2.2556 |
| BeH | 2.5469 | 162.9 | 137.3 | +0.8323 | −1.4387 | −0.5497 | +2.1198 | +0.1314 |
| BH | 2.3163 | 173.3 | 92.1 | +0.6679 | −1.8201 | −0.3630 | +1.5471 | −0.6360 |
| CH | 2.0941 | 176.4 | 53.0 | −0.0235 | −0.7327 | +0.1685 | −0.4921 | −1.0563 |
| NH | 1.9347 | 160.7 | 37.1 | −0.3036 | −0.2456 | +0.1943 | −0.5874 | −0.6387 |
| OH | 1.8111 | 149.5 | 24.4 | −0.5427 | +0.1192 | +0.1684 | −0.9829 | −0.6953 |
| FH | 1.7211 | 135.0 | 15.7 | −0.7073 | +0.3373 | +0.1261 | −1.2174 | −0.7540 |
| NaH | 3.6176 | 109.0 | 177.4 | +0.7084 | −0.0568 | −0.1234 | +2.5627 | +2.3825 |
| MgH | 3.3043 | 261.6 | 145.7 | +0.6870 | −1.4812 | −0.2424 | +2.2701 | +0.5465 |
| AlH | 3.1222 | 267.9 | 126.4 | +0.7677 | −2.0537 | −0.2900 | +2.3969 | +0.0532 |
| SiH | 2.8634 | 271.4 | 102.0 | +0.6903 | −1.7362 | −0.3003 | +1.9766 | −0.0599 |
| PH | 2.6655 | 259.6 | 77.7 | +0.4858 | −1.3047 | −0.1887 | +1.2949 | −0.1985 |
| SH | 2.5134 | 251.4 | 52.8 | +0.0342 | −0.5100 | +0.0713 | +0.085 | −0.353 |
| ClH | 2.3928 | 241.6 | 39.4 | −0.2420 | −0.0360 | +0.1253 | −0.5791 | −0.4898 |

a.u., which leads to suggest it is an ionic molecule. This net charge is to be compared with that coming from a Mulliken analysis, which is 0.543 a.u. In the QTAIM, the volume of the H basin triples that of H in dihydrogen, while that of the Li basin is extremely small. These facts are compatible with the differences in the polarizabilities of a hydride and a lithium cation, as well as with the large discrepancy in their atomic dipoles. The hydride in the QTAIM LiH is heavily polarized towards the $Li^+$, and the latter polarizes also along the same direction, although almost a hundred times smaller in magnitude. We can also check how the CT dipole moment is rather similar to the molecular dipole. We can see that, across the full table, this is only a good approximation in LiH and NaH, showing how useless it is to try to rationalize molecular dipoles from net atomic charges. Finally, the effect of correlation in $H_2$ deserves special attention. If we compare the results in Table 3.4 (B3LYP) with those in Table 3.2 (Hartree-Fock), we see that the effectively correlated B3LYP results offer a slightly larger dipole moment. This is basically due to the change in equilibrium distance, $R_e$.

As we move along the second period, the volume of the H atom decreases in a monotonous way, clearly correlated to its net charge. The H atom is negatively charged ($Q_H = -Q_A$) up to BH, being close to neutral (although positively charged) in CH, then passing to be positively charged. It is also in CH that the volumes of both basins become comparable. These two results are compatible with the similar electronegativities of both atoms. In BH, the polarizability of the boron atom has increased so much that its counterpolarization is big enough to invert the dipole moment.

As we run over the third period, all these features are reproduced once again, although slightly displaced toward atoms with higher $Z$ values, something that can be rationalized from the smaller electronegativity of the non-metal atoms. For instance, the SH molecule displays smaller net charges than its equivalent OH counterpart. The H atomic volume in this case is similar to that in dihydrogen. This is another manifestation of the transferability of atomic properties, that will be analyzed in the next subsection.

### *3.4.7 Transferability of Atomic Properties*

The union set of any number of atomic basins is a superbasin that complies with all the requisites of a QTAIM region. Over the years, the science of Chemistry has shown that some groups of atoms provide a system with constant and predictable properties. These groups of atoms are called **functional groups**. Formally,

$$\mathcal{G} = \bigcup_A \Omega_A. \tag{3.136}$$

One of the great successes of the QTAIM was to show that **group properties**,

$$O_{\mathcal{G}} = \sum_A O_{\Omega_A}, \tag{3.137}$$

correspond faithfully to those obtained from experiments, and that their constancy is based on that of their atomic basins.

Rigorously speaking, there exist two types of transferable groups. We have on the one side, what we call **rigid groups**, whose separatrices, populations, and multipole moments are fixed and independent of the environment in which we place them. On the other, we find **buffer groups**, that acquire or donate electrons rather easily, with properties that vary more or less linearly within the limits set by the corresponding charges they hold. The latter are those groups used to classify the **inductive effects** of organic chemistry. In both cases, we can predict with high fidelity their behavior as a function of their environment.

The classical example of transferable groups comes from the empirical study of the physico-chemical properties along homologous series. In linear chain saturated hydrocarbons $CH_3$–$(CH_2)_n$–$CH_3$ it is possible to fit the experimental enthalpies of formation of these compounds to the following form:

$$\Delta H_f^\circ = 2\alpha + n\beta. \tag{3.138}$$

**Table 3.5** Some group properties for the methyl and methylene groups in the first alkanes. Changes are reported with respect to the standard methyl group in ethane and the standard methylene in hexane. Populations in electrons and energies in kcal/mol. Data taken from Ref. [7] with permission

| Molecule | $\Delta N(CH_2)$ | | $\Delta E(CH_2)$ | | $\Delta N(CH_3)$ | $\Delta E(CH_3)$ |
|---|---|---|---|---|---|---|
| Ethane | | | | | 0.000 | 0.0 |
| Propane | −0.034 | | 21.7 | | 0.017 | −10.9 |
| Butane | −0.018 | | 10.8 | | 0.018 | −10.8 |
| Pentane | −0.018 | +0.001 | 10.9 | 0.0 | 0.017 | −9.9 |
| Hexane | −0.018 | +0.000 | 10.6 | −1.1 | 0.018 | −9.9 |

where $\alpha$ and $\beta$ are the contribution of the methyls and methylene groups, respectively. The accepted values at 298.15 K are $\alpha = 10.12$ and $\beta = -4.29$ kcal/mol, respectively.

Back in 1980, HF calculations performed by Krishnan et al. [44] showed the same correlations, with average deviations smaller than the experimental errors. Bader and coworkers [7] analyzed such results from the QTAIM point of view in 1987. Table 3.5 show some of their results. There is no perfect transferability. All groups are slightly different, in agreement with the holographic theorem [59]. Notwithstanding, we clearly observe two gross methyl groups: that in methane, and the methyl in the rest of alkanes; and three types of methylene: in propane and two in butane. This classification is actually a chemically guided taxonomy. We are isolating methyl or methylene groups flanked by two methyls, methyl+methylne or two methylene groups.

It is surprising that the energy changes of these objects lies in the chemical scale, although the group energies themselves are measured in the tens or hundreds of atomic units. Even more interesting is to check how the energy changes balance out. For instance, the energy of the methyl group in propane is about 11 kcal/mol lower than the energy of the standard methyl. This change is coupled with a gain of about 0.017 electrons. Simultaneously, the methylene looses twice this charge and about 22 kcal/mol with respect to the standard methylene. At the end of this process, the energy of propane can still be written in terms of standard methyl and methylene groups as $2\alpha + \beta$. We come to the same conclusions if we examine the rest of Table 3.5. In other words, $\Delta E/\Delta N$ is almost a constant quantity, and the electrons transferred seem to possess a characteristic energy, independent of the group in which they finally reside.

The physical basis for the transferability of energies lies in the so-called virial energy partition [6]. Within the QTAIM atomic partition, at equilibrium (or stationary) geometries on the potential energy surface the virial theorem is fulfilled, and $2\langle T\rangle = -\langle V\rangle$. Provided that $E = \langle T\rangle + \langle V\rangle$, it follows that $\langle E\rangle = -\langle T\rangle$, so that after an atomic partition of the kinetic energy $E_\Omega = -T_\Omega$ . Since kinetic energies are one-center (not two-center) quantities, this means that we can ascribe energies to atoms in stationary states and justifies the transferability of group energies (or enthalpies) in Chemistry. It assigns a proper energy to each QTAIM atom or group.

We will dwell further on a more general way of using these atomic energies in Sect. 4.3.4.

Not only the energy is transferable. Other quantities, such as atomic dipoles, atomic volumes or even the correlation energy show an additive behavior that includes cancellations like those we have just commented above.

The correct behavior of the QTAIM group properties as regards their comparison to group experimental data has been shown once and again, with a vast literature leaving few unexplored territories [7]. It has been proposed, for instance, to use the QTAIM like a procedure to theoretically synthesize compounds. One would start from a fragment library, that would include nuclear positions, separatrices and density values over a 3D grid. Those groups would be then superimposed by appropriate translations and rotations to build the density of the molecule. Finally, the molecular observables would be obtained by a scalar or vector addition process from the different observables kept on the library. These techniques have been applied to oligopeptides, small fragments of DNA and, small polymers of rather rigid conformation [8].

## 3.5 Reduced Density Matrices and Multielectron Descriptors

The electron density is just one out of many possible probability densities that can be used in theoretical chemistry. In fact, it is the first element in a hierarchy of objects that determine the probability of joint statistical events regarding the position of electrons. Starting with Born's interpretation, we define the $N$-th order reduced density matrix (nRDM) as

$$\rho_N(\boldsymbol{x}_1, \ldots, \boldsymbol{x}_N; \boldsymbol{x}'_1, \ldots, \boldsymbol{x}'_N) = N!\Psi^*(\boldsymbol{x}'_1, \ldots, \boldsymbol{x}'_N)\Psi(\boldsymbol{x}_1, \ldots, \boldsymbol{x}_N). \quad (3.139)$$

A spinless version $\rho_N(\boldsymbol{r}_1, \ldots, \boldsymbol{r}_N; \boldsymbol{r}'_1, \ldots, \boldsymbol{r}'_N)$ also exists if the spin coordinates of all electrons are integrated out. Its diagonal part, $\rho_N(\boldsymbol{r}_1, \ldots, \boldsymbol{r}_N; \boldsymbol{r}_1, \ldots, \boldsymbol{r}_N) = \rho_N(\boldsymbol{r}_1, \ldots, \boldsymbol{r}_N)$ is an $N$-th order reduced density (nRD) that measures the probability density of finding $N$-tuples of electrons in a $3N$ dimensional volume element $d\boldsymbol{r}_1 \ldots d\boldsymbol{r}_N$.

We can nevertheless be interested in more compact structures in which the position of several electrons is averaged out. For instance, we might question about the probability density of finding $m < N$ electrons at $\boldsymbol{r}_1 \ldots \boldsymbol{r}_m$. These reduced densities can be easily obtained by integrating out the coordinates of the non-interesting electrons, taking into account indistinguishability[4]

[4] Recall from Combinatory Logic that: $\binom{N}{m} = \frac{N!}{m!(N-m)!}$ , gives us the number of $n$-tuples in the system, so that the number of pairs is given by $\binom{N}{2} = \frac{N!}{2!(N-2)!} = \frac{N(N-1)}{2}$. If we know consider ordered pairs (e.g. $AB \neq BA$), we have twice as many, i.e. $N(N-1)$. More generally, for ordered

$$\rho_m(\boldsymbol{x}_1, \ldots, \boldsymbol{x}_m; \boldsymbol{x}'_1, \ldots, \boldsymbol{x}'_m) = \frac{N!}{(N-m)!} \int \Psi^*(\boldsymbol{x}'_1, \ldots, \boldsymbol{x}'_N)\Psi(\boldsymbol{x}_1, \ldots, \boldsymbol{x}_N) d\boldsymbol{x}_{m+1} \ldots d\boldsymbol{x}_N \,, \tag{3.140}$$

with spinless and diagonal versions defined in analogous ways. In the above equation we assume that the limit $\boldsymbol{x}'_k \to \boldsymbol{x}_k$ has been taken before integrating all $k > m$ coordinates. We will do that in the following to avoid cumbersome notation. Notice that the nRDs are normalized to $n!/(N-n)!$, the number of ordered $n$-tuples of electrons. This is the so-called McWeeny's normalization [57].

These simple definitions allow us to understand that the electron density, $\rho(\boldsymbol{r})$ is nothing but the diagonal part of the first order density matrix, 1RDM.

It is easy to show (we refer the reader to the excellent presentation of McWeeny [57]) that in an $n$-body Hamiltonian the energy depends only on, at most, the nRDM. Actually, we do not need the full wavefunction, but only the nRDM to obtain the expectation value of an $n$-electron operator. In the standard Coulombic Hamiltonian of non-relativistic quantum mechanics, this means that the energy can be obtaiend with the help of the 1-RDM and the 2RD. The 2RD is known as the pair density, and will be examined below. Unfortunately, the dream of solving the electronic structure problem using only the 2RDM instead of the full wavefunction faces enormous problems. We will not dwelve into these subtleties in this monograph [23].

### 3.5.1 *The Pair Density*

If the one-particle density is key to provide a quantum theory of the atom in the molecule, the pair density plays an equivalent role in building a theory of **the electron pair**. Let us recall some properties and notation regarding the second order density matrix,

$$\rho_2(\boldsymbol{x}'_1, \boldsymbol{x}'_2; \boldsymbol{x}_1, \boldsymbol{x}_2) = N(N-1) \int \Psi^*(\boldsymbol{x}'_1, \boldsymbol{x}'_2, \ldots, \boldsymbol{x}_N)\Psi(\boldsymbol{x}_1, \boldsymbol{x}_2, \ldots, \boldsymbol{x}_N)\, d\boldsymbol{x}_3 \ldots \boldsymbol{x}_N, \tag{3.141}$$

which we normalize to the ordered pairs of electrons in the system, $N(N-1)$. The diagonal part of this object is the pair density, $\rho_2(\boldsymbol{x}_1, \boldsymbol{x}_2)$, a positive definite quantity on $\mathbb{R}^3 \otimes \mathbb{R}^3$. Were the electrons non-correlated objects, the pair density would immediately be obtained from the one-body density,

$$\rho_2^{\text{ind}}(\boldsymbol{x}_1, \boldsymbol{x}_2) = \frac{N-1}{N}\rho(\boldsymbol{x}_1)\rho(\boldsymbol{x}_2), \tag{3.142}$$

where the $N$–dependent factor takes into account the renormalization of the density of the second electron, that should integrate to $N-1$, not to $N$, provided that the first electron has been localized (and located) at $\boldsymbol{x}_1$. The difference between $\rho_2$ and

$n$-tuples (which allow $m!$ permutations), we will have: $\frac{N!}{(N-m)!}$. Note that different normalizations have been proposed depending on the use of ordered or unordered pairs.

$\rho_2^{\text{ind}}$ is that the former contains all correlation effects. The antisymmetry principle also demands that

$$\rho_2(\boldsymbol{x}_1', \boldsymbol{x}_2'; \boldsymbol{x}_1\boldsymbol{x}_2) = -\rho_2(\boldsymbol{x}_2', \boldsymbol{x}_1'; \boldsymbol{x}_1, \boldsymbol{x}_2), \tag{3.143}$$

so $\rho_2(\boldsymbol{x}_1, \boldsymbol{x}_1) = 0$, and *the density probability of finding two same-spin electrons at the same point in space vanishes*. This implies the existence of very important deviations of the true pair density with respect to that of independent particles. We thus define the so-called **correlation function** or **correlation factor** $f$:

$$\rho_2(\boldsymbol{x}_1, \boldsymbol{x}_2) = \rho(\boldsymbol{x}_1)\rho(\boldsymbol{x}_2)\left\{1 + f(\boldsymbol{x}_1, \boldsymbol{x}_2)\right\}. \tag{3.144}$$

We should not forget that the correlation factor exists even for non-correlated electrons, due to the above mentioned normalization function: $\rho(\boldsymbol{x}_1)\rho(\boldsymbol{x}_2)$ integrates to $N^2$ instead of $N(N-1)$, so a given electron would interact with itself. This is intimately linked to self-interaction corrections: if we accept that $f = 0$ is the independent electron reference, then our electron system is affected by **self-interaction** and an electron feels the Coulombic interaction due to its own density. Appropriately dealing with self-interaction terms is extremely important in density functional theory.

It is also interesting to analyze the case of HF. $\rho_1(\boldsymbol{x}_1, \boldsymbol{x}_2)$ can be expressed in terms of orbitals:

$$\rho_1(\boldsymbol{x}_1, \boldsymbol{x}_2) = \sum_i \Psi_i^*(\boldsymbol{x}_1)\Psi_i(\boldsymbol{x}_2), \tag{3.145}$$

where $\Psi_i(\boldsymbol{x})$ are spinorbitals. $n$th-order reduced densities can be reconstructed from their $n-1$ analogues. For example, for $\rho_2(\boldsymbol{x}_1, \boldsymbol{x}_2)$ we can see that the final expression will depend on the spin of the electrons under study. For electrons of different spin, there will be no correlation, so that we recover independent events as in Eq. 3.142:

$$\rho_2^{\alpha\beta}(\boldsymbol{x}_1, \boldsymbol{x}_2) = \rho^\alpha(\boldsymbol{x}_1)\rho^\beta(\boldsymbol{x}_2) \tag{3.146}$$

For same spin electrons, we have:

$$\rho_2^{\alpha\alpha}(\boldsymbol{x}_1, \boldsymbol{x}_2) = \rho^\alpha(\boldsymbol{x}_1)\rho^\alpha(\boldsymbol{x}_2) - |\rho_1^{\alpha\alpha}(\boldsymbol{x}_1, \boldsymbol{x}_2)|^2 = \rho^\alpha(\boldsymbol{x}_1)\rho^\alpha(\boldsymbol{x}_2) - \sum_i^\alpha |\Psi_i^*(\boldsymbol{x}_1)\Psi_i(\boldsymbol{x}_2)|^2, \tag{3.147}$$

so that if $\boldsymbol{x}_2 \to \boldsymbol{x}_1$, the probability tends to zero. Thus, the probability of finding electrons with the same spin on the same point is zero. This is called Fermi correlation or exchange, and results from the Pauli principle.

In the study of joint probability distribution functions, it has been found very useful to pay attention to conditional probabilities.[5] When it is known that a refer-

[5] Recall that given two events $A$ and $B$, the conditional probability of $A$ given $B$, $P(A|B)$, is given by:

ence electron (RE) is located at $\boldsymbol{x}_1$, the conditional probability density to find other electrons at $\boldsymbol{x}_2$ becomes

$$\rho(\boldsymbol{x}_2|\boldsymbol{x}_1) = \frac{\rho_2(\boldsymbol{x}_1, \boldsymbol{x}_2)}{\rho(\boldsymbol{x}_1)}. \tag{3.149}$$

It is clear that this object integrates to $N - 1$, and that

$$\begin{aligned} \rho(\boldsymbol{x}_2|\boldsymbol{x}_1) &= \rho(\boldsymbol{x}_2)\left\{1 + f(\boldsymbol{x}_1, \boldsymbol{x}_2)\right\}, \\ \rho_2(\boldsymbol{x}_1, \boldsymbol{x}_2) &= \rho(\boldsymbol{x}_1)\rho(\boldsymbol{x}_2|\boldsymbol{x}_1). \end{aligned} \tag{3.150}$$

Finally, it is also useful to define the difference between the conditional probability and the standard density. This new function is known as the **exchange-correlation hole**, $h_{xc}$ [57] :

$$h_{xc}(\boldsymbol{x}_2|\boldsymbol{x}_1) = \rho(\boldsymbol{x}_2|\boldsymbol{x}_1) - \rho(\boldsymbol{x}_2) = \rho(\boldsymbol{x}_2) f(\boldsymbol{x}_1, \boldsymbol{x}_2). \tag{3.151}$$

We call it a hole due to its two basic properties valid at any $\boldsymbol{x}_1$,

$$\begin{aligned} \int h_{xc}(\boldsymbol{x}_2|\boldsymbol{x}_1) d\boldsymbol{x}_2 &= -1, \\ h_{xc}(\boldsymbol{x}_2 \rightarrow \boldsymbol{x}_1|\boldsymbol{x}_1) &= -\rho(\boldsymbol{x}_1). \end{aligned} \tag{3.152}$$

It is known that the hole displays a cusp when $\boldsymbol{x}_1 \rightarrow \boldsymbol{x}_2$ as a consequence of the Coulombic singularity in $r_{12}^{-1}$. This is reminiscent of Kato's theorem [42] in Sect. 3.1 for the electron density cusp at nuclear positions. We refer the reader to Ref. [23] for details on the electron-electron coalescence cusp.

With these expressions, the electron repulsion energy, $V_{ee}$, may be written in the following ways:

$$\begin{aligned} V_{ee} &= \frac{1}{2} \int \frac{\rho_2(\boldsymbol{x}_1, \boldsymbol{x}_2)}{r_{12}} d\boldsymbol{x}_1 d\boldsymbol{x}_2 = \frac{1}{2} \int \frac{\rho(\boldsymbol{x}_1)\rho(\boldsymbol{x}_2|\boldsymbol{x}_1)}{r_{12}} d\boldsymbol{x}_1 d\boldsymbol{x}_2 \\ &= \frac{1}{2} \int \frac{\rho(\boldsymbol{x}_1)\rho(\boldsymbol{x}_2)}{r_{12}} d\boldsymbol{x}_1 d\boldsymbol{x}_2 + \frac{1}{2} \int \frac{\rho(\boldsymbol{x}_1) h_{xc}(\boldsymbol{x}_2|\boldsymbol{x}_1)}{r_{12}} d\boldsymbol{x}_1 d\boldsymbol{x}_2 \\ &= V_{\mathrm{C}} + V_{xc}, \end{aligned} \tag{3.153}$$

where we have introduced the classical Coulombic repulsion (with self-interaction) and the **exchange-correlation** energy.

We have already seen for $\rho_2^{HF}(\boldsymbol{x}_, \boldsymbol{x}_2)$ that it is can be interesting to properly distinguish between correlation effects due to same-spin electrons, which we call **Fermi correlation**, from those due to opposite-spin electrons, named **Coulomb correlation**. To do that, we separate the exchange-correlation hole into a part due to

---

$$P(A|B) = \frac{P(A \cap B)}{P(B)} \tag{3.148}$$

where $P(A \cap B)$ is the probability of the joint event $A$ and $B$.

same-spin electrons, the **Fermi hole**, and a part coming from opposite-spin electrons, the **Coulomb hole**:

$$h_{xc}(\boldsymbol{x}_2|\boldsymbol{x}_1) = h_{xc}(\boldsymbol{r}_2(s_2 = s_1)|\boldsymbol{x}_1) + h_{xc}(\boldsymbol{r}_2(s_2 \neq s_1)|\boldsymbol{x}_1) = h_F(\boldsymbol{r}_2|\boldsymbol{r}_1) + h_C(\boldsymbol{r}_2|\boldsymbol{r}_1). \tag{3.154}$$

With this partition and the antisymmetry principle, it comes out that only the Fermi hole is responsible for electron exclusion, so that

$$\begin{aligned} &\int h_F(\boldsymbol{r}_2|\boldsymbol{r}_1)\, d\boldsymbol{r}_2 = -1, \\ &h_F(\boldsymbol{r}_2 \to \boldsymbol{r}_1|\boldsymbol{r}_1) = -\rho(\boldsymbol{r}_1), \\ &\int h_C(\boldsymbol{r}_2|\boldsymbol{r}_1)\, d\boldsymbol{r}_2 = 0. \end{aligned} \tag{3.155}$$

The specific shape of the Coulomb and Fermi holes depends strongly on the system under scrutiny. In general, the Fermi hole is deepest close to the RE, decaying quickly as we move away from it. It can be shown to be negative definite in single determinant descriptions. Notice that if the RE is far from the molecular frame, Eq. 3.151 shows that $h_F$ will be rather shallow in its nearbies and, since it integrates to -1, it will necessarily display non-negligible values relatively far from the probe electron, **leaving it behind**.

It is common use to study the behavior of the holes using the dissociation of the dihydrogen molecule, as originally done by Baerends y Gritsenko [11]. Since in this case there are only two electrons, the only purpose of the Fermi hole is to correct for the self-interaction error, eliminating the repulsion of the RE with itself. The hole density is, therefore, that of an electron with the same spin as that of the RE, with opposite density sign: $h_F = -\rho/2$. This means that the Fermi hole does not depend, in this case, on the position of the RE, displaying two well developed contributions peaking around the two nuclei even in the separated atom limit.

Let us consider a molecule close to its dissociation limit, with one electron at the hydrogen atom labeled as $H_A$ and the other at the atom that we label as $H_B$. The behavior of the Fermi and Coulomb holes is qualitatively shown in Fig. 3.11. Nonetheless, do not forget that it is only $h_{xc}$ that has true physical meaning.

Let us firstly place a RE in the vicinity of nucleus A (Fig. 3.11 top). This electron should feel a practically bare $A$ nucleus, and a nucleus $B$ almost completely shielded by its own electron. The purpose of $h_{xc}$ is to guarantee that the interelectron repulsion tends properly towards its asymptotic limit, $R_{AB}^{-1}$. $h_{xc}$ must therefore delete the density of electron 2 in the neighborhood of nucleus $A$, leaving that density untouched close to nucleus $B$. Under these circumstances the Fermi hole is a very bad approximation to the true behavior of the full exchange-correlation hole. In particular, in the vicinity of our RE, $h_F$ partially shields nucleus $A$, and does not shield nucleus $B$ sufficiently. As a consequence, the electron responds building a rather diffuse density, that gives rise to a too small kinetic energy, a not large enough nuclear attraction and, of course, a too large interelectron repulsion. All these facts are well known in the failure of the Hartree-Fock method to correctly dissociate dihydrogen.

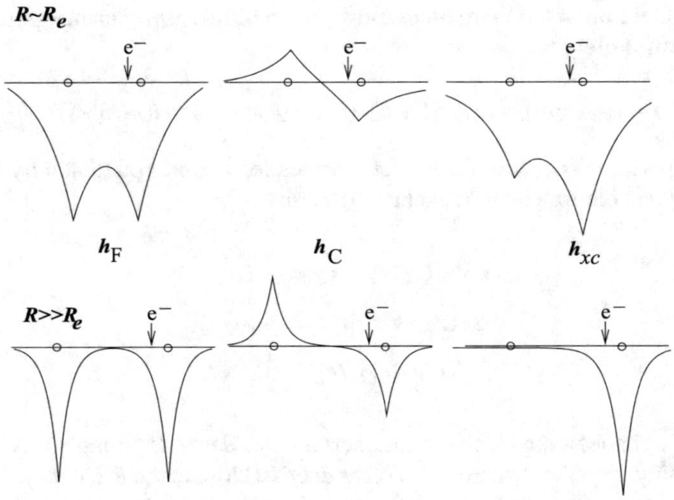


**Fig. 3.11** Qualitative behavior of the Fermi (left), Coulomb (middle), and full (right) correlation holes in the $H_2$ molecule at two representative internuclear distances. The position of the RE is shown by an arrow. Adapted from Ref. [11]

The task of the Coulomb hole is to build a density depletion around the RE. Since its integrated density vanishes, the charge that is taken out from one region has to accumulate in others. This is the expected behavior of the Coulombic repulsion. In the present case, the task of $h_C$ at large internuclear distances is many-fold (Fig. 3.11 bottom). On the one hand, it guarantees the cancellation of the Fermi hole at nucleus $B$. On the other hand, it makes the hole deeper in the vicinity of nucleus $A$. Since the Fermi hole is static, the Coulomb hole must be dynamic, and if we move the RE toward nucleus $A$, $h_C$ must jump towards it when we cross the midpoint of the internuclear axis.

In the next sections, we will see how using these higher order densities gives access from a QTAIM point of view to well rooted concepts in Quemistry such energy partitions (Sect. 3.5.2), electron localization (Sect. 3.5.3) and orbitals (Sect. 3.5.4).

### 3.5.2 Energy: Interacting Quantum Atoms

Any approach to chemical bonding must include an energetic viewpoint. Since the QM energy of a molecular system can only be truly partitioned into those terms that make up the model Hamiltonian chosen (e.g. the kinetic energy, the electron-nucleus attraction, the interelectron repulsion, and the nuclear-nuclear interaction in the Coulomb approximation), a large number of chemically appealing decompositions of the energy have been proposed over the years [48]. These can be grossly partitioned into intermolecular and intramolecular energy decomposition analyses (EDAs).

Summarizing a vast field, inter-EDAs have been dominated by perturbation theory, while most intra-EDAs have inhereted, in one way or another, the spirit of Morokuma's decomposition [43].

In the former, researchers take profit of the multipolar expansion of the interaction Hamiltonian to express directly the interaction energy between two (or more) approaching fragments in terms of the wavefunctions of the isolated systems. At large distances, when the antisymmetry of the interacting wavefunctions can be neglected and the multipolar expansion converges, this approach (which is known as the polarization approximation) gives rise to a plethora of terms. Consider two molecules or fragments A and B with Hamiltonians $H^A$ and $H^B$ so that $H^A|i^A\rangle = E_i^A|i^A\rangle$ and $H^B|j^B\rangle = E_j^B|j^B\rangle$, with $i, j = 0, 1, 2, \ldots$. The sum $H^0 = H^A + H^B$ and the simple products $|i^A j^B\rangle = |i^A\rangle|j^B\rangle$ are the Hamiltonian and eigenstates of the unperturbed A+B supermolecule:

$$H^0|i^A j^B\rangle = (E_i^A + E_j^B)|i^A j^B\rangle = E_{ij}^0|i^A j^B\rangle. \tag{3.156}$$

When A and B interact with each other, the interaction Hamiltonian is as the (standard) electrostatic expression (in atomic units)

$$H' = -\sum_{i\in A}\sum_{\beta\in B} Z_\beta r_{i\beta}^{-1} - \sum_{j\in B}\sum_{\alpha\in A} Z_\alpha r_{j\alpha}^{-1} + \sum_{i\in A}\sum_{j\in B} r_{ij}^{-1}. \tag{3.157}$$

It is straightforward to show that the first order energy correction takes the form

$$E^{(1)} = E_{\text{elstat}} = \iint d\boldsymbol{r}_1 d\boldsymbol{r}_2 r_{12}^{-1} \rho_t^A(\boldsymbol{r}_1)\rho_t^B(\boldsymbol{r}_2), \tag{3.158}$$

which is nothing but the electrostatic interaction between the two undeformed total (electronic plus nuclear) densities $\rho^t = -\rho(\boldsymbol{r}) + \sum_\alpha Z_\alpha \delta(\boldsymbol{r} - \boldsymbol{R}_\alpha)$ of the fragments. The first order energy correction conveys the electrostatic interaction between the molecules, that includes the well-known ion-ion, ion-dipole, dipole-dipole and, in general, multipole-multipole terms. The second order corrections involve electronic excitations in the interacting molecules, and include induction (if excitation is restricted to one fragment) and dispersion (if both are simultaneously excited).

This language has completely shaped the way we understand condensed phases. When antisymmetry is non-negligible, technical difficulties arise that have no definite solution. Over the years, one formalism, symmetry adapted perturbation theory (SAPT), has finally dominated the field, introducing other terms related to exchange, like exchange-repulsion, exchange-dispersion, etc. [41].

As the distance between the fragments becomes smaller, perturbation theory ceases to converge, and a switch to intra-EDAs is needed. Most EDAs chemically decompose the expectation value of the Hamiltonian by separating two interacting fragments (more or less arbitrarily) and building the wavefunction of the total system by combining those of the fragments in a series of steps. Typically, the fragments are first prepared in the geometry that they will meet in the final molecule, giving rise to preparation energy ($E_{prep}$). Then their standard electrostatic interaction is obtained, $E_{elstat}$. In a second step, an antisymmetric HF or Kohn-Sham determinant is formed from that of the fragments and its energy, measured with respect to the prepared fragments. It is ascribed to the joint effect of Pauli repulsion and the previously computed $E_{elstat}$. Finally, the wavefunction is allowed to relax, and the drop in energy assigned to orbital interactions, $E_{orb}$. The total energetic change after a given chemical process is thus written as

$$\Delta E = E_{prep} + E_{elstat} + E_{Pauli} + E_{orb}. \tag{3.159}$$

A recurrent problem of these intra-EDAs is the need of appropriate reference states.

In real space, the energy enclosed by a region can also be obtained using different prescriptions. We have seen in the previous section (Sect. 3.4.7) that the virial energy partition allows to obtain additive atomic energies. Unfortunately, we have also seen in the generalized open system case that extra terms appear for non-stationary states (Sect. 3.4.3.3). Hence, the virial energy partitioning can not be applied to general points on energy surfaces, falling way short as a general tool in reactivity. Moreover, it does not facilitate the interpretation of binding energies, which become a sum of atomic contributions that sum up all the interactions taking place upon binding.

A different approach to energy partitions in real space comes from the fact that in the QTAIM the union of all atomic domains completely fills the space: $\bigcup_A \Omega_A = R^3$, so that as we saw in Sect. 3.4.6, the terms in the standard Coulomb Hamiltonian have well defined expectation values over QTAIM basins or pair of basins. This partition is known as the Interacting Quantum Atoms (IQA) approach [13, 30, 67]. It starts from recognizing that only the 1- and 2-particle reduced density matrices (RDMs) are needed to recover the energy of a molecular system:

$$\begin{aligned} E = h + V_{ee} + V_{nn} = & \int_\infty \hat{h}\rho_1(\boldsymbol{r}_1;\boldsymbol{r}_1')_{|\boldsymbol{r}_1'\to\boldsymbol{r}_1} d\boldsymbol{r}_1 \\ & + \frac{1}{2}\int_\infty\int_\infty \rho_2(\boldsymbol{r}_1,\boldsymbol{r}_2) r_{12}^{-1} d\boldsymbol{r}_1 d\boldsymbol{r}_2 \\ & + \frac{1}{2}\sum_A\sum_{A\neq B} \frac{Z^A Z^B}{R_{AB}}. \end{aligned} \tag{3.160}$$

Here $h$ is the one-electron Hamiltonian, $h = t - \sum_A Z_A/r_A$, with kinetic and electron-nucleus terms, and $\rho_1$ and $\rho_2$ are the spinless 1- and 2RDMs.

IQA thus partitions all the integrals above into one- (for one-electron operators) and two-basin contributions (in the case of interelectron repulsion):

$$E = \sum_A T_A + \sum_{A,B} V_{en}^{AB} + \sum_{A>B} V_{nn}^{AB} + \sum_{A>B} V_{ee}^{AB}, \tag{3.161}$$

in a self-explanatory notation. All intra-atomic components are gathered together to form the atomic self-energy, $E^A_{self}$ or $E^A_{intra}$:

$$E^A_{self} = T^A + V^{AA}_{en} + V^{AA}_{ee}, \tag{3.162}$$

while all the inter-basin ones form the interbasin interaction energy, $E^{AB}_{int}$,

$$E^{AB}_{int} = V^{AB}_{en} + V^{BA}_{en} + V^{AB}_{ee} + V^{AB}_{nn}. \tag{3.163}$$

For a set of isolated, non-interacting atoms, the atomic self-energies tend to the free atomic energies, and interaction energies vanish.

The total energy takes then a familiar expression,

$$E = \sum_A E^A_{self} + \sum_{A>B} E^{AB}_{int}. \tag{3.164}$$

The above equations are immediately generalized when $A$ is a group of basins, the group-self-energy now containing all the energetic terms in which only the atomic labels of the atoms in the group are found, and the interaction energy between two groups gathering all terms with mixed atomic labels.

Atomic or group self-energies measure the proper energy of a component of the system in-the-molecule. These groups are deformed with respect to their *in vacuo* states, so a deformation energy may be defined given such reference states,

$$E^A_{def} = E^A_{self} - E^A_{vac}. \tag{3.165}$$

$E_{int}$ takes then the usual form

$$E_{int} = \sum_A E^A_{def} + \sum_{A>B} E^{AB}_{int}. \tag{3.166}$$

Notice that the introduction of a reference is not at all needed. IQA provides a reference-free energy partition, being one element of a very limited category of methods with that property. References are only used to decompose interaction energies. As a by-product, basis set superposition errors, which plague supermolecular partitioning methods, are immediately counterpoise corrected [17]. They only affect the deformation terms, and never the interaction ones.

A further point is needed. The electronic repulsion terms contain all types of contributions, Coulombic, exchange, and correlation. It is thus profitable to separate them. This can be done, as shown in the Sect. 3.5.1, by separating the Coulombic part in $\rho_2$ , $\rho_2(\boldsymbol{r}_1, \boldsymbol{r}_2) = \rho(\boldsymbol{r}_1)\rho(\boldsymbol{r}_2) + \rho_{xc}(\boldsymbol{r}_1, \boldsymbol{r}_2)$, where $\rho_{xc}$ is the exchange-correlation density. Exchange itself is not well-defined in general correlated cases, but can be obtained in an invariant manner from the Fock-Dirac expression, $\rho_X(\boldsymbol{r}_1, \boldsymbol{r}_2) = \rho_1(\boldsymbol{r}_1; \boldsymbol{r}_2)\rho_1(\boldsymbol{r}_2; \boldsymbol{r}_1)$, so that $\rho_{xc}(\boldsymbol{r}_1, \boldsymbol{r}_2) = \rho_X(\boldsymbol{r}_1, \boldsymbol{r}_2) + \rho_{corr}(\boldsymbol{r}_1, \boldsymbol{r}_2)$. Notice that $\rho_{corr}$ is equivalent to the cumulant density previously defined. This partition allows us to write

$$V_{ee}^{AB} = V_C^{AB} + V_X^{AB} + V_{corr}^{AB}, \tag{3.167}$$

where $V_C^{AB}$ is the classical Coulombic electrostatic electron-electron interaction. If we gather all the classical interaction terms and define a classical interaction, $V_{cl}^{AB} = V_{nn}^{AB} + V_{ne}^{AB} + V_{en}^{AB} + V_C^{AB}$, then

$$E_{int}^{AB} = V_{cl}^{AB} + V_{xc}^{AB}. \tag{3.168}$$

In terms of these quantities, $E_{int} = E_{def}^{A} + E_{def}^{B} + V_{xc}^{AB} + V_{cl}^{AB}$ for a two-fragment complex.

All IQA energy terms are the expectation values of physical interactions once the atomic partition has been defined. There is no ambiguity in their interpretation, nor reference to a fictitious state. All the energetic components are derived from the supramolecular wavefunction, isolated fragments being only required if the binding energy is to be calculated. Electrons within the IQA partition are indistinguishable, they are not attributed to a given fragment. This allows performing the partition with physical meaning all along a dissociation curve: from long range, to covalent, to the repulsive regime (where perturbative approaches fail). From the charge point of view, fragments hold fractionary number of electrons that can cover a wide range, from zero for perfect covalent cases to nominal ionic charges in very ionic ones. No perturbative approach is able to describe this type of situation either.

As caveats of the method we can mention the fact that the electron repulsion in any invariant RDMs method faces the problem of electrostatic self-interaction. As an example, the intermolecular Coulomb repulsion, $\int \rho^A(\boldsymbol{r}_1)\rho^B(\boldsymbol{r}_2)/r_{12}\, d\boldsymbol{r}_1\, d\boldsymbol{r}_2$ assumes that electrons in different fragments are distinguishable. However, only the full $\rho(\boldsymbol{r})$ is available in orbital-free techniques (i.e. $A$, $B$ labels are dropped). This means that an electron at $\boldsymbol{r}_1$ will interact with itself at $\boldsymbol{r}_2$. Indeed, a large part of the Fermi hole that builds around an electron is due to the self-interaction correction. In other words, using invariant densities implies self-interaction issues. Comparison with results obtained from Pauli violating densities (i.e. those from methods where fragment densities are defined) needs care. The $H_2$ molecule is a very good example of self-interaction: $V_X^{AB}$ will be non-zero even though an equal spin pair does not exist. In strong bonding interactions, $V_{xc}^{AB}$, being the non-classical interaction energy, measures covalency. Similarly, $V_{cl}$ describes ionicity.

To exemplify IQA, Table 3.6 shows some IQA data for a few systems, computed at the aug-cc-pVTZ//CAS[valence] level [48]. IQA can deal equally well with strong (intramolecular) and weak (inter- or intramolecular) interactions, at variance with SAPT, and its results are generally easy to interpret. Take for instance the singlet (S) state of dihydrogen. We measure the deformation energy of each H atom taking the *in vacuo* energy as a reference. This deformation is small, about 9.6 kcal/mol. This means that, in this case, the formation of the molecule has a very limited impact on the self-energy of each of the fragments. The total classical electrostatic interaction between the two moieties is destabilizing, reaching 25 kcal/mol. It can be shown that

**Table 3.6** Some IQA results obtained at the CAS level. All units in kcal/mol. The I label means an ionic reference

| | $E_{def}^{A}$ | $E_{def}^{B}$ | $V_{cl}^{AB}$ | $V_{xc}^{AB}$ | $E_{int}^{AB}$ |
|---|---|---|---|---|---|
| $H_2$ (Singlet) | 9.6 | 9.6 | 25.1 | −139.6 | −114.4 |
| $H_2$ (Triplet) | 73.4 | 73.4 | 60.0 | −71.0 | −11.0 |
| H-F (I) | −124.1 | 56.1 | −224.7 | −81.8 | −306.5 |
| H-Cl (I) | −240.4 | 37.0 | 21.8 | −168.0 | −146.2 |
| H-F | 189.5 | −22.6 | −224.7 | −81.8 | −306.5 |
| H-Cl | 73.2 | −22.5 | 21.8 | −168.0 | −146.2 |
| Li-H (I) | −6.1 | 14.5 | −141.2 | −27.1 | −168.4 |
| Li-F (I) | 22.7 | −29.5 | −164.8 | −29.8 | −194.6 |
| Li-H | 117.0 | 6.6 | −141.2 | −27.1 | −168.4 |
| Li-F | 145.7 | −100.4 | −164.8 | −29.8 | −194.6 |
| $F_2$ | 53.0 | 53.0 | 48.2 | −191.6 | −143.4 |
| $Cl_2$ | 50.3 | 50.3 | 28.9 | −174.6 | −145.8 |
| HF-HF | 4.0 | 6.4 | −4.4 | −9.6 | −14.0 |
| $H_2O$-$H_2O$ | 5.0 | 7.2 | −5.2 | −10.8 | −16.0 |

two symmetric non-interpenetrating neutral densities lead necessarily to destabilizing electrostatics. Standard EDAs do provide small, but non-negligible stabilizing electrostatic interactions that may become very large in strongly binded systems, like $N_2$. All the IQA stabilizing energy comes from the exchange-correlation contribution, which demonstrates that $H_2$ is covalently bound. If we now examine the first excited state (triplet, dissociative, computed at the singlet's equilibrium distance), now the deformation of each atom is around eight times larger than in the ground state, the electrostatic destabilization is more than twice as big, and the covalent contribution has almost halved. The IQA answer to the unstable character of the dihydrogen triplet lies in the large atomic deformation energy of each of the H atoms, that can be traced back to Pauli's principle (or to Pauli's repulsion).

On the opposite binding extreme we can take the LiH molecule. Now the deformation energy is very large (123.6 kcal/mol), and the covalent contribution much smaller than the electrostatic one, which is stabilizing. This is a typically ionic interaction. Actually, the large total deformation energy of the system is not symmetrically distributed between the two atoms. It is 117 kcal/mol for the Li moiety, while 6.6 kcal/mol for H. This is easy to understand if we take into account the large QTAIM charges of both atoms ($\approx \pm 0.90\ e$): it takes 117 kcal/mol to take the neutral Li atom to its in-the-molecule (almost ionized) $Li^+$ state. If we take the ions, not the neutral atoms, as energetic reference, the deformation energy is much smaller. Thus, in systems where large charge transfers occur, we expect important deformation energies due to ionization costs. These do only occur when such an ionization is over-compensated by the stabilizing electrostatic interaction between the oppositely charged ions left behind. Since ionization is normally very demanding from the energetic point of view, truly ionic interactions are scarce in nature. Surely, other systems lie in between, as shown in Table 3.6, and more or less polar interactions abound, as expected from chemical grounds. For instance, the HF molecule has considerably large both ionic and covalent contributions, pointing toward a polar-covalent interaction.

Finally, IQA can also be profitably used in intermolecular interactions. The HF and water hydrogen bonded dimers teach us that the molecular deformation of the hydrogen bond donor and acceptor are not symmetric, and that although the electrostatic interaction can be used to sense the strength of the complex, the exchange-correlation contribution is far from negligible. Electrons are clearly delocalized to some extent over the two interacting fragments [46].

### 3.5.3 Pair Density and Electron Localization

The study of the pair density has been rather fruitful to back up the concept of **spatial localization** in molecular quantum chemistry. The initial ideas were set up in the 1950s and 1960s when Daudel and coworkers [25] became interested in the $P_n(\Omega)$ function, that determines the probability of finding a given integer number of electrons, $n$, in a given region of space, $\Omega$.

Let us call $\Omega' = \mathbb{R}^3 - \Omega$. Then,

$$P_n(\Omega) = \binom{N}{n} \int_\Omega d\boldsymbol{x}_1 \dots \int_\Omega d\boldsymbol{x}_n \int_{\Omega'} d\boldsymbol{x}_{n+1} \dots \int_{\Omega'} d\boldsymbol{x}_N \Psi^*(\boldsymbol{x}_1, \dots, \boldsymbol{x}_N)\Psi(\boldsymbol{x}_1, \dots, \boldsymbol{x}_N), \tag{3.169}$$

where we have used a non-ordered paired normalization. Since the total number of electrons is $N$,[6]

$$\sum_{n=0}^{n=N} P_n(\Omega) = 1, \quad N_\Omega = \sum_{n=0}^{n=N} n P_n(\Omega), \tag{3.170}$$

independently of $\Omega$.

Daudel's original proposal was to find the regions, or **loges**, that would minimize Shannon's entropy of the distribution $P$.[7] Searching for those regions has

[6] The first equality refers to the fact that the total probability must be equal to one by definition, which in our case, means summing up all the occupation probabilities. The second expression gives us the average occupation of a given $\Omega$ from the statistics of a random variable $\bar{x} = \sum_i x_i P(X_i)$.

[7] The concept of entropy in information theory provides the "uncertainty" in a given event. For example, a normal coin has maximum entropy (we have maximum probability of having either heads or tails). At the other end, a two-face coin has minimum entropy. Mathematically, this is measured by Shannon's entropy:

$$H(X) = -\sum_{i=1} P(x_i) \log P(x_i) \qquad (3.171),$$

where the variable X can have $x_1 \dots x_n$ values with $P(X_1) \dots P(x_n)$ probabilities.

several drawbacks. It is necessary to know the full $n$-th order density and requires optimization of 3D objects. Both characteristics lead to cumbersome algebra and to computational complexity. Only recent computational advance has allowed some authors, like A. Savin, to recover these ideas, now focused on the probability of finding a pair of electrons in a region (the so-called maximum probability domains) [19].

The variance of the electron population in a region, defined in Eq. 3.120, may be expressed in terms of the $P_n$ as follows[8]:

$$\sigma^2(\Omega) = \sum_{n=0}^{n=N} (n - N_\Omega)^2 P_n(\Omega) = \sum_{n=0}^{n=N} n^2 P_n(\Omega) - \left[ \sum_{n=0}^{n=N} n P_n(\Omega) \right]^2 . \tag{3.172}$$

If the variance in a region vanishes, then only one non-zero term may appear in the last expression, let us say $n = l$, so that $P_l(\Omega) = 1$, and $N_\Omega = l$. The $l$ electrons are completely localized in the region, for which the number of electrons is well-defined.

Let us go back to Eq. 3.169 and define specifically the **pair population** of a region $\Omega$,

$$D_{\Omega,\Omega} = \frac{1}{2} \int_\Omega \int_\Omega \rho_2(\boldsymbol{x}_1, \boldsymbol{x}_2) d\boldsymbol{x}_1 d\boldsymbol{x}_2 = \sum_{n=0}^{n=N} \frac{n(n-1)}{2} P_n(\Omega), \tag{3.173}$$

where as it is to be expected, the sum actually starts at $n = 2$, which is the minimum number of electrons for which there is a pair.

A bit of algebra leads us to[9]

$$\sigma^2(\Omega) = 2D_{\Omega,\Omega} - N_\Omega(N_\Omega - 1), \tag{3.174}$$

and, therefore, if the variance vanishes, $D_{\Omega,\Omega} = l(l-1)/2$, where $l$ is an integer. We say that the region displays a **pure pair population**. As Bader and Stephens showed, [10], we can also speak in that case of localization of the electrons in region $\Omega$, since the number of electrons as well as the number of their pairs, is determined by $\rho_1$. In this case, the full state vector is an antisymmetrized tensor product of independent state vectors in $\Omega$ and $\Omega'$. If the variance vanishes, then only one of the $P_n$ is non-vanishing, and this immediately implies through Eq. 3.174 that $N_\Omega = n$ is an integer number.[10] Otherwise,

$$2D_{\Omega,\Omega} > N_\Omega(N_\Omega - 1). \tag{3.175}$$

---

[8] You are probably acquainted with the variance of a discrete variable measured $n$ times: $S^2 = \frac{(\sum x - \bar{x})^2}{n}$. For a random variable $X$ taking values $x_1 \ldots x_n$ with probabilities $P_1 \ldots P_n$, $S^2 = (\sum_i x_i - \bar{x})^2 P(X_i)$. Hence, using Eq. 3.170: $\sigma^2(\Omega) = \sum_i (n - N_\Omega)^2 P_n(\Omega) = \sum_i (n^2 P_n(\Omega)) + N_\Omega^2 - 2N_\Omega \sum_i (n P_n(\Omega) = \sum_i (n^2 P_n(\Omega)) - N_\Omega^2$.

[9] Developing Eq. 3.173 and introducing Eq. 3.172: $2D_{\Omega,\Omega} = \sum_n n^2 P_n(\Omega) - \sum_n n P_n(\Omega) = \sigma^2(\Omega) + N_\Omega^2 - N_\Omega$. Reorganizing, we obtain Eq. 3.174.

[10] From Eqs. 3.173 and 3.174: $D_{\Omega,\Omega} = N_\Omega(N_\Omega - 1) = n(n-1)$.

We can establish a clean relation between these quantities and the exchange-correlation hole. Using Eq. 3.151 we easily find that[11]

$$\sigma^2(\Omega) = N_\Omega + \int_\Omega \int_\Omega \rho(\boldsymbol{x}_1) h_{xc}(\boldsymbol{x}_2|\boldsymbol{x}_1) d\boldsymbol{x}_1 d\boldsymbol{x}_2 = N_\Omega + F_{\Omega,\Omega}, \quad (3.176)$$

where we have defined the $F_{\Omega,\Omega}$ function, that turns out to be a measure of the electron correlation in region $\Omega$.

Suppose now that $h_{xc}$ is completely contained in the region $\Omega$. Namely, that $h_{xc}(\boldsymbol{x}_2 \notin \Omega | x_1 \in \Omega) = 0$. Under these conditions, it follows that

$$\int_\Omega h_{xc}(\boldsymbol{x}_2|\boldsymbol{x}_1) d\boldsymbol{x}_2 = -1,$$
$$\sigma^2(\Omega) = 0, \quad (3.177)$$

and the complete localization of the hole implies necessarily a pure pair population. It is in principle possible that $F_{\Omega,\Omega}$ vanishes and that $h_{xc}$ does not fulfil Eq. 3.177. In this case a residual correlation would still exist, but we will not consider it here. The criterion of localization here introduced can be attained if the $\Omega$ region maximized the intra-basin electron correlation, i.e. **if** the *exchange-correlaton hole is completely localized in the region, independently of the position of the RE*. This establishes the sought link between the localization criteria and the formation of electron pairs.

We can go into greater details if we turn back to Eqs. 3.127 and 3.128, we reinterpret term using Eq. 3.153, and we suppose that the space is divided into regions bearing completely localized holes. Doing so, we come to the conclusion that the inter-basin repulsion does not possess correlation contributions, being described exclusively by the classical term. Thus, *the localization of the hole in a region is equivalent to the localization of the electrons inside*.

If the holes are completely localized in two regions $\Omega_A$ and $\Omega_B$, and we apply Eq. 3.176 to the joint region $\Omega_A \cup \Omega_B$, we find that

$$\sigma^2(\Omega_A \cup \Omega_B) = 2F_{\Omega_A,\Omega_B} = 0,$$
$$F_{\Omega_A,\Omega_B} = \int_{\Omega_A} \rho(\boldsymbol{x}_1) d\boldsymbol{x}_1 \int_{\Omega_B} h_{xc}(\boldsymbol{x}_2|\boldsymbol{x}_1) d\boldsymbol{x}_2, \quad (3.178)$$

where we have used the $A \leftrightarrow B$ symmetry. $F_{\Omega_A,\Omega_B}$ can be interpreted, then, as a measure of the interbasin electron correlation strength. In this way, we can introduce the **interbasin pair population**, $D_{\Omega_A,\Omega_B}$. When $F_{\Omega_A,\Omega_B}$ is negligible $D_{\Omega_A,\Omega_B} = N_{\Omega_A} N_{\Omega_B}$, is the number of ordered pairs between two disjoint sets.

It has been common practice to transform these functions into a set of localization indices with simple interpretation. In order to avoid a cumbersome notation, we will drop the $\Omega$ subscript in what follows, understanding that $A \equiv \Omega_A$. The diagonal part

[11] $\sigma^2(\Omega) = \int_\Omega \int_\Omega (\rho(\boldsymbol{x}_1)\rho(\boldsymbol{x}_2) + \rho(\boldsymbol{x}_1) h_{xc}(\boldsymbol{x}_1, \boldsymbol{x}_2)) d\boldsymbol{x}_1 d\boldsymbol{x}_2 - N_\Omega^2 + N_\Omega$ with $\int_\Omega \int_\Omega \rho(\boldsymbol{x}_1) \rho(\boldsymbol{x}_2) d\boldsymbol{x}_1 d\boldsymbol{x}_2 = N_\Omega^2$.

of $F$ is a measure of the intrabasin localization. Taking into account that, in general, $|F_{AA}| < N_A$, we now introduce the **atomic localization index, LI**, $\lambda_A = |F_{AA}|$. Similarly, $F_{AB}$ measures the delocalization among the electrons of the $A$ region into the $B$ domain, and vice versa [3]. We call it the **delocalization index, DI**, $\delta_{AB} = 2F_{AB}$. Localization and delocalization indices are subjected to a sum rule,[12]

$$F_{AA} + \sum_{B \neq A} F_{AB} = -N_A. \tag{3.179}$$

It should be noted that expression Eq. 3.178 is closely related to Mayer's definitions of bond order [51, 52] in those single-determinant cases where the 2RDM can be expressed in terms of the 1RDM (see Eqs. 3.145 and 3.147). In HF, the fluctuation can be expressed as a function of the domain overlaps, $S_{ij}(\Omega)$, between the spinorbitals $\psi_i$, $S_{ij}(\Omega) = \int_\Omega \psi_i^*(\boldsymbol{x})\psi_j(\boldsymbol{x})d\boldsymbol{x}$[13]:

$$\sigma^2(\Omega) = N(\Omega) - \sum_i^N S_{ii}^2(\Omega) - \sum_{i \neq j}^N S_{ij}^2(\Omega), \tag{3.181}$$

and the $F_{AB}$ functions become:

$$F_{AB} = \sum_i \sum_j S_{ij}(\Omega_A) S_{ij}(\Omega_B) \tag{3.182}$$

The relevance of the fluctuation in the number of particles contained in a region of space has been addressed in depth by Diner, Claverie and Malrieu [27]. Among the most important applications of this analysis is undoubtedly the study of delocalisation in conjugate systems, for which Lewis' theory is not capable of providing a defined structure.

Figure 3.11 allows us to understand the behavior of these indices in the dihydrogen molecule. If we start with a HF model, thus with a static Fermi hole, the integral of $h_F$ over all space is $-1$, and this quantity is shared equally between the two basins. Obviously, $F_{AA} = -1/2$ and $2F_{AB} = -1$. This mean that half of the average electron dwelling in each basin is delocalized over the other center, and that *the equally shared pair of electrons* gives rise to $\delta$ equal to one. These results are independent of the basis set and of the internuclear distance, and should be interpreted both as a success as well as pointing to the failures of the HF model: By allowing electron delocalization efficiently, the HF mean-field answers deal rather well with normal chemical bonds. On the contrary, by letting delocalization occur at any distance, the HF solution fails to

---

[12] $\sum_B F_{\Omega_A,\Omega_B} = \int_{\Omega_A} \rho(\boldsymbol{x}_1)d\boldsymbol{x}_1 \sum_B \int_{\Omega_B} h_{xc}(\boldsymbol{x}_2|\boldsymbol{x}_1)d\boldsymbol{x}_2 = N_A \int_{\mathbb{R}^3} h_{xc}(\boldsymbol{x}_2|\boldsymbol{x}_1)d\boldsymbol{x}_2 = -N_A$

[13]

$$\int_\Omega \int_\Omega \rho(\boldsymbol{x}_1)\rho(\boldsymbol{x}_2)d\boldsymbol{x}_1 d\boldsymbol{x}_2 - \sum_i \sum_j \int_\Omega \int_\Omega (\phi_i^*(\boldsymbol{x}_1)\phi_i(\boldsymbol{x}_2))(\phi_j^*(\boldsymbol{x}_1)\phi_j(\boldsymbol{x}_2))d\boldsymbol{x}_1 d\boldsymbol{x}_2 = N_\Omega^2 - \sum_i S_{ij}^2(\Omega) \tag{3.180}$$

localize each of the electrons of a bonding pair on a different atom at large distances, precluding dissociation. If we introduce now the Coulomb hole, in the dissociative limit $F_{AA} = -1$, $F_{AB} = 0$. The effects of $h_C$ at distances close to equilibrium are easy to sense: an increase of $\lambda$ coupled to a decrease of $\delta$ or, in other words, an increase in intrabasin electron localization, that decreases the number of shared pairs [49].

Similar rules can be used to predict the behavior of these indices using chemical intuition. In the dinitrogen molecule, for instance, the Lewis rules establish that core and non-bonding electrons would be localized, with three shared pairs of electrons. This model leads to predict $\lambda = 5.5$ and $\delta = 3.0$. The results of a Hartree-Fock calculation provide $\lambda = 5.48$, $\delta = 3.04$, indicating that the non-bonding pairs are slightly delocalized [29]. If we include correlation, these figures change to $\lambda = 5.89$, $\delta = 2.22$. Similarly, an ionic molecule like LiF should ideally show $\delta = 0$ with $\lambda$ values corresponding to the ionic electron populations, 2.0 and 10.0 for $Li^+$ and $F^-$, respectively. The Hartree-Fock data are compatible with this picture: $\delta = 0.178$. The inclusion of electron correlation now increases this delocalization up to 0.193. Fradera and coworkers [29] undertook years ago a study of the meaning of the delocalization indices for a set of standard simple molecules. Table 3.7 shows some results. It is today clear that delocalization indices encode cleanly what chemists understand as electron delocalization.

It is important to take into account that the formation of a bond is accompanied by a decrease in the number of inter-atomic electron pairs and of an increase in the number of intra-atomic electron pairs. This is the meaning of **electron pair sharing**. In the ionic limit, a direct relation between the reduction of the number of intrabasin pairs and the number of Lewis shared pairs does not exist. The link is established as we approach the homopolar bond. Also notice that delocalization indices are to be understood as covalent bond orders, lacking information about *ionic* contributions, and that they should not be be confused with traditional integer bond orders.

### *3.5.4 Domain-Averaged Fermi Holes and Natural Adaptive Orbitals*

Topological bond orders between two real space regions have been introduced in the previous section. By using the pair density, the measure of the shared pairs of electrons provided by the delocalization index $\delta$ gives us a generalization of the covalent bond order. Note that this localization/delocalization concepts do not show an intrinsic relationship to the existence of a BCP. Semi-quantitative aspects of this connection will be introduced in Sect. 4.3.3.1. The way we have introduced localization/delocalization indices comes directly from a partition of the number of electrons into one and two basin contributions. This definition paves the way to a multicenter generalization. A detailed discussion may be found in Ref. [31], and a

**Table 3.7** Localization and delocalization indices for some diatomic molecules at the HF 6-311++G(2d, 2p) (top), and CISD 6-311++G(2d, 2p) (bottom) levels. The noble gas results are CISD 6-311G(2d, 2p). Data taken from Ref. [29]

| Molecule | | $N_A$ | $D_{AA}$ | $\lambda_A$ | $D_{AB}$ | $\delta_{AB}$ | $\sum D_{AB}/2$ | $\sum F_{AB}/2$ |
|---|---|---|---|---|---|---|---|---|
| $H_2$ | H | 1.000 | 0.250 | 0.500 | 0.500 | 1.000 | 1.000 | −2.000 |
| $N_2$ | N | 7.000 | 21.761 | 5.479 | 47.479 | 3.042 | 91.000 | −14.000 |
| $F_2$ | F | 9.000 | 36.321 | 8.358 | 80.358 | 1.283 | 153.000 | −18.000 |
| LiF | Li | 2.060 | 1.136 | 1.971 | 20.387 | 0.178 | 66.001 | −12.000 |
| | F | 9.940 | 44.447 | 9.851 | | | | |
| CO | C | 4.647 | 8.865 | 3.860 | 42.677 | 1.574 | 91.006 | −14.000 |
| | O | 9.354 | 39.463 | 8.567 | | | | |
| $CN^-$ | C | 5.227 | 11.598 | 4.121 | 44.748 | 2.210 | 90.995 | −13.999 |
| | N | 8.773 | 34.649 | 7.668 | | | | |
| $NO^+$ | N | 5.525 | 13.102 | 4.323 | 45.622 | 2.405 | 91.000 | −14.000 |
| | O | 8.475 | 32.276 | 7.273 | | | | |
| Molecule | | $N_A$ | $D_{AA}$ | $\lambda_A$ | $D_{AB}$ | $\delta_{AB}$ | $\sum D_{AB}/2$ | $\sum F_{AB}/2$ |
| $H_2$ | H | 1.000 | 0.212 | 0.575 | 0.573 | 0.849 | 1.000 | −2.000 |
| $N_2$ | N | 7.000 | 21.555 | 5.891 | 47.890 | 2.219 | 91.000 | −14.001 |
| $F_2$ | F | 9.000 | 36.251 | 8.498 | 80.497 | 1.005 | | |
| LiF | Li | 2.067 | 1.151 | 1.973 | 20.440 | 0.193 | 65.999 | −12.005 |
| | F | 9.932 | 44.409 | 9.838 | | | | |
| CO | C | 4.794 | 9.454 | 4.072 | 43.410 | 1.443 | 90.997 | −14.001 |
| | O | 9.206 | 38.133 | 8.484 | | | | |
| $CN^-$ | C | 5.434 | 12.519 | 4.490 | 45.601 | 1.888 | 90.995 | −14.000 |
| | N | 8.566 | 32.874 | 7.621 | | | | |
| $NO^+$ | N | 5.803 | 14.421 | 4.837 | 46.605 | 1.934 | 91.008 | −14.001 |
| | O | 8.197 | 29.982 | 7.231 | | | | |
| $He_2$ | He | 2.000 | 1.001 | 1.998 | 3.998 | 0.004 | 6.000 | −4.000 |
| $Ar_2$ | Ar | 18.000 | 153.003 | 17.993 | 323.994 | 0.013 | 630.000 | −36.000 |

set of related ideas were progressively developed by Giambiagi and coworkers [37], Bultinck and collaborators [18], and Matito et al. [50].

What is needed is a general quantity that can be integrated over one, two, three, etc, number of basins, and that always adds to the number of electrons, $N$. Fortunately, this type of object exists in statistics, and it is called a cumulant. An $n$-th order cumulant density contains information about simultaneous $n$-object fluctuations. The second order cumulant density is directly the exchange-correlation density, $\rho_{xc}$, whose integration provides the localization and delocalization descriptors. From the general $n$-th order reduced density (nRD, $\rho_n(\boldsymbol{r}_1, \boldsymbol{r}_2, \ldots, \boldsymbol{r}_n)$), which is defined as the $n$-tuple probability density at points $\boldsymbol{r}_1, \ldots, \boldsymbol{r}_n$, it is legitimate to extract the irreducible part which in general cannot be written explicitly in terms of lower order

RDs (remember that we saw that this can be done within the HF method). This is the cumulant. For instance:

$$\rho_2(\boldsymbol{r}_1, \boldsymbol{r}_2) = \rho(\boldsymbol{r}_1)\rho(\boldsymbol{r}_2) - \rho_{xc}(\boldsymbol{r}_1, \boldsymbol{r}_2), \tag{3.183}$$

defines the second order cumulant, $\rho_2^c = \rho_{xc}$. All cumulants integrate to $N$, and can be recursively obtained from those of further order:

$$\int \rho_n^c(\boldsymbol{r}_1, \ldots, \boldsymbol{r}_n)\, d\boldsymbol{r}_1 \ldots d\boldsymbol{r}_n = N,$$
$$\int \rho_n^c(\boldsymbol{r}_1, \ldots, \boldsymbol{r}_n)\, d\boldsymbol{r}_n = \rho_{n-1}^c(\boldsymbol{r}1, \ldots, \boldsymbol{r}_{n-1}). \tag{3.184}$$

Expressions for cumulants of order greater than two are relatively cumbersome, and may be found in Ref. [31]. An $n$-basin partition of the $n$-th order cumulant density (nCD) provides an $n$-basin partition of the number of electrons:

$$N = \underbrace{\sum_a \ldots \sum_n}_{n} N_{a \ldots n}. \tag{3.185}$$

In this way, for $n = 1$, the number of electrons is divided into basins, each contribution being nothing but the QTAIM atomic population. For $n = 2$, the $N_{aa}$ terms are the localization indices, describing how many electrons are localized in a basin, while the $N_{ab}$ (and their equal $N_{ba}$ counterparts) define the number of shared or delocalized electrons. Recall that the DI is related to the 2-center covariance of the electron distribution. Generalizing this algebra, for $n > 2$ the $N_{ab \ldots n}$ populations are, except for possibly normalization factors like $n!$, $n$-center delocalization indices $\delta^{ab \ldots n}$, that measure the mutual fluctuation of the electron populations among the different centers. Several definitions have been given over the years [14, 15].

In this way, a natural generalization of the 2-center bond order is readily available in real space, providing orbital invariant descriptors for otherwise difficult concepts. Notice also that if we integrate all but one electron coordinate of a nCD over a set of basins we obtain a $(n-1)$th basin partition of the electron density:

$$\int_a d\boldsymbol{r}_2 \ldots \int_n d\boldsymbol{r}_n\, \rho_c^{n+1}(\boldsymbol{r}_1, \boldsymbol{r}_2, \ldots, \boldsymbol{r}_n) = \rho_{ab \ldots n}(\boldsymbol{r}_1),$$
$$\sum_{ab \ldots n} \rho_{ab \ldots n}(\boldsymbol{r}) = \rho(\boldsymbol{r}),$$
$$\int \rho_{ab \ldots n}(\boldsymbol{r}) d\boldsymbol{r} = N_{ab \ldots n}. \tag{3.186}$$

This can also be done with the density matrices instead of the densities themselves, so that non-diagonal densities can also be obtained. The electron density can thus be understood at any point $\boldsymbol{r}$ as the sum of 1, 2, ..., $n$ contributions. Diagonalization of the $n$-center non-diagonal densities give rise to a natural orbital decomposition of the $n$-center densities. The occupation numbers (eigenvalues) of each of these orbitals add to $N_{ab...n}$, thus decomposing the $n$-center bond order into effective one-electron contributions. These orbitals are known as natural adaptive orbitals (NAdOs) [58].

The one-basin decomposition of the density (obtained through the second order cumulant) was proposed by Ponec [62, 63], who called it the domain-averaged Fermi hole (DAFH). The associated one-electron functions are called domain natural orbitals (DNOs). The DAFH technique has found a large number of applications. When the DAFH of a given basin is diagonalized, the DNOs with close to zero occupation number $n$ are completeley localized outside the basin, those with occupations close to one (or two if doubly occupied orbitals are used instead of spin orbitals) are almost fully localized in the basin, and those with occupations in between delocalize outside the basin. It can be shown that in the case of a single determinant wavefunction, the contribution of each spin-DNO to the DI between the basin and the rest of the system is given by $4n(1 - n)$. DNOs display the local symmetry of the basin in which they are obtained. Techniques to fully localize them in cases of high symmetry have been devised, as in the case of the isopycnic transformation [20].[14]

A two-basin decomposition gives orbitals describing directly two-center delocalizations. These 2-NAdOs have occupations that add directly to the DI, providing a decomposition of a "bond" into one-electron components. They have been used to show how traditional bonding ideas appear from real space considerations [58]. Although, strictly speaking, multicenter bond indices and general NAdOs can only be obtained from electronic structure methods with a true wavefunction (like HF or any multiconfigurational approach), it has become customary to use Kohn-Sham pseudo-determinants to obtain approximate descriptors. Care has to be taken if a single reference is qualitatively incorrect, as in homolytic bond dissociations.

### 3.5.5 Other Density and Reduced Density-Based Descriptors

A wealth of both local and integrated quantities related to the various reduced densities or density matrices that we have presented have been explored over the years. Since the field is vast, we will only consider a couple of examples to show the versatility of the topological approach: the source function and the local spin.

#### 3.5.5.1 The Source Function

The source function (SF) analysis provides an interesting tool in real space descriptions of chemical bonding by allowing to dissect the density at a point into additive

[14] Isopycnic transformations are linear orbital transformations that leave the first-order density matrix invariant. They make it possible to localize not only the Hartree-Fock spin-orbitals, but also the natural spin-orbitals.

components that come from an exhaustive partition of space, usually the atomic one provided by the QTAIM. Actually, it allows to partition any scalar field into such an additive sum, since it is based on Green's identity:

$$f(\boldsymbol{r}) = -\frac{1}{4\pi}\int \frac{\nabla^2 f(\boldsymbol{r}')}{|\boldsymbol{r}-\boldsymbol{r}'|}d\boldsymbol{r}'. \tag{3.187}$$

Notice that this expression is general. When applied to $f = \rho$, the density at a point is written as a result (effect) of how the Laplacian field behaves over all the space (a cause). This is the basis of the SF analysis [5, 32, 33]:

$$\rho(\boldsymbol{r}) = \int \mathrm{LS}(\boldsymbol{r}, \boldsymbol{r}')d\boldsymbol{r}', \tag{3.188}$$

where

$$\mathrm{LS}(\boldsymbol{r}, \boldsymbol{r}') = -\frac{\nabla^2\rho(\boldsymbol{r}')}{4\pi|\boldsymbol{r}-\boldsymbol{r}'|}. \tag{3.189}$$

Here, $(4\pi|\mathbf{r}-\mathbf{r}'|)^{-1}$ is a kind of *influence* function [4] that measures how important $\nabla^2\rho(\boldsymbol{r}')d\boldsymbol{r}'$ is in determining the *effect* $\rho(\mathbf{r})$. Given a reference point $\boldsymbol{r}$ (RP), the Laplacian at other points divided by the distance to the RP is the *cause* of the *effect* density at the RP. Now we can clearly use a topological partition to divide the integral of the local source (LS) into atomic contributions. Thus, labelling our atoms as $A$,

$$\rho(\boldsymbol{r}) = \int \mathrm{LS}(\boldsymbol{r}, \boldsymbol{r}')d\boldsymbol{r}' = \sum_A \int_A \mathrm{LS}(\boldsymbol{r}, \boldsymbol{r}')d\boldsymbol{r}' = \sum_A \mathrm{SF}(\boldsymbol{r}, A). \tag{3.190}$$

Each of these terms are the so-called atomic SF contributions to the density at the RP. The source function analyses thus provides a way to transform an apparently local scalar like the density into a non-local one, so that each atom participates in building up the density value at a given point. As it has been shown, chemical intuition is well maintained by this construction, and it is usually found that the density at a point within an atomic basin $A$ is dominated by the atomic SF of that basin, followed by contributions from basins with which that atom shares interatomic surfaces (its bonded contributions). The SF analysis then becomes a very useful tool in understanding what atoms, and to what extent, influence density values. We must mention, nevertheless, that the SF contributions are not positive definite, and that in some cases negative values can be found. This casts doubts over the true meaning of these contributions, but does not invalidate the overall appealing nature of the procedure, which has gained popularity in the last years. Depending only on the electron density and its derivatives, the SF is amenable to experimental determination with the accurate $\nabla^2\rho$ distribution that are now routinely obtained with high-quality single-crystal X-ray diffraction experiments [32, 36, 71].

An important insight that may help to understand the significance of the SF is clear if Eq. 3.190 is integrated again over a given QTAIM region, since this provides a decomposition of the atomic population:

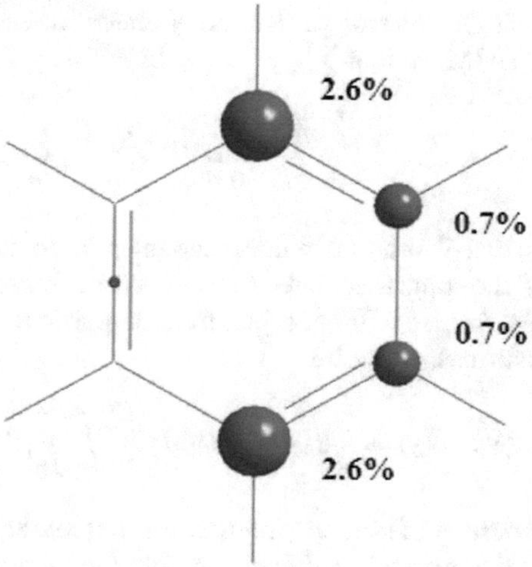


**Fig. 3.12** SF%(A) atomic values for some atoms at the C-C BCP in benzene, computed at the B3LYP/DZVP2 level. Adapted with permission from Ref. [61]. Copyright (2011) American Chemical Society

$$\int_A \rho(\boldsymbol{r})d\boldsymbol{r} = N_A = \sum_B \int_B \mathrm{SF}(\boldsymbol{r}, A)\, d\boldsymbol{r} = N_{AA}^{SF} + \sum_B N_{AB}^{SF}. \qquad (3.191)$$

This shows how the atomic population becomes a sum of intra- and interbasin pair contributions, just as we found when examining the cumulant expansion. In this sense, $N_{AA}^{SF}$ and $N_{AB}^{SF}$ play a role equivalent to those of the localization and delocalization indices, respectively. Thus, the SF may be used to grasp delocalization information from a purely one-particle property.

Traditionally, SF data are offered relative to the density at the RP through Source Function percentage contributions, SF%, defined as $\mathrm{SF}\%(A) = \mathrm{SF}(\boldsymbol{r}, A)/\rho(\boldsymbol{r}) \times 100)$ [32, 34]. Usually, SF% values are visualized using a ball and stick molecular model where each ball has a volume proportional to its associated SF% value and where the location of the chosen RP. As an example, Fig. 3.12 shows how 6.6% the density at the C-C BCP in benzene is due to non-adjacent C atomic contributions.

#### 3.5.5.2 Local Spin

The atomic partition provided by the QTAIM can also be used to write the expectation values of other operators, like the total squared spin, $S^2$. The importance of this so-called local spin cannot be overlooked. Usually, the distribution of the radical-like character of $S \neq 0$ species is examined through the spin density, $\rho_s(\boldsymbol{r}) = \rho^\alpha(\boldsymbol{r}) - \rho^\beta(\boldsymbol{r})$, which vanishes at every point in space for singlets. This means that there is no easy way to examine this problem in, let us say, singlet diradicals.

In 2001, Clark and Davidson [21] showed how to partition the expectation value of $S^2$ into intra- and interatomic contributions, much as in IQA. To that end, they used the vector spin operator of an $N$ electron system, $\boldsymbol{S} = \sum_i^N \boldsymbol{S}_i$, and considered a single-electron projection operator for each atom $A$, $P^A$ like that provided by the QTAIM, so that $\sum_A P^A = 1$. In this way $\boldsymbol{S}_A = \sum_i^N \boldsymbol{S}_i P_i^A$, and

$$S^2 = \sum_{A,B} \boldsymbol{S}_A \cdot \boldsymbol{S}_B = \sum_A S_A^2 + \sum_A \sum_{B \neq A} \boldsymbol{S}_A \cdot \boldsymbol{S}_B, \tag{3.192}$$

so that $S^2$ is a sum of intra- and inter-fragment terms, the latter measuring the coupling of the spins associated to each pair of fragments. If we further take into account that the former will need just the one-particle, and the second the two-particle density matrices, we write,

$$\langle \boldsymbol{S}_A \cdot \boldsymbol{S}_B \rangle = \frac{3}{4} \delta_{AB} \int_A \rho(\boldsymbol{r}) d\boldsymbol{r} + \int_A \int_B \boldsymbol{S}_1 \cdot \boldsymbol{S}_2 \, \rho_2(\boldsymbol{r}_1, \boldsymbol{r}_2; \boldsymbol{r}_1', \boldsymbol{r}_2') d\boldsymbol{r}_1 d\boldsymbol{r}_1, \tag{3.193}$$

where in the second integral we first let the operator act and then equate the primed and unprimed coordinates before the integration takes place.

When this path is taken, the atomic local spin $\langle S_A^2 \rangle$ of atoms in restricted singlets become non-zero. For instance, Clark and Davidson [21] showed that in a Hartree-Fock description of $H_2$, $\langle S_A^2 \rangle = 3/8$. This situation has been heavily criticized by many (see Refs. [1, 26, 38, 53–56, 66]) and has led to propose other expressions for the local atomic spin that provide zero values in restricted singlets. A very fruitful approach has been proposed by Ramos-Cordoba and coworkers [64, 65], who built a one-parameter family of local spins,

$$\begin{aligned} \langle S^2 \rangle_{\mathrm{RC}_a} = {} & a \int u(\boldsymbol{r}_1) \, d\boldsymbol{r}_1 + (2a-1) \iint \Lambda(\boldsymbol{r}_1, \boldsymbol{r}_2; \boldsymbol{r}_1, \boldsymbol{r}_2) \, d\boldsymbol{r}_1 d\boldsymbol{r}_2 \\ & - \frac{1}{2} \iint \Lambda(\boldsymbol{r}_1, \boldsymbol{r}_2; \boldsymbol{r}_2, \boldsymbol{r}_1) \, d\boldsymbol{r}_1 d\boldsymbol{r}_2. \end{aligned} \tag{3.194}$$

where $\Lambda$ is a spinless two-particle cumulant and $u(\boldsymbol{r}_1; \boldsymbol{r}_1') = 2\rho(\boldsymbol{r}_1; \boldsymbol{r}_1') - \int \rho(\boldsymbol{r}_1; \boldsymbol{r}_2)\rho(\boldsymbol{r}_2; \boldsymbol{r}_1') d\boldsymbol{r}_2$ is the effectively unpaired density matrix $u$ introduced by Takatsuka, Fueno and Yamaguchi [70]. Ramos-Cordoba and coworkers have found that by taking $a = 3/4$ the resulting local spin satisfies all the desirable properties, both at single determinant as well as correlated levels of theory.

This local spin has been found very useful to identify singlet diradicals or polyradicals, and is becoming a popular way to recognize them.

## References

1. Alcoba, D.R., Torre, A., Lain, L., Bochicchio, R.C.: Determination of local spins by means of a spin-free treatment. J. Chem. Theory Comput. **7**(11), 3560–3566 (2011)
2. Angulo, J.C., Dehesa, J.S., Galvez, F.J.: Atomic-charge convexity and the electron density at the nucleus. Phys. Rev. A **42**(1), 641 (1990)

3. Angyan, J.G., Loos, M., Mayer, I.: Covalent bond orders and atomic valence indices in the topological theory of atoms in molecules. J. Phys. Chem. **98**(20), 5244–5248 (1994)
4. Arfken, G.: Mathematical Methods for Physicists, 3rd edn. Academic, San Diego, California (1985)
5. Bader, R.F., Gatti, C.: A green's function for the density. Chem. Phys. Lett. **287**(3–4), 233–238 (1998)
6. Bader, R.F.W.: Atoms in Molecules. Oxford University Press, Oxford (1990)
7. Bader, R.F.W., Carroll, M.T., Cheeseman, J.R., Chang, C.: Properties of atoms in molecules: atomic volumes. J. Am. Chem. Soc. **109**, 7968 (1987)
8. Bader, R.F.W., Chang, C.: Properties of atoms in molecules: energetic of electrofilic aromatic substitution. J. Phys. Chem. **93**, 5095 (1989)
9. Bader, R.F.W., Popelier, P.L.A.: Atomic theorems. Int. J. Quantum Chem. **45**, 189 (1993)
10. Bader, R.F.W., Stephens, M.E.: Spatial localization of the electronic pair and number distributions in molecules. J. Am. Chem. Soc. **97**, 7391 (1975)
11. Baerends, E.J., Gritsenko, O.V.: A quantum chemical view of density functional theory. J. Phys. Chem. A **101**, 5383 (1997)
12. Bingel, W.A.: The behaviour of the first-order density matrix at the coulomb singularities of the Schrödinger equation. Z. Naturforsch. A **18**, 1249 (1963)
13. Blanco, M.A., Martín Pendás, A., Francisco, E.: Interacting quantum atoms: a correlated energy decomposition scheme based on the quantum theory of atoms in molecules. J. Chem. Theory Comput. **1**, 1096 (2005)
14. Bochicchio, R., Ponec, R., Torre, A., Lain, L.: Multicenter bonding within the AIM theory. Theor. Chem. Acc.: Theory, Comput. Model. (Theoretica Chimica Acta) **105**(4–5), 292–298 (2001)
15. Bollini, C.G., Giambiagi, M., de Giambiagi, M.S., de Figueiredo, A.P.: Graphical linking of a mo multicenter bond index to vb structures. Struct. Chem. **12**(2), 113–120 (2001)
16. Born, M.: Zur Quantenmechanik der Stoßvorgänge. Z. Phys. **37**(12), 863 (1926)
17. Boys, S., Bernardi, F.: The calculation of small molecular interactions by the differences of separate total energies. Some procedures with reduced errors. Mol. Phys. **19**(4), 553–566 (1970)
18. Bultinck, P., Ponec, R., Van Damme, S.J.: Multicenter bond indices as a new measure of aromaticity in polycyclic aromatic hydrocarbons. J. Phys. Org. Chem. A **18**, 706 (2005)
19. Cancès, E., Keriven, R., Lodier, F., Savin, A.: How electrons guard the space: shape optimization with probability distribution criteria. Theor. Chem. Acc. **111**, 373 (2004)
20. Cioslowski, J.: Isopycnic orbital transformations and localization of natural orbitals. Int. J. Quant. Chem. **S24**, 15 (1990)
21. Clark, A.E., Davidson, E.R.: Local spin. J. Chem. Phys. **115**(16), 7382–7392 (2001)
22. Cohen, L.: Local kinetic energy in quantum mechanics. J. Chem. Phys. **70**, 788 (1979)
23. Coleman, A.J., Yakulov, V.I.: Reduced Density Matrices. Coulson's Challenge. Springer, New York, USA (2000)
24. Coppens, P.: Electron density from x-ray diffraction. Annu. Rev. Phys. Chem. **43**, 663–692 (1992)
25. Daudel, R.: The Fundamentals of Theoretical Chemistry. Pergamon Press, Oxford (1968)
26. Davidson, E.R., Clark, A.E.: Analysis of wave functions for open-shell molecules. Phys. Chem. Chem. Phys. **9**(16), 1881 (2007)
27. Diner, S., Malrieu, J.P., Claverie, P.: Localized bond orbitals and the correlation problem. Theoret. Chim. Acta (Berl.) **13**, 1 (1969)
28. Finzel, K.: How does the ambiguity of the electronic stress tensor influence its ability to serve as bonding indicator. Int. J. Quant. Chem. **114**(9), 568 (2014)
29. Fradera, X., Austen, M.A., Bader, R.F.W.: The Lewis model and beyond. J. Phys. Chem. A **103**, 304 (1999)
30. Francisco, E., Martín Pendás, A., Blanco, M.A.: A molecular energy decomposition scheme for atoms in molecules. J. Chem. Theory Comput. **2**, 90 (2006)
31. Francisco, E., Martín Pendás, A., García-Revilla, M., Boto, R.A.: A hierarchy of chemical bonding indices in real space from reduced density matrices and cumulants. Comput. Theor. Chem. **1003**, 71 (2013)

32. Gatti, C.: The source function descriptor as a tool to extract chemical information from theoretical and experimental electron densities. Struct. Bond. **147**, 193–286 (2012)
33. Gatti, C.: Challenging chemical concepts through charge density of molecules and crystals. Physica Scripta **87**(4), 048,102 (2013)
34. Gatti, C., Cargnoni, F., Bertini, L.: Chemical information from the source function. J. Comput. Chem. **24**(4), 422–436 (2003)
35. Gatti, C., Macchi, P. (eds.): Modern Charge-density Analysis. Springer, Dordrecht (2012)
36. Gatti, C., Saleh, G., Presti, L.L.: Source function applied to experimental densities reveals subtle electron-delocalization effects and appraises their transferability properties in crystals. Acta Crystallogr. Sect. B Struct. Sci. Cryst. Eng. Mater. **72**(2), 180–193 (2016)
37. Giambiagi, M., de Giambiagi, M.S., Mundim, K.C.: Definition of a multicenter bond index. Struct. Chem. **1**, 423 (1990)
38. Herrmann, C., Reiher, M., Hess, B.A.: Comparative analysis of local spin definitions. J. Chem. Phys. **122**(3), 034,102 (2005)
39. Hoffmann-Ostenhof, M., Hoffmann-Ostenhof, T.: "Schrödinger inequalities" and asymptotic behavior of the electron density of atoms and molecules. Phys. Rev. A **16**(5), 1782 (1977)
40. Itzykson, C., Zuber, J.B.: Quantum Field Theory, international edn. McGraw-Hill, New York (1985)
41. Jeziorski, B., Moszynski, R., Szalewicz, K.: Perturbation theory approach to intermolecular potential energy surfaces of van der Waals complexes. Chem. Rev. **94**, 1887 (1994)
42. Kato, W.A.: On the eigenfunctions of many-particle systems in quantum mechanics. Commun. Pure Appl. Math. **10**, 151 (1957)
43. Kitaura, K., Morokuma, K.: A new energy decomposition scheme for molecular interactions within the Hartree-Fock approximation. Int. J. Quant. Chem. **10**, 325 (1976)
44. Krishnan, R., Frisch, M.J., Pople, J.A.: Contribution of triple substitutions to the electron correlation energy in fourth order perturbation theory. J. Chem. Phys. **72**, 4244 (1980)
45. Landau, L.D., Lifshitz, E.M.: Mechanics. In: Course of Theoretical Physics, vol. 1, 3rd edn. Butterworth-Heinemann, Oxford (1993)
46. Martín Pendás, A., Blanco, M.A., Francisco, E.: The nature of the hydrogen bond: a synthesis from the interacting quantum atoms picture. J. Chem. Phys. **125**, 184,112 (2006)
47. Martín Pendás, A., Francisco, E., Blanco, M.A., Gatti, C.: Bond paths as privileged exchange channels. Chem. Eur. J. **13**, 9362 (2007)
48. Martín Pendás, A., Francisco, E., Casals-Sainz, J.: Quantitative determination of the nature of intermolecular bonds by EDA analysis, chapter 6. In: Novoa, J.A. (ed.) Intermolecular Interactions in Crystals: Fundamentals of Crystal Engineering, 1st edn. Royal Society of Chemistry, London (2018)
49. Matito, E., Duran, M., Solà, M.: A novel exploration of the Hartree-Fock homolytic bond dissociation problem in the hydrogen molecule by means of electron localization measures. J. Chem. Educ. **83**(8), 1243 (2006)
50. Matito, E., Solà, M., Salvador, P., Duran, M.: Electron sharing indexes at the correlated level. application to aromaticity calculations. Faraday Discuss. **135**, 9904 (2007)
51. Mayer, I.: Bond orders and valences in the SCF theory: a comment. Theor. Chim. Acta **67**, 315 (1985)
52. Mayer, I.: On bond orders and valences in the ab initio quantum chemical theory. Int. J. Quant. Chem. **29**, 73 (1986)
53. Mayer, I.: Local spins: An alternative treatment for single determinant wave functions. Chem. Phys. Lett. **440**(4–6), 357–359 (2007)
54. Mayer, I.: Local spins: an improved treatment for correlated wave functions. Chem. Phys. Lett. **478**(4–6), 323–326 (2009)
55. Mayer, I.: Local spins: Improving the treatment for single determinant wave functions. Chem. Phys. Lett. **539–540**, 172–174 (2012)
56. Mayer, I., Matito, E.: Calculation of local spins for correlated wave functions. Phys. Chem. Chem. Phys. **12**(37), 11,308 (2010)
57. McWeeny, R.: Methods of Molecular Quantum Mechanics, 2nd edn. Academic, London (1992)

58. Menéndez, M., Boto, R.A., Francisco, E., Martín Pendás, A.: One-electron images in real space: natural adaptive orbitals. J. Comput. Chem. **36**, 833 (2015)
59. Mezey, P.G.: The holographic electron density theorem and quantum similarity measures. Mol. Phys. **96**, 169 (1999)
60. Mohajeri, A., Ashrafi, A.: Aromaticity in terms of ring critical point properties. Chem. Phys. Lett. **458**(4–6), 378 (2008)
61. Monza, E., Gatti, C., Presti, L.L., Ortoleva, E.: Revealing electron delocalization through the source function. J. Phys. Chem. A **115**(45), 12864–12878 (2011)
62. Ponec, R.: Electron pairing and chemical bonds. Chemical structure, valences and structural similarities from the analysis of the Fermi holes. J. Math. Chem. **21**, 323 (1997)
63. Ponec, R.: Electron pairing and chemical bonds. molecular structure from the analysis of pair densities and related quantities. J. Math. Chem. **23**, 85 (1998)
64. Ramos-Cordoba, E., Matito, E., Mayer, I., Salvador, P.: Toward a unique definition of the local spin. J. Chem. Theory Comput. **8**(4), 1270–1279 (2012)
65. Ramos-Cordoba, E., Matito, E., Salvador, P., Mayer, I.: Local spins: improved Hilbert-space analysis. Phys. Chem. Chem. Phys. **14**(44), 15,291 (2012)
66. Reiher, M.: On the definition of local spin in relativistic and nonrelativistic quantum chemistry. Faraday Discus. **135**, 97–124 (2007)
67. Salvador, P., Mayer, I.: One- and two-center physical space partitioning of the energy in the density functional theory. J. Chem. Phys.**126**(23), 234,113 (2007)
68. Shanon, R.D.: Revised effective ionic radii and systematic studies of interatomic distances in halides and chalcogenides. Acta Crystallogr. llogr. A **32**, 751 (1976)
69. Steiner, E.: Charge densities in atoms. J. Chem. Phys. **39**, 689 (1963)
70. Takatsuka, K., Fueno, T., Yamaguchi, K.: Distribution of odd electrons in ground-state molecules. Theor. Chim.Acta **48**(3), 175–183 (1978)
71. Thomsen, M.K., Gatti, C., Overgaard, J.: Probing cyclic $\pi$-electron delocalization in an imidazol-2-ylidene and a corresponding imidazolium salt. Chem. - A Eur. J. **24**(19), 4973–4981 (2018)

# Chapter 4
# Electron Pairing Descriptors

**Abstract** The concept of electron (de)localization that has been explored in the previous chapter has fascinated chemists for decades. There are situations in which electrons, pairs of electrons, or more generally $n$-electron-tuples behave as localized entities, which extend their influence in a very limited spatial region. Atomic core electrons, for instance, can in most chemically relevant problems be ignored, merely contributing to shielding the nuclear charge. In other cases, like in alternate polyenes, electrons act as highly delocalized objects with long-range potential effects. Chemists have long recognized this difference in, for instance, separating inductive from mesomeric effects. Here we describe some of the theoretical efforts to construct scalars in real space that sense these different behaviors. We pay especial attention to the electron localization function (ELF), which has been widely used in the literature. Several interpretations of ELF are put forward, and its relation to Gillespie's valence shell electron pair repulsion (VSEPR) model examined.

## 4.1 Introduction

The number of ideas that were related to molecular geometry and properties with a quantum basis but put forward before the advent of Schrödinger equation is amazing [67]. At the beginning of the twentieth century, the understanding of crystalline salts had achieved a fairly reasonable state: it was already known that they were governed by electrostatic forces between contrary charged ions.

However, trying to bring these ideas to the molecular realm had failed miserably. By the end of the nineteenth century, there was a major consensus on the existence of forces that held atoms together mainly by pairs. In 1858, Archibald S. Couper represented this "force" as a (dotted) line for the first time [19], a convention that lasts until today (and looks like it will still be around for a long time!). It was also admitted that electrons were responsible for bonding (but note that those were the electrons of Thomson's model!), so that a given atom formed a characteristic number of bonds (the so-called "valence").



Á. Martín Pendás and J. Contreras-García, *Topological Approaches to the Chemical Bond*, Theoretical Chemistry and Computational Modelling,
https://doi.org/10.1007/978-3-031-13666-5_4

This was the state of the art when G.N. Lewis foresaw in his ground-breaking 1916 paper that electron pairing was fundamental for chemical bonding [36]. And this was way before Pauli's proposal of the exclusion principle! How did he come to this conclusion? Lewis observed that most stable molecules had an even number of electrons, so that their composition could be understood if it was assumed that each bond corresponded to the sharing of two electrons between the bonded atoms (i.e. the rule of two). Moreover, he observed that following this line of reasoning, most central atoms had eight electrons around them (aka. the rule of eight, known nowadays as the octet rule). He even suggested that these eight electrons were placed around the atom forming a cube, which set the basis for ulterior studies of molecular geometry.

Within these postulated rules, he was able to explain the structure of $CX_4$ compounds and introduce the concept of lone pairs to explain the geometry of $NX_3$ and $OX_2$ compounds. He also put forward the basis for double and triple bonds in ethylene and acetylene. This was the birth of molecular geometry, which culminated with the Valence Shell Electron Pair Repulsion (VSEPR) theory [25].

Already at that time the existence of the exceptions that we have all studied in high school (hypovalent compounds such as $BF_3$ and hypervalent moieties such as $PCl_5$ and $SF_6$) was known. Nevertheless, hypo- and hypervalent structures represented a very small number of systems at that time, so the rule of eight was still very useful. We will analyze these cases in Chap. 6, but here it is more important to highlight the fact that these rules seemed to clash with well-known and accepted theories.

On the one hand, Bohr's model had been introduced in 1913 [6], and due to its ability to describe the hydrogen spectrum, it had soon been accepted by the community. However, how could the pair rule be understood within an orbit model? Lewis tried to reconcile both atomic viewpoints in his book [37], where he proposed that each electron should have somehow fixed orbits, which gave rise to preferential positions.

On the other hand, since electrostatic forces were already well known, it was far from clear how the proposed pairing of particles with the same charge could be explained. This apparent contradiction between the Lewis model and the Coulombic repulsion among the electrons was solved by Lennard-Jones [35], who proposed that the most probable spatial disposition of eight electrons is at the vertices of two tetrahedra of different spin (and that is where the existence of two "types of electrons" —i.e. spin—comes in).

Following up on this study, Linnett [38] came to the conclusion that electrons were not always associated with forming pairs, as Lewis had proposed, but rather that the two tetrahedra proposed by Lewis stayed as far away as possible due to spin correlation, yielding a cubic arrangement with alternated spins (i.e. the double quartets). This cube could rotate freely, explaining the spatial arrangement of electrons in atoms, accounting for electrostatic repulsion and taking spin into account. Recall that electrostatic interactions are generally one order smaller in magnitude than those induced by the exclusion principle.

However, if electrons are freely rotating, why do we observe Lewis pairs? Linnett also gave an answer to this question: the atomic delocalized electronic arrangement

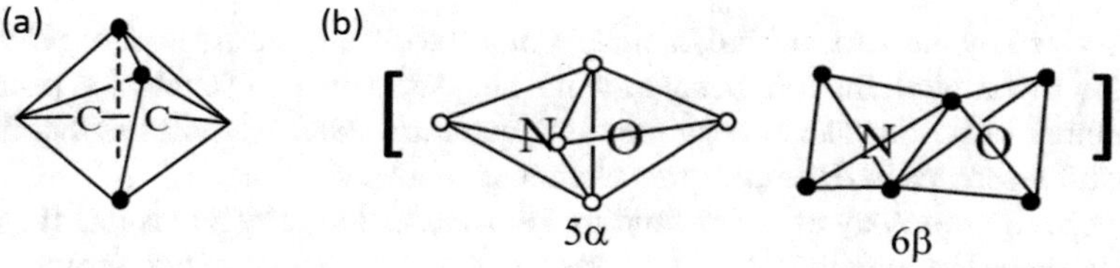


**Fig. 4.1** Linnett's double quartet structures for **a** ethyne **b** nitrogen monoxide. Reproduced with permission from Ref. [28]

evolved upon molecule formation. When atoms bind, nuclear potentials from the other atoms lead to electronic localization. For example, the two tetrahedra should coalesce on a vertex when the bonding pair is formed to yield a molecule such as A-H. The remaining six electrons remain delocalized. In the water molecule, the hydrogen cores induce the formation of four localized pairs (two bonding and two non-bonding). Another example where electrons remain delocalized is acetylene. The electrons in the triple bond are delocalized within a circle around the molecular axis. Keep this picture in mind (Fig. 4.1a) because it will be extremely useful to understand ELF topology in acetylene! (Sect. 4.3.3.1). Note that the mere pairing of electrons would never be able to explain unpaired electrons, nor paramagnetic molecules such as NO. Linnet's simple approach is able to explain the paramagnetic nature of NO in real space. Figure 4.1b shows the paramagnetic orientations of $\alpha$ and $\beta$ electrons separately. This approach is however limited, and finds trouble to explain energetics, which are easier to understand from bond orders and molecular orbitals.

Anyhow, the persistence of the electron pairing concept and its use for understanding molecular geometry and properties is extremely notorious. Moreover when we realize that these concepts have survived the delocalized picture provided by quantum mechanics. We can understand and marry these two viewpoints if we take a loose Lewis approach: electrons are not strictly localized in pairs, but rather the most probable positions of opposed spin electrons are generally those predicted by Lewis and Linnett's model (or VSEPR).

In this Chapter, we will review several scalar functions that reveal this average localization, thus providing a real space viewpoint that merges localization and quantum mechanics. But first of all, we will review how this "most probable" position can be understood from statistics to yield the so-called electronic Pauli repulsion and with it, the Pauli principle.

## 4.2 The Pauli Principle

The Pauli exclusion principle was formulated by Wolfgang Pauli in 1925 [48]. It states that two fermions cannot occupy the same quantum state simultaneously, that is, two electrons cannot have all the same quantum numbers. Hence, if two electrons are described by the same orbital, their spin must be different. And no more electrons can enter that orbital. Instead, bosons are not subject to the Pauli exclusion principle: any number of bosons can occupy the same quantum state. We will see that bosons are useful reference in defining some chemical concepts.

A more rigurous way of understanding Pauli exclusion principle is that the wavefunction should be antisymmetric with respect to the exchange of two fermions (and symmetric for the exchange of two bosons). Hence, if the coordinates of two electrons are exchanged within the system, its wavefunction should change sign (and stay the same for bosonic particles).

With the simple approach of Linnett, we have seen that Pauli exclusion principle helps understanding a good number of features of the electronic structure in atoms, molecules and solids. It enables to understand the organization of atoms into shells, the sharing of electrons in molecules and the periodic nature of their properties. However, it is important to deeply understand what the Pauli principle physically implies. When talking about electron pairing, the first concept that is recalled in most textbooks is Pauli repulsion. But there is no such thing as a Pauli force, so what exactly do we mean when we use the Pauli principle to rationalize electronic structure?

From a physical point of view, Pauli force refers to an effective attraction between bosons or to an effective repulsion between fermions. This is induced by a differential behavior of same or different spin electrons which lead to different statistics that keep them closer or further away, respectively.

As an example, let us use a model toy of two particles 1 and 2 described by orthonormal orbitals $\psi_a$ and $\psi_b$ [26]. If these particles are distinguishable $(dist)$, the system wavefunction is given by $\Psi_{dist}(\boldsymbol{x}_1, \boldsymbol{x}_2) = \psi_a(\boldsymbol{x}_1)\psi_b(\boldsymbol{x}_2)$, so that the mean-square separation between particles is given by[1]

$$\langle (x_1 - x_2)^2 \rangle_{dist} = \langle x^2 \rangle_a + \langle x^2 \rangle_b - 2\langle x \rangle_a \langle x \rangle_b, \tag{4.1}$$

where $\langle x \rangle_i = \int x|\psi_i(x)|^2 dx$. If instead, particles are non-distinguishable, the system wavefunction needs to be symmetric if they are bosons (+), and antisymmetric (−) if they are fermions:

$$\Psi(x_1, x_2)_\pm = \frac{1}{\sqrt{2}}[\psi_a(x_1)\psi_b(x_2) \pm \psi_a(x_2)\psi_b(x_1)] \tag{4.2}$$

---

1

$$\int\int (\boldsymbol{x}_1 - \boldsymbol{x}_2)^2 |\Psi(\boldsymbol{x}_1, \boldsymbol{x}_2)|^2 d\boldsymbol{x}_1 d\boldsymbol{x}_2 = \int\int \boldsymbol{x}_1^2 |\psi_a(\boldsymbol{x}_1)\psi_b(\boldsymbol{x}_2)|^2 d\boldsymbol{x}_1 d\boldsymbol{x}_2$$
$$+ \int\int \boldsymbol{x}_2^2 |\psi_a(\boldsymbol{x}_1)\psi_b(\boldsymbol{x}_2)|^2 d\boldsymbol{x}_1 d\boldsymbol{x}_2 - 2\int\int \boldsymbol{x}_1\boldsymbol{x}_2 |\psi_a(\boldsymbol{x}_1)\psi_b(\boldsymbol{x}_2)|^2 d\boldsymbol{x}_1 d\boldsymbol{x}_2$$
$$= \int \boldsymbol{x}_1^2 |\psi_a(\boldsymbol{x}_1)|^2 d\boldsymbol{x}_1 + \int \boldsymbol{x}_2^2 |\psi_b(\boldsymbol{x}_2)|^2 d\boldsymbol{x}_2 - 2\int \boldsymbol{x}_1 |\psi_a(\boldsymbol{x}_1)|^2 d\boldsymbol{x}_1 \int \boldsymbol{x}_2 |\psi_b(\boldsymbol{x}_2)|^2 d\boldsymbol{x}_2,$$

where we have used $\int |\psi_a(\boldsymbol{x})|^2 d\boldsymbol{x} = 1$.

so that the mean-square separation is now given by[2]

$$\langle (x_1 - x_2)^2 \rangle_{\pm} = \langle (x_1 - x_2)^2 \rangle_{dist} \mp 2|\langle x \rangle_{ab}|^2, \tag{4.3}$$

where $\langle x \rangle_{ab} = \int x \psi_a^*(x) \psi_b(x) dx$. Comparing Eq. 4.3 with Eq. 4.1,[3] we can see that bosons tend to be closer than distinguishable particles, whereas fermions tend to be further apart with respect to the same reference.

In a "loose" way, we can say that there is an effective attraction force between bosons that tends to collapse them, and an effective repulsive force between fermions that tends to put them apart. However, in the strict sense, we must bear in mind that such force **does not exist**, that we observe a differential behavior due to the symmetry requirements of the wavefunction [44]. Along this Chapter, we will see how different quantities, such as the conditional pair probability for same spin electrons, or the kinetic energy densities, can be used to construct local properties that reveal this "Pauli repulsion" and also how bosons and non-interacting electrons can be used to highlight the differential behavior.

## 4.3 The Electron Localization Function (ELF)

### *4.3.1 Definition*

#### 4.3.1.1 Theoretical Background: The Homogeneous Electron Gas

The homogeneous electron gas, or Thomas Fermi (TF) gas, is a widely used reference when analyzing electronic distributions in $\mathbb{R}^3$. It comes from the analysis of periodic independent electrons (i.e. electrons under zero potential). The solution for

---

2

$$\langle (\boldsymbol{x}_1 - \boldsymbol{x}_2)^2 \rangle_{\pm} = \frac{1}{2} \int \int (\boldsymbol{x}_1 - \boldsymbol{x}_2)^2 |\psi_a(\boldsymbol{x}_1)\psi_b(\boldsymbol{x}_2) \pm \psi_a(\boldsymbol{x}_2)\psi_b(\boldsymbol{x}_1)|^2 d\boldsymbol{x}_1 d\boldsymbol{x}_2$$
$$= \frac{1}{2} \int \int (\boldsymbol{x}_1 - \boldsymbol{x}_2)^2 \left[ |\psi_a(\boldsymbol{x}_1)\psi_b(\boldsymbol{x}_2)|^2 + |\psi_a(\boldsymbol{x}_2)\psi_b(\boldsymbol{x}_1)|^2 \pm 2|\psi_a(\boldsymbol{x}_1)^* \psi_b^*(\boldsymbol{x}_2)\psi_a(\boldsymbol{x}_2)\psi_b(\boldsymbol{x}_1)| \right] d\boldsymbol{x}_1 d\boldsymbol{x}_2$$
$$= \frac{1}{2} \int \int [\boldsymbol{x}_1^2 |\psi_a(\boldsymbol{x}_1)\psi_b(\boldsymbol{x}_2)|^2 + \boldsymbol{x}_2^2 |\psi_a(\boldsymbol{x}_1)\psi_b(\boldsymbol{x}_2)|^2 - 2\boldsymbol{x}_1\boldsymbol{x}_2 |\psi_a(\boldsymbol{x}_1)\psi_b(\boldsymbol{x}_2)|^2$$
$$+ \boldsymbol{x}_1^2 |\psi_a(\boldsymbol{x}_2)\psi_b(\boldsymbol{x}_1)|^2 + \boldsymbol{x}_2^2 |\psi_a(\boldsymbol{x}_2)\psi_b(\boldsymbol{x}_1)|^2 - 2\boldsymbol{x}_1\boldsymbol{x}_2 |\psi_a(\boldsymbol{x}_2)\psi_b(\boldsymbol{x}_1)|^2$$
$$\pm 2\boldsymbol{x}_1^2 \psi_a^*(\boldsymbol{x}_1)\psi_b^*(\boldsymbol{x}_2)\psi_a(\boldsymbol{x}_2)\psi_b(\boldsymbol{x}_1) \pm 2\boldsymbol{x}_2^2 \psi_a^*(\boldsymbol{x}_1)\psi_b^*(\boldsymbol{x}_2)\psi_a(\boldsymbol{x}_2)\psi_b(\boldsymbol{x}_1) \mp 4\boldsymbol{x}_1\boldsymbol{x}_2 \psi_a^*(\boldsymbol{x}_1)\psi_b^*(\boldsymbol{x}_2)\psi_a(\boldsymbol{x}_2)$$
$$\psi_b(\boldsymbol{x}_1)] d\boldsymbol{x}_1 d\boldsymbol{x}_2 = \langle (\boldsymbol{x}_1 - \boldsymbol{x}_2)^2 \rangle_{dist} \mp |2\langle \boldsymbol{x} \rangle_{ab}|^2 \pm 4S^2 \langle x^2 \rangle_{ab},$$

where $S = \int \psi_a(\boldsymbol{x})\psi_b(\boldsymbol{x}) d\boldsymbol{x}$. Since we have assumed orthonormal orbitals:

$$\langle (\boldsymbol{x}_1 - \boldsymbol{x}_2)^2 \rangle_{\pm} = \langle (\boldsymbol{x}_1 - \boldsymbol{x}_2)^2 \rangle_{dist} \mp |2\langle \boldsymbol{x} \rangle_{ab}|^2$$

.

3 Note that there is a typo in the original articles for Eq. 4.3.

this system is given by plane waves $\phi_{\mathbf{k}}(\mathbf{r})$, where $\mathbf{k}$ is the wavevector and $\lambda$ is the wavelength $\lambda = 2\pi/|\mathbf{k}|$.

Assuming a cubic cell of $a$ lattice parameter, wavevectors along each direction (x, y, z) are given by $k_i = 2\pi n_i/a$, where $n_i$ are quantum numbers. Hence, the volume of each state (i.e. for a given $k_x, k_y, k_z$) is given by $8\pi^3/V$ where $V = a^3$. Following the usual occupation of states, from smaller to bigger energy, this allows to obtain the number of occupied states, given by $Vk_F^3/6\pi^2$ where $k_F$ is known as the Fermi wavevector. Since we have two electrons per state, the number of electrons is given by

$$N = \frac{Vk_F^3}{3\pi^2}. \tag{4.4}$$

Hence, the electron density of the homogeneous electron gas can also be related to $k_F$ by

$$\rho = \frac{k_F^3}{3\pi^2}. \tag{4.5}$$

Assuming a sphere of radius $r_s$, known as the Seitz radius, we have a measure of the "compactedness" of the system:

$$r_s = \left(\frac{3}{4\pi\rho}\right)^{1/3}. \tag{4.6}$$

Note that in this model, the Fermi momentum $p_F$ is a spatial function, instead of a quantum mechanical operator, given by

$$p_F(\mathbf{r}) = \hbar k_F = \hbar\left(3\pi^2\rho(\mathbf{r})\right)^{1/3}. \tag{4.7}$$

Similarly, we can obtain the kinetic energy as

$$T_h = \frac{p_F^2}{2m} = \frac{\hbar^2\left(3\pi^2\rho(\mathbf{r})\right)^{2/3}}{2m} \tag{4.8}$$

which allows defining a local kinetic energy density so that $T_h = \int t_h(\mathbf{r})d\mathbf{r}$ as

$$t_h(\mathbf{r}) = c_F(\rho(\mathbf{r}))^{5/3} \tag{4.9}$$

where $c_F = \frac{3\hbar^2\left(3\pi^2\right)^{2/3}}{10m}$ is known as the Fermi constant. Or in atomic units:

$$c_F = \frac{3\left(3\pi^2\right)^{2/3}}{10}. \tag{4.10}$$

We can also calculate the Fermi hole curvature, $\nabla'^2 \rho_2^{\alpha\alpha}$:

$$\nabla'^2 \rho_2^{\alpha\alpha}(\boldsymbol{r}) = \frac{3}{5}(6\pi^2)^{2/3} \rho^{\alpha}(\boldsymbol{r})^{5/3}. \tag{4.11}$$

#### 4.3.1.2 Original Definition

The geometrical patterns proposed in the VSEPR model are based on the properties of the same spin pair density, $\rho_2^{\alpha\alpha}$ or $\rho_2^{\beta\beta}$, which were studied by Lennard–Jones in 1949 [34]. Following the same principle, Becke and Edgecombe [5] also used the conditional probability of finding same spin electrons to build probably the most widely used local function that reveals electron localization, known as the Electron Localization Function (ELF).

The conditional probability of finding an electron at point $\boldsymbol{r}_2$ when another same spin electron is located at $\boldsymbol{r}_1$ was introduced in Sect. 3.5.1. In a closed-shell system, we can limit ourselves to one of the spins, $\alpha$ for example, leading to Eq. 3.149, which for the HF approximation reads

$$\rho^{\alpha\alpha}(\boldsymbol{r}|\boldsymbol{r}') = \frac{\rho_2^{\alpha\alpha}(\boldsymbol{r}, \boldsymbol{r}')}{\rho^{\alpha}(\boldsymbol{r})} = \rho^{\alpha}(\boldsymbol{r}') - \frac{|\rho^{\alpha}(\boldsymbol{r}, \boldsymbol{r}')|^2}{\rho^{\alpha}(\boldsymbol{r})}. \tag{4.12}$$

Equation 4.12 can be simplified through a Taylor expansion if we consider the conditional probability of finding an electron of the same spin at a distance $s$ from the reference electron, $\rho_{av}^{\alpha\alpha}(\boldsymbol{r}, \boldsymbol{r} + s)$, irrespective of direction.[4]

---

[4] We perform a Taylor expansion in $s$ around $\boldsymbol{r}$ [4]. To do this we notice that the operator $e^{s\cdot\nabla}$ provides the conventional Taylor expansion of a function, since $e^{s\cdot\nabla} f(\boldsymbol{r}) \approx f(\boldsymbol{r}) + \boldsymbol{s} \cdot \nabla f(\boldsymbol{r}) + \ldots$:

$$\rho^{\alpha\alpha}(\boldsymbol{r}|\boldsymbol{r} + s) = e^{s\cdot\nabla'} \rho^{\alpha\alpha}(\boldsymbol{r}|\boldsymbol{r}')|_{\boldsymbol{r}=\boldsymbol{r}'}, \tag{4.13}$$

where $\nabla'$ acts on $\boldsymbol{r}'$. If we now build a spherical coordinate system such that the $\nabla'$ vector coincides with the polar axis, then $\boldsymbol{s} \cdot \nabla'$ is given by $\cos\theta\, |s\nabla'|$, so that changing $\cos\theta = u$ the spherical average is then given by

$$\langle e^{s\cdot\nabla'} \rangle = \frac{1}{4\pi} \int_{-1}^{+1} \int_0^{2\pi} e^{u|s\cdot\nabla'|} du\, d\phi = \frac{\sinh(|s \cdot \nabla'|)}{|s \cdot \nabla'|}, \tag{4.14}$$

where we have used $\sinh = (e^x - e^{-x})/2$. Now using the Taylor expansion of sinh:

$$\langle e^{s\cdot\nabla'} \rangle = 1 + \frac{1}{3!} s^2 (\nabla')^2 + \frac{1}{5!} s^4 (\nabla')^4 + \frac{1}{7!} s^6 (\nabla')^6 + \cdots \tag{4.15}$$

Note that the first non-vanishing term in the expansion is the quadratic one. The $s$ independent term vanishes because the conditional probability density of finding two electrons of the same spin at the same position is zero due to the Pauli principle. Hence, we have

$$\rho_{av}^{\alpha\alpha}(\boldsymbol{r}|\boldsymbol{r} + s) = \langle e^{s\cdot\nabla'} \rho^{\alpha\alpha}(\boldsymbol{r}|\boldsymbol{r} + s) \rangle \simeq \frac{1}{6} s^2 (\nabla')^2 \rho^{\alpha\alpha}(\boldsymbol{r}|\boldsymbol{r} + s)|_{s=0} \tag{4.16}$$

$$\rho_{av}^{\alpha\alpha}(\boldsymbol{r}|\boldsymbol{r}+s) \simeq \frac{1}{6}s^2(\nabla')^2\rho^{\alpha\alpha}(\boldsymbol{r}|\boldsymbol{r}+s)|_{s=0}. \tag{4.17}$$

Equation 4.17 can be further simplified if we calculate $(\nabla')^2\rho^{\alpha\alpha}(\boldsymbol{r}|\boldsymbol{r}+s)|_{s=0}$ for real HF orbitals:[5]

$$\begin{aligned}
&\nabla' \cdot \nabla' \frac{|\rho^{\alpha}(\boldsymbol{r},\boldsymbol{r}')|^2}{\rho(\boldsymbol{r})} \\
&= \frac{1}{\rho(\boldsymbol{r})}\sum_i^{\alpha}\sum_{j\neq i}^{\alpha}\phi_i(\boldsymbol{r})\phi_j(\boldsymbol{r})[\nabla^2\phi_i(\boldsymbol{r}')\phi_j(\boldsymbol{r}') + 2\nabla\phi_i(\boldsymbol{r}')\nabla\phi_j(\boldsymbol{r}') + \phi_i(\boldsymbol{r}')\nabla^2\phi_j(\boldsymbol{r}')] \\
&= 2\frac{\sum_i^{\alpha}\sum_{j\neq i}^{\alpha}\phi_i(\boldsymbol{r})\phi_j(\boldsymbol{r})\nabla\phi_i(\boldsymbol{r}')\nabla\phi_j(\boldsymbol{r}')}{\rho(\boldsymbol{r})} + 2\sum_i^{\alpha}\phi_i(\boldsymbol{r}')\nabla^2\phi_i(\boldsymbol{r}').
\end{aligned}$$

So that for $\boldsymbol{r} = \boldsymbol{r}'$:[6]

$$\nabla' \cdot \nabla' \frac{|\rho^{\alpha}(\boldsymbol{r},\boldsymbol{r}')|^2}{\rho(\boldsymbol{r})}\Bigg|_{r=r'} = \frac{1}{2}\frac{(\nabla\rho(\boldsymbol{r}))^2}{\rho^{\alpha}(\boldsymbol{r})} + \nabla^2\rho^{\alpha}(\boldsymbol{r}) - 2\sum_i^{\alpha}|\nabla\phi_i(\boldsymbol{r})|^2. \tag{4.24}$$

Using Eq. 4.24 we can obtain a simple expression for $\nabla'^2\rho^{\alpha\alpha}$:

[5] The following expressions are used:

$$|\rho^{\alpha}(\boldsymbol{r},\boldsymbol{r}')|^2 = \sum_i^{\alpha}\sum_j^{\alpha}\phi_i(\boldsymbol{r})\phi_j(\boldsymbol{r})\phi_i(\boldsymbol{r}')\phi_j(\boldsymbol{r}') \tag{4.18}$$

$$\nabla'|\rho^{\alpha}(\boldsymbol{r},\boldsymbol{r}')|^2 = \sum_i^{\alpha}\sum_j^{\alpha}\phi_i(\boldsymbol{r}')\phi_j(\boldsymbol{r}')[\nabla\phi_i(\boldsymbol{r})\phi_j(\boldsymbol{r}) + \phi_i(\boldsymbol{r})\nabla\phi_j(\boldsymbol{r})] \tag{4.19}$$

$$\nabla'\nabla'|\rho^{\alpha}(\boldsymbol{r},\boldsymbol{r}')|^2 = \sum_i^{\alpha}\sum_j^{\alpha}\phi_i(\boldsymbol{r}')\phi_j(\boldsymbol{r}')[\nabla^2\phi_i(\boldsymbol{r})\phi_j(\boldsymbol{r}) + 2\nabla\phi_i(\boldsymbol{r})\nabla\phi_j(\boldsymbol{r}) + \phi_i(\boldsymbol{r})\nabla^2\phi_j(\boldsymbol{r})] \tag{4.20}$$

.

[6] It suffices to use the following HF expression for real orbitals:

$$\nabla\rho^{\alpha}(\boldsymbol{r}) = 2\sum_i^{\alpha}\nabla\phi_i(\boldsymbol{r})\phi_i(\boldsymbol{r}) \tag{4.21}$$

$$(\nabla\rho(\boldsymbol{r}))^2 = 4\sum_i^{\alpha}\sum_j^{\alpha}\phi_i(\boldsymbol{r})\phi_j(\boldsymbol{r})\nabla\phi_i(\boldsymbol{r})\nabla\phi_j(\boldsymbol{r}) \tag{4.22}$$

$$\nabla^2\rho^{\alpha}(\boldsymbol{r}) = 2\sum_i^{\alpha}\phi_i(\boldsymbol{r})\nabla^2\phi_i(\boldsymbol{r}) + |\nabla\phi_i(\boldsymbol{r})|^2 \tag{4.23}$$

.

$$\nabla'^2 \rho^{\alpha\alpha}(\boldsymbol{r}, \boldsymbol{r}') = \nabla^2 \rho(\boldsymbol{r}) - \nabla' \cdot \nabla' \frac{|\rho^\alpha(\boldsymbol{r}, \boldsymbol{r}')|^2}{\rho(\boldsymbol{r})} = 2\sum_i^\alpha |\nabla\phi_i(\boldsymbol{r})|^2 - \frac{1}{2}\frac{(\nabla\rho(\boldsymbol{r}))^2}{\rho^\alpha(\boldsymbol{r})}. \tag{4.25}$$

Substituting back into Eq. 4.17, we finally obtain

$$\rho_{av}^{\alpha\alpha}(\boldsymbol{r}|\boldsymbol{r}+s) = \frac{1}{3}\left[\sum_i^\alpha |\nabla\phi_i(\boldsymbol{r})|^2 - \frac{|\nabla\rho^\alpha(\boldsymbol{r})|^2}{4\rho^\alpha(\boldsymbol{r})}\right] s^2 + \cdots \tag{4.26}$$

The expression in brackets:

$$D^\alpha(\boldsymbol{r}) = \sum_i^\alpha |\nabla\phi_i(\boldsymbol{r})|^2 - \frac{|\nabla\rho^\alpha(\boldsymbol{r})|^2}{4\rho^\alpha(\boldsymbol{r})} \tag{4.27}$$

is at the core of the Electron Localization Function. Note that $D^\alpha(\boldsymbol{r})$ is then directly related to the Fermi hole curvature, $C = \nabla'^2 \rho_2^{\alpha\alpha}(\boldsymbol{r}, \boldsymbol{r}')$ at the same level (HF with real-value orbitals [21]):

$$D^\alpha(\boldsymbol{r}) = \frac{1}{2}\nabla'^2 \rho^{\alpha\alpha}(\boldsymbol{r}, \boldsymbol{r}') = \frac{1}{2}\nabla'^2 \frac{\rho_2^{\alpha\alpha}(\boldsymbol{r}, \boldsymbol{r}')}{\rho^\alpha(\boldsymbol{r})} = \frac{\nabla'^2 \rho_2^{\alpha\alpha}(\boldsymbol{r}, \boldsymbol{r}')}{2\rho^\alpha(\boldsymbol{r})}. \tag{4.28}$$

This expression will be especially useful for developing the ELF version beyond HF (Sect. 4.3.1.4).

In spite of the fact that the physical meaning of pairing should be contained in $D^\alpha(\boldsymbol{r})$, this is not the case. Figure 4.2 left shows the results for the nitrogen molecule. The core density hides the desired effects. This is further highlighted with the solid black line inf Fig. 4.2 right. A renormalization is needed in order to get rid of the electron density dependence of $D^\sigma(\boldsymbol{r})$. The value for the homogeneous electron

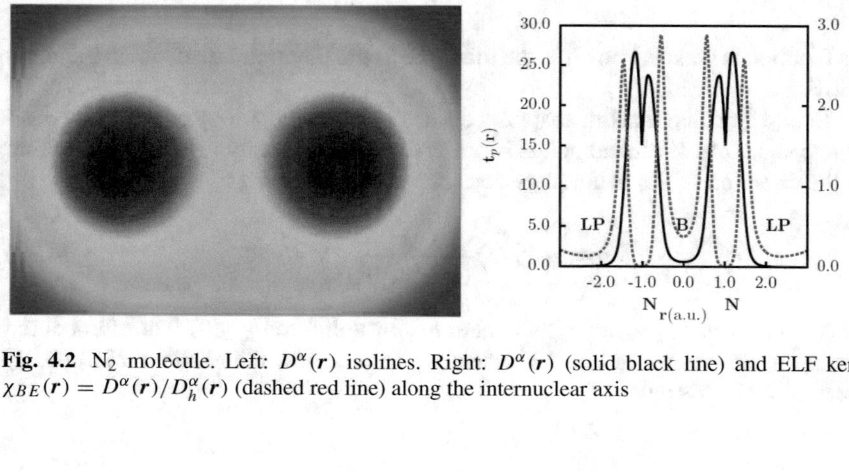


**Fig. 4.2** $N_2$ molecule. Left: $D^\alpha(\boldsymbol{r})$ isolines. Right: $D^\alpha(\boldsymbol{r})$ (solid black line) and ELF kernel $\chi_{BE}(\boldsymbol{r}) = D^\alpha(\boldsymbol{r})/D_h^\alpha(\boldsymbol{r})$ (dashed red line) along the internuclear axis

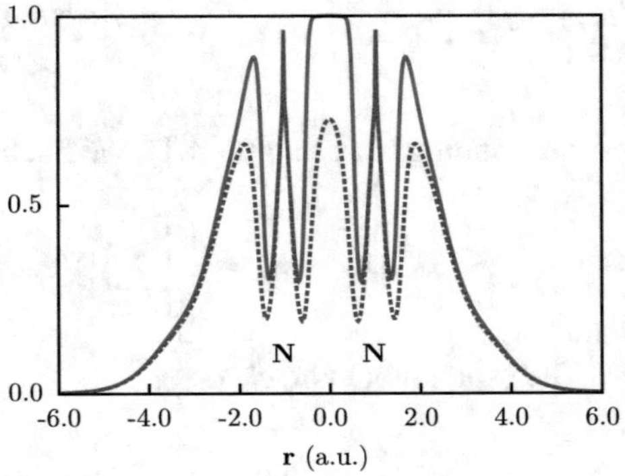


**Fig. 4.3** Mapped ELF (blue) along internuclear axis for $N_2$. The zero is set at the BCP

gas is used as a reference $D_h^\alpha(\boldsymbol{r}) = \frac{3}{5}(6\pi^2)^{2/3}\rho^\alpha(\boldsymbol{r})^{5/3}$ (see Eq. 4.11), leading to an adimensional ELF kernel:

$$\chi_{BE}^\alpha(\boldsymbol{r}) = \frac{D^\alpha(\boldsymbol{r})}{D_h^\alpha(\boldsymbol{r})}. \tag{4.29}$$

As shown in Fig. 4.2 right (red dashed line), once we use the homogeneous electron gas as a reference, the kernel $\chi_{BE}$ does reveal chemical structure. We expect $\chi_{BE}$ to be low in those areas where Pauli repulsion is low. For the nitrogen molecule, these are the nuclei, the bond and the lone pairs. In Fig. 4.2 right we see indeed that $\chi_{BE}$ shows low values in these areas, whereas it is high in between, where the probability of finding the same spin electrons increases.

Although $\chi_{BE}$ contains all the chemical information, it is not bounded from above and it is inversely proportional to localization. Hence, a Lorentzian mapping is used so that the final form, $\eta_{BE}(\boldsymbol{r})$, runs from 0 to 1 [5]:

$$\eta_{BE}(\boldsymbol{r}) = \frac{1}{1 + \chi_{BE}^2(\boldsymbol{r})}. \tag{4.30}$$

This function is what is known in the literature as the Electron Localization Function or ELF.

It should be noted that this mapping does not alter the topology of $\chi(\boldsymbol{r})$ (the position and signature of critical points is preserved). This is easily verified by checking that the direction of $\nabla\eta$ is uniquely determined by that of $\nabla\chi$:

$$\nabla\eta = -2\chi(\boldsymbol{r})\frac{1}{(1 + \chi(\boldsymbol{r})^2)^2}\nabla\chi(\boldsymbol{r}) \tag{4.31}$$

since all the terms are positive (remember that by definition $\chi \geq 0$). The effect of the mapping is also shown in Fig. 4.3, where we can see how the maxima (closer to 1) appear at the Lewis pairs.

ELF has two well-defined physical limits:

- ELF = 1 corresponds to perfectly localized regions, where $D^\alpha \to 0$:

$$\chi_{BE} = \frac{D_h^\alpha}{D_h^\alpha} = 0, \quad ELF = \frac{1}{1+0^2} = 1. \tag{4.32}$$

- ELF = 1/2 corresponds to the homogeneous electron gas ($D^\alpha = D_h^\alpha$):

$$\chi_{BE} = \frac{D_h^\alpha}{D_h^\alpha} = 1, \quad ELF = \frac{1}{1+1^2} = \frac{1}{2}. \tag{4.33}$$

The meaning of values close to zero, which are found in between perfectly localized regions, is less clear and we will analyze them in section "Saddle points: $f$-localization domains and bifurcation trees".

Note that up to now we have defined ELF for a given spin. Nonetheless, calculations are usually done in many occasions for closed-shell systems for the total density. In those cases, a total ELF can be defined as

$$ELF = \frac{1}{1+\chi_{BE}^2}, \tag{4.34}$$

where

$$\chi_{BE}(\boldsymbol{r}) = \frac{D(\boldsymbol{r})}{D_h(\boldsymbol{r})}. \tag{4.35}$$

$$D(\boldsymbol{r}) = D^\alpha(\boldsymbol{r}) + D^\beta(\boldsymbol{r}) = \sum_i |\nabla\phi_i(\boldsymbol{r})|^2 - \frac{|\nabla\rho(\boldsymbol{r})|^2}{4\rho(\boldsymbol{r})} \tag{4.36}$$

$$D_h(\boldsymbol{r}) = c_F\rho(\boldsymbol{r})^{5/3}. \tag{4.37}$$

The constant $c_F = \frac{3}{10}(3\pi^2)^{2/3}$ already seen in Sect. 4.3.1.1 (Eq. 4.10) takes into account that now $\rho(\boldsymbol{r}) = 2\rho^\alpha(\boldsymbol{r})$.[7] Unless otherwise stated we will stick from now on to the closed-shell formulation (Eq. 4.34).

Let's see what this function leads to when used in a molecule we know. Figure 4.4 shows a 2D cut and 3D isosurface of ELF in ethanol. If we think about the Lewis structure of this compound one would expect a total of $(6 \times 2) + (1 \times 6) + (8 \times 1)$ electrons = 26 electrons. We have three cores (two carbons and one oxygen) holding 3 pairs, so that we have 20 electrons (10 pairs) to distribute in the valence. These

7

$$D_h(\boldsymbol{r}) = \frac{3}{5}(6\pi^2)^{2/3}\rho^\alpha(\boldsymbol{r})^{5/3} = \frac{3}{5}2^{2/3}(3\pi)^{2/3}\frac{\rho(\boldsymbol{r})^{5/3}}{2^{5/3}} = \frac{3}{10}(3\pi^2)^{2/3}\rho(\boldsymbol{r})^{5/3} = c_F\rho(\boldsymbol{r})^{5/3} \tag{4.38}$$

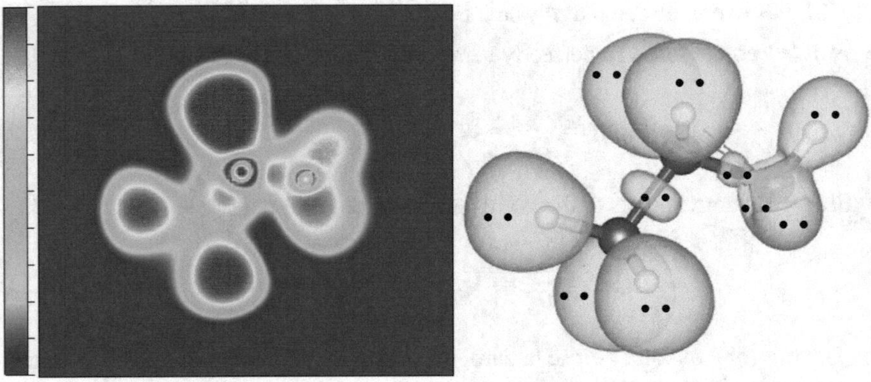

**Fig. 4.4** Left: ethanol ELF 2D cut (red, high value; blue, low value). Right: 3D isosurface. VSEPR and core electron pairs have been added for the sake of clarity

have been represented on top of the 3D isosurfaces as usually done in Chemistry textbooks for the sake of clarity. We can see how ELF surfaces surround the groups of electron pairs: the C–O bond, the lone pairs of oxygen and the C–H bonds. We will see that the C and O cores would have also been visible if we had not used the ball-and-stick representation. It is also interesting to note that the two lone pairs of oxygen appear together, we will show later (section "Saddle points: $f$-localization domains and bifurcation trees") that they appear separately at higher ELF values and how this reflects the fact that they are very delocalized among them like we saw for Linnett's pairs. We also see that hydrogen atoms occupy very high volumes. This is general for all hydrogen basins. Since Hydrogen only has one electron, the "repulsion" due to the Pauli exclusion principle is very low.

#### 4.3.1.3 DFT Definition

The original definition of ELF, $\text{ELF}_{BE}$, was introduced for HF wavefunctions. However, the pair density and its curvature are not explicitly defined in DFT, which is the workhorse of big systems and solid state calculations. This drawback was overcome thanks to the introduction of an ELF definition for DFT by Savin et al. [57].

Within Kohn–Sham DFT, the single-particle pseudo-density matrix $\rho_1(\boldsymbol{r}; \boldsymbol{r}')$ is determined by energy minimization for a given electron density $\rho(\boldsymbol{r})$. Hence, it seems intuitive to look for quantities containing $\rho_1$, that is kinetic energies. More specifically, if we look closer at the expression for the positive definite kinetic energy density introduced in Sect. 3.4.2 (Eq. 3.45):

$$t(\boldsymbol{r}) = \frac{1}{2}(\nabla \cdot \nabla')\rho_1(\boldsymbol{r}; \boldsymbol{r}')|_{\boldsymbol{r}'=\boldsymbol{r}} \tag{4.39}$$

which can be developed in terms of orbitals $\phi_i$:

$$t(\boldsymbol{r}) = \frac{1}{2}\sum_i |\nabla\phi_i(\boldsymbol{r})|^2 \tag{4.40}$$

we can easily see the similarities with the first term in the expression of $D^\alpha$ (Eq. 4.27), now interpreted within the Kohn–Sham DFT framework. Now the pertinent information is directly contained in a three-dimensional function, and the six-dimensional $\rho_1(\boldsymbol{r}, \boldsymbol{r}')$ is not required.

The positive definite kinetic energy density has a lower bound given by the Weizäcker kinetic energy density, $t_w$. This is the kinetic energy density of a fictitious bosonic system which has the same density as the reference system. This can be easily verified by analyzing the kinetic energy density for a system whose wavefunction can be assimilated to a single orbital (all electrons occupy the same orbital), $\psi = \phi$. Assuming a real space orbital, the wavefunction and its gradient can be easily expressed in terms of the electron density as $\psi = \sqrt{\rho}$ and $\nabla\psi = \frac{\nabla\rho}{2\sqrt{\rho}}$ so that the bosonic kinetic energy density is given by

$$t_w = \frac{1}{2}\nabla\phi^*(\boldsymbol{r})\nabla\phi(\boldsymbol{r}) = \frac{1}{8}\frac{|\nabla\rho(\boldsymbol{r})|^2}{\rho(\boldsymbol{r})}. \tag{4.41}$$

For those points in space where electrons can be described by a unique orbital, $t(\boldsymbol{r}) = t_w(\boldsymbol{r})$. Instead, when the exchange hole becomes multicenter in nature, the kinetic energy density rises. The above difference is also known as the Pauli kinetic energy density, $t_P$, since it is the excess of kinetic energy density due to the fermionic nature of electrons:

$$t_P(\boldsymbol{r}) = t(r) - \frac{1}{8}\frac{|\nabla\rho(\boldsymbol{r})|^2}{\rho(\boldsymbol{r})}. \tag{4.42}$$

Following the same recipe as in $\chi_{BE}$ we now need to divide by the same quantity for the homogeneous electron gas reference, i.e. by the homogeneous electron gas kinetic energy density (Eq. 4.9).

$$\chi_S(\boldsymbol{r}) = \frac{t_P(\boldsymbol{r})}{t_h(\boldsymbol{r})} \tag{4.43}$$

where $c_F = \frac{3}{10}(3\pi^2)^{2/3}$. Note that the DFT interpretation of ELF is directly done for closed-shell systems, which explains the use of $c_F$ as was done in Eq. 4.37. Introducing again the mapping, we arrive at $\eta_S$:

$$\eta_S = \frac{1}{1 + \chi_S^2} \tag{4.44}$$

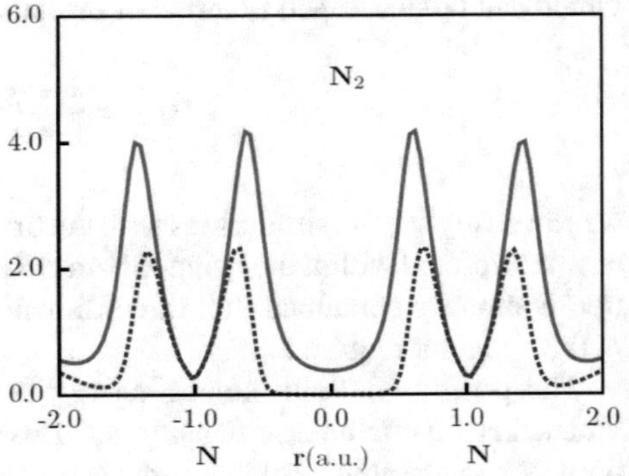


**Fig. 4.5** Scaled kinetic energy densities, $t/t_h$ (red) and $t_w/t_h$ (blue), along the internuclear axis in $N_2$

which is the equivalent to Eq. 4.34 within the DFT framework. Figure 4.5 illustrates the two terms that now give rise to ELF, $t/t_h$ and $t_w/t_h$ and how they reveal chemical structure in $N_2$ as well as the fact that $t_w$ is a lower bound to $t$.

Within this interpretation, ELF is conceived as a measure of the influence of the exclusion principle on the kinetic energy density. Pauli principle has a small influence on electrons alone or forming pairs, so their excess of kinetic energy density is small and ELF approaches unit. Instead, in the regions in between electron pairs, the probability of finding the same spin electrons rises, and so does the kinetic energy density, leading to lower ELF values.

It is interesting to note that an expression equivalent to $\chi_S(\boldsymbol{r})$ was found earlier by Deb and Ghosh [20]. They proposed the following formulation of the kinetic energy density:

$$K(\boldsymbol{r}) = -\frac{1}{4}\nabla^2\rho(\boldsymbol{r}) + \frac{1}{8}\frac{|\nabla\rho(\boldsymbol{r})|^2}{\rho(\boldsymbol{r})} + c_F f(\boldsymbol{r})\rho^{5/3}. \tag{4.45}$$

If we substitute $K(\boldsymbol{r})$ by Eq. 3.47 under the monodeterminantal approximation, we easily see that $f(\boldsymbol{r}) = \chi_S$.[8]

Finally, notice that (independently of the ELF interpretation adopted) ELF tries to pinpoint electron localization through a one-particle descriptor. Two-particle behavior is at most included locally, with the curvature of the same spin conditional pair density. This is advantageous, since scalar densities are easily visualized in real space, but hides much complexity that would remain in a truly two-particle, i.e. six-dimensional, localization or delocalization descriptor. Provided that the analysis of

---

8

$$\frac{1}{2}\sum_i |\nabla\phi_i(\boldsymbol{r})|^2 - \frac{1}{4}\nabla^2\rho(\boldsymbol{r}) = -\frac{1}{4}\nabla^2\rho(\boldsymbol{r}) + \frac{1}{8}\frac{|\nabla\rho(\boldsymbol{r})|^2}{\rho(\boldsymbol{r})} + c_F f(\boldsymbol{r})\rho^{5/3} \tag{4.46}$$

$$f(\boldsymbol{r}) = \frac{1}{c_E\rho^{5/3}}\left[\frac{1}{2}\sum_i |\nabla\phi_i(\boldsymbol{r})|^2 - \frac{1}{8}\frac{|\nabla\rho(\boldsymbol{r})|^2}{\rho(\boldsymbol{r})}\right] = \chi_S(\boldsymbol{r}) \tag{4.47}$$

.

high dimensional objects is extremely difficult, ELF as well as other electron localization or localizability indicators have never left the one-particle or local two-particle picture.

#### 4.3.1.4 Correlated Definition

Given the two main interpretations of ELF, those of Becke & Edgecombe and Savin, the inclusion of correlation is not unique.

**Wavefunction—$\chi_{BE}$**

A correlated version of $\chi_{BE}$ can be developed through the connection with the curvature of the pair density presented in Eq. 4.17 [42]. This allows to develop the ELF definition for post-HF methods in terms of the correlated pair density curvature:[9]

$$D(\mathbf{r}) = \frac{\nabla'^2 \rho_2^{\beta\beta}(\mathbf{r}, \mathbf{r}')|_{\mathbf{r}=\mathbf{r}'} + \nabla'^2 \rho_2^{\alpha\alpha}(\mathbf{r}, \mathbf{r}')|_{\mathbf{r}=\mathbf{r}'}}{2c_F \rho^{8/3}(\mathbf{r})}, \tag{4.48}$$

where $\rho_2$ is now correlated and given by

$$\nabla^2 \rho_2^{\sigma\sigma}(\mathbf{r}, \mathbf{r}')|_{\mathbf{r}=\mathbf{r}'} = \tag{4.49}$$

$$\sum_{ijkl} \rho_{ij}^{kl}\, \phi_i^*(\mathbf{r})\phi_k(\mathbf{r})[\phi_l(\mathbf{r})\nabla^2\phi_j^*(\mathbf{r}) + \phi_j^*(\mathbf{r})\nabla^2\phi_l(\mathbf{r}) + 2\nabla\phi_j^*(\mathbf{r})\nabla\phi_l(\mathbf{r})],$$

where $\phi_i$ are the molecular orbitals and $\rho_{ij}^{kl}$ are the first-order density matrix elements.

With this expression at hand, we can compare the effects of correlation on chemical bonding [43]. One of the typical expected changes upon inclusion of correlation is the weakening of covalent bonds. In general, this is indeed the case. As an example, the population of the C–O bond in the CO molecule decreases from 3.32 electrons to 2.99 electrons when we go from a HF to a CISD calculation with a 6-311+G (2df, 2p) basis set.

**DFT—$\chi_S$**

The DFT formulation includes already some correlation from the approximate exchange-correlation functional. Beyond that, Savin's definition in terms of the kinetic energy density can also be used for post-HF methods in the configuration interaction (CI) framework. It suffices to express the correlated wavefunction in terms of natural orbitals and fractional occupations:

$$t(\mathbf{r}) = \frac{1}{2}\sum_i n_i |\nabla\psi_i(\mathbf{r})|^2 - \frac{1}{4}\nabla^2\rho(\mathbf{r}), \tag{4.50}$$

[9] Note that this expression will be recalled in the Sect. 4.3.1.7 due to potential problems for the spin separation.

where $\psi_i(\boldsymbol{r})$ are the Kohn–Sham orbitals with $n_i$ occupations. We will see in Sect. 7.4 that this formulation is also adapted for experimental densities.

#### 4.3.1.5 Time-Dependent ELF

With the recent advances in ultra-short lasers, capable of measuring electronic movement during a reaction, it is essential to have a tool to analyze the chemical changes for time-dependent processes.

Just like in the case of correlation, the expression of the time-dependent ELF (TDELF) [10] can be derived from the relationship with the curvature of the Fermi hole (Eq. 4.28), $D^\alpha(\boldsymbol{r}) = C(\boldsymbol{r})/2\rho^\alpha(\boldsymbol{r})$ in the case of non-zero current densities [21]. We have that for a static current-carrying single-determinant state:[10]

$$C(\boldsymbol{r}) = 2\rho^\alpha(\boldsymbol{r})\sum^{\alpha}|\nabla\phi_i(\boldsymbol{r})|^2 - \frac{1}{2}|\nabla\rho^\alpha(\boldsymbol{r})|^2 - (j^\alpha(\boldsymbol{r}))^2 = C_{HF}(\boldsymbol{r}) - (j^\alpha(\boldsymbol{r}))^2. \quad (4.54)$$

The main difference is the additional term proportional to the current density, $j_\alpha^2$ (remember we already analyzed density currents in Chap. 3). This term naturally arises when the analysis we carried out for $\chi_{BE}$ (Eqs. 4.18–4.25) is carried out without assuming real-valued orbitals, i.e. without assigning vanishing orbital currents [21].

It is easy to see that this derivation also applies to time-dependent monodeterminantal wavefunctions built from time dependant orbitals, $\phi_i^\alpha(\boldsymbol{r}, t)$, leading to

$$D^\alpha(\boldsymbol{r}, t) = \sum_i^{\alpha}|\nabla\phi_i^\alpha(\boldsymbol{r}, t)|^2 - \frac{1}{4}\frac{(\nabla\rho^\alpha(\boldsymbol{r}, t))^2}{\rho^\alpha(\boldsymbol{r}, t)} - \frac{(j^\alpha(\boldsymbol{r}, t))^2}{\rho^\alpha(\boldsymbol{r}, t)}. \quad (4.55)$$

---

10 Using the nomenclature $\{k \leftrightarrows j\}$ meaning substituting $k$ and $j$ indexes [21]:

$$C(\boldsymbol{r}) = \sum_k\sum_j(\phi_k^*\phi_k\nabla\phi_j^*\nabla\phi_j + \{k \leftrightarrows j\}) - \sum_k\sum_j(\phi_k^*\phi_j\nabla\phi_j^*\nabla\phi_k + \{k \leftrightarrows j\}) = \quad (4.51)$$

$$2\rho(\boldsymbol{r})\sum_j|\nabla\phi_j|^2 - 2\sum_k\sum_j\phi_k^*\phi_j\nabla\phi_j^*\nabla\phi_k$$

Developing the last term in this equation:

$$\sum_k\sum_j\phi_k^*\phi_j\nabla\phi_j^*\nabla\phi_k = \left|\sum_j\phi_j^*\nabla\phi_j\right|^2 = \quad (4.52)$$

$$\left|\frac{1}{2}\sum_j(\phi_j^*\nabla\phi_j + \phi_j\nabla\phi_j^*) + \frac{1}{2}\sum_j(\phi_j^*\nabla\phi_j - \phi_j\nabla\phi_j^*)\right|^2 = \frac{1}{4}|\nabla\rho|^2 + \frac{1}{2}(j(\boldsymbol{r}))^2$$

with $j(\boldsymbol{r}) = \sum_j(\phi_j^*\nabla\phi_j - \phi_j\nabla\phi_j^*)$. So that the curvature is given by:

$$C(\boldsymbol{r}) = 2\rho(\boldsymbol{r})\sum_j|\nabla\phi_j|^2 - \frac{1}{2}|\nabla\rho(\boldsymbol{r})|^2 - (\boldsymbol{j}(\boldsymbol{r}))^2 \quad (4.53)$$

.

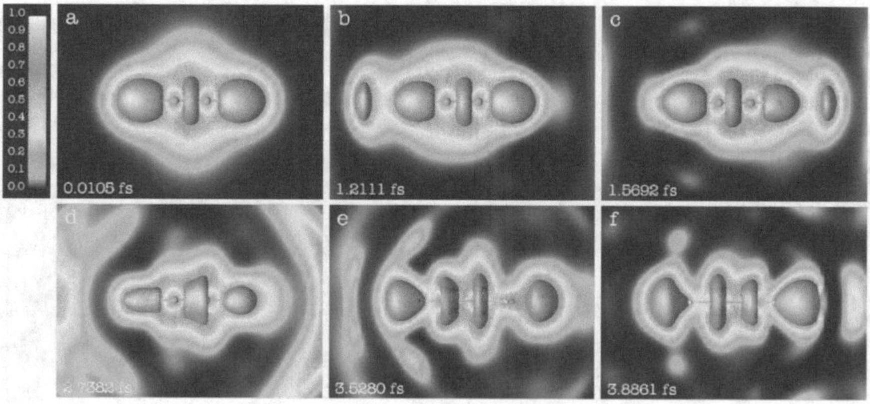


**Fig. 4.6** Snapshots of TDELF for the excitation of acetylene by a laser pulse polarized along the molecular axis. Reprinted figure with permission from [10]. Copyright 2005 by the American Physical Society

The core $\chi_{TDELF} = D^{\alpha}(\boldsymbol{r}, t)/D_h^{\alpha}(\boldsymbol{r}, t)$ can then be easily calculated constructing (see Eq. 4.11):

$$D_h^{\alpha}(\boldsymbol{r}, t) = \frac{3}{5}(6\pi^2)^{2/3}\rho^{\alpha}(\boldsymbol{r}, t)^{5/3}. \tag{4.56}$$

The analysis of the TDELF allows to visualize bonding changes along a photochemical reaction, providing also information on the time scales at which the reaction occurs. Figure 4.6 provides an example for the excitation of acetylene by a laser pulse where the transition from the bonding $\pi$ state to the antibonding $\pi^*$ state is clearly revealed [10].

#### 4.3.1.6 ELF in the Reciprocal Space

It is possible to define ELF in the reciprocal space, in order to understand the momenta distributions in solid state.

The Fourier transform of the wavefunction defined in real space, $\phi(\boldsymbol{x}_1, \ldots, \boldsymbol{x}_N)$, allows to obtain the wavefunction in reciprocal space, which thus depends on the momenta, $\boldsymbol{p}_i$:

$$\phi(\boldsymbol{p}_1, \boldsymbol{p}_2, \ldots, \boldsymbol{p}_N) = \tag{4.57}$$

$$= (2\pi)^{-3N/2} \int \phi(\boldsymbol{x}_1, \boldsymbol{x}_2, \ldots, \boldsymbol{x}_N) e^{ir_1 p_1} e^{ir_2 p_2} \ldots e^{ir_N p_N} d\boldsymbol{r}_1 d\boldsymbol{r}_2 \cdots d\boldsymbol{r}_N$$

Integrating this function gives access to the electron momentum density (EMD), which can be measured experimentally from Compton profiles. Although EMD does

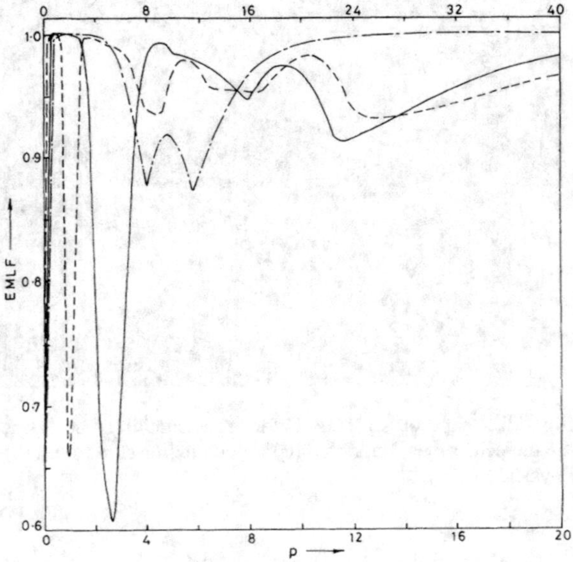


**Fig. 4.7** EMLF versus momentum for Ne (dashed-dot line), Ar (continuous line) and Kr (dashed line). The top scale is for Kr. Atomic units. Reproduced with permission from Ref. [33]

show an internal structure, it is not able, just like the electron density in real space, to reveal atomic structure. Instead, it is again necessary to resort to derivatives. It is possible to resort to the radial distributions or the Laplacian, or to ELF in the reciprocal space (EMLF for Electron Momentum Localization Function) [32, 33].

Figure 4.7 shows EMLF for several noble gases. The regions of low momentum correspond to the valence, where electrons are slower. As momentum increases, we are moving towards the inner shells of the atoms. Similar localization measures in reciprocal space (see Sect. 4.4.3) have been shown to be able to recover the Aufbau principle not only qualitatively, but also quantitatively, with shell populations close to the expected ones. As an example, 2.2 and 7.8 electrons are obtained for the K and L shells of Neon by Kohout et al. [32].

#### 4.3.1.7 Separation into Different Contributions

Whereas the electron density can be easily divided into additive terms (e.g. spin, orbital contributions), doing the same with ELF is not straightforward. Nonetheless, additive separations can be formulated with a bit of mathematical playing [49].

Starting from the partition of the electron density into additive components $\rho(\boldsymbol{r}) = \rho_a(\boldsymbol{r}) + \rho_b(\boldsymbol{r})$, ELF core is then defined in terms of additive, $a$ and $b$, and non-additive, $c$, components:[11]

[11] The non-additive terms can be easily obtained by developing for $\rho(\boldsymbol{r}) = \rho_a(\boldsymbol{r}) + \rho_b(\boldsymbol{r})$

$$\chi(\boldsymbol{r}) = \frac{t_{P,a}(\boldsymbol{r}) + t_{P,b}(\boldsymbol{r}) + t_{w,c}(\boldsymbol{r})}{t_{h,a}(\boldsymbol{r}) + t_{h,b}(\boldsymbol{r}) - t_{h,c}(\boldsymbol{r})} \geq \frac{t_{P,a}(\boldsymbol{r}) + t_{P,b}(\boldsymbol{r})}{t_{h,a}(\boldsymbol{r}) + t_{h,b}(\boldsymbol{r})} = \chi_{add}(\boldsymbol{r}) \tag{4.62}$$

where the terms $t_{w,c}$ and $t_{h,c}$ are the non-additive Weizsäcker and HEG, respectively.

**By Spin**

Becke and Edgecombe [5] developed their formulation for spin densities. The sum of both spin components leads to

$$\chi_S^{(1)} = \frac{t_P^\alpha + t_P^\beta}{t_h^\alpha + t_h^\beta} = \chi_{add} \tag{4.63}$$

which is equivalent to the additive part in Eq. 4.62 if we make $\rho_a = \rho_\alpha$ and $\rho_b = \rho_\beta$. Note that the additive part is more localized than the total ELF. Thus, the total ELF is lower than the additive ELF.

In the case of closed-shell ($x$ =1/2, $\rho_a(\boldsymbol{r}) = \rho_b(\boldsymbol{r}) = \rho(\boldsymbol{r})/2$), the relation $\chi_S = \chi_{add}$ remains exactly verified since the non-additive terms become exactly zero over the space. However, the definition of a total ELF for open-shell cases is not unique. Within the DFT formulation, the scaling with the homogeneous gas constant is given as for Becke's approach by Eq. 4.38:

$$\chi_S^{(1)} = \frac{t_P^\alpha + t_P^\beta}{t_h^\alpha + t_h^\beta} = \frac{t_P^\alpha + t_P^\beta}{2^{2/3} c_F [(\rho^\alpha)^{5/3} + (\rho^\beta)^{5/3}]} \tag{4.64}$$

where

$$t_P^\alpha(\boldsymbol{r}) = t^\alpha(r) - \frac{1}{8}\frac{|\nabla\rho^\alpha(\boldsymbol{r})|^2}{\rho^\alpha(\boldsymbol{r})} \tag{4.65}$$

$$t_h^\alpha(\boldsymbol{r}) = \frac{3}{5}(6\pi^2)^{2/3}(\rho^\alpha)^{5/3}. \tag{4.66}$$

Instead, if we start from the Fermi hole curvature formulation used for the correlated cases (Eq. 4.48), the splitting leads to

---

$$t_w(\boldsymbol{r}) = \frac{1}{8}\frac{|\nabla(\rho_a(\boldsymbol{r}) + \rho_b(\boldsymbol{r})|^2}{\rho_a + \rho_b} = \frac{x}{8}\frac{|\nabla\rho_a(\boldsymbol{r})|^2}{\rho_a(\boldsymbol{r})} + \frac{1-x}{8}\frac{|\nabla\rho_b(\boldsymbol{r})|^2}{\rho_b(\boldsymbol{r})} + \frac{2}{8}\frac{\nabla\rho_a(\boldsymbol{r})\nabla\rho_b(\boldsymbol{r})}{\rho_a + \rho_b} \tag{4.58}$$

$$t_h(\boldsymbol{r}) = c_F\rho(\boldsymbol{r}) = c_F(x^{5/3}\rho^{5/3}(\boldsymbol{r}) + (1 - x^{5/3})\rho^{5/3}(\boldsymbol{r}) = c_F(\rho_a^{5/3}(\boldsymbol{r}) + g(x)\rho_b^{5/3}(\boldsymbol{r})) \tag{4.59}$$

where $x = \rho_a/\rho$, $1 - x = \rho_b/\rho$ and $g(x) = (1 - x^{5/3})/(1 - x)^{5/3}$. Hence, the non-additive terms are given by

$$t_{w,c} = \frac{2}{8}\frac{\nabla\rho_a(\boldsymbol{r})\nabla\rho_b(\boldsymbol{r})}{\rho_a + \rho_b} \tag{4.60}$$

$$t_{h,c} = c_F(2^{2/3} - 1)\rho_a^{5/3}(\boldsymbol{r}) + c_F(2^{2/3} - g(x))\rho_b^{5/3}(\boldsymbol{r}) \tag{4.61}$$

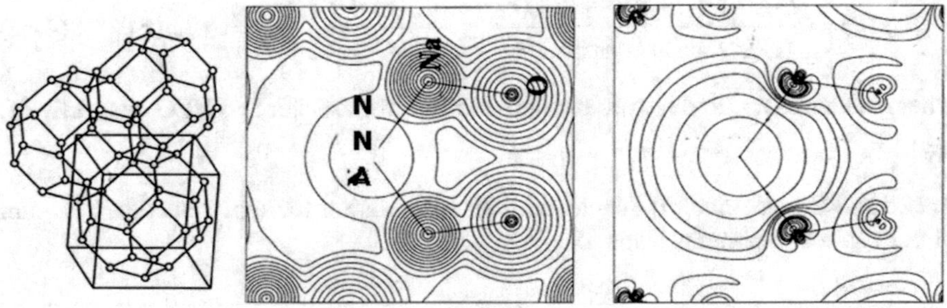


**Fig. 4.8** Spin density distribution of electrosodalite in the plane (1, 1, 0). Contour lines at 2, 4 and $8\times10^n$ a.u., $n$ from 0 to –4. Adapted with permission from Ref. [41]

$$\chi_S^{(2)} = \frac{2(\rho_\alpha t_P^\alpha + \rho^\beta t_P^\beta)}{c_F(\rho^\alpha + \rho^\beta)^{8/3}} \tag{4.67}$$

Both $\chi^{(1)}$ and $\chi^{(2)}$ converge to the spin-independent ELF kernel in Eq. 4.43 in a closed-shell case, but only Eq. 4.63 respects the spin scaling in the HEG.[12] In an open shell system the measurement of localization is different for both approaches, which becomes apparent locally in one electron regions:

$$\chi^{(1)}[\rho_\beta \to 0] = \frac{t_P}{2^{2/3}c_F(\rho^\alpha)^{5/3}} \tag{4.68}$$

$$\chi^{(2)}[\rho^\beta \to 0] = \frac{t_P}{c_F(\rho^\alpha)^{5/3}}. \tag{4.69}$$

As evidenced by Eqs. 4.68 and 4.69, $\chi^{(1)}$ will be noticeably smaller than $\chi^{(2)}$ in single-electron regions, arising purely from the different references.

Thus, in highly spin polarized systems, it is highly advised to use separately $ELF_\alpha$ and $ELF_\beta$ rather than the total ambiguous-definite total ELF [30]. Moreover, this allows to retrieve interesting spin information. As an example, the electrosodalite crystal, $Na_8(AlSiO_4)_6$, has a missing electron associated with its F center [41]. The representation of $ELF^\alpha$ shows a maximum at the position of that center, while that of $ELF^\beta$ shows a minimum (Fig. 4.8).

### By Symmetry: $\sigma$ and $\pi$

Since ELF depends formally only on the density it has no information about $\pi$ and $\sigma$ bonds. It is possible to carry out a partition similar to the one done for the spin ($\chi^{(1)}$, Eq. 4.63) and construct separate ELF functions for $\sigma$ and $\pi$ orbitals [54]. The following components are used:

---

[12] Note that formulation $\chi^{(1)}$ is implemented in DGRID whereas formulation $\chi^{(2)}$ is implemented in TopMod.

$$\rho^{\pi}(\boldsymbol{r}) = \sum_{i \in \pi} |\phi_i(\boldsymbol{r})|^2 \tag{4.70}$$

$$t_p^{\pi} = \sum_{i \in \pi} |\nabla \phi_i(\boldsymbol{r})|^2 - \frac{1}{8} \frac{|\nabla \rho^{\pi}(\boldsymbol{r})|^2}{\rho^{\pi}(\boldsymbol{r})} \tag{4.71}$$

so that we have

$$\chi_S^{(1)} = \frac{t_P^{\sigma} + t_P^{\pi}}{t_h^{\sigma} + t_h^{\pi}} = \frac{t_P^{\sigma} + t_P^{\pi}}{2^{2/3} c_F [(\rho^{\sigma})^{5/3} + (\rho^{\pi})^{5/3}]}. \tag{4.72}$$

Note nevertheless that once again, $\chi \neq \chi^{\sigma} + \chi^{\pi}$ even if $\rho = \rho^{\sigma} + \rho^{\pi}$.

Due to the term $c_F \rho^{5/3}$ in the denominator of $\chi$, it should be noted that the decomposition of the ELF into contributions as in Eq. 4.72 is only reasonable if the orbitals considered do not interfere much in the same regions of space. Unlike the electron density and the kinetic energy density, $c_F \rho^{5/3}$ cannot be separated into orbital contributions [31]. Since the $\sigma$ and $\pi$ contributions are considerably large in common regions of space, this approach should be used with caution [31].

To avoid this problem, a second formulation is possible. The problematic term of the denominator can be divided into two density-dependent contributions, of which only one is decomposed as a function of the orbitals:

$$\chi_S^{(3)} = \frac{t_P^{\sigma} + t_P^{\pi}}{c_F \rho^{2/3} [\rho^{\sigma} + \rho^{\pi}]} \tag{4.73}$$

thus obtaining a splitting of $\chi$ into additive orbital contributions.

The partition into components can again shed light into subtle bonding effects such as aromaticity. An example of the use of ELF$^{\pi}$ is presented in Sect. 6.3.2.1 for the classification of aromatic compounds. It is also interesting in those cases where the understanding of the system properties escapes the basic VSEPR or crystal-field theories. Transition metal complexes are a good example. The study of their topologies allows to verify that their geometries, which escape common chemical rules, correspond to interactions between ligands and a structured external core [22].

### 4.3.2 Prototype Bond Types

ELF isosurfaces reveal in a very intuitive manner the type of bonding present in a system. It suffices to analyze the general shape of the surfaces. Let's see some examples.

ELF is naturally conceived to identify covalent interactions. In those cases, a basin appears in between the covalently bonded atoms. This is shown for several $CH_3X$ (X = $CH_3$, $NH_2$, OH, F) compounds in Fig. 4.9. We can see in the picture how polarity is also revealed. Homopolar covalent bonds are centrosymmetric, whereas

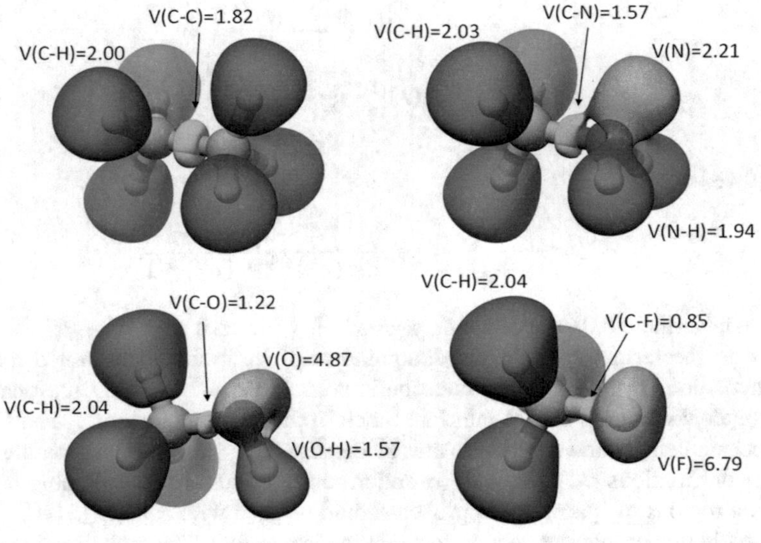


**Fig. 4.9** $CH_3$-X ELF basins and populations at the B3LYP/6-31G* level. V(C-H) basins are shown in blue, V(C-C) and V(C-X) basins are shown in green, and V(X) basins are shown in orange. Isovalue = 0.8

they become smaller and more deformed towards the more electronegative atom as the bond becomes more polar. Do not worry for the notation at this stage, we will see it in detail in the following section.

Starting from this covalent picture, let's visually analyze other bonding types. Stretched covalent bonds usually lead to two different ELF basins instead of one (see Sect. 6.2 on reactivity); and this can also be observed in weak covalent bonds (recently renamed as protocovalent [40] or charge-shift bonds [62]). The typical case of $F_2$ is shown in Fig. 4.10b.

When polarity of the bond increases so much that it becomes ionic (e.g. Li in LiF in Fig. 4.10c), the valence basin completely migrates to one atom, so that no ELF basin lies in between the atoms and at least the hardest atom looks spherical-like. This effect is further enhanced in solid state (see LiF in Fig. 4.10d). We will see in the coming sections that ionic compounds are also easy to identify by looking at the charges (Sect. 4.3.3.2) and the fluctuations (Sect. 4.3.3.3).

Finally, let's have a look at simple models of delocalized bonds. Figure 4.10e shows the $Li_2$ molecule. $Li_2$ is remarkable due to the huge volume of the bond basin (compare for example with the ethane!). Also, the isosurface is very flat along the bond. This provides a good idea of how metallic periodic systems look like. In a 3D metal, we expect a flat ELF [66]. This is shown in Fig. 4.10f, where the valence of Fe extends throughout the crystal. More details will be given in the chapter on solid state about the periodic systems (Sect. 7.2.2.1).

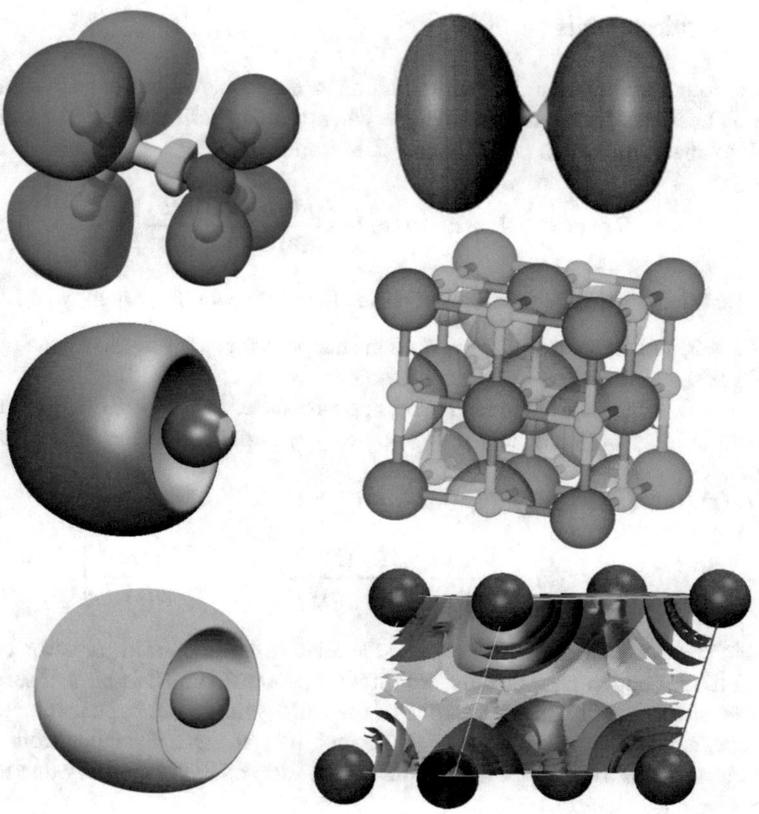

**Fig. 4.10** ELF for **a** $CH_3BH_3$; **b** $F_2$; **c** LiH (molecular); **d** LiF (solid); **e** $Li_2$; **f** Fe (solid)

### *4.3.3 Analyzing ELF Results*

As we already stated, the topological method is not restricted to the $\rho$ scalar. Any derivable field defined over $\mathbb{R}^3$ can be subjected to such an analysis. The usefulness of such a partitioning depends on the physical meaning that we assign to the scalar, on the existence of theoretical results warranting a meaning for such an induced topology or, more frequently, on the empirical results that are obtained after such a procedure.

We have seen that ELF picture provides an intuitive characterization of bond types. However, much richer interpretations can be done if we use extra tools from the topological analysis framework. In this section, we will review these tools and how to apply them correctly.

#### 4.3.3.1 Critical Points

ELF provides an intuitive image with maxima located where electrons are localized. Hence, it becomes interesting to analyze when the ELF critical points occur. It is enough to check for which points $\nabla\eta = 0$. Differentiating $\eta$:

$$\nabla\eta(\mathbf{r}) = -2(\eta(\mathbf{r}))^2\chi(\mathbf{r})\left(\frac{\nabla t_P(\mathbf{r})}{t_P(\mathbf{r})} - \frac{5\nabla\rho(\mathbf{r})}{3\rho(\mathbf{r})}\right) \tag{4.74}$$

we see that there are three possible scenarios that delineate the topology of ELF:

1. ELF = 0, and therefore $\chi(\mathbf{r}) \to \infty$. This happens for $t_P(\mathbf{r}) \to \infty$ (at the minima of $\chi(\mathbf{r})$) and for $t_h(\mathbf{r}) \to 0$ (at $\mathbf{r} \to \infty$).
2. $\chi(\mathbf{r}) = 0$, which corresponds to the opposite case, in which ELF = 1 due to a bosonic ($t(\mathbf{r}) = t_w(\mathbf{r})$) and/or high density situation, as is the case of atomic nuclei.
3. $\nabla t_P(\mathbf{r})/t_P(\mathbf{r}) = 5\nabla\rho(\mathbf{r})/(3\rho(\mathbf{r}))$, with

$$\nabla t_P(\mathbf{r}) = \nabla t(\mathbf{r}) - \frac{\nabla\rho(\mathbf{r})}{8}\left[\frac{2\nabla\rho(\mathbf{r})\cdot\nabla\rho(\mathbf{r})}{\rho(\mathbf{r})} - \frac{(\nabla\rho(\mathbf{r}))^2}{\rho(\mathbf{r})}\right], \tag{4.75}$$

   which reflects those situations where the relative gradient of the electron density and the kinetic energy density are directly proportional. It is also observed in those situations in which electrons of the same spin do not appear in a zone of space, as can be the case of the $H_2$ molecule or at great distances from atoms of elements of the *s* block (alkali show saturation of ELF at infinity due to their lone valence electron).

The three cases listed highlight the relationship between the ELF and Laplacian critical points, and explain the homomorphism found in most cases between the two topologies [3].

Overall, most of the chemically relevant critical points come from case (2), where electrons behave like bosons (this was already seen in Fig. 4.5). These maxima can be separated from the chemical point of view in two well-differentiated groups: core and valence. The first order critical points in between those basins will also help us to develop the chemical image of the system. We will review how to use this information in the next sections.

**ELF Maxima and VSEPR**

The concept of atomic shell and the *core*-valence separation are at the very basis of chemistry: atomic shells are built up from orbitals with the same principal quantum number and filled with electrons following the Aufbau principle, so that their charge densities are interpenetrated. The goal of revealing the atomic structure of atoms from some local property has been developed over many years with varying results. The density *per se* as a monotonically decreasing exponential function does not provide information about this. However, derived properties such as the radial

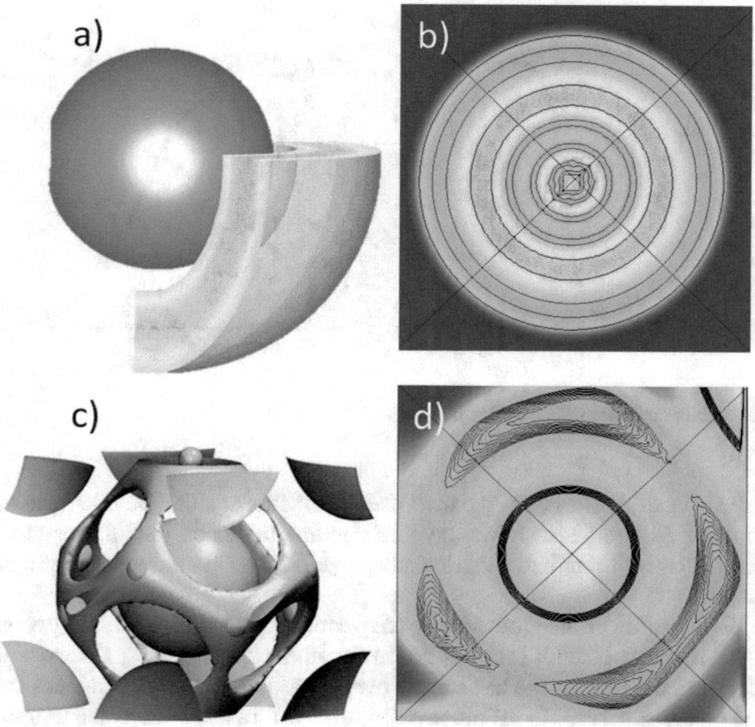


**Fig. 4.11** **a** 3D cut of isolated Xe core, **b** 2D contour map of isolated Xe core, **c** 3D isosurface of Xe core in $XeF_8^{2-}$, **d** contour map of Xe core in $XeF_8^{2-}$

distribution $4\pi^2\rho$, the Laplacian $-\nabla^2\rho$, the logarithm of electron density or the function $-|\nabla^2\rho|/\rho$ are able to at least partially resolve the atomic structure. ELF works better than many other methods for detecting shells, for example, the Laplacian of the density (see Sect. 4.4.1) does not yield a shell structure in some heavier atoms, while ELF does.

ELF is able to recover the number of shells expected from the Aufbau principle. Core maxima are found around nuclei with $Z > 2$ (building a structure analogous to that expected for the inner shells of atoms). This is shown in Fig. 4.11. A maximum of ELF always appears at the nucleus position. For isolated atoms, the following shells are spherically degenerated. That is, each of the radial maximum sees no alteration in the angular directions $\theta$, $\phi$ if we use an spherical coordinate system. This means that there always exist two perpendicular directions to the radial vector with Lyapunov index $\lambda = 0$, so that the CPs are of the $(1, +1)$, $(1, -1)$ or $(0, 0)$ kind, the latter in the case of a radial inflection point. The formation of a molecule breaks the SO(3) symmetry of the atoms. The basins of the atomic shells are broken down into several smaller basins connected by the corresponding separatrix. However, upon analyzing

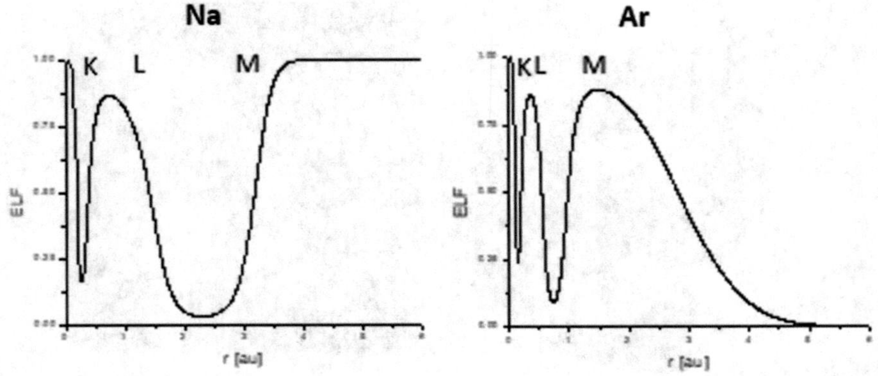


**Fig. 4.12** Shell structure of **a** Sodium **b** Argon

the ELF surfaces, it is easy to see they usually still form a unique chemical unit: an atomic shell. Figure 4.11c shows the core maxima in $XeF_8^{2-}$ (small green spheres), whereas the overall retained spherical shape (and its deformation) is highlighted in Fig. 4.11d.

The ELF profile of the atoms along the periodic table is divided into two groups, according to their behavior for $r \rightarrow \infty$. In the atoms having only $s$ electrons in their valence shell, the ELF of the last shell grows towards 1.0 at large distances from the nucleus (see Fig. 4.12a). In all other cases, the ELF reaches the valence maximum and then declines to zero at distances away from the nucleus (Fig. 4.12b). The first result is due to the fact that $D_\sigma$ fades asymptotically in finite systems, as it does in perfectly localized systems, so that the ELF tends to unity at distances far from the nucleus. The opposite behavior is determined by a faster decreasing denominator $D_\sigma^0 \propto \rho^{5/3}$ for $l > 0$.

There is no absolute scale for ELF. For example, not all maximal values of ELF are close to 1 for the atomic shells. Only pure s-shells show values close to 1. Shells where s and p orbitals coexist, yield values closer to 0.8, while those where s, p and d orbitals are present yield values closer to 0.6 [55]. This also implies that the sequence of maxima has decreasing ELF values. Furthermore, inner shells penetrate the valence shell and lower the value of ELF. Note that this can have unwanted effects if pseudopotentials are used.

Once the core structure is understood, we can have a look at the valence. Valence maxima fill the rest of the space. These regions are largely assimilated to Lewis entities, typically bonds and lone pairs. Figure 4.13 shows an example of each. We can see that two maxima are found for the lone pairs of water, as well as one maximum for the N-N bond in the $N_2$ molecule (as well as the respective nitrogen lone pairs).

The existence of degenerate critical points is still possible in the case of linear molecules ($D_{\infty h}$, $C_{\infty v}$), where due to symmetry it is possible to find ring attractors centered on the internuclear axis. This is illustrated in Fig. 4.13c for HF. This situation

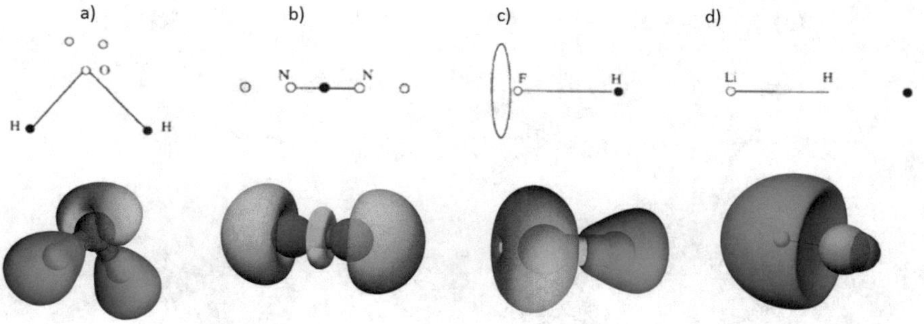


**Fig. 4.13** ELF critical points (top) and localization domains (bottom) for **a** $H_2O$, **b** $N_2$, **c** HF, **d** LiH. Core attractors are represented by full circles, non-bonding ones by empty ones. Degenerate attractors are represented by lines

is highly unstable, so that small perturbations (e.g. an approaching molecule) give rise to a symmetry breaking and these degenerate points are resolved in at least one punctual maximum.

It can be seen from Fig. 4.13 that hydrogen constitutes an outlier within the ELF framework. Since H has only one electron, Pauli repulsion is very small (actually it is non-existent in the atom) and it shows up as a full valence without core (i.e. its associated volumes are very large like in the $H_2O$ molecule). This also leads to anomalies with respect to the position of the ELF maximum. The position of the hydrogen atom can change according to the bond in which it takes part. In the case of electron-sharing interactions involving hydrogen, the bonding attractors are very close to the protons. However, in cases where the hydrogen is practically ionized, its location is not very precise because it extends over a large region of practically constant ELF value, and tends to unity (see LiH in Fig. 4.13d).

Bond multiplicity can also be studied graphically in terms of the number and distribution of the bond attractors. Figure 4.14 represents the ELF maxima of ethane, ethylene and acetylene. It can be seen that ethane has a single bond attractor, corresponding to the single C–C bond, that the double C=C bond of ethylene is reflected by two bond attractors symmetric with respect to the molecular plane and that in acetylene, the revolution symmetry is responsible for the presence of a circular disynaptic attractor. Recall at this point Linnett's representation of acetylene (Fig. 4.1a). It is similar to the acetylene localization image provided by ELF, a single toroidal bond basin containing the bonding electrons (not the single plus two double bonds provided by MOs).

Within the ELF framework, there is a widespread notation in order to characterize the different types of maxima. We use $C(A)$ to identify the core basin of an atom $A$, $V(A)$ to name its non-shared valence basin(s) (i.e. the lone pairs) and $V(A, B)$ for the shared valence basins between the $A$ and $B$ atoms (i.e. a bond in between atoms $A$ and $B$). This notation is used in Fig. 4.9. For those cases where several attractors appear that correspond to the same atom(s), an additional counter can be

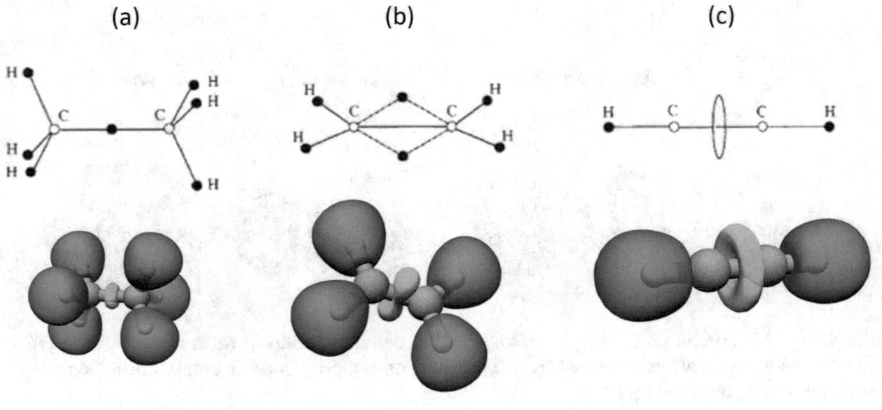


**Fig. 4.14** ELF critical points (top) and isosurfaces (bottom) for **a** $C_2H_6$, **b** $C_2H_4$, **c** $C_2H_2$. Core attractors are represented by full circles, non-bonding ones by empty ones. Degenerated attractors are represented by lines

**Table 4.1** ELF(**r**) basin classification according to synapticity

| Synaptic order | Name | Symbol | Example |
|---|---|---|---|
| 0 | Asynaptic | | F-centers |
| 1 | Monosynaptic | V(A) | Lone pairs |
| 2 | Disynaptic | V(A, B) | Covalent bonds |
| 3 | Trisynaptic | V(A, B, C) | 3-center-2-electron bonds |
| >3 | Polysynaptic | V(A, B, C, …) | Polycenter bonding |

added: $V_i(A)$. For example, in the water molecule, the two lone pairs will be noted as $V_1(O)$ and $V_2(O)$.

The number of atoms to which a given valence belongs is known as the synaptic order [63]. Hydrogen basins are counted formally as cores when we assign synapticities. Similarly, if the core of a given atom (when it has several shells) shows several maxima, they are only counted as one (like the Xe case above). The possible cases are summarized in Table 4.1. Monosynaptic valence basins are attributed to atomic shells when they surround the core as we have seen for Xe, and to lone electron pairs otherwise. Disynaptic basins are attributed to common chemical bonds. However, other less common synapticities are also found which enrich chemical insight. For example, asynaptic (synaptic order zero) basins are associated with F centers in solid state and polysynaptic basins (synapticity greater than 2) are associated with multicenter bonds. This allows a clear division between the conventional two-center-two-electron bonds (2c-2e) and more complex bonding schemes such as 3c-2e bonds (see Sect. 6.1.2.1). It is common use to color basins according to their synapticity for a more visual characterization (e.g. hydrogen basins in grey, bonds in green, cores in magenta and lone pairs in orange).

The analysis of the distribution of the ELF valence attractors provides a quantum-mechanical confirmation of Gillespie's Valence Shell Electron Pair Repulsion Model (VSEPR) [25]. The relationship between the representation of systems without lone pairs in the central atom ($ab_x$, $x = 2$–$6$) found in all books of basic structural chemistry and the image provided by ELF is overwhelming (see Fig. 4.15). Furthermore, the ELF analysis is able to provide even more detailed descriptions of the chemical structure, such as the position of the lone pairs in systems of type $ab_3e_2$ or the cylindrical shape of the attractors in molecules with three lone pairs ($ab_2e_3$).

### Saddle Points: $f$-localization Domains and Bifurcation Trees

We have already seen several examples where several electron pairs appear as a unique Lewis entity. For instance, we have seen the external shell of Xe with 8 electrons. We have also seen that the valence of metals can be seen as a full surface flat ELF surface that includes several maxima, mimicking the homogeneous electron gas model. This highlights the fact that it is not only interesting to localize the attractors as we have done in the previous section, but also to analyze the relationship between them, sometimes to consider them an ensemble.

A correlation can be established between the ELF values in the limits 0 and 1 and the Heisenberg and Pauli principles [52, 53]. The points at which the ELF approaches one correspond to an infinite error in the assignment of momentum and therefore a perfect localization of electrons, whereas in cases where it becomes zero, the momentum can be determined with great precision but the electronic position is affected by a large uncertainty.

Hence, looking at the low values of ELF can also reveal a great deal of information. Based on the orbital interpretations of ELF seen in Sect. 4.3.5.1 [9, 46, 55], the value of ELF at the the first order saddle points can be directly related to electron delocalization [17]: the greater the ELF value at the saddle point in between two basins, the greater the delocalization among them. This is why we will call them from now on *bips*, for bond interaction points. Core-valence *bips* typically show values close to 0.05. The valence-valence basins can show rather high values: V(O)-V(O) separate at nearly 0.9 (Fig. 4.18). We will see in the next section that variance can be calculated in a more accurate manner. However, it requires the integration of pair density quantities, which are not easily accessible in solid-state calculations. Hence, *bip* values can become an interesting source of information in the analysis of bonding in the periodic state (see some examples in Sect. 7).

In those cases where we do not have access to saddle point values, *bips* can be obtained visually thanks to the concept of localization domains. A localization domain is the volume enclosed within an isosurface $\eta(\boldsymbol{r}) = f$ in which $\eta(\boldsymbol{r}) > f$ (see Fig. 4.16). This volume may contain one or more attractors. In the latter case, the domain is called reducible since it is composed of several basins that will be revealed as the ELF value of the limiting isosurface increases. For low values of $f$, there is just one domain that encapsulates the whole system. As this value increases, there is usually a first separation between the core and the valence domains. For even higher ELF values, we see the separation of the global valence domain into sub-domains located around the cores of the most electronegative atoms. Finally the irreducible

**Fig. 4.15** ELF isosurfaces for molecules with structures typically predicted by VSEPR

| VSEPR reference | Molecule | ELF |
|---|---|---|
| $ab_2$ | $BeCl_2$ |  |
| $ab_2e_3$ | $XeCl_2$ | 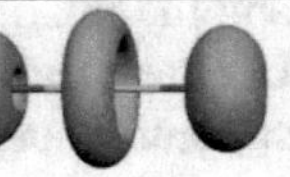 |
| $ab_3$ | $BCl_3$ | 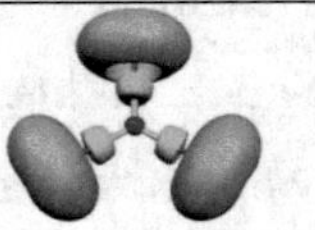 |
| $ab_3e$ | $NH_3$ | 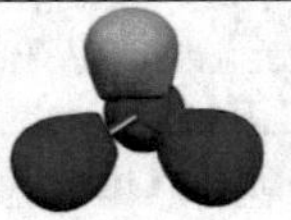 |
| $ab_3e_2$ | $ClF_3$ |  |
| $ab_4$ | $CH_4$ |  |
| $ab_4e$ | $SF_4$ | 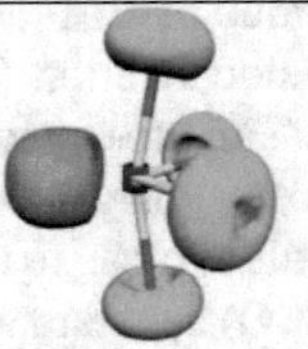 |
| $ab_5$ | $PCl_5$ | 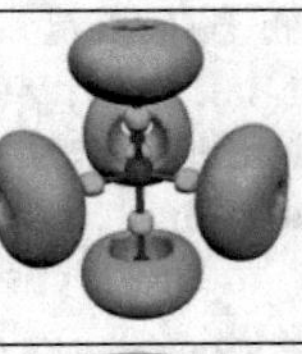 |
| $ab_6$ | $SCl_6$ | 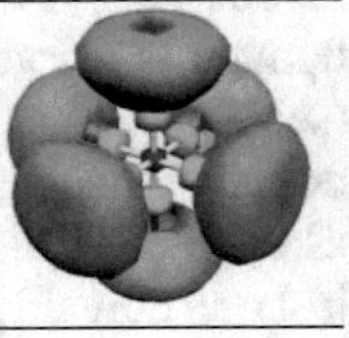 |

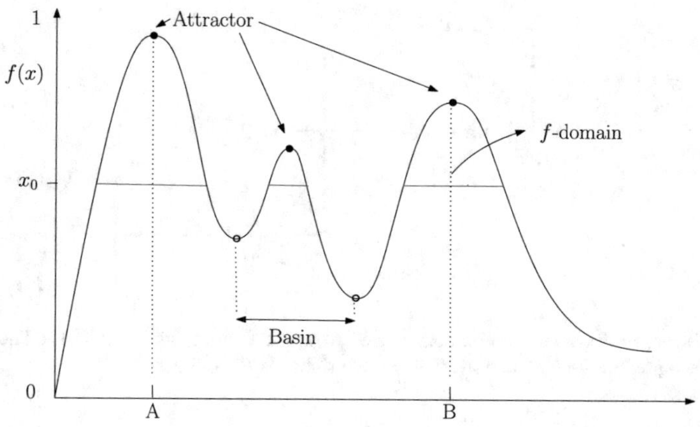


**Fig. 4.16** Localization domains of the function $f(x)$ for $\eta = x_0$, value at which the three system basins are already revealed

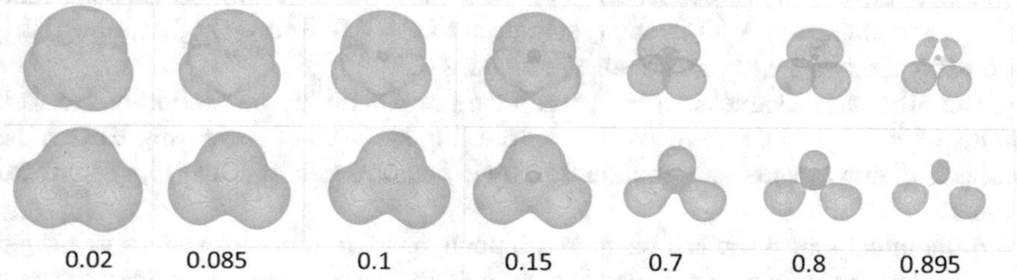


**Fig. 4.17** Localization domains for water at $f = 0.02, 0.085, 0.1, 0.15, 0.7, 0.8, 0.895$ illustrating the reduction of domains. Two different views are shown in the top and bottom rows

valence domains corresponding to the mono-, di- and polysynaptic attractors are separated so that the full basin structure emerges. This is exemplified for water in Fig. 4.17. In a complex, such as when we deal with systems with weak hydrogen bridges, the first reduction gives rise to two compound domains that are associated with the interacting fragments. Similarly, the first separation in an ionic compound provides the reducible domains of each of the ions.

The monitoring of this process is carried out graphically through a tree structure. These are the bifurcation diagrams (known in topology as Reeb graphs). The bifurcation diagram is a tree-shaped graph showing the values of the attractors and the highest values of the saddle points of each basin, from which the irreducible domains are obtained.

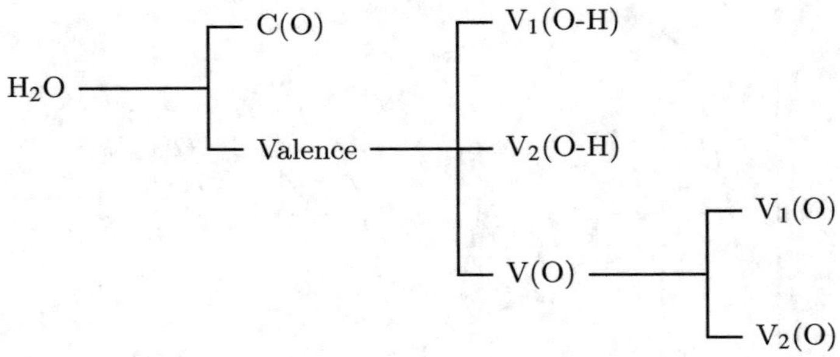


**Fig. 4.18** Bifurcation tree for water, constructed from the $f$-domains in Fig. 4.17. The following values are assigned to the bifurcations from left to right: 0.085, 0.8 and 0.895

An example of bifurcation diagram is given in Fig. 4.18 for $H_2o$. It has been constructed thanks to the $f$-domain analysis in Fig. 4.17. ELF domains are first separated into core and valence. The valence domain includes the V(O-H) bonds and the oxygen lone pairs, V(O), which separate at ELF = 0.8. At even higher value (ELF = 0.895), the two lone pairs are also reduced.

The bifurcation values can be determined graphically by inspection of the fork points of the reducible domains (i.e. by plotting surfaces of increasing ELF value and noting down when they separate) or by exhaustive listing of the (3, –1) critical points.

A chemical unit is the last basin of a branch, as long as it corresponds to a filled volume. However, as we have seen for example for the Xe atoms, it is often the case that irreducible domains that correspond to basins of the same type are not clearly separated at the end of a branch because the value of the function at the characteristic point and at the attractor are extremely similar. In these cases, the division will be more chemically significant if they are all considered together, all the more so as the number of these basins is usually quite dependent on the quality of the base. This is how the concept of the super-basin (aka. *cwm*) is born: a group of basins at the end of the branch are taken as a whole. The values of the ELF at the branching, along with chemical knowledge or synapticity can be used to assign the chemical nature of the domains in non-intuitive cases.

#### 4.3.3.2 Basin Integration

Just like with the density, the topology induced by the ELF gradient allows the space to be divided into basins, each of which is made up of the set of points that lead to the same maximum of the function. Among the most interesting properties to analyze by integration are the population, the population fluctuation and higher moments, which we will introduce in the coming sections.

#### 4.3.3.3 Populations

The atomic charge is probably one of the most widely used concepts in chemistry as it is a basic pillar in defining the type of bonding of a system and its reactivity. The ELF function not only qualitatively predicts the atomic structure, the number of shells coinciding with those predicted by the Aufbau principle, but the integration of the electronic density within these layer basins also leads to a population in the layer very close to that expected according to this same principle (Table 4.2).

As expected from chemical knowledge, cores are very little affected when molecules are formed. Hence, the core populations remain very close to the ones in Table 4.2. ELF populations in the valence follow a series of rules, directly related to VSEPR theory:

- The population of a disynaptic basin is generally less than two electrons. Higher values can be found in cases where multiple (or partially multiple) bonds are described by a single basin. This is illustrated for the ethane and ethylene, where the bond charges are respectively 1.81 and 3.42 (see Exercise 9.4.2.1).
- The population of monosynaptic basins is generally larger than that of disynaptic basins, exceeding two electrons. For example, the oxygene lone pairs in methanol hold ca. 2.3 electrons (see Exercise 9.4.2.2).
- The population of a protonated disynaptic basins generally ranges between 1.3 and 2.5 electrons.

These rules result from the competition between Pauli's repulsion and the electron-core attraction potential. The latter tends to concentrate the electron density around the nucleus and to give the atoms its radial structure. On the other hand, the minimization of the Pauli repulsion favors the partitioning into basins, either large enough so that the electrons can avoid each other, or into basins whose population does not exceed two electrons. Bond populations also reflect polarization, this was already seen in the $CH_3X$ series in Fig. 4.9, where the series $X$ = C, N, O, F leads bonding populations of 1.82, 1.57, 1.22 and 0.85, respectively.

Unlike other population analyses, topological partitioning provides populations that are not generally dependent on the basis set or method [47]. It can be seen in Table 4.3 that small variations ($\simeq$ 0.2 electrons) result from the addition of polarization functions. Consistent with chemical intuition, the addition of polarization functions increases the populations of disynaptic basins (bonds) to the detriment of monosynaptics (lone pairs), increasing the dissociation energies and elongation force constants. As far as the method is concerned, the inclusion of the correlation gives rise to the opposite trend (described in Sect. 4.3.1.4). As correlation is included, the ELF basin population decreases, along with the corresponding force constant. Moreover, as expected from chemical knowledge, the relative population of these basins reveals some correlation with the bond order [12].

**Fluctuation**

We can resort to the fluctuation calculations introduced in Sect. 3.5.3 to calculate the delocalization in ELF basins, it suffices to use ELF basins instead of QTAIM ones.

**Table 4.2** Radii and shell populations determined from Clementi and Roetti basis-sets using the ELF topology. $q^K$ is the K-shell charge in electrons, $r^K$ its radius in a.u., and $Q_0^{10}$ the total charge in electrons within a sphere of radius equal to 10 bohr. Data taken from Ref. [30]

| Átomo | $q^K$ | $r^K$ | $q^L$ | $r^L$ | $q^M$ | $r^M$ | $q^O$ | $Q^{10}$ |
|---|---|---|---|---|---|---|---|---|
| Li($^2$S) | 2.0 | 1.62 | 1.0 | – | – | – | – | 3.0 |
| Be($^1$S) | 2.0 | 1.02 | 2.0 | – | – | – | – | 4.0 |
| B($^2$P) | 2.0 | 0.75 | 3.0 | – | – | – | – | 5.0 |
| C($^3$P) | 2.1 | 0.58 | 3.9 | – | – | – | – | 6.0 |
| N($^4$S) | 2.1 | 0.48 | 4.9 | – | – | – | – | 7.0 |
| O($^3$P) | 2.1 | 0.40 | 5.9 | – | – | – | – | 8.0 |
| F($^2$P) | 2.1 | 0.34 | 6.9 | – | – | – | – | 9.0 |
| Ne($^1$S) | 2.2 | 0.30 | 7.8 | – | – | – | – | 10.0 |
| Na($^2$S) | 2.2 | 0.2641 | 7.9 | 2.263 | 1.0 | – | – | 11.0 |
| Mg($^1$S) | 2.2 | 0.2363 | 7.9 | 1.686 | 2.0 | – | – | 12.0 |
| Al($^2$P) | 2.2 | 0.2138 | 7.9 | 1.397 | 2.9 | – | – | 13.0 |
| Si($^3$P) | 2.2 | 0.1951 | 7.9 | 1.189 | 3.9 | – | – | 14.0 |
| P($^4$S) | 2.2 | 0.1793 | 7.9 | 1.034 | 4.9 | – | – | 15.0 |
| S($^3$P) | 2.2 | 0.1683 | 7.9 | 0.914 | 5.9 | – | – | 16.0 |
| Cl($^2$P) | 2.2 | 0.1541 | 7.9 | 0.817 | 6.9 | – | – | 17.0 |
| Ar($^1$S) | 2.2 | 0.1439 | 7.9 | 0.739 | 7.9 | – | – | 18.0 |
| K($^2$S) | 2.2 | 0.1350 | 7.9 | 0.673 | 8.0 | 3.27 | 0.9 | 19.0 |
| Ca($^1$S) | 2.2 | 0.1270 | 7.9 | 0.618 | 8.0 | 2.55 | 1.9 | 20.0 |
| Sc($^2$D) | 2.2 | 0.1200 | 7.9 | 0.573 | 8.8 | 2.39 | 2.1 | 21.0 |
| Ti($^3$F) | 2.2 | 0.1138 | 7.9 | 0.535 | 9.7 | 2.29 | 2.1 | 22.0 |
| V($^4$F) | 2.2 | 0.1082 | 8.0 | 0.501 | 10.7 | 2.20 | 2.2 | 23.0 |
| Cr($^7$S) | 2.2 | 0.1032 | 8.0 | 0.472 | 12.5 | 2.47 | 1.3 | 24.0 |
| Mn($^6$S) | 2.2 | 0.0986 | 8.1 | 0.445 | 12.5 | 2.07 | 2.2 | 25.0 |
| Fe($^5$D) | 2.2 | 0.0944 | 8.1 | 0.421 | 13.4 | 2.00 | 2.2 | 26.0 |
| Co($^4$F) | 2.2 | 0.0903 | 8.2 | 0.399 | 14.3 | 1.94 | 2.2 | 27.0 |
| Ni($^3$F) | 2.2 | 0.0870 | 8.2 | 0.379 | 15.3 | 1.89 | 2.2 | 28.0 |
| Cu($^2$S) | 2.2 | 0.0837 | 8.3 | 0.362 | 17.4 | 2.40 | 1.1 | 29.0 |
| Zn($^1$S) | 2.2 | 0.0807 | 8.4 | 0.345 | 17.2 | 1.81 | 2.2 | 30.0 |
| Ga($^2$P) | 2.2 | 0.0778 | 8.4 | 0.329 | 17.2 | 1.60 | 3.1 | 31.0 |
| Ge($^3$P) | 2.2 | 0.0751 | 8.4 | 0.315 | 17.2 | 1.43 | 4.1 | 32.0 |
| As($^4$S) | 2.2 | 0.0726 | 8.5 | 0.301 | 17.2 | 1.30 | 5.1 | 33.0 |
| Se($^3$P) | 2.2 | 0.0702 | 8.5 | 0.289 | 17.2 | 1.19 | 6.1 | 34.0 |
| Br($^2$P) | 2.2 | 0.0680 | 8.5 | 0.278 | 17.1 | 1.10 | 7.1 | 35.0 |
| Kr($^1$S) | 2.2 | 0.0659 | 8.5 | 0.267 | 17.1 | 1.02 | 8.1 | 36.0 |
| Rb($^2$S) | 2.2 | 0.0639 | 8.6 | 0.257 | 17.1 | 0.95 | 8.2 | 37.0 |
| Sr($^1$S) | 2.2 | 0.0621 | 8.6 | 0.248 | 17.0 | 0.89 | 8.3 | 38.0 |
| Cd($^1$S) | 2.2 | 0.0481 | 8.8 | 0.183 | 17.1 | 0.58 | 17.9 | 48.0 |

**Table 4.3** Effect of the basis set on the calculation of populations ($N(\Omega)$), variance ($\sigma^2$) and fluctuations ($\lambda$) of the basins ($\Omega$) of $H_2O$ at the B3LYP level. Data taken from Ref. [11]

| Base | $\Omega$ | $N(\Omega)$ | $\sigma^2$ | $\lambda$ |
|---|---|---|---|---|
| STO-3G | C(O) | 2.03 | 0.30 | 0.15 |
| | V(H1,O) | 1.37 | 0.65 | 0.47 |
| | V(O) | 2.61 | 1.05 | 0.47 |
| 6-31G(d) | C(O) | 2.09 | 0.36 | 0.17 |
| | V(H1,O) | 1.53 | 0.73 | 0.48 |
| | V(O) | 2.43 | 1.09 | 0.47 |
| 6-31+G(d) | C(O) | 2.10 | 0.36 | 0.17 |
| | V(H1,O) | 1.59 | 0.75 | 0.47 |
| | V(O) | 2.36 | 1.09 | 0.46 |
| 6-311++G(d, p) | C(O) | 2.10 | 0.36 | 0.17 |
| | V(H1,O) | 1.69 | 0.79 | 0.47 |
| | V(O) | 2.25 | 1.06 | 0.47 |
| 6-311++G(3df,2p) | C(O) | 2.10 | 0.36 | 0.17 |
| | V(H1,O) | 1.65 | 0.78 | 0.47 |
| | V(O) | 2.30 | 1.08 | 0.47 |

For the variance in ELF populations within a given basin, $\Omega$ [65] (Eq. 3.174), we have

$$\sigma^2(\Omega) = \int\limits_{\Omega}\int\limits_{\Omega} \rho_2(\boldsymbol{r}_1, \boldsymbol{r}_2) d\boldsymbol{r}_1 d\boldsymbol{r}_2 - [N(\Omega)]^2 + N(\Omega) \tag{4.76}$$

where $N(\Omega)$ is the population of the basin and $\rho_2(\boldsymbol{r}_1, \boldsymbol{r}_2)$ is the spinless pair function.

Similarly, we can resort to the covariance, $cov(\Omega_i, \Omega_j)$, to obtain information of the population fluctuation between two basins, $\Omega_i$ and $\Omega_j$:

$$\mathrm{cov}(\Omega_i, \Omega_j) = N(\Omega_i)N(\Omega_j) - \int\limits_{\Omega_i}\int\limits_{\Omega_j} \rho_2(\boldsymbol{r}_1, \boldsymbol{r}_2) d\boldsymbol{r}_1 d\boldsymbol{r}_2. \tag{4.77}$$

Note that the fluctuation is an extensive quantity, making difficult the comparison among molecules with different numbers of basins. Hence, it is convenient to introduce the relative fluctuation, $\lambda(\Omega)$

$$\lambda(\Omega) = \frac{\sigma^2(\Omega)}{N(\Omega)} \tag{4.78}$$

which is a positive quantity and lower than one.[13]

[13] We use $\lambda$ also for the localization index. Normally the context clearly allows to know which concept it refers to.

**Table 4.4** Comparison between ELF and minimum electron fluctuation domains (MinFLuc). Data taken from Ref. [51]

| Molecule | Partitioning | $\sum \sigma^2$ |
|---|---|---|
| LiH | ELF | 0.13 |
| | MinFluc | 0.18 |
| $BeH_2$ | ELF | 0.41 |
| | MinFluc | 0.42 |
| $BH_4^-$ | ELF | 1.99 |
| | MinFluc | 2.10 |

Looking at the results for $H_2O$ from Table 4.3 (e.g. for the 6-31G(d) basis set) we can see that the core, populated roughly by 2.09 electrons, exhibits a fluctuation of 0.36, and $\lambda = 0.17$. The population of the disynaptic V(H1O) domain associated with the OH bond is 1.53, while the population of the monosynaptic V(O) domains associated with lone pairs are equal to 2.43. These values deviate from Lewis' prediction, which is also reflected in the exalted values of electron fluctuation (0.73 and 1.09, respectively) and relative fluctuation (0.48 and 0.47, respectively).

These values also serve to illustrate general trends in electron fluctuations: core basins are well separated from valence basins, with values of $\lambda$ typically of 0.1. Protonated disynaptic basins are in the order of 0.3–0.5 for OH and CH bonds, respectively. It is also around 0.5 for delocalized CC bonds, and slightly lower (ca. 0.4) for single and double CC bonds (see Table 4.3). Note that these values respect the hierarchy provided by *bip* values in the previous section.

It is interesting to note that the analysis of *loges* minimizing electron fluctuation is very close to the ELF results.[14] Table 4.4 shows that differences between the fluctuations in both types of domains are only marginal. In other words, ELF partitioning is reminiscent of the partitioning induced by the direct optimization of the minimum electron fluctuation domains for these simple systems [51].

ELF Moments

As it was done with QTAIM basins, in addition to the computation of the population, it is possible to define higher order moments, $M_i$, within ELF basins [50]. We use $M$ instead of $N$ to distinguish them from those used for atomic basins. Hence, the population would just be the charge of the basin ($q(\Omega) = M^{00}(\Omega)$). First moments correspond to the dipolar polarization of the charge distribution within the basin:

[14] Since a general optimization procedure for these domains is not available, a two step approach was taken. Firstly, the core domains were assumed to be spherical, the radius being determined by the minimum electron fluctuation condition. The remaining valence space was partitioned to give the bond domains on the symmetry-based considerations.

$$M^{1x}(\Omega) = -\int_{\Omega} (x - X_c)\rho(\mathbf{r})d\mathbf{r} \tag{4.79}$$

$$M^{1y}(\Omega) = -\int_{\Omega} (y - Y_c)\rho(\mathbf{r})d\mathbf{r} \tag{4.80}$$

$$M_{1z}(\Omega) = -\int_{\Omega} (z - Z_c)\rho(\mathbf{r})d\mathbf{r} \tag{4.81}$$

where $X_c$,$Y_c$ and $Z_c$ are the cartesian coordinates of the basin attractors. The first moment basin magnitude is then defined as

$$|M^1(\Omega)| = \sqrt{\sum_i M^{1i}(\Omega)^2}. \tag{4.82}$$

These quantities allow to compare the dipolar polarization as the environment of the basin changes. Higher order terms would allow to analyze quadrupoles, etc.

ELF basin's dipole contributions have been used to rationalize inductive polarization effects and hydrogen bond interactions whereas bond quadrupole polarization moments enable to discuss bond multiplicities. ELF multipoles allow to introduce the very common partitions into electrostatic and electronic contributions. For example, the analysis of $M^0$, $M^1$ and $M^2$ moments in aminoacids showed that they share a common electronic and electrostatic common ground, which explains their transferability [50].

### 4.3.4 *IQA for ELF*

To go further in the analysis of energetic terms beyond ELF moments, we can introduce a partition similar to the one we used for the electron density in the Interacting Quantum Atoms Section (Sect. 3.5.2). Before doing so, it should be recalled that the kinetic energy density is not well defined (any terms $a\nabla^2\rho$ for $a$ = constant leads to the same total kinetic energy) except within the QTAIM partition. Admitting this indetermination, we can apply the IQA energetic partition to other topological partitions, such as the one induced by ELF. This has the advantage of providing insight into the energetics involved in electron pairs. Some small differences appear from the formal point of view. When considering bonding basins (and valence in general) within the IQA partition, the electron-nuclear attractions vanish ($V_{en}^{BB} = 0$, $V_{nn}^{AB} = 0$, $V_{en}^{AB} = 0$). If we take a bonding basin that we label "Bond" which interacts with the core basin of atom $A$, then we can express the different energy terms as

$$E_{intra}^{Bond} = T_{Bond} + V_{Coul}^{Bond} \tag{4.83}$$

$$E_{inter}^{A-Bond} = V_{Coul}^{A-Bond} + V_{XC}^{A-Bond}. \tag{4.84}$$

Changing the core of atom $A$ by another bond or a lone pair will easily allow us to calculate interactions between two "bonds" and between "bonds" and "lone pairs".

#### 4.3.4.1 Energetic Terms

With this in mind, it is possible to analyze the changes upon elongation in the terms that contribute to $E_{intra}^{Bond}$ and $E_{inter}^{A-Bond}$. We will analyze $CH_3$-$NH_2$ in detail, as a representative example of a molecule showing bonds and lone pairs (you can check the ELF picture for $CH_3$NH2 in Fig. 4.9). Note that similar results are obtained for the other molecules of the series $CH_3$-X (X=$CH_3$, OH, F) shown in Figure [45].

**Electrostatic Energy**

The electrostatic energy in these systems is reasonably well approximated by classical monopole electrostatics terms between point charges centered at the ELF maximum (which can be shown to lay very close to the charge barycentre of the basin) [45]:

$$E_{elec}^{AB} \propto \frac{q^A q^B}{R} \tag{4.85}$$

where $q^i$ stands for the charge at basin $i$ (core, bond or lone pair). Thus, the total electrostatic interaction should follow the classical Coulomb potential $Eelec = C_1/R + C_2$, where $C_1$ and $C_2$ are constants.

This can be verified by calculating the classical components of $E_{intra}^{Bond}$ and $E_{inter}^{A-Bond}$ from the IQA-ELF partition (see Eqs. 4.83 and 4.84) for the interactions between the bond, V(C,N) and both cores (C(N), C(C)), as well as the interactions between the N lone pairs (V(N)) and C(N). Results are shown in Fig. 4.19. $R^2$ is higher than 0.97 in all the $CH_3$X molecules, for both intra and inter-basin contributions. Note that intrabasin terms appear as repulsive interactions, since they correspond to electron-electron repulsions, whereas interbasin Bond-core terms are positive, due to the extra attractive contribution from the nucleus. Overall, this indicates that zeroth order electrostatics is a reasonably accurate approach for ELF basins.

**Kinetic Energy**

The electron density falling out exponentially, we can assume a zeroth order expansion of the kinetic energy density in the region far away from the cores, i.e. we can approximate the kinetic energy density by the homogeneous electron gas, $t(r) = c_F \rho^{5/3}$. Assuming that the charges follow a linear behavior upon stretching (Fig. 4.20a), we can obtain an easy relationship between the kinetic energy for valence basins and the distance:

$$T \propto \frac{q^{5/3}}{R_B^2} = \frac{q^{5/3}}{R - R_{cores}} \tag{4.86}$$

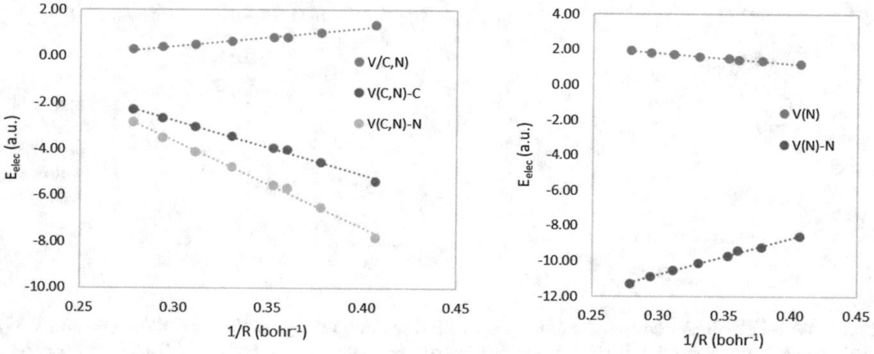


**Fig. 4.19** ELF-IQA computed classical electrostatic interaction energy (intra and inter) against 1/R for a) V(C-N) and, b) V(N) in the $CH_3NH_2$ molecule. Regression coefficients: $R^2 = 0.978$ for $E_{intra}^{V(C,N)}$, 0.995 for $E_{inter}\, V(C-N)-C$, 0.996 for $E_{inter}\, V(C-N)-N$, 0.989 for $E_{intra}\, V(N)$, and 0.994 for $E_{inter}\, V(N)-N$. Reprinted with permission from Ref. [45]

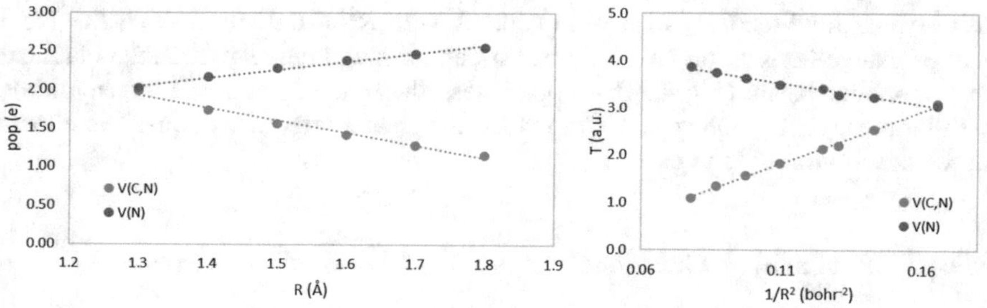


**Fig. 4.20** **a** V(C, N) and V(N) populations as a function of the C–N bond distance. Linear regression coefficients: $R^2 = 0.982$ for V(C, N) and 0.989 for V(N). **b** ELF-IQA computed bond kinetic energy against $1/R^2$ for V(C-N) and V(N) basins in $CH_3NH_2$. Regression coefficients: $R^2 = 0.997$ for V(C, N) and 0.984 for V(N). Reprinted with permission from Ref. [45]

Note that although Eq. 4.86 depends on $R_B$, the bond distance, $T$ can easily be expressed in terms of the total radius by taking into account that the total radius is the sum of the core radii, $R_{cores}$, (which are constant) and the bond distance, $R_B$. This is shown in the second part of Eq. 4.86. The validity of the equation derived, that is of $T \propto R^{-2}$, is presented in Fig. 4.20b. Overall, very good agreements ($R^2 =$ 0.984–1.00) are found in the full $CH_3X$ series in both bonds and lone pairs.

### Exchange-Correlation Energy

The exchange-correlation energy within the QTAIM approach is given by:

$$E_{xc}^{AB} \simeq -\frac{\delta^{AB}}{2R} \simeq -\frac{k}{R^2}. \tag{4.87}$$

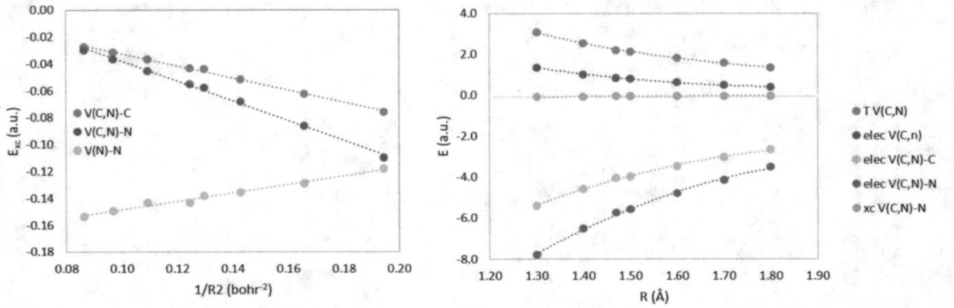


**Fig. 4.21** **a** ELF-IQA computed exchange-correlation energies against $1/R^2$ for V(C-N) and V(N) basins with C and N in $CH_3$-$NH_2$. Regression coefficients: $R^2 = 0.998$ for V(C, N)-C, 0.997 for V(C, N)-N and 0.988 for V(N)-N. **b** ELF-IQA energy terms for C–N bond in methanamine. Reprinted with permission from Ref. [45]

Figure 4.21a shows that this relationship with respect to $R$ also holds for ELF in $CH_3NH_2$ (and the $CH_3X$ series in general, with $R^2 > 0.99$ in most cases). Note that exchange-correlation terms are two orders of magnitude smaller than classical terms for bond basins (Fig. 4.21b). Nonetheless, they are only one order of magnitude smaller in the case of lone pairs. This reveals the greater relevance of quantum effects in the description of lone pairs.

#### 4.3.4.2 Similarity to Old Models

According to the previous discussion, the total bond energy would be the sum of kinetic, electrostatic and exchange-correlation contributions, even though in most cases the latter will play a minor role. Interestingly, this formulation is formally equivalent to the one provided by the Bond Charge Model (BCM) [7, 8]. In the BCM, the energy is the sum of three different contributions (in atomic units):

$$E(q, R) = E_0 + E_1 + E_2 = \sum_A E_A \frac{q}{2} - C\frac{q^2}{R} + D'\frac{q}{R_B^2}, \tag{4.88}$$

where $E_0$ is the total core energy. That is, the sum of the core energies of each atom, $E_A$. $E_1$ accounts for classical electrostatic interactions and $E_2$ represents the kinetic energy of the valence electrons. The BCM model has the same energy terms as within the ELF-IQA approach (with the exception of the exchange-correlation term, which is non-classical). Moreover, they also show the same dependencies with $R$. We can thus reformulate the Bond Charge Model within IQA as follows:

$$E_{core} = E_A \tag{4.89}$$

$$E_{intra}^{val} + E_{inter}^{(core-val)} = -C' \frac{q^2}{R_B} \tag{4.90}$$

$$T_{val} = D' \frac{q}{R_B^2} \tag{4.91}$$

$$E_{xc} \simeq 0 \tag{4.92}$$

where *core* and *val* stand, respectively, for core and valence. Overall, applying ELF-IQA enables to retrieve the energetics between ELF basins in an analytical manner, enabling to justify the success of the BCM model, but also its failures (e.g. we have seen that the behavior of lone pairs is less adapted to this description).

### 4.3.5 ELF, Orbitals and Others

The arbitrary reference to the electron gas in the ELF definition, present in both Becke's and Savin's definition, is definitely the biggest source of doubt about ELF true physical meaning [27, 29, 58, 64]. In their 1990 definition, Becke and Edgecombe stressed that the spin conditional probability contains the information necessary for the recovery of the shell structure of atoms. However, we want to stress once more that this function is inoperative unless it is calibrated against the conditional probability in the homogeneous electron gas. The relevance of the quotient becomes even more evident if we consider that this formulation determined the success of the ELF in comparison to the extremely similar localization function proposed by Luken and Culberson, in which the value of the homogeneous gas is subtracted from $D(\mathbf{r})$ (instead of used in the denominator). This apparently fortuitous and arbitrary reference is what finally allows the quantitative recovery of Lewis' structure and the gain of chemical vision to the detriment of physical meaning.

Numerous interpretations have been proposed throughout the years for the seminal definition of the ELF. The ELF *kernel*, originally defined as a local measure of the curvature of the Fermi hole calculated at the HF level, was supplemented by the DFT formulation we have already seen in terms of local excess of kinetic energy density. We have also briefly reviewed Dobson's development in terms of the curvature of the pair density. This option was also later developed independently by Silvi and Kohout. Orbital interpretations also came to light from Burdett's hand [9], as well as from Nalewajski's [46]. A detailed account of these interpretations is presented below.

#### 4.3.5.1 Orbital Nodes

Burdett y McCormick [9] proposed a very interesting interpretation of ELF in terms of orbitals nodes. Starting from the observation that the ELF picture is extremely stable with respect to the method of calculation, to such an extent that even extended Hückel calculations, where the pair function has no meaning, are able to recover the pairing picture, they suggested that ELF must be intrinsically related to orbitals. More specifically, they proposed that ELF is related to the orbital nodes of the occupied orbitals in the system.

Since within the orbital definition, the kinetic energy density is limited from below by that of a localized system, the crucial term in the ELF expression is $\sum_i |\nabla\phi_i|^2\rho^{-5/3}$. Thus, in those regions where the gradient (or the kinetic energy density) increases, there is a decrease in the localization and in the ELF, a fact that is closely related to the occupation of nodal orbitals. For example, as electrons are added to a metal, these must occupy orbitals with a greater number of nodes, so that this effect exceeds that of an increase in electronic density and the possibility of electrons being in bonding regions is reduced until they reach values below 1/2. This result allows associating ELF values lower than 1/2 to regions of very low density (such as those far from the nuclei) or where the contribution of a large number of nodes compensates for that of the density (such as the regions between the atomic shells)

Figure 4.22-top shows the classical bonding vs antibonding picture of $H_2$ wavefunctions. The slope in the bonding region is considerably larger in the antibonding case. Since the kinetic energy density of the electrons is directly related to the slope, or gradient, $\nabla\phi_i$, it is easy to see that large contributions to the kinetic energy density occur in the regions between atoms when orbitals with nodes are populated. This

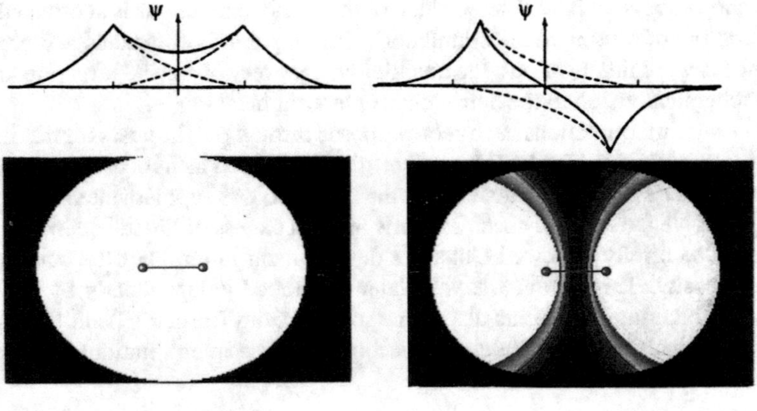


**Fig. 4.22** Top: bonding (left) and antibonding (right) wavefunctions for $H_2$. Note the larger slope in the bonding region for the antibonding case. Bottom: ELF plot for bonding (left) and antibonding $H_2$ (right). Note the construction of localization in the bonding case. Adapted with permission from Ref. [9]. Copyright 1998 American Chemical Society

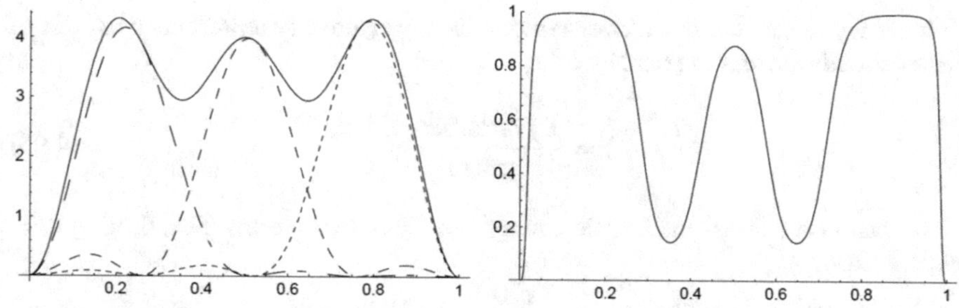


**Fig. 4.23** Left: The localized orbitals (squared), with dashed curves, and their sum, the density (full curve), as a function of the position. Right: ELF as a function of the position, showing the regions where the orbitals localize by values close to 1. Reprinted with permission from Ref. [55]

increases the ELF denominator, leading to lower values of the function, i.e. to less localization (Fig. 4.22-bottom). Moreover, since the orbital is antibonding we have a combined effect of $D$ increasing and $D_h$ decreasing.

A more quantitative relationship between ELF and orbitals can be established if we assume localized orbitals, $\phi_i$ [55]. A toy system with three localized orbitals[15] is shown in Fig. 4.23-left. It can be seen that in the region where $\phi_i$ is localized, $|\phi_i|^2 \simeq \rho$, which are the regions where ELF$\simeq$1 (Fig. 4.23-right). Instead, ELF goes to very low values in the regions in between localized orbitals.

#### 4.3.5.2 Information Measure

The conditional spin probability that serves as the basis for the seminal definition of the ELF can be related to the Fisher interorbital information measure contained in the electronic distribution function [46]. This fact is not surprising once we recall that Fisher information is a measure of the amplitude of the electron distribution, and therefore of localization.

The probability density to find a $\sigma$ electron at $\boldsymbol{r}$ is given by

$$p_\sigma(\boldsymbol{r}) = \frac{\rho_\sigma(\boldsymbol{r})}{N_\sigma}, \tag{4.93}$$

where $N_\sigma$ is the total number of $\sigma$ electrons.

Recall that in Sect. 3.5.3 we referred to Shannon entropy (Eq. 3.171) as a way to recover information on probabilities. Here we will resort to the information measure, $I[p]$. $I[p]$ characterizes the localization of the electron distribution as follows:

$$I[p] = \int p(\boldsymbol{r})[\nabla \ln p(\boldsymbol{r})]^2 d\boldsymbol{r} = \int \frac{|\nabla p(\boldsymbol{r})|^2}{p(\boldsymbol{r})} d\boldsymbol{r} = \int f[p; \boldsymbol{r}] d\boldsymbol{r} \tag{4.94}$$

---

[15] Six non-interacting particles in a one-dimensional box.

It is easy to see that the Weizsäker kinetic energy density represents a quarter of Fisher's total information density $(f_t)$:

$$t_w(\boldsymbol{r}) = \frac{1}{4}\frac{|\nabla\rho^\alpha(\boldsymbol{r})|^2}{\rho^\alpha(\boldsymbol{r})} = \frac{1}{4}f^\alpha. \tag{4.95}$$

We can also express the kinetic energy density in these terms. Substituting $\rho_i = |\phi_i|^2$ in Eq. 4.94:

$$f[\rho_i; \boldsymbol{r}] = \frac{|\nabla\rho_i(\boldsymbol{r})|^2}{\rho_i(\boldsymbol{r})} = 4|\nabla\phi_i(\boldsymbol{r})|^2, \tag{4.96}$$

we see that the kinetic energy density represents the sum over the Fisher information densities associated with the orbitals:

$$t^\alpha(\boldsymbol{r}) = \sum_\alpha |\nabla\phi_i(\boldsymbol{r})|^2 = \frac{1}{4}\sum f_i^\alpha[\rho_i^\alpha; \boldsymbol{r}] = \frac{1}{4}f_a^\alpha[\rho^\alpha; \boldsymbol{r}]. \tag{4.97}$$

The subscript $a$ makes reference to its additive nature.

Diving the total Fisher information in Eq. 4.95 into an additive (Eq. 4.97) and a non-additive part:

$$t_w(\boldsymbol{r}) = \frac{1}{4}\left[f_a^\alpha + f_n^\alpha\right] = t(\vec{r}) + \frac{1}{4}f_n^\alpha \tag{4.98}$$

where $f_n^\alpha$ is the non-additive part of the Fisher information.

We can then see that the $D_\alpha(\boldsymbol{r}) = t(\boldsymbol{r}) - t_w(\boldsymbol{r})$ function is proportional to the negative non-additive contribution to the Fisher information defined in terms of the electron densities:

$$D_\alpha(\boldsymbol{r}) = -\frac{1}{4}f^\alpha. \tag{4.99}$$

In the KS theory the MO densities refer to the hypothetical non-interacting system, while the overall density corresponds to the interacting system. Hence, the source of the non-additivity of Eq. 4.98 is the electron interaction in the real system. Furthermore, given the fact that the additive part combines all intra-orbital contributions, the non-additive part can be alternatively interpreted as a measure of the interorbital Fisher information density.

#### 4.3.5.3 Probabilities

We have seen the definition of the probabilities of finding $n$ electrons in a region $\Omega$, $P_n(\Omega)$ in Sect. 3.5.3. Maximizing the probability, $P_n(\Omega)$, of finding $n$ electrons in a region $\Omega$ gives rise to regions $\Omega_y$ that correspond to the atomic shells, bonds, and bonding pairs [23, 51].

Starting from a HF wavefunction whose overlap matrix in the $\Omega$ region, $S_{ij}(\Omega)$, has eigenvalues $\lambda_1, \lambda_2, \ldots$, the probabilities of finding zero, one and two electrons in $\Omega$ are given by

$$P_0(\Omega) = \Pi_i(1 - \lambda_i) \tag{4.100}$$

$$P_1(\Omega) = \sum_j \lambda_j \Pi_{i \neq j}(1 - \lambda_i) \tag{4.101}$$

$$P_2(\Omega) = \sum_{j>k} \lambda_j \lambda_k \Pi_{i(\neq j,k)}(1 - \lambda_i). \tag{4.102}$$

While the calculation of these probabilities does not require much computational time, the location of the $\Omega_\nu$ regions does, and since they can be expected to resemble the ELF basins for $\nu = 2$, it is reasonable to ask ourselves to what extent one can be approximated by the other.

Following our toy model in Sect. 4.3.5.1, let's imagine a system with two doubly occupied localized orbitals, so that spin orbitals $\phi_1$ and $\phi_2$ describe the two electrons localized in $\Omega_1$ and $\phi_3$ and $\phi_4$ describe two electrons localized in $\Omega_2$. If we focus on region $\Omega_1$, we have

$$\begin{aligned}
\lambda_4 &= S_{11}(\Omega_1) = 1 \\
\lambda_1 &= S_{12}(\Omega_1) = 1 \\
\lambda_2 &= S_{13}(\Omega_1) = 0 \\
\lambda_3 &= S_{14}(\Omega_1) = 0
\end{aligned}$$

Hence:

$$\begin{aligned}
P_0(\Omega_1) &= 0 \\
P_1(\Omega_1) &= 0 \\
P_2(\Omega_1) &= 1
\end{aligned}$$

More generally, if we consider a Slater determinant built from perfectly located orbitals in $L_k$ regions, it can be seen from a quick inspection of the above equations that only $p(2; L_k)$ will be non-zero, and that its value will be one. This is the best possible probability, so $L_k$ will be one of the $\Omega_{\nu=2}$. Assuming a deformation of the latter so that it does not cover all of the $L_k$, the overlaps will not all be 0 or 1, but some zero value will increase and some 1 value will decrease, resulting in an overall decrease of $p(2; \Omega)$. An analogous reasoning from what we say in Sect. 4.3.5.1 allows to demonstrate that these same regions correspond to the ELF basins. It is sufficient to first consider a $\boldsymbol{r} \in L_k$. Since the two spin orbitals located in $L_k$ have the same spatial part, the density will simply be given by $\rho(\boldsymbol{r}) = 2|\psi_k|^2$, which gives ELF $= 1$ within the region. Beyond the basin boundary, $\psi_k$ becomes infinite, so that

ELF = 0. It is sufficient to assume a slight smoothing of the curve to obtain the ELF profile.

It is interesting to consider the cases where localized orbitals cannot be defined unequivocally. For example, the localized orbitals and $\Omega_{\nu=2}$ in the Ne atom divide the space into a region of *core* and four valence regions reminiscent of the $sp^3$ orbitals, while the ELF, which is adapted to symmetry, leads to atomic shells with spherical symmetry and 2 and 8 electrons for the K and L shells, respectively. In this case, it is the $\Omega_{\nu=8}$ probability which leads to the L shell.

#### 4.3.5.4 Constrained Electron Population

The concept of electron density localization at a given point **r** can be understood from a simple statistical point of view as the charge standard deviation enclosed in a sampling volume centered at that point. This approach is known as constrained population approach. It is based on interconnecting two different local quantities: the control property is used to divide the space in compact mutually excluding microcells which fill up the total volume. The volume of these cells is determined so as to lead to the same integral value of the control property. Then, the other property, known as sampling property, is integrated within these cells. This leads to a discrete distribution of the sampling property integral. This approach will be further exploited in Sect. 4.4.3 in the development of the ELI functions.

If we want to focus on electron localization, we have to keep in mind that the smaller the standard deviation, the higher the localization [65]. It is common to use the square of the standard deviation (i.e. the variance, $\sigma^2$), which is the expectation value of the variance operator in Eq. 3.174:[16]

$$\begin{aligned}\sigma^2(V(\boldsymbol{r})) &= = 2D(V(\mathbf{r}), V(\mathbf{r})) - N(V(\mathbf{r}))\,(N(V(\mathbf{r})) - 1)\\ &= D(V(\mathbf{r}), V(\mathbf{r})) - q^2 + q\end{aligned} \tag{4.103}$$

$q = N(V(\mathbf{r}))$ and $D(V(\mathbf{r}), V(\mathbf{r}))$ stand for the one particle and two-particle densities integrated over the sample $V(\mathbf{r})$ (Eq. 3.173), respectively. Note that only $D(V(\mathbf{r}), V(\mathbf{r}))$ depends on the position, $-q^2 + q$ can be regarded as a constant and ignored.

The integrated pair density can be separated in two terms, the same spin contribution and the opposite spin contribution (almost proportional to $q^2$):

$$D(V(\mathbf{r}), V(\mathbf{r})) = D^{\alpha\alpha}(V(\mathbf{r})V(\mathbf{r})) + D^{\beta\beta}(V(\mathbf{r}), V(\mathbf{r})) + 2D^{\alpha\beta}(V(\mathbf{r}), V(\mathbf{r})). \tag{4.104}$$

[16] Recall that $\sigma^2(i, j) =< ij > - < i >< j >$ so that for the number of electrons in a system with basins $A$ and $B$, we have $\sigma^2(A, B) = \hat{N}(A)\hat{N}(B) - \bar{N}(A)\bar{N}(B) = \hat{\pi}(A, B) + \delta_{AB}\bar{N}(A) - \bar{N}(A)\bar{N}(B)$. Hence, for different basins, we have: $\sigma^2(A, B) = \hat{\pi}(A, B) - \bar{N}(A)\bar{N}(B)$, whereas for same basins: $\sigma^2(A, A) = \hat{\pi}(A, A) - \bar{N}(A)(\bar{N}(A) - 1)$.

The expectation value of the variance of the integrated opposite spin pair density is given by:[17]

$$\begin{aligned}\langle\sigma^2(D^{\alpha\beta}(V(\mathbf{r}),V(\mathbf{r}))\rangle &= D^{\alpha\alpha}(V(\mathbf{r}),V(\mathbf{r}))D^{\beta\beta}(V(\mathbf{r}),V(\mathbf{r}))\\ &+N^{\alpha}(V(\mathbf{r}))D^{\beta\beta}(V(\mathbf{r}),V(\mathbf{r}))\\ &+N^{\beta}(V(\mathbf{r}))D^{\alpha\alpha}(V(\mathbf{r}),V(\mathbf{r}))+N^{\alpha}(V(\mathbf{r}))N^{\beta}(V(\mathbf{r}))\\ &-(D^{\alpha\beta}(V(\mathbf{r})))^2. \end{aligned} \tag{4.105}$$

Note that the integrated opposite spin pair variance also depends on the same spin pair densities. On its turn, the integrated same spin pair density has been numerically shown to be proportional to $q^{5/3} \times c_\pi(\mathbf{r})$, where $c_\pi(\mathbf{r})$ is a function of the position. Hence, we can use $c_\pi(\mathbf{r})$ as a local function independent of the size of the sample [64]:

$$c_\pi(\mathbf{r}) = lim_{q\to 0}\frac{D(V(\mathbf{r}),V(\mathbf{r}))}{q^{5/3}}. \tag{4.106}$$

This function allows to localize "electronic groups". This can be easily illustrated by a simple example in which two $\alpha$ and two $\beta$ spin electrons are confined in a box of volume $\Omega$ (see Fig. 4.24). In we assume the electron density probability to be uniform and without spin polarization, we see that the spin pair functions are also constant and given by

$$\begin{aligned}\rho(\mathbf{r}) &= 4/\Omega\\ \rho^{\alpha}(\mathbf{r}) = \rho^{\beta}(\mathbf{r}) &= 2/\Omega\\ D^{\alpha\beta}(\mathbf{r}_1,\mathbf{r}_2) = D^{\beta\alpha}(\mathbf{r}_1,\mathbf{r}_2) &= 4/\Omega^2.\end{aligned}$$

This model enables to consider two localization cases:

- The opposite spin pairs are delocalized over the box and the same spin pair functions are constant. In this case, $D^{\alpha\alpha}(\mathbf{r}_1,\mathbf{r}_2) = D^{\beta\beta}(\mathbf{r}_1,\mathbf{r}_2) = 2/\Omega^2$ and therefore $c_\pi(\mathbf{r})$ is constant. This is a model for delocalized electrons: the localization function is constant.
- Each opposite spin pair occupies one half of the box. In this case, $c_\pi(\mathbf{r})$ will change depending on the sampling point:

---

[17] Recall from Eq. 3.146 that $\rho_2^{\alpha\beta}(\boldsymbol{r},\boldsymbol{r}') = \rho^{\alpha}(\boldsymbol{r})\rho^{\beta}(\boldsymbol{r}')$, hence we can express

$$\begin{aligned}\langle\sigma^2(D^{\alpha\beta}(V(\mathbf{r}),V(\mathbf{r}))\rangle &= \hat{N}^{\alpha}(V(\mathbf{r}))\hat{N}^{\beta}(V(\mathbf{r})) - (D^{\alpha\beta}(V(\mathbf{r})))^2\\ &= D^{\alpha\alpha}(V(\mathbf{r})) + \bar{N}^{\alpha}(V(\mathbf{r}))(D^{\beta\beta}(V(\mathbf{r}))) + \bar{N}^{\beta}(V(\mathbf{r})) - (D^{\alpha\beta}(V(\mathbf{r})))^2\\ &= D^{\alpha\alpha}(V(\mathbf{r}),V(\mathbf{r}))D^{\beta\beta}(V(\mathbf{r}),V(\mathbf{r}))\\ &+ N^{\alpha}(V(\mathbf{r}))D^{\beta\beta}(V(\mathbf{r}),V(\mathbf{r})) + N^{\beta}(V(\mathbf{r}))D^{\alpha\alpha}(V(\mathbf{r}),V(\mathbf{r}))\\ &+ N^{\alpha}(V(\mathbf{r}))N^{\beta}(V(\mathbf{r})) - (D^{\alpha\beta}(V(\mathbf{r})))^2\end{aligned}$$

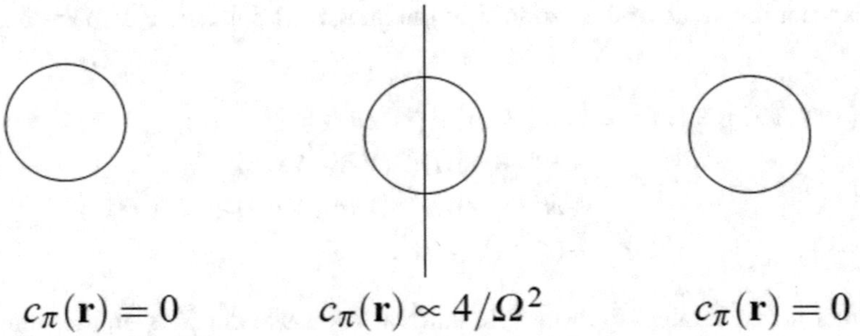


**Fig. 4.24** Model system with $2\alpha$ and $2\beta$ electrons. Each half of space contains an electron pair (i.e. $1\alpha$,$1\beta$). Density is assumed to be constant. The control property is the charge $q$. The antiparallel pair population of the sample is also constant and equal to $q^2/2$. The spin pair composition is the sampling property and it is zero when the sampling volume is entirely contained on the same half of space. It increases when contained in both parts and it reaches its maximum when placed in between them (see text)

- $\mathbf{r}_1, \mathbf{r}_2 \in$ the same half box: $D^{\alpha\alpha}(\mathbf{r}_1, \mathbf{r}_2) = D^{\beta\beta}(\mathbf{r}_1, \mathbf{r}_2) = 0$. Hence $c_\pi(\mathbf{r}) = 0$ ($\mathbf{r} \notin$ the boundary).
- $\mathbf{r}_1, \mathbf{r}_2 \in$ the different half boxes: $D^{\alpha\alpha}(\mathbf{r}_1, \mathbf{r}_2) = D^{\beta\beta}(\mathbf{r}_1, \mathbf{r}_2) = 4/\Omega^2$. Hence $c_\pi(\mathbf{r}) \propto 4/\Omega^2$ ($\mathbf{r} \in$ the boundary).

This is a good model for localized electrons: the localized function is zero within each basin and rises in between them. The function enables to locate the boundary between the two opposite spin electron pairs.

It can be easily demonstrated for Hartree–Fock wavefunctions that [64]:

$$c_\pi(\mathbf{r}) \approx \chi_\sigma(\mathbf{r}). \tag{4.107}$$

After the Lorentzian transformation of $\chi_\sigma(\mathbf{r})$, we see that ELF tends to 1 in those regions where the localization is high and to small values at the boundaries between such regions.

## 4.4 Other Localization Descriptors

Many seemingly unrelated scalar fields have been derived over the years to sense the spatial degree of electron pairing. Let's review the knowledge that has been usually retrieved from them as well as the physical connection among them.

### 4.4.1 The Laplacian of the Electron Density

The topology of $\nabla^2\rho$ is also intimately linked to the VSEPR model. Since the VSEPR model works nicely, and deals with the geometrical disposition of electron pairs, it is reasonable to look after some type of link between $\rho_2$ and $\nabla^2\rho$. A partial answer to this alleged correspondence can be found in the partial condensation of electron pairs. In valence regions, Fermi's correlation is not strong enough to force a complete localization of electron pairs (in section "ELF maxima and VSEPR" we saw that ELF was very close to 1 in the first shell, but typically lower in the valence). However, it may well be sufficiently intense so as to induce a significant decrease in the number of electron pairs involved (with respect to random pairing). We call these regions, regions of partial pair condensation.

In order to develop this idea, we have to go back to the properties of the same spin pair density (Sect. 3.5.1). In *closed-shell systems*, according to Eq. 3.151,

$$\rho^\sigma(\boldsymbol{r}_2|\boldsymbol{r}_1) = \rho(\boldsymbol{r}_2) + h_{xc}^{\alpha\alpha}(\boldsymbol{r}_2|\boldsymbol{r}_1) + h_{xc}^{\beta\beta}(\boldsymbol{r}_2|\boldsymbol{r}_1). \tag{4.108}$$

In most molecular systems at geometries close to equilibrium, the correlation hole will be dominated by the Fermi contribution, and the minimum value for $\rho^\sigma(\boldsymbol{r}_2|\boldsymbol{r}_1)$ will appear when $\boldsymbol{r}_2 \to \boldsymbol{r}_1$. Whenever a region in which the Fermi hole of a given electron is well localized, the rest of the same spin electrons will be excluded from it. By using $\alpha \leftrightarrow \beta$ symmetry, the same would happen to the opposite spin hole. If the region fully encapsulates the hole, we will have excluded an electron from it, and created a zone occupied by a density corresponding to a single pair of electrons. Under these conditions, the hole hardly contributes to $\rho^\sigma(\boldsymbol{r}_2|\boldsymbol{r}_1)$ in regions far apart from the reference electron, so that its conditional probability will approach the total electron density.

If the condensation of the electron pair is not complete (i.e. whenever the Fermi hole is not completely localized) the spatial regions where $\rho^\sigma(\boldsymbol{r}_2|\boldsymbol{r}_1)$ is a minimum for a given position of the reference electron will give the location where the partial pair condensation occurs. Usually $\rho(\boldsymbol{r})$ has no other minima than those at infinity, and it is necessary to turn to its Laplacian to find its fine structure. Calculations show that, in general, there exists a homeomorphism between $\nabla^2\rho(\boldsymbol{r}_2)$ and $\rho^\sigma(\boldsymbol{r}_2|\boldsymbol{r}_1)$ [2] that turns out to be relatively independent of the position of the reference electron. Normally, the pattern of maxima and minima of $\nabla^2$ and $\rho^\sigma(\boldsymbol{r}_2|\boldsymbol{r}_1)$ recovers the shell structure when $\boldsymbol{r}_1$ is located in the valence, particularly when $\boldsymbol{r}_1$ is located in the atomic shells. Similarly, the bonds and lone pairs can be reconstructed by locating the reference electron at the nucleus.

These ideas, together with the success of $\nabla^2\rho$ on recovering the number, geometry, shape and size of the pairs of the VSEPR theory allows us to state that:

1. Fermi holes are rather localized in simple molecular systems.
2. The analysis of $\nabla^2\rho$ allows to qualitatively grasp the behavior of the conditional pair density in a simplified manner.

#### 4.4.1.1 General Features of $\nabla^2\rho$

The Laplacian field is related to local charge accumulations or depletions. A point characterized by $\nabla^2\rho < 0$ is identified with an accumulation of density with respect to its nearbies, while if $\nabla^2\rho > 0$, we understand that the density escapes from the point toward its surroundings. This interpretation is grounded in the existence of a principal system of curvature at each point of space. Provided that the trace of a matrix is invariant under an orthogonal transformation,

$$\nabla^2\rho(\boldsymbol{r}) = \lambda_1 + \lambda_2 + \lambda_3, \tag{4.109}$$

where, as usual, $\lambda_i$ are the principal curvatures of the density at a point $\boldsymbol{r}$ (which is not necessarily a CP). A positive Laplacian corresponds, thus, to one or more positive principal curvature directions with total curvature larger to the sum of all those with negative or zero curvature directions. We condensate the complexity of the 3D scalar in a single index that provides information on whether the accumulation(depletion) along one or more directions is more important than the depletion(accumulation) along the rest.

Let us examine, first, the general features of $\nabla^2\rho$ in simple systems. We start by looking at the structure of the Laplacian in isolated spherically symmetric (or in spherically averaged) atoms. Since in a spherical coordinates system $\rho(r, \theta, \phi) = \rho(r)$, we have

$$\nabla^2\rho(r) = \rho''(r) + 2\frac{\rho'(r)}{r}, \tag{4.110}$$

where $\rho'$ and $\rho''$ are the first and second radial derivatives of the density, and the sign of the Laplacian depends on a subtle balance between the radial curvature of the density and its slope.

Admitting an exponential behavior, $\rho(r) = Ne^{-\zeta r}$, an approximation which is rather valid both close to the nucleus and in the large distance asymptotic limit, $\nabla^2\rho(r) = Ne^{-\zeta r}(\zeta^2 - 2\zeta/r)$. Examining this expression we find that a cutoff radius exists, $r_c = 2/\zeta$ at which a sign change occurs. $\nabla^2\rho < 0$ when $r < r_c$ and vice versa. Therefore, $\lim_{r\to 0}\nabla^2\rho = -\infty$ and $\lim_{r\to\infty}\nabla^2\rho = 0^+$. Actually, atomic densities are well approximated not by an exponential but by a set of exponential segments of decreasing exponents as we move out from the nucleus (note that we will further use this in Sect. 5.3.3.4). This simple argument shows that the Laplacian should show a set of oscillations, with a negative and positive window in each of these segments. This is exactly the behavior observed, and $\nabla^2\rho$ in atoms is found as a set of pairs of density accumulation and density depletion regions.

Figure 4.25 shows a couple of examples. In the Helium atom, for instance, we observe how the Laplacian starts at a very negative value (actually a divergence for an exponential) at the nucleus, then it touches zero at a distance $r = 0.53$ bohr to increase up a maximum value $\nabla^2\rho = 0.703$ a.u. at $r = 0.732$ bohr, to decay exponentially to zero at large distances. Correspondingly, a logarithmic map of the electron density shows a single segment with a behavior closely characterized

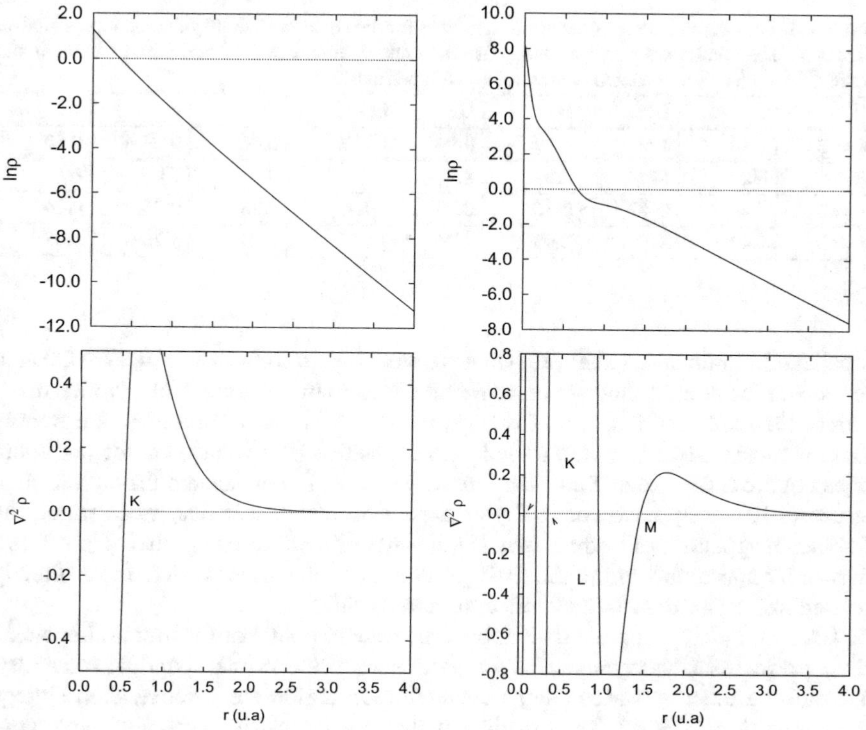


**Fig. 4.25** $\ln\rho$(top) and $\nabla^2\rho$ (bottom) along any radial direction $i$ for the He (left) and Ar (right) atoms. Densities from the quasi-HF exponential Clementi and Roetti data set. All data in a.u

by a single exponent. This single segment is tripled in Ar, where the logarithmic map of the density shows three different exponents and the Laplacian shows three sets of regions of charge accumulation and charge depletion. The Laplacian cuts the $r$ axis five times, at $r$ equal to 0.055, 0.205, 0.345, 0.951, 1.463 bohr, displaying maxima and minima at $r = 0.074, 0.239, 0.439, 1.078, 1.775$ a.u. with values $90500, -791, 191, -1.211, 0.216$ a.u., respectively.

It is tempting to identify each of these pairs of accumulation and depletion regions with an atomic shell. These are labeled employing standard chemical notation in Fig. 4.25: $K$, $L$, $M$ for 1st, 2nd and 3rd shells, respectively. It is well known that the map between the Laplacian and the atomic shell structure is faithful until the fourth period. The Laplacian of the calcium atom shows only three shells instead of four (both at the HF and more accurate correlated levels of theory). This situation worsens as we run across the periodic table [1]. It has been debated whether this is a pitfall of the Laplacian scalar field to recover the atomic shell structure or if it is the latter that smooths out for large enough nuclear charges. The discussion has very likely come to an end, since as we have seen, other scalar fields like ELF or the electronic

**Table 4.5** Maximum charge concentration radii $r_i$ for the $i$th atomic shells of some representative elements. The results are based on the quasi-HF multi-$\zeta$ atomic wavefunctions of Clementi and Roetti. All data in a.u. Adapted from ref. [1], with permission

| R | Li | Be | B | C | N | O | F | Ne |
|---|---|---|---|---|---|---|---|---|
| $i = 2$ | 2.49 | 1.59 | 1.19 | 0.94 | 0.78 | 0.66 | 0.57 | 0.50 |
| R | Na | Mg | Al | Si | P | S | Cl | Ar |
| $i = 2$ | 0.44 | 0.40 | 0.36 | 0.33 | 0.30 | 0.28 | 0.26 | 0.24 |
| $i = 3$ | 3.44 | 2.55 | 2.08 | 1.76 | 1.52 | 1.34 | 1.20 | 1.08 |

localizability indicator (ELI) [29] (that we will soon introduce, Sect. 4.126) are able to recover the atomic shell structure for all atoms in the periodic table. Notice that a plot of the Laplacian in Ca, for instance, really shows four oscillations. It is simply that the most outwards minimum is found at positive $\nabla^2\rho$ values, i.e. the minimum does not cross the $r$ axis. Since the link between the Laplacian and the atomic shell structure is mainly based on a Taylor expansion of the pair density in which the Laplacian appears as a leading term [1], it is not difficult to accept that $\nabla^2\rho$ offers a first-order approximation to the shell structure, and that other scalars more directly related to the pair density may reflect it more faithfully.

Like we already saw for ELF above, the critical points of the atomic Laplacian field are necessarily degenerate as a consequence of the atomic spherical symmetry. We call the most outwards charge concentration region the **valence shell charge concentration** (VSCC). The position of the radial zeros, minima and maxima in atoms has been successfully correlated to a number of measures of atomic size, like the orbital radii defined by Alex Zunger [70]. Table 4.5 gathers some of these values.

#### 4.4.1.2 Topological Analysis of $\nabla^2\rho$

We will analyze the reshaping that the spherically symmetric $\nabla^2\rho$ field suffers when a system of atoms interacts to form different types of molecules through simple examples. This will be easily generalized to more complex cases.

We have chosen the formation of the $Cl_2$ molecule (Fig. 4.26) to illustrate the topology of the Laplacian in a covalent molecule. The topology of $\rho$ is simple, with just a single bond critical point. The Laplacian is mostly atomic close to the nuclei, showing $K$, $L$ and $M$ shells with clear core charge accumulation and charge depletion regions that are basically conserved from the free atoms. On the contrary, the valence shell has suffered a considerable distortion. In the first place, the regions of charge accumulation of both atoms have fused together, and we can appreciate a continuous region of charge accumulation (negative Laplacian) that encapsulates the whole molecule. We will see that this behavior is not general, but rather characteristic of certain interaction types.

The topology of $\nabla^2\rho$ is considerably more complex, since axial symmetry forces the appearance of a one-dimensional degeneracy for any CP not located on the

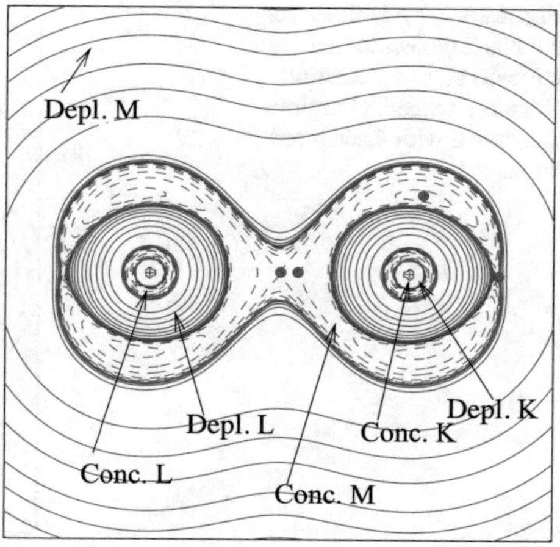


**Fig. 4.26** Isolines of $\nabla^2\rho$ in the $Cl_2$ molecule at the HF/6-311G** level. Negative Laplacian isolines are shown dashed in red. Some representative critical points have been highlighted

**Fig. 4.27** $\nabla^2\rho = \pm 0.6$ a.u. (green and blue, respectively) in the $Cl_2$ molecule

internuclear axis. In order to keep the discussion simple, and focusing on that axis, we find a (3, +1) CP at the midpoint between the Cl atoms, with two very close symmetrically located (3, −1) CPs on its left and on its right, respectively. Each Cl atom also displays a pair of (3, −1) and (3, −3) CPs located very close to the point where the Laplacian vanishes in the valence shell. It is however more interesting to check how the VSCC generates two toroidal charge concentration distributions surrounding each of the chlorine atoms. Figure 4.27 shows the $\nabla^2\rho = -0.6$ a.u. isosurface in $Cl_2$. These toroidal regions can be successfully associated with those of preferred electrophilic attack. Similarly, in the charge depletion regions along the internuclear axis, in the rear part of each Cl, we find the positions prone to a nucleophilic attack.

A similar analysis will now be offered for another archetypal diatomic molecule, NaCl. In Fig. 4.28, a diagram similar to that in Fig. 4.26 is presented. At first sight, several outstanding differences are clear. At variance with the $\rho$ field, the Laplacian seems to be sensitive to the type of interaction. In NaCl, the VSCCs of the isolated atoms forming the molecule do not fuse together. Actually, the bonding region is not

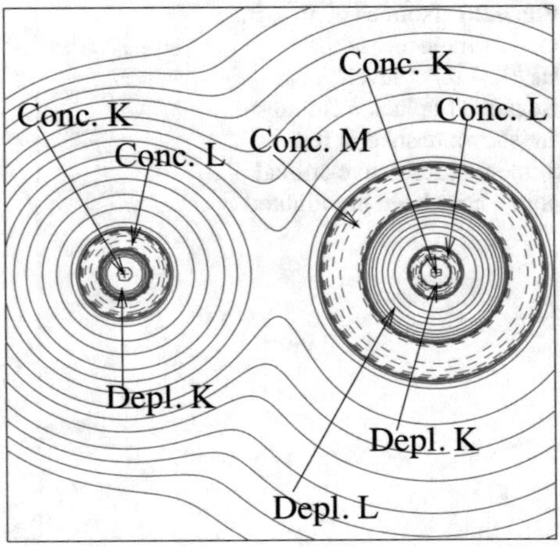


**Fig. 4.28** $\nabla^2\rho$ isolines for the NaCl molecule in a HF/6-311G** calculation. Negative Laplacian regions are shown with dashed red lines

a region of charge accumulation but, on the contrary, of charge depletion. If we focus on the number of atomic shells around each of the atoms, we immediately see that the chlorine conserves as many shells as the free atom, while the sodium atom has lost the valence shell charge concentration corresponding to the $M$ shell. As expected, this type of interaction is completely different from that observed in $Cl_2$: NaCl is formed by ions that are kept together thanks to electrostatic forces.

The analysis of the Laplacian also allows to find second order effects in the density redistributions that accompany bond formation. In NaCl we can check how a small, but clear polarization of the VSCC of the chloride anion towards the el $Na^+$ cation appears. This type of small distortions are important from the structural point of view, since they account for the final atomic dipoles and further electrostatic multipoles that influence the total forces exerted over the nuclei.

To complete this survey, we explore other types of classical bonds. We have chosen the BeH and the $Ne_2$ molecules. Their Laplacian fields are shown in Fig. 4.29. In BeH we see how the position of the separatrix assigns to the Be atom a small part of the VSCC originally belonging to the H atoms. In this way, the major part of the Be VSCC has counterpolarized and that of H has distorted considerably. The region close to the BCP displays a negative Laplacian, and between both nuclei there exists a large charge accumulation. This corresponds to a very polarized bond, with a non-negligible charge transfer, but nonetheless smaller than in the extreme NaCl case. Finally, the interaction between the two closed shells of the Ne atoms practically does not distort their original electron distribution. No accumulation regions are shared and the Laplacian at the bond critical point is positive.

It is usual to classify chemical interactions into two great classes according to the sign of the Laplacian in the bonding region (normally at the BCP). We talk about shared shell interactions when, like in the $Cl_2$ molecule, the VSCC shells of the two bonded atoms have fused together or at least when, like in BeH, the internuclear region close to the BCP lies in a region of charge accumulation. To simplify, when

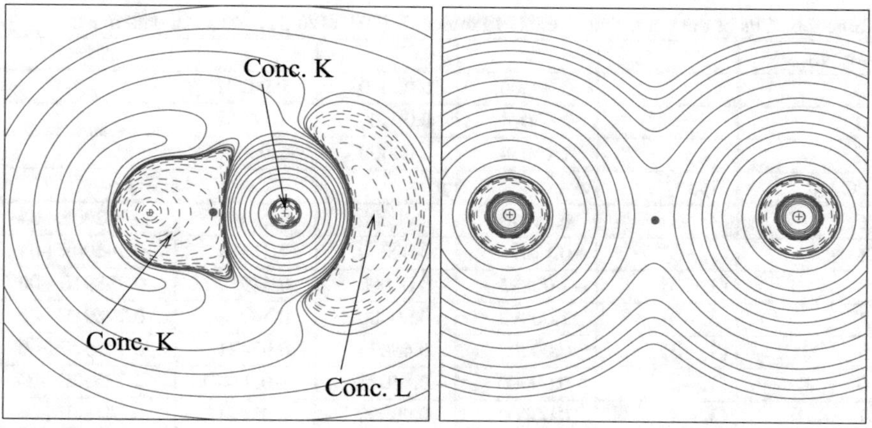


**Fig. 4.29** $\nabla^2\rho$ isolines for BeH (B3LYP/6-311G** density, left) and for $Ne_2$ (CISD/6-311G** density, right). Negative Laplacian regions are identified in red dashed lines, and the position of the BCP is marked by a red dot

$\nabla^2\rho(\boldsymbol{r}_b) < 0$. On the contrary, we talk about closed-shell interactions if the two bonded atoms do not share their VSCCs and the bonding region is a depletion zone, with $\nabla^2\rho(\boldsymbol{r}_b) > 0$. This latter type of interaction is found in systems that we would normally classify as ionic (NaCl), or weakly bound ($Ne_2$).

The topological analysis of the Laplacian becomes much more interesting in non-linear molecules, in which the CPs of the scalar fields are not degenerate anymore, save in exceptional cases. Let us start by looking at the Laplacian structure of the water molecule. It displays 33 different non-degenerate CPs, which are listed in Table 4.6. A pictorial representation of the position of these CPs is shown in Fig. 4.30. Notice the different complexity of the density and Laplacian fields, since the former just displays five CPs. This complexity stems from the rupture of the shell structure, to which the Laplacian is sensitive. For the isolated oxygen atom we find a radial Laplacian maximum for the K shell at $r = 0.17$ bohr ($\nabla^2\rho = 1325$ a.u.), and a radial maximum and a minimum for the L shell at $r = 0.66$ y 1.21 bohr ($\nabla^2\rho = -3.48$ and 0.53 a.u.), respectively; for the H atom, there is just a K maximum at $r = 1.37$ bohr ($\nabla^2\rho = 0.02$ a.u.). The $\nabla^2\rho$ isolines, in the now familiar format we have been using, are found in Fig. 4.31. In the first place, we find that the O–H link is of the shared shell type, and a region of negative Laplacian encapsulates the whole molecule.

As the CPs themselves are regarded, we find two maxima in the molecular plane in the vicinities of the oxygen atom, separated by a $(3, -1)$ CP, at a distance of about 0.17 a.u. from the nucleus. They correspond to points 1, 2 and 7 in Table 4.6, they can be easily distinguished by the large value of their $\nabla^2\rho$.

The spherical symmetry of the K shell of O has been broken, and the charge depletion is maximum in the directions opposite to the bonds. There exists, thus, a polarization of the oxygen core that, otherwise, remains largely unaffected. As the L shell is regarded, we find four minima of $\nabla^2\rho$ distributed approximately at the

**Table 4.6** CPs of the Laplacian of the $H_2O$ molecule (B3LYP/6-311G**). All data in a.u

| Atomo | | $x$ | $y$ | $z$ | |
|---|---|---|---|---|---|
| O | | 0.00000 | 0.00000 | 0.08201 | |
| H | | −1.43094 | 0.00000 | −1.04058 | |
| H | | 1.43094 | 0.00000 | −1.04058 | |
| PC | Tipo | $x$ | $y$ | $z$ | $\nabla^2\rho$ |
| 1 | (3, −3) | 0.11073 | 0.00000 | 0.21489 | 0.14200E+04 |
| 2 | (3, −3) | −0.11073 | 0.00000 | 0.21489 | 0.14200E+04 |
| 3 | (3, −3) | −1.00893 | 0.00000 | 0.54976 | 0.79891E+00 |
| 4 | (3, −3) | 1.00893 | 0.00000 | 0.54976 | 0.79891E+00 |
| 5 | (3, −3) | 0.00000 | 0.45075 | −0.97471 | 0.72552E+00 |
| 6 | (3, −3) | 0.00000 | −0.45075 | −0.97471 | 0.72552E+00 |
| 7 | (3, −1) | 0.00000 | 0.00000 | 0.25481 | 0.14186E+04 |
| 8 | (3, −1) | 0.68655 | 0.00000 | 0.22125 | −0.14526E+01 |
| 9 | (3, −1) | −0.68655 | 0.00000 | 0.22125 | −0.14526E+01 |
| 10 | (3, −1) | 0.00000 | 0.00000 | −0.63692 | −0.11753E+01 |
| 11 | (3, −1) | 0.00000 | 0.00000 | −1.06575 | 0.71720E+00 |
| 12 | (3, −1) | 0.53824 | 1.01022 | −0.19212 | 0.58614E+00 |
| 13 | (3, −1) | −0.53824 | 1.01022 | −0.19212 | 0.58614E+00 |
| 14 | (3, −1) | −0.53824 | −1.01022 | −0.19212 | 0.58614E+00 |
| 15 | (3, −1) | 0.53824 | −1.01022 | −0.19212 | 0.58614E+00 |
| 16 | (3, −1) | 0.00000 | 0.00000 | 1.23368 | 0.59280E+00 |
| 17 | (3, 1) | 0.00000 | 0.00000 | 0.73277 | −0.41436E+01 |
| 18 | (3, 1) | 0.86464 | 0.00000 | −0.60906 | −0.91233E+00 |
| 19 | (3, 1) | −0.86464 | 0.00000 | −0.60906 | −0.91233E+00 |
| 20 | (3, 1) | 0.00000 | 1.11771 | 0.45383 | 0.51301E+00 |
| 21 | (3, 1) | 0.00000 | −1.11771 | 0.45383 | 0.51301E+00 |
| 22 | (3, 1) | 0.52147 | 0.28469 | −0.30594 | −0.25390E+01 |
| 23 | (3, 1) | −0.52147 | 0.28469 | −0.30594 | −0.25390E+01 |
| 24 | (3, 1) | −0.52147 | −0.28469 | −0.30594 | −0.25390E+01 |
| 25 | (3, 1) | 0.52147 | −0.28469 | −0.30594 | −0.25390E+01 |
| 26 | (3, 1) | 1.37153 | 0.00000 | −1.77630 | 0.13861E+00 |
| 27 | (3, 1) | −1.37153 | 0.00000 | −1.77630 | 0.13861E+00 |
| 28 | (3, 3) | 0.00000 | 0.60049 | 0.30555 | −0.52912E+01 |
| 29 | (3, 3) | 0.00000 | −0.60049 | 0.30555 | −0.52912E+01 |
| 30 | (3, 3) | 0.57320 | 0.00000 | −0.36436 | −0.26459E+01 |
| 31 | (3, 3) | −0.57320 | 0.00000 | −0.36436 | −0.26459E+01 |
| 32 | (3, 3) | 1.42559 | 0.00000 | −1.03679 | −0.26728E+02 |
| 33 | (3, 3) | −1.42559 | 0.00000 | −1.03679 | −0.26728E+02 |

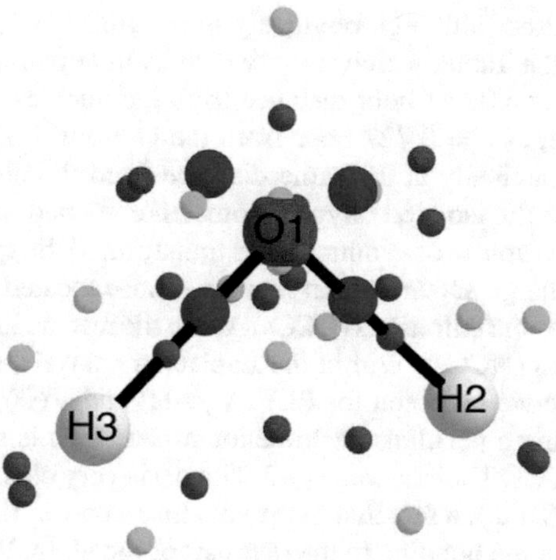


**Fig. 4.30** Position of the CPs of $\nabla^2\rho$ in the water molecule (B3LYP/6-311G**). Oxygen and hydrogen nuclei in white and black, respectively. In red, green, magenta and blue the (3, −3), (3, −1), (3, +1) y (3, +3) CPs, respectively. The latter has been highlighted, as it corresponds to charge concentrations

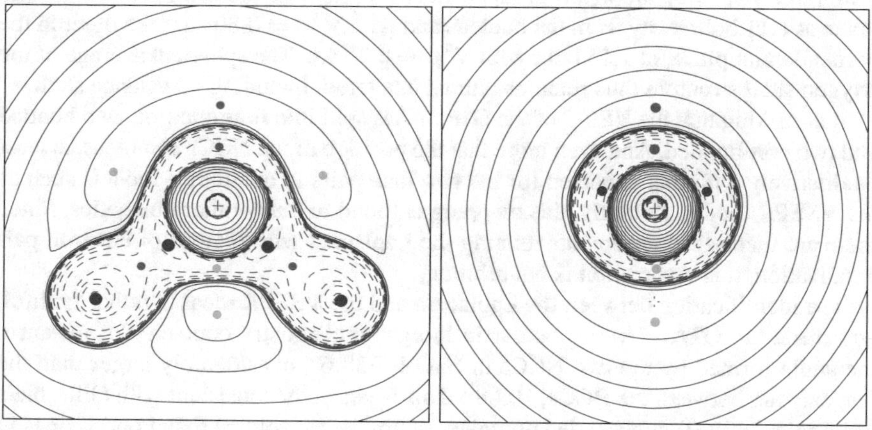

**Fig. 4.31** $\nabla^2\rho$ isolines for the $H_2O$ molecule (B3LYP/6-311G** density) in the molecular (left) and perpendicular to the molecular (right) planes that contain the $C_{2v}$ axis. The CPs are presented using the same color code used in the previous figure

corners of a tetrahedron, although obviously preserving the $C_{2v}$ symmetry. These are points 28–31 in the Table, which form two equivalent pairs. Two of them along the bond directions, at a 0.641 bohr distance from the nucleus and the other two in the orthogonal plane, at $r = 0.727$ bohr from the O atom. Let us stress that these points are located practically at the same distance from the nucleus as the L shell associated minima in the isolated oxygen atom. Like we had seen for ELF in Sect. 4.3.3, we see how the rupture of symmetry has transformed the spherical atomic shell in a set of discrete charge accumulation regions. Those located along the bonds are called bonded charge concentrations (BCCs), while the rest are known as non-bonded charge concentrations (NCCs). Within the Laplacian analysis, this nomenclature is equivalent to the one we had seen for ELF: V(O-H) and V(O), respectively. In the case we are presenting a peculiar phenomenon arises: $\nabla^2\rho$ is more negative in the NCCs than in the BCCs. Their average (−3.97 a.u.) is very close to the VSCC in the isolated atom (−3.48 a.u.), a fact that highlights that a part of the atomic charge has been transferred from the bonding to the rear part of the atom. We should not forget, however, that the H atoms pump electron charge towards the oxygen.

In a similar way we recognize symmetry broken depletion regions in the valence shell. They give rise to four maxima of the Laplacian, also in a quasi-tetrahedral disposition, although rotated 90° with respect to the VSCCs. These are CPs 3 to 6 in Table 4.6. They are again grouped in two pairs, one located in the molecular plane at 1.11 bohr away from the nucleus and with $\nabla^2\rho = 0.80$ a.u.; the other in the perpendicular plane, at 1.15 bohr with $\nabla^2\rho = 0.73$ a.u. The spherical average of the oxygen shell structure thus remains almost unaltered, including its valence shell.

It is striking that the VSCC of the O atom has split into four maxima, two bonded and two non-bonded. And even more that the position of the NCCs coincides, at least qualitatively with that expected for the two lone pairs in qualitative models such as the VSEPR model. Actually, this mapping is found in many other molecules. Since there are theoretical arguments relating the Laplacian with regions of electron pair localization, this association is not arbitrary.

The identification between the Laplacian and the VSEPR ideas is rather faithful, and made the QTAIM very popular in inorganic chemistry courses. For instance, the angle formed by the two NCCs in $H_2O$ is 138.6°, considerably larger than the one existing between the BCCs, 103.8°. This is perfectly consistent with Gillespie's basic rules [25]: (i) non-bonded domains occupy larger volume than bonded ones in the same valence shell; bonded domains decrease their size as the electronegativity of the ligands increases, and increase it with the electronegativity of the central atom; (iii) bonded domains corresponding to double and triple bonds are larger than those of single bonds. In our case, the angle subtended by the non-bonded domains is larger than the one between the bonded domains.

We will end this section by presenting the Laplacian of another classical example, the $ClF_3$ molecule. Its geometry, well known, is the prototype of a T-shaped molecule and, according to the VSEPR model, it corresponds to a trigonal bipyramidal disposition of the five pairs of valence electrons around the Cl atom. The topology of

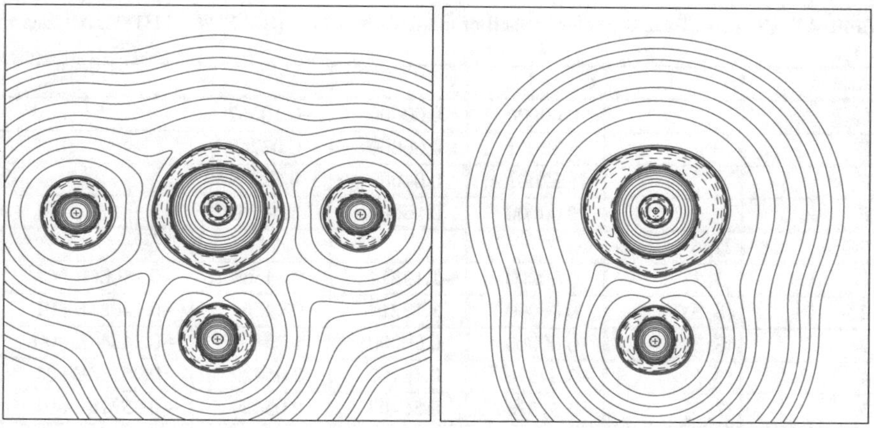

**Fig. 4.32** $\nabla^2\rho$ isolines in the $ClF_3$ molecule (B3LYP/6-311G** density) in the molecular (left) and perpendicular to molecular (right) planes containing the $C_{2v}$ symmetry axis

$\nabla^2\rho$ displays 75 different CPs. A large number of them correspond to the symmetry breaking of the inner shells of the Cl and F atoms. We will skip them here.

In Fig. 4.32, we show the isolines of the Laplacian in the two interesting molecular planes. Let us observe, in the first place, that the classification of the bond type according to the Laplacian is different from that in the water molecule. In $ClF_3$, the distortion suffered by the valence shells of both the Cl and the F atoms is very small and $\nabla^2\rho$ in the three BCPs is positive. From the point of view of the trigonal distribution of the electron pairs, the axial bonds are longer than the equatorial one, thus the Laplacian at the equatorial bond critical point, 0.017 a.u., is much smaller than in the axial ones, 0.215 a.u.

Table 4.7 shows the valence (3, +3) CPs in this system. The first three correspond to the Cl, and clearly show the existence of two NCCs in trigonal configuration at 1.15 bohr away from the Cl atom, in excellent agreement with the VSCC accumulation found in the free atom, in which they are found at 1.19 a.u. They form an angle of 155.4°. The third maximum is the BCC of the equatorial bond at a similar distance from the central atom. It shows the same asymmetry in the $\nabla^2\rho$ value that was already discussed in $H_2O$. CPs 4–9 are NCCs of the F atoms. No BCC appear for the axial bonds, although some scar-like features, like ring or bond critical points do appear. This fact is usual in many cases. When we face closed-shell links, the BCCs may be absent, since the density accumulation is not large enough to force the appearance of a minimum in the Laplacian. On the contrary, NCCs are almost always present, for non-bonded electrons try to get as far away from each other as possible, creating minima of $\nabla^2\rho$ close to the positions expected by the VSEPR theory.

**Table 4.7** (3, +3) CPs in the valence shell of the $ClF_3$ molecule (B3LYP/6-311G**). All data in a.u

| Atomo | | $x$ | $y$ | $z$ | |
|---|---|---|---|---|---|
| Cl | | 0.00000 | 0.00000 | −0.10067 | |
| F | | −3.32061 | 0.00000 | 0.02749 | |
| F | | 3.32061 | 0.00000 | 0.02749 | |
| F | | 0.00000 | 0.00000 | 3.04568 | |
| PC | Tipo | $x$ | $y$ | $z$ | $\nabla^2\rho$ |
| 1 | (3, 3) | 0.00000 | −1.12214 | −0.34564 | −0.14088E+01 |
| 2 | (3, 3) | 0.00000 | 1.12214 | −0.34564 | −0.14088E+01 |
| 3 | (3, 3) | 0.00000 | 0.00000 | 1.15807 | −0.34848E+00 |
| 4 | (3, 3) | −3.37576 | 0.56407 | 0.05429 | −0.95318E+01 |
| 5 | (3, 3) | 3.37576 | −0.56407 | 0.05429 | −0.95318E+01 |
| 6 | (3, 3) | −3.37576 | −0.56407 | 0.05429 | −0.95318E+01 |
| 7 | (3, 3) | 3.37576 | 0.56407 | 0.05429 | −0.95318E+01 |
| 8 | (3, 3) | 0.00000 | 0.56031 | 3.09451 | −0.10485E+02 |
| 9 | (3, 3) | 0.00000 | −0.56031 | 3.09451 | −0.10485E+02 |

#### 4.4.1.3 Comparison with ELF

The topologies of the Laplacian and ELF are intimately related. The main differences that arise are:

1. Atomic radii are bigger within ELF than within $-\nabla^2\rho$.
2. Covalent bonds within ELF lead to a basin, whereas they do not for the Laplacian.

According to Bader and Heard [2] most of these differences come from the normalization with respect to the homogeneous electron gas in the ELF definition. Indeed, we have already seen that a connection exists between the spin conditional pair density (the sum of both $\alpha\alpha$ and $\beta\beta$ contributions) and the Laplacian (Sect. 4.4.1). It should be noted that the differences between both functions have been exploited in the past to obtain a richer chemical insight. Llusar et al. [39] exploited the advantages of each technique to analyze metal-metal interactions in dimers and complexes. Chesnut and Bartolotti [13] did the same for discussing aromaticity in several substituted cyclopentadienyls. For a more extensive review of the simultaneous use of ELF and the Laplacian, the reader is directed to Ref. [22] and references therein.

### *4.4.2 The Pair Density*

It would be reasonable to think at this point, why not applying the topological analysis to the pair density itself? The high dimensionality of this object makes it difficult to analyze and interpret. However, several lower dimensionality quantities have been defined over the years contracting the pair density, $\rho_2(\boldsymbol{r}_1, \boldsymbol{r}_2)$. An implicit possibility that has already been explored is expanding $\boldsymbol{r}_2$ in a Taylor series around $\boldsymbol{r}_1$, as done in the ELF, but the most widespread explicit approach is probably to split the spinless pair density into the intracular and extracular densities [16]. This involves introducing a new system of coordinates $\boldsymbol{r} = \boldsymbol{r}_1 - \boldsymbol{r}_2$, and $\boldsymbol{R} = (\boldsymbol{r}_1 + \boldsymbol{r}_2)/2$. By doing this, the intracular, $I(\boldsymbol{r})$, and the extracular, $E(\boldsymbol{r})$, densities can be written as

$$I(\boldsymbol{r}) = \int \rho_2(\boldsymbol{r}_1, \boldsymbol{r}_2)\delta\left((\boldsymbol{r}_1 - \boldsymbol{r}_2) - \boldsymbol{r}\right) d\boldsymbol{r}_1 d\boldsymbol{r}_2,$$

$$E(\boldsymbol{R}) = \int \rho_2(\boldsymbol{r}_1, \boldsymbol{r}_2)\delta\left(\frac{\boldsymbol{r}_1 + \boldsymbol{r}_2}{2} - \boldsymbol{R}\right) d\boldsymbol{r}_1 d\boldsymbol{r}_2. \tag{4.111}$$

$I$ and $E$ are nothing but the distribution functions for the interelectron separation and the center of mass coordinate for electron pairs, respectively. Performing the above-mentioned isometric change of variables,[18] we can simplify these expressions:

$$I(\boldsymbol{r}) = \int \rho_2(\boldsymbol{R} + \boldsymbol{r}/2, \boldsymbol{R} - \boldsymbol{r}/2) d\boldsymbol{R} = \int \rho_2(\boldsymbol{R}, \boldsymbol{R} - \boldsymbol{r}) d\boldsymbol{R},$$

$$E(\boldsymbol{R}) = \int \rho_2(\boldsymbol{R} + \boldsymbol{r}/2, \boldsymbol{R} - \boldsymbol{r}/2) d\boldsymbol{r} = 2\int \rho_2(\boldsymbol{r}, 2\boldsymbol{R} - \boldsymbol{r}) d\boldsymbol{r}. \tag{4.117}$$

Let us state some of the basic properties of these functions. If $\mathcal{R}$ is a symmetry operation acting on the nuclear system, then the pair density is invariant under such operation: $\rho_2(\mathcal{R}\boldsymbol{r}_1, \mathcal{R}\boldsymbol{r}_2) = \rho_2(\boldsymbol{r}_1, \boldsymbol{r}_2)$. Provided that the symmetry operators induce isometric coordinate changes, simple algebraic manipulations show that $I(\boldsymbol{r}) = I(\mathcal{R}\boldsymbol{r})$ and that $E(\boldsymbol{R}) = E(\mathcal{R}\boldsymbol{R})$. *The intra- and extracular densities are also invariant under the symmetry operations of the nuclear frame.* Taking profit of the symmetry of $\rho_2$, it is also easy to show that $I(\boldsymbol{r}) = I(-\boldsymbol{r})$. The intracular density has, therefore, an additional inversion center, a fact that increases its point group symmetry when the molecule is non-centrosymmetric. Moreover, $I$ is also invariant against molecular translations, since the center of mass of the electron pair is the variable of integration.

---

[18] We just need to recall that:

$$\int f(x)\delta(x - a)dx = f(a) \tag{4.113}$$

and make use of the change of variable relationships:

$$\boldsymbol{r}_1 = \boldsymbol{r} + 2\boldsymbol{R} - \boldsymbol{r}_1 = \frac{\boldsymbol{r}}{2} + \boldsymbol{R} \tag{4.114}$$

$$\boldsymbol{r}_2 = 2\boldsymbol{R} - \boldsymbol{r}_1 = \boldsymbol{R} - \frac{\boldsymbol{r}}{2} \tag{4.115}$$

$$\left|\frac{\partial(\boldsymbol{r}_1, \boldsymbol{r}_2)}{\partial(\boldsymbol{r}, \boldsymbol{R})}\right| = 1 \tag{4.116}$$

The second set of equalities are obtained after another simple change of variables.

Both densities integrate to $N(N-1)$. Finally, it is known [16] that $I(\boldsymbol{r})$ shows a cusp at $\boldsymbol{r}=\boldsymbol{0}$ (let us recall that at this point there is a Coulombic singularity), and that this coalescence cusp is very difficult to simulate, particularly when Gaussian basis sets are used.

At this point, it is appropriate to examine a simple model for the pair density that allows us to gain some intuition on the behavior of these functions. To that end, we will imagine a Hartree–Fock scenario with closed-shell atoms. Under these conditions, the second order density is given by Eq. 3.146, and $\rho_2(\boldsymbol{r}_1,\boldsymbol{r}_2)$ can be approximated by the following expression:

$$\rho_2(\boldsymbol{r}_1,\boldsymbol{r}_2) \simeq \sum_{A,B} \{\rho_A(\boldsymbol{r}_1-\boldsymbol{R}_A)\rho_B(\boldsymbol{r}_2-\boldsymbol{R}_B) - \rho_A(\boldsymbol{r}_2-\boldsymbol{R}_A;\boldsymbol{r}_1-\boldsymbol{R}_A)\rho_B(\boldsymbol{r}_1-\boldsymbol{R}_B;\boldsymbol{r}_2-\boldsymbol{R}_B)\}, \quad (4.118)$$

where $\rho_A$ is the first-order density matrix of atom $A$, referred to its nucleus, located at $\boldsymbol{R}_A$. The positive terms in the above expression are Coulomb contributions, and the negative exchange ones. Now, for instance, we can check that the $A, B$ part of the intracular density can be written as $\int \rho_A(\boldsymbol{r}_1-\boldsymbol{R}_A)\rho_B(\boldsymbol{r}_1-\boldsymbol{r}-\boldsymbol{R}_B)d\boldsymbol{r}_1$, so that changing $\boldsymbol{r}_1-\boldsymbol{R}_A$ to a new $\boldsymbol{R}$ variable, we obtain $\int \rho_A(\boldsymbol{R})\rho_B(\boldsymbol{R}-\boldsymbol{r}+\boldsymbol{R}_A-\boldsymbol{R}_B)d\boldsymbol{R}$.

Defining now $\boldsymbol{R}_{AB}=\boldsymbol{R}_B-\boldsymbol{R}_A$, $\boldsymbol{S}_{AB}=(\boldsymbol{R}_A+\boldsymbol{R}_B)/2$, and $\mathcal{J}_{AB}$ ad $\mathcal{K}_{AB}$ as the functions defined as follows:

$$\begin{aligned}
\mathcal{J}^I_{AB}(\boldsymbol{r}) &= \int \rho_A(\boldsymbol{R})\rho_B(\boldsymbol{R}-\boldsymbol{r})d\boldsymbol{R},\\
\mathcal{K}^I_{AB}(\boldsymbol{r}) &= \int \rho_A(\boldsymbol{R}-\boldsymbol{R}_{AB}-\boldsymbol{r};\boldsymbol{R})\rho_B(\boldsymbol{R}+\boldsymbol{R}_{AB};\boldsymbol{R}-\boldsymbol{r})d\boldsymbol{R},\\
\mathcal{J}^E_{AB}(\boldsymbol{R}) &= 2\int \rho_A(\boldsymbol{r})\rho_B(2\boldsymbol{R}-\boldsymbol{r})d\boldsymbol{r},\\
\mathcal{K}^E_{AB}(\boldsymbol{R}) &= 2\int \rho_A(2\boldsymbol{R}-\boldsymbol{r}-\boldsymbol{R}_{AB};\boldsymbol{r}/2)\rho_B(\boldsymbol{r}+\boldsymbol{R}_{AB};2\boldsymbol{R}-\boldsymbol{r})d\boldsymbol{r}. \quad (4.119)
\end{aligned}$$

We get, after some algebra,

$$\begin{aligned}
I(\boldsymbol{r}) &= \sum_{A,B}\left\{\mathcal{J}^I_{AB}(\boldsymbol{r}-\boldsymbol{R}_{AB})-\mathcal{K}^I_{AB}(\boldsymbol{r}-\boldsymbol{R}_{AB})\right\},\\
E(\boldsymbol{R}) &= \sum_{A,B}\left\{\mathcal{J}^E_{AB}(\boldsymbol{R}-S_{AB})-\mathcal{K}^E_{AB}(\boldsymbol{R}-S_{AB})\right\}. \quad (4.120)
\end{aligned}$$

In the limit of large interatomic distances, the exchange contributions, $\mathcal{K}^I$, $\mathcal{K}^E$ will be negligible with respect to the Coulombic ones, and the intra- and extracular densities will be dominated by the latter. If we suppose, additionally, that the atomic

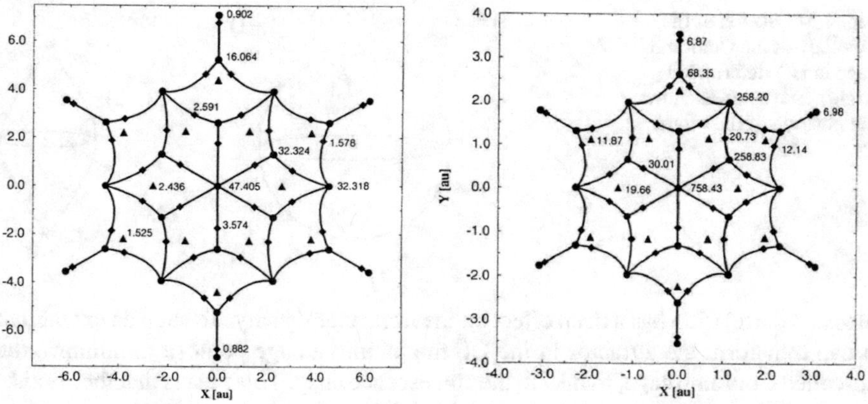


**Fig. 4.33** Topology of the intracular (left) and extracular (right) densities in the molecular plane of a HF/6-31G* benzene molecule. Attractors are marked as points, bond critical points as diamond, and ring critical points as triangles. From reference [14]

densities are spherical, then it is straightforward to prove that $\mathcal{I}^E_{AB}(\boldsymbol{r}) = 2\mathcal{I}^I_{AB}(2\boldsymbol{r})$, and the gross features of both $I$ and $E$ will be determined by the correlation functions of the atomic densities,

$$C_{AB}(\boldsymbol{r}) = \int \rho_A(\boldsymbol{R})\rho_B(\boldsymbol{R} - \boldsymbol{r})d\boldsymbol{R}. \tag{4.121}$$

In the $A = B$ case, this is just an autocorrelation or Patterson function [24], well known in Crystallography. It shows a maximum at $\boldsymbol{r} = \boldsymbol{0}$ which is equal to $\int \rho_A^2(\boldsymbol{r})d\,\boldsymbol{r}$, and tends asymptotically to zero very quickly. In the case of a $A \neq B$ pair, we must compute the overlap between two densities that are progressively displaced one with respect to the other. Usually a maximum is found at the origin, but if the two densities have very different magnitudes this rule can be violated. Turning back to Eq. 4.120, we expect that

1. $I$ displays attractors at the origin, at position vectors that coincide with some $\boldsymbol{R}_{AB}$. This would imply a $A \leftrightarrow B$ symmetry and, therefore the presence of the inversion center we have previously referred to.
2. $E$ shows attractors at $\boldsymbol{R}_A$ and for position vectors equal to some $\boldsymbol{S}_{AB}$.

In Fig. 4.33, we show the topology of the intra- and extracular densities in the molecular plane of the $C_6H_6$ molecule [14] for a HF calculation. The location of the critical points of these functions is considerably more cumbersome than in the case of $\rho$. Notwithstanding, we can clearly observe how the basic features of the model we have just discussed appear in an actual calculation. Recent range-separations of the intracular density have also been proposed [69].

We expect that exchange and correlation effects on $I$ y $E$ should be more intense around the origin and position vectors $\boldsymbol{R}_A$, respectively. It is important to notice that

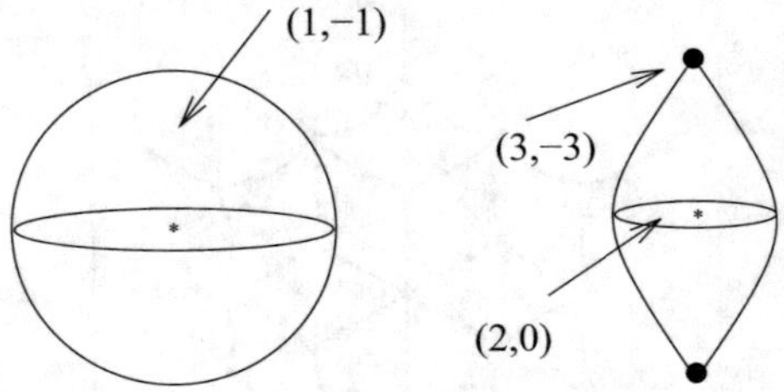


**Fig. 4.34** Sketch of the topology of the Coulomb cage in $H^-$ (left) and $H_2$ (right). The asterisk shows the position of the origin

electron correlation has a deep effect on the intracular density. To such an extent that it can transform the attractor in the HF model into a cage point (a minimum), the so-called Coulomb cage, thanks to the coalescence cusp. This means that there exists a finite region around the origin in $I(\boldsymbol{r})$ in which there exists a qualitative difference between the true correlated and the Hartree–Fock behavior. The effective radius of such a region has been used as an effective measure of the size of the Coulomb hole [18]. Although the coalescence cusp cannot be obtained from gaussian functions, and we even need basis sets that depend explicitly on the interelectron coordinates, the standard quantum mechanical methods to approximate electron correlation are able to simulate these phenomena reliably.

The presence of the Coulomb cage induces important changes in the critical point pattern of $I$, as well as in their connectivity. Given that the topology of $I$ at distances larger than the hole's radius is basically non-correlated, a set of critical points around $\boldsymbol{r} = \boldsymbol{0}$ must appear that warrant the large distance Hartree–Fock structure. If we look at the cage from a distance, it must appear like an attractor. In isolated atoms, this long-range behavior is usually attained by means of a degenerate sphere of $(1, -1)$ CPs located at a distance $R_I$ from the origin. For instance, $R_I \simeq 1.1$ a.u. in the $H^-$ anion, and the cage contains about 0.0025 electron pairs.

In molecules, the cage possesses the symmetry of the system plus the inversion center that characterizes $I(\boldsymbol{r})$. If we place the atoms of a $H_2$ molecule along the $z$ axis, a (2, 0) ring of CPs with radius close to 0.65 a.u. appear in the $xy$ plane, with two attractors at approximately $(0, 0, \pm 1.0)$ a.u. Figure 4.34 shows a sketch of how these CPs are located. A detailed analysis by Cioslowski and Liu [15] shows that the size of the cages in two-electron systems is about 0.6 a.u. in the directions orthogonal to all the $\boldsymbol{R}_{AB}$ vectors, and of about 1.0 au for the rest. The classification and physical meaning of the Coulomb cages in more complex molecules is not yet well known.

### 4.4.3 ELI

The development of ELI-d [32] and ELI-q [29] (where ELI stands for Electron Localizability Indicator) is based on the so-called "constrained population" approach introduced in Sect. 4.3.5.4. Recall that this approach is based on interconnecting two different local quantities: the control property (used to divide the space into cells), and the sampling property (which determines the volume of the cell so that the integral of the sampling constant is constant) thus leading to a discrete distribution of the sampling property integral.

The properties that are used in the construction of ELI are the electron density and the same spin pair density. If the sampling property is the pair density for cells with a fixed charge, we are calculating ELI-q, $\Upsilon_q^\sigma$. If instead, we sample the electron density for volumes with a fix spin pair density integral, we are calculating ELI-d, $\Upsilon_D^\sigma$.

Let's analyze ELI-d further. For small enough cells, $D^{\sigma\sigma}$, can be approximated as [29]:

$$D^{\sigma\sigma}(\Omega) = \int_\Omega \int_\Omega \rho_2^{\sigma\sigma}(\vec{r}_1, \vec{r}_2) d\vec{r}_1 d\vec{r}_2 \simeq \frac{1}{12} V(\Omega)^{8/3} g(\Omega), \tag{4.122}$$

where $g(\Omega)$ represents the curvature of the Fermi hole in $\Omega$.

Integrating the sampling property, the electron density, over these cells we have

$$Q^\sigma(\Omega) = \int_\Omega \rho^\sigma(\vec{r}) d\vec{r} \simeq \rho^\sigma(\Omega) V(\Omega), \tag{4.123}$$

and substituting the value of $V(\Omega)$ derived from Eq. 4.122 into Eq. 4.123, we have:

$$Q^\sigma(\Omega) \simeq (D^{\sigma\sigma})^{3/8} \rho(\Omega) \left( \frac{12}{g(\Omega_i)} \right)^{3/8} \tag{4.124}$$

ELI-d is then defined as

$$\Upsilon_D^\sigma(\Omega) = \rho(\Omega) \left( \frac{12}{g(\Omega)} \right)^{3/8} = \frac{Q^\sigma(\Omega)}{(D^{\sigma\sigma})^{3/8}}, \tag{4.125}$$

which is proportional to the charge and the volume of the microcell.

For $D^{\sigma\sigma}$ infinitely small, the discrete positions can be considered continuous, so that the index can be defined in terms of local properties:

$$\Upsilon_D^\sigma(\Omega)|_{D^{\sigma\sigma}\to 0} \longrightarrow \Upsilon_D^\sigma(\vec{r}) = \rho(\vec{r}) \left( \frac{12}{g(\vec{r})} \right)^{3/8}. \tag{4.126}$$

It can be shown that the relationship between the indices when the control and sampling properties are interchanged is given by

$$\Upsilon_D^\sigma(\vec{r}) = \left( \frac{1}{\Upsilon_q^\sigma(\vec{r})} \right)^{3/8}, \tag{4.127}$$

which provides the corresponding expression for $\Upsilon_q^\sigma(\vec{r})$.

ELF and ELI-d [32] are intimately related at the HF level. Recalling the expression for the curvature of the Fermi hole within the HF approximation, $C_{HF}(\vec{r})$ (Eq. 4.28):

$$C_{HF}(\vec{r}) = \sum_{i<j}^{occ,\sigma} |\psi_i(\vec{r})\nabla\psi_j(\vec{r}) - \psi_j(\vec{r})\nabla\psi_i(\vec{r})|^2 \tag{4.128}$$

$$= 2\rho_\sigma(\vec{r}) \left( \sum_i^{occ,\sigma} |\nabla\psi_i(\vec{r})|^2 - \frac{1}{4}\frac{|\nabla\rho_\sigma(\vec{r})|^2}{\rho_\sigma(\vec{r})} \right) \tag{4.129}$$

and substituting Eq. 4.128 into the ELI-d expression (Eq. 4.126), we have

$$\Upsilon_{D,HF}^{\sigma}(\vec{r}) = \left[ \frac{12\rho_\sigma(\vec{r})^{8/3}}{2\rho_\sigma(\vec{r}) \left( \sum_i^{occ,\sigma} |\nabla\psi_i(\vec{r})|^2 - \frac{1}{4}\frac{|\nabla\rho_\sigma(\vec{r})|^2}{\rho_\sigma(\vec{r})} \right)} \right]^{3/8} = \tag{4.130}$$

$$= \left[ \frac{6\rho_\sigma(\vec{r})^{5/3}}{\sum_i^{occ,\sigma} |\nabla\psi_i(\vec{r})|^2 - \frac{1}{4}\frac{|\nabla\rho_\sigma(\vec{r})|^2}{\rho_\sigma(\vec{r})}} \right]^{3/8}. \tag{4.131}$$

Making use of Eq. 4.127, we have the following relationships between ELI-d, ELI-q and the ELF kernel:

$$\Upsilon_{D,HF}^{\sigma}(\vec{r}) = [\chi_{BE}(\mathbf{r})]^{-3/8} \tag{4.132}$$

$$\Upsilon_{q,HF}^{\sigma}(\vec{r}) = \chi_{BE}(\mathbf{r}). \tag{4.133}$$

### 4.4.4 LOL

Since the kinetic energy density is able to reflect non-local changes in the first-order density matrix, Schmider and Becke [60, 61, 68] proposed using the Localized Orbital Locator as a bonding indicator. Alike in the case of ELF, the positive definite local kinetic energy, $t(\mathbf{r})$ is used as a starting point (recall Eq. 4.39):

$$t(\vec{r}) = \frac{1}{2}(\nabla \cdot \nabla')\rho_1(\vec{r}; \vec{r}')|_{\vec{r}'=\vec{r}}. \tag{4.134}$$

However, this function depends on the electron density, the main dependency needs to be taken away in order to reveal electron pairing. Just like in the ELF definition, it suffices to compare the kinetic energy density with the homogeneous electron gas of the same density:

$$t_h(\mathbf{r}) = \frac{3}{10}(3\pi^2)^{2/3}\rho^{5/3}(\mathbf{r}). \tag{4.135}$$

The LOL kernel, $v(\boldsymbol{r})$ is thus given by

$$v(\boldsymbol{r}) = \frac{t_h(\boldsymbol{r})}{t(\boldsymbol{r})}. \tag{4.136}$$

It is bounded from below by 0, since by definition the positive definite kinetic energy density cannot achieve negative values. It is not bounded from above. Hence it is also common to remap it in the [0,1] interval:

$$\mathrm{LOL} = \frac{v(\boldsymbol{r})}{1 + v(\boldsymbol{r})}. \tag{4.137}$$

The general features are very similar to the ones from ELF. This quantity takes high values where the kinetic energy density is low with respect to the homogeneous electron gas. The value $v(\boldsymbol{r})$ =1/2 corresponds to the regions where the kinetic energy density is the same as in a homogeneous electron gas with the same density. The distribution of maxima and minima of LOL reveals the shell structure of atoms [60] and is capable of identifying covalent bonds, lone pairs and regions of electron localization in general.

The connection between ELF and LOL comes out straightforward from their corresponding kernels:

$$\chi(\boldsymbol{r}) = \frac{t(\boldsymbol{r}) - t_w(\boldsymbol{r})}{t_h(\boldsymbol{r})} = \frac{t(\boldsymbol{r})}{t_h(\boldsymbol{r})} - \frac{t_w(\boldsymbol{r})}{t_h(\boldsymbol{r})} = v + t_{bose}(\boldsymbol{r}) \tag{4.138}$$

where $t_{bose}$ is the kernel of NCI that we will see in the next Chapter.

Both ELF and LOL are able to recover the position of localized orbitals in a system. Figure 4.35 shows how the maxima of localized orbitals and LOL are located at the same place. Following the interpretations of ELF we saw in Sect. 4.3.5.1, these points correspond to the maxima of the orbitals in those regions that can be described by a unique localized orbital (i.e. where electrons are slower) [55, 56].

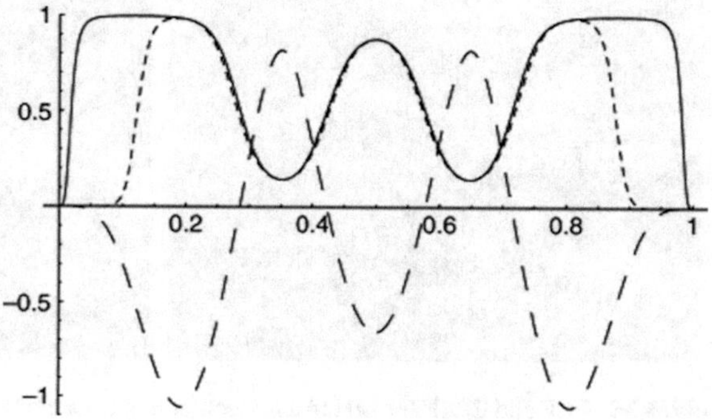


**Fig. 4.35** ELF (full curve) and functions related to it: LOL (shortdashed curve), and the difference between the Fermi hole mobility function for the given system and the uniform electron gas (divided by 100 to fit into the same graph, long-dashed curve), as a function of the position. Reprinted with permission from Ref. [55]

It should be noted that in spite of ELF and LOL having a similar shape, the chemical content of these two functions is not equivalent. Integration within LOL domains does not lead to the expected occupation numbers.

### 4.4.5 Eletron Pair Localization Function (EPLF)

This function allows the localization of electron pairs from the statistical measurement of the electronic positions across Monte Carlo configurations [59]:

$$d_{\sigma\sigma}(\vec{r}) = \langle\langle \sum_{i=1}^{N} \delta(\vec{r} - \vec{r}_i)\ \min_{j;\sigma_j=\sigma_i} |\vec{r}_i - \vec{r}_j|\ \rangle\rangle \tag{4.139}$$

$$d_{\sigma\sigma'}(\vec{r}) = \langle\langle \sum_{i=1}^{N} \delta(\vec{r} - \vec{r}_i)\ \min_{j;\sigma_j\neq\sigma_i} |\vec{r}_i - \vec{r}_j|\ \rangle\rangle \tag{4.140}$$

$$\text{EPLF} = \frac{d_{\sigma\sigma}(\vec{r}) - d_{\sigma\sigma'}(\vec{r})}{d_{\sigma\sigma}(\vec{r}) + d_{\sigma\sigma'}(\vec{r})}, \tag{4.141}$$

where $\vec{r}_i$ are the positions of the $N$ electrons for a given configuration, $\sigma_i$ is the spin of electron $i$, $\langle\langle ... \rangle\rangle$ is the stochastic average over the Monte Carlo configurations, and $d_{\sigma\sigma}(\vec{r})$ is the average distance between an electron at $\vec{r}$ and the closest same spin electron. It should be noted that there should be at least a pair of electrons of each spin so that the function is defined.

The EPLF function has values in the range $[-1, 1]$. In the regions of space where electrons are not paired, the distances $d_{\sigma\sigma}(\vec{r})$ and $d_{\sigma\sigma'}(\vec{r})$ are very similar, so that EPLF $\simeq 0$. In those regions where the electron density goes to zero, EPLF is also close

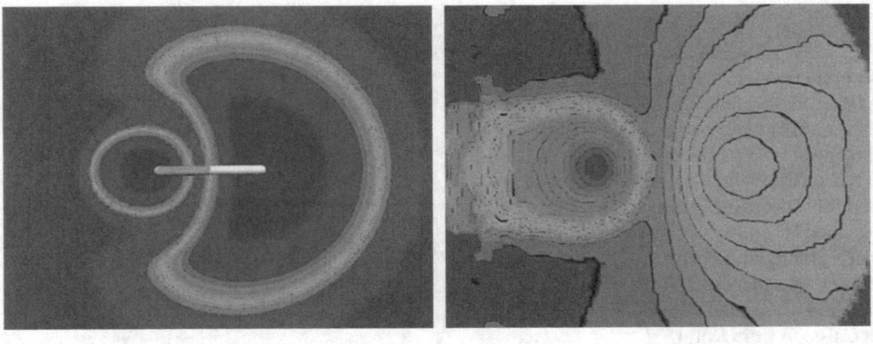

**Fig. 4.36** ELF and EPLF for LiH in a molecular plane. Li atom on the left side, H on the right side. Reproduced with permission from Ref. [59]

to zero. If instead opposite spin electrons are paired, we have that $d_{\sigma\sigma'}(\vec{r}) << d_{\sigma\sigma}(\vec{r})$, so that EPLF tends to 1. The $-1$ limits identify anomalous situations in which same spin electrons are closer than opposite spin ones (Fig. 4.36).

## References


1. Bader, R.F.W.: Atoms in Molecules. Oxford University Press, Oxford (1990)
2. Bader, R.F.W., Heard, G.L.: The mapping of the conditional pair density onto the electron density. J. Chem. Phys. **111**, 8789 (1999)
3. Bader, R.F.W., Johnson, S., Tang, T.H., Popelier, P.L.A.: The electron pair. J. Phys. Chem. **100**, 15,398 (1996)
4. Becke, A.D.: Hartree-fock exchange energy of an inhomogeneous electron gas. Int. J. Quantum Chem. **23**, 1915–1922 (1983)
5. Becke, A.D., Edgecombe, K.E.: A simple measure of electron localization in atomic and molecular systems. J. Chem. Phys. **92**, 5397 (1990)
6. Bohr, N.: I. On the constitution of atoms and molecules. Lond., Edinb., Dublin Philos. Mag. J. Sci. **26**(151), 1–25 (1913)
7. Borkman, R.F., Parr, R.G.: Toward an understanding of potential-energy functions for diatomic molecules. J. Chem. Phys. **48**, 1116 (1968)
8. Borkman, R.F., Simons, G., Parr, R.: Simple bond-charge model for potential-energy curves of heteronuclear diatomic molecules. J. Chem. Phys. **50**, 58 (1969)
9. Burdett, J.K., McCormick, T.A.: Electron localization in molecules and solids: the meaning of ELF. J. Phys. Chem. A **102**, 6366 (1998)
10. Burnus, T., Marques, M.A.L., Gross, E.K.U.: Time-dependent electron localization function. Phys. Rev. A **71**, 010,501 (2005)
11. Chamorro, E., Duque, M., Cárdenas, C., Santos, J., Tiznado, W., Fuentealba, P.: Condensation of the highest occupied molecular orbital within the electron localization function domains. J. Chem. Sci. **117**, 419 (2005)
12. Chesnut, D.B.: The use of parameter ratios to characterize the formal order of chemical bonds. Chem. Phys. **271**, 9 (2001)
13. Chesnut, D.B., Bartolotti, L.J.: The pair density description of aromaticity in some substituted cyclopentadienyl systems: a comparison of AIM and ELF bonding descriptors. Chem. Phys. **257**, 175 (2000)
14. Cioslowski, J., Liu, G.: Topology o electron–electron interactions in atoms and molecules i. the hartree–fock approximation. J. Chem. Phys. **105**, 8187 (1996)
15. Cioslowski, J., Liu, G.: Topology of electron–electron interactions in atoms and molecules ii. the correlation cage. J. Chem. Phys. **110**, 1882 (1999)
16. Coleman, A.J.: Density matrices in the quantum theory of matter: energy, intracules and extracules. Int. J. Quant. Chem. **S1**, 457 (1967)
17. Contreras-García, J., Recio, J.M.: Electron delocalization and bond formation under the elf framework. Theor. Chem. Acc. **128**, 411 (2011)
18. Coulson, C.A., Neilson, A.H.: Electron correlation in the ground state of helium. Proc. Phys. Soc. London **78**, 831 (1961)
19. Couper, A.S.: On a new chemical theory. Phil. Mag. **16**, 104–116 (1858)
20. Deb, B., Gosh, S.K.: Direct calculation of electron density in many-electron systems. i. application to closed shell atoms. Int. J. Quantum Chem. **23**, 1–26 (1983)
21. Dobson, J.F.: Interpretation of the Fermi hole curvature. J. Chem. Phys. **94**, 4328 (1991)
22. http://www.nucleomileniocuantica.cl/~personal/pfuentealba/4 ELFBook.pdf
23. Gallegos, A., Carbó-Dorca, R., Lodier, F., Cancés, E., Savin, A.: Maximal probability domains in linear molecules. J. Comput. Chem. **26**, 455 (2005)

24. Giacovazzo, C.: Fundamentals of Crystallography. International Union of Crystallography Oxford University Press, Chester, England Oxford New York (1992)
25. Gillespie, R.J.: The valence-shell electron-pair repulsion (VSEPR) theory of directed valency. J. Chem. Educ. **40**, 295 (1963)
26. Griffiths, D.J.: Introduction to Quantum Mechanics. Prentice Hall, Englewood Cliffs, NJ (1995)
27. http://www.cpfs.mpg.de/ELF:
28. Khaniani, Y., Badiei, A.: Linnett double quartet theory, challenging the pairing electrons. J. Chem. **6**, 489,153 (2008)
29. Kohout, M.: A measure of electron localizability. Int. J. Quantum Chem. **97**, 651 (2004)
30. Kohout, M., Savin, A.: Atomic shell structure and electron numbers. Int. J. Quantum Chem. **60**, 875 (1996)
31. Kohout, M., Savin, A.: Influence of core-valence separation of electron localization function. J. Comp. Chem. **18**, 1431 (1997)
32. Kohout, M., Wagner, F.R., Grin, Y.: Atomic shells from the electron localizability in momentum space. Int. J. Quantum Chem. **106**, 1499 (2006)
33. Kulkarni, S.A.: Localization of electron momentum in atomic and molecular systems. Phys. Rev. A **50**, 2202 (1994)
34. Lennard-Jones, J.E.: Equivalent orbitals in molecules of known symmetry. Proc. Roy. Soc. London, Ser. A **198**, 14 (1949)
35. Lennard-Jones, J.K.: The spatial correlation of electrons in molecules. J. Chem. Phys. **20**, 1024 (1952)
36. Lewis, G.N.: The atom and the molecule. J. Am. Chem. Soc. **38**, 762 (1916)
37. Lewis, G.N.: Valence and the Structure of Atoms and Molecules. Chemical Catalog Co., New York, US (1923)
38. Linnett, J.W.: A modification of the Lewis-Langmuir octet rule. J. Am. Chem. Soc. **83**(12), 2643–2653 (1961)
39. Llusar, R., Beltrán, A., Andrés, J., Fuster, F., Silvi, B.: Topological analysis of multiple metal–metal bonds in dimers of the $M_2$(Formamidinate)$_4$ type with M = Nb, Mo, Tc, Ru, Rh, and Pd. J. Phys. Chem. A **105**, 9460 (2001)
40. Llusar, R., Beltrán, A., Andrés, J., Noury, S., Silvi, B.: Topological analysis of electron density in depleted homopolar chemical bonds. J. Comput. Chem. **20**(14), 1517 (1999)
41. Madsen, G.K.H., Gatti, C., Iversen, B.B., Damjanovic, L., Stucky, G.D., Srdanov, V.I.: F center in sodium electrosodalite as a physical manifestation of a non-nuclear attractor in the electron density. Phys. Rev. B **59**, 12,359 (1999)
42. Matito, E., Silvi, B., Duran, M., Solà, M.: Electron localization function at the correlated level. J. Chem. Phys. **125**, 024,301 (2006)
43. Matito, E., Solà, M., Salvador, P., Duran, M.: Electron sharing indexes at the correlated level. Application to aromaticity calculations. Faraday Discuss. **135**, 9904 (2007)
44. Mullin, W.J., Blaylock, G.: Quantum statistics: is there an effective fermion repulsion or boson attraction? Am. J. Phys. **71**, 1223 (2003)
45. Munárriz, J., Laplaza, R., Martín Pendás, A., Contreras-García, J.: A first step towards quantum energy potentials of electron pairs. Phys. Chem. Chem. Phys. **1**, 4215 (2019)
46. Nalewajski, R.F., Köster, A.M., Escalante, S.: Electron localization function as information measure. J. Phys. Chem. A **44**, 10,038 (2005)
47. Noury, S., Colonna, F., Savin, A., Silvi, B.: Analysis of the delocalization in the topological theory of chemical bond. J. Molec. Struct. **450**, 59 (1998)
48. Pauli, W.: Über den einfluß der geschwindigkeitsabhängigkeit der elektronenmasse auf den zeemaneffekt. Z. Phys. **31**(1), 373–385 (1925)
49. Pilmé, J.: Electron localization function from density components. J. Comp. Chem. **38**, 204–210 (2017)
50. Pilmé, J., Piquemal, J.: Advancing beyond charge analysis using the electronic localization function: chemically intuitive distribution of electrostatic moments. J. Comp. Chem **29**, 1440 (2008)

51. Ponec, R., Chaves, J.: Electron pairing and chemical bonds. electron fluctuation and pair localization in ELF domains. J. Comput. Chem. **26**, 1205 (2005)
52. Putz, M.V.: Markovian approach of the electron localization functions. Int. J. Quantum Chem. **105**, 1 (2005)
53. Putz, M.V.: Density functionals of chemical bonding. Int. J. Mol. Sci. **9**, 1050 (2008)
54. Santos, J.C., Tiznado, W., Contreras, R., Fuentealba, P.: Sigma-$\pi$ separation of the electron localization function and aromaticity. J. Chem. Phys. **120**, 1670 (2004)
55. Savin, A.: The electron localization function (ELF) and its relatives: interpretations and difficulties. J. Molec. Struct. **727**, 127 (2005)
56. Savin, A.: On the significance of ELF basins. J. Chem. Sci. **117**, 473 (2005)
57. Savin, A., Jepsen, O., Flad, J., Andersen, O., Preuss, H., von Schneiring, H.G.: Electron localization in solid-state structures of the elements: the diamond structure. Angew. Chem. Int. Ed. Engl. **31**, 187 (1997)
58. Savin, A., Nesper, R., Wengert, S., Fässler, T.F.: ELF: the electron localization function. Angew. Chem. Int. Ed. Engl. **36**, 1809 (1997)
59. Scemama, A., Chaquin, P., Caffarel, M.: Electron pair localization function: a practical tool to visualize electron localization in molecules from quantum monte carlo data. J. Chem. Phys. **121**, 1725 (2004)
60. Schmider, H., Becke, A.: Chemical content of the kinetic energy density. J. Molec. Struct. **527**, 51 (2000)
61. Schmider, H.L., Becke, A.D.: Two functions of the density matrix and their relation to the chemical bond. J. Chem. Phys. **116**, 3184 (2002)
62. Shaik, S., Danovich, D., Silvi, B., Lauvergnat, D.L., Hiberty, P.C.: Charge-shift bonding-a class of electron-pair bonds that emerges from valence bond theory and is supported by the electron localization function approach. Chem. Eur. J. **11**(21), 6358 (2005)
63. Silvi, B.: The synaptic order: a key concept to understand multicenter bonding. J. Molec. Struct. **614**, 3 (2002)
64. Silvi, B.: The spin-pair compositions as local indicators of the nature of the bonding. J. Phys. Chem. A **107**, 3081 (2003)
65. Silvi, B.: How topological partitions of the electron distributions reveal delocalization. Phys. Chem. Chem. Phys. **6**, 256 (2004)
66. Silvi, B., Gatti, C.: Direct space representation of the metallic bond. J. Phys. Chem. **104**, 947 (2000)
67. Sutcliffe, B.T.: The development of the idea of a chemical bond. Int. J. Quantum Chem. **58**, 645–655 (1996)
68. Tsirelson, V., Stash, A.: Determination of the electron localization function from electron density. Chem. Phys. Lett. **351**, 142 (2002)
69. Via-Nadal, M., Rodríguez-Mayorga, M., Ramos-Cordoba, E., Matito, E.: Singling out dynamic and nondynamic correlation. J. Phys. Chem. Lett. **10**(14), 4032–4037 (2019)
70. Zunger, A.: Systematization of the stable crystal structure of all ab-type binary compounds: a pseudopotential orbital-radii approach. Phys. Rev. B **22**, 5839 (1980)

# Chapter 5
# Weak Interactions

**Abstract** Non-covalent interactions dominate soft matter, and have been particularly difficult to model theoretically. Given their importance in supramolecular chemistry, be it in biomolecular problems or in crystal engineering, much effort has been devoted to visualize them in real space. In this Chapter, we will review several scalar fields devised to bridge this gap, paying particular attention to the Non-Covalent Interactions index (NCI). More specifically, we will review the relationship of NCI to the bosonic kinetic energy density, its ability to recover electron localization as well as weak interactions (even beyond QTAIM features). We will show some examples where the use of NCI is specially advised and we will introduce the tools for a more detailed NCI analysis (integrals, bifurcation points, etc.). We will finish the chapter with some other existing tools for weak interactions and their mutual relationship.

## 5.1 Introduction

Weak interactions span a wide range of binding energies, and encompass hydrogen bonding, dipole–dipole interactions [34] and London dispersion, as well as more up-to-date interactions such as halogen bonds [42], $\text{CH}\cdots\pi$ [9] and $\pi\cdots\pi$ [41] interactions. Repulsive interactions, also known as steric clashes, should not be disregarded either since they can be at the heart of many reactivity and conformational questions.

It is commonly understood that bonds refer to strong covalent interactions, and that their rearrangements determine chemical reactions. However, in spite, or rather because of their weakness, weak interactions are of uttermost importance to our world as we conceive it. Chemical interactions between a protein and a drug, or a catalyst and its substrate, self-assembly of nanomaterials [23, 36], and even some chemical reactions [19, 51], are dominated by non-covalent interactions.

More specifically, non-covalent interactions are of paramount importance in chemistry and especially in bio-disciplines [12, 39], since they set up the force field scenario through which chemical species interact with each other without a significant electron sharing between them. They represent, in fact, the machinery through which molecules recognize themselves and establish how molecules will



Á. Martín Pendás and J. Contreras-García, *Topological Approaches to the Chemical Bond*, Theoretical Chemistry and Computational Modelling,
https://doi.org/10.1007/978-3-031-13666-5_5

approach and eventually pack together. This is so precisely because of their weak nature, which enables them to be labile and versatile when needed.

During the last decade, non-covalent interactions have also raised a great deal of interest in the context of materials through self-assembly [35] and crystallization [21], whose underlying general rules are at the moment too faraway to be fully rationalized and understood [26]. Knowledge of such rules would, in principle, allow to build from scratch (even complex) materials exhibiting the desired properties [18, 23, 31]. Non-covalent interactions are also extremely relevant at the intramolecular level. Although it cannot be ignored that a given observed structure is generally the outcome of a "drawing" among a plethora of energetically similar, but structurally dissimilar options [16], understanding intermolecular non-covalent interactions and their mutual interplay in the supramolecular assemblies are nonetheless a fundamental step in making progress in structural prediction and evolution.

In spite of the relevance of these interactions in chemistry and related disciplines, identifying and quantifying non-covalent interactions remains a difficult issue. Given how useful and central this concept is to chemistry, we will devote this whole chapter to the identification of weak interactions in real space.

## 5.2 Insight from Previous Tools

### 5.2.1 QTAIM

The bond paths we considered in Chap. 3 are indicative of all types of interatomic interactions. This means that they not only appear for closed-shell interactions (such as ionic), but also for weak interactions such as hydrogen bonds or van der Waals. Moreover, we have seen in Sect. 4.4.1 that we can differentiate these two types of interactions by means of $\nabla^2\rho(\mathbf{r})$. Whereas shared interactions (covalent and polar bonds) have $\nabla^2\rho(\mathbf{r}) < 0$ at their BCP, closed-shell interactions have positive $\nabla^2\rho(\mathbf{r})$ at their BCP. This is due to the fact that they are stabilized by the contraction of the electronic charge within the atomic basins, rather than shared between them. Since the density is contracting away from the interatomic surface, the density at the BCP for closed-shell interactions is also rather low.

This is a generalization of what we saw in Sect. 4.4.1 for $Cl_2$ and NaCl: whereas $Cl_2$ shows a charge accumulation, NaCl shows a charge depletion. These charge depletions are also found in weak non-covalent interactions. Figure 5.1 shows the critical points in water and methane dimers. Nuclei are in blue and bond critical points in white. These dimers illustrate that BCPs appear both in covalent and intermolecular interactions. We have a BCP for the hydrogen bond in Fig. 5.1a, and another one for the van der Waals interaction in Fig. 5.1b. $\nabla^2\rho(\mathbf{r}) < 0$ for the O-H and C-H covalent bonds, whereas $\nabla^2\rho(\mathbf{r}) > 0$ for the hydrogen bond and the van der Waals interaction. Differences also appear when looking at the electron densities: Whereas shared shell interactions (O-H and C-H covalent bonds) appear at ca. $\rho(\mathbf{r}) = 0.2$ a.u., closed-shell

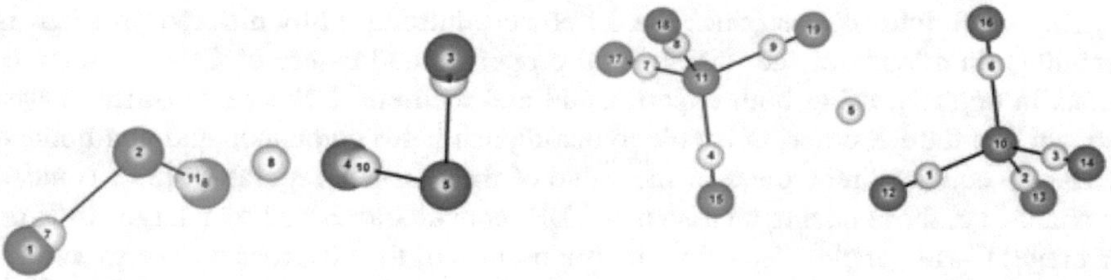

**Fig. 5.1** QTAIM critical points in water dimer (left) and methane dimer (right). (3, −3) points in blue and (3, −1) points in white

**Table 5.1** Some scalar densities at the BCP in a.u.: electron density $\rho$, Laplacian $\nabla^2\rho$, positive definite kinetic energy density, potential $V$ and energy density $H$

| Dimer | $\rho$ | $\nabla^2\rho$ | $t$ | $V$ | $H$ |
|---|---|---|---|---|---|
| $(CH_4)_2$ | 0.002513 | 0.009766 | 0.001838 | −0.00123 | 0.009766 |
| $(H_2O)_2$ | 0.017589 | 0.068866 | 0.015206 | −0.01319 | 0.002011 |
| $(HF)_2$ | 0.016924 | 0.080081 | 0.017432 | −0.01484 | 0.002588 |

interactions appear at much lower densities: $\rho(\boldsymbol{r}) = 0.03$ a.u. for the hydrogen bond and $\rho(\boldsymbol{r}) = 0.003$ a.u. for methane dimer. Hence, the electron density itself also allows to differentiate between non-covalent interactions: hydrogen bonds appear at higher densities than van der Waals interactions. It becomes then interesting to formalize these differences between closed-shell interactions themselves.

In order to do so, the Laplacian dichotomous classification can be softened to identify the "transit region" associated with incipient covalent bond formation [22]. Whereas shared interactions are characterized by $|V(\boldsymbol{r})|/t(\boldsymbol{r}) > 2$ and negative $\nabla^2\rho(\boldsymbol{r})$, closed-shell interactions with positive $\nabla^2\rho(\boldsymbol{r})$ can be divided into purely closed shell, with $|V(\boldsymbol{r})|/t(\boldsymbol{r}) < 1$ and incipient covalent bonding with $1 < |V(\boldsymbol{r})|/t(\boldsymbol{r}) < 2$. The latter give rise to $H(\boldsymbol{r}) = V(\boldsymbol{r}) + t(\boldsymbol{r}) < 0$. Hence, QTAIM provides a clear criterion to characterize different closed-shell interactions, separating stronger interactions involving orbital overlap, such as hydrogen bonds, from correlation effects, such as van der Waals interactions.

Table 5.1 shows the variation of the electron density and energy densities for several dimers. Since all of them are closed-shell interactions, the Laplacian of the electron density is positive for all of them. However, the energy ranges they span are considerably different. Methane dimer involves dispersive interactions, so the electron density and $H(\boldsymbol{r})$ are very small. Both quantities increase in the hydrogen bond of water dimer. We have also added HF dimer as an example of a very strong hydrogen bond: here the $|V(\boldsymbol{r})|/t(\boldsymbol{r})$ ratio increases and $H(\boldsymbol{r})$ becomes negative.

It is very important to recall that rigorously speaking, comparison between BCP properties from one system to another is only meaningful when the nuclei participating in the bond belong to the same family. This taken into account, the information extracted from energy densities can be extremely useful for analyzing and generalizing weak interactions (since the differences introduced by different overlaps is smaller). In a study of the fundamental properties and nature of CH··· O interactions in crystals, using both experimental and theoretical charge densities, it was shown that the BP criterion is able to distinguish between bonded and non-bonded CH··· O contacts, regardless of the value of their O··· H separation [32]. Bonded contacts, i.e. those having an associated BP, were characterized by a large (140° on average) C-H-O angle, disclosing the importance of the electrostatic energy contribution (monopole–dipole and dipole–dipole interactions) that favors linear or close to linear geometries over bent ones. Conversely, the non-bonded ones, i.e. those with no BP, were better classified as van der Waals non-bonded contacts because they could be easily folded down to C-H-O angles close to 90°, being their interaction independent of the C-H-O angle value. This was a clear example of how the electron density topology allows to gain further precious information relative to that obtained from conventional structural analysis.

## 5.2.2 *ELF*

### 5.2.2.1 Hydrogen Bonds with ELF

ELF has been developed to identify chemical bonds due to electron pairing. Thus, it provides very valuable information in the analysis of shared-electron interactions such as covalent or metallic bonding. However, it becomes a difficult tool in the description of weak interactions. Let us use the same examples already commented when presenting the difficulties of QTAIM with weak interactions to illustrate it. Figure 5.2 shows the ELF isosurfaces for water dimer (hydrogen bond) and methane dimer (van der Waals). It can be seen how Lewis' features clearly stand out: C–H bonds in methane, O–H bonds and oxygen lone pairs in water. But nothing appears for the weak interaction. Is ELF completely blind to weak interactions? Not really. As shown in the identification of weak interactions by QTAIM, hydrogen bonding includes a certain degree of molecular orbital overlap which can be indirectly casted by ELF thanks to the bifurcation diagrams (Sect. 4.3.3.1).

It is possible to define the core–valence bifurcation (CVB) index [25] as

$$\mathrm{CVB} = \mathrm{ELF}_{cv}(AH) - \mathrm{ELF}_{vv}(AHB), \tag{5.1}$$

where $\mathrm{ELF}_{cv}(AH)$ is the value of ELF at the $(3, -1)$ point that separates core and valence in the AH molecule and $\mathrm{ELF}_{vv}(AHB)$ is the value of ELF at the $(3, -1)$ point that separates the valence basin of AH from that of B. In other words, this index enables us to answer the question: Is a hydrogen bonded complex a molecule or an assembly of molecules? CVB is positive in the case of weak complexes and negative in stronger ones.

Table 5.2 shows the ELF plot of the HF$\cdots N_2$ and HF$\cdots NH_3$ complexes. The CVB indices, the counterpoise-corrected complexation energies and the $\nu_{FH}$ harmonic fre-

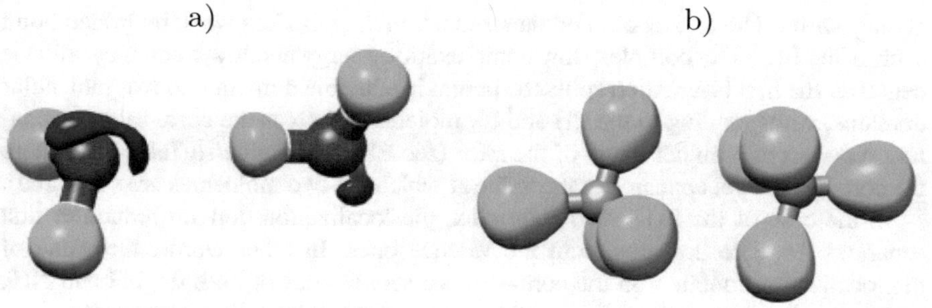


**Fig. 5.2** ELF = 0.85 isosurfaces in the water dimer (left) and the methane dimer (right). Hydrogen basins in green and lone pairs in blue

**Table 5.2** Complex data and the corresponding ELF in 3D and 1D profiles for $HF \cdots N_2$ and $HF \cdots NH_3$

| | $HF \cdots N_2$ | $HF \cdots NH_3$ |
|---|---|---|
| ELF 3D | 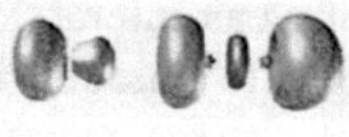 | 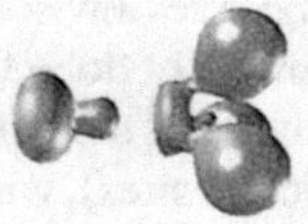 |
| ELF 1D | 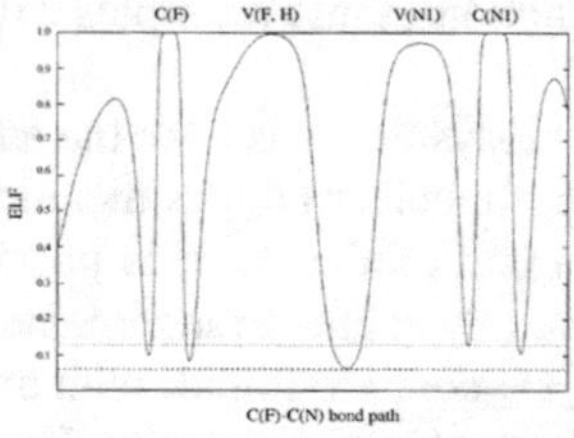 | 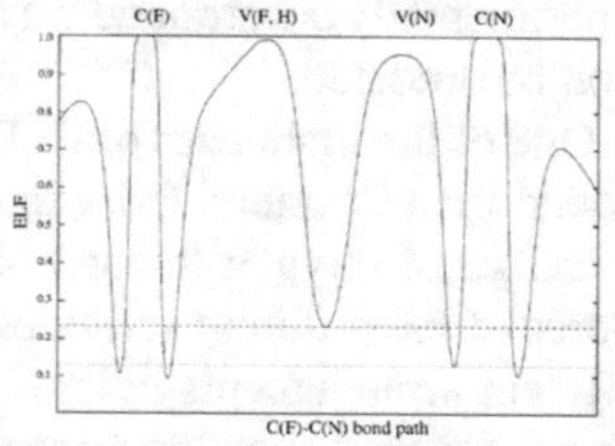 |
| CVB | -0.063 | 0.083 |
| De | 7.6 | 54.6 |
| Δω | -33 | -617 |

quency shifts. The localization domain reduction diagram in a weak hydrogen bond such as the FH$\cdots N_2$ complex (low complexation energy and low frequency shift) is negative: the first bifurcation splits the parent irreducible domain into two molecular domains corresponding to the FH and $N_2$ moieties, whereas the core–valence separation occurs at a higher value of the ELF (see ELF-1D Figure in Table 5.2, where the core has not yet appeared at the value at which the two molecules are separated).

In the case of the FH$\cdots NH_3$ complex, the localization domain reduction first separates the core domains from the valence ones. In other words, the value of the localization function on the core–valence separatrices of both the FH and $NH_3$ moieties is lower than that of the critical point of the V(N)-V(F, H) boundary.

It is interesting to note that for a given proton donor moiety, almost linear correlations have been found between the core–valence bifurcation indexes and the complexation energies or AH stretching frequencies [25]. Therefore, it provides an additional basis for a sub-classification of hydrogen bonds.

#### 5.2.2.2 Biosystems and Core Orientation

We will here review some representative applications to provide a guideline of what can be currently done for the analysis of weak interactions in biosystems with ELF. Due to the size of these systems, smaller toy models are usually introduced (e.g. capped active sites), which give then access to the wavefunction. With the increase in computational power, as well the access to highly accurate experimental densities and experimentally constrained wavefunctions, these examples will probably (hopefully) soon be outdated.

One of the main uses of ELF in biosystems has been the analysis of coordination around metal centers. For example, dinuclear copper metalloproteins are enzymes in charge of oxygen transportation and oxidation. It is possible to understand the source of the different activities given by subtle arrangements. Most commonly two different arrangements in their $Cu_2O_2$ core are found: with and without O–O bond, the one without showing a much shorter Cu–Cu distance. The ELF picture confirms the O–O bond in the first case, with a very small Cu$\rightarrow$O charge transfer. Instead, the second form shows two V(O) basins instead of a bonding basin, and a much larger Cu$\rightarrow$O charge transfer.

Similar approaches have been used for chelating agents related to poisoning. It is now accepted that one of the main problems of saturnism (plomb poisoning) is that $Pb^{2+}$ displaces the naturally occurring chelating agents in a number of metalloenzymes, disturbing their function. $Pb^{2+}$ can chelate in two different forms, holodirected (VSEPR $AL_n$ shape) or hemidirected (VSEPR $AL_nE$ shape). The ability of ELF to localize lone pairs has enabled to proof that the hemidirected form appears up to $n = 6$. From then on, the lone pair merges with the outer core and the holodirected form is present [27]. Understanding how this chelation occurs is thus relevant for the design of potential chelating agents that would displace $Pb^{2+}$ itself.

## 5.3 The Non-covalent Interactions Index (NCI)

From the previous sections, we can see that QTAIM and ELF can be limited when looking at weak interactions, or at least not-so-straight-forward to analyze. Delocalized interactions are mainly reflected as 2-body interactions in the bond path image of QTAIM, so that deviations from linearity or IQA terms need to be invoked for delocalized interactions. ELF is primarily reflecting electron localization, so that $(3, -1)$ points and bifurcation trees are needed to analyze hydrogen bonds.

Along this section, we will introduce the use of NCI, another scalar field specially developed to reveal non-covalent interactions. It is based on the electron density and its derivatives, allowing simultaneous analysis and visualization of a wide range of non-covalent interaction types as real space surfaces, adding an important tool to the chemist's arsenal we have seen up to now [14, 15, 29].

### 5.3.1 *Definition*

NCI is based on the analysis of the reduced density gradient, $s(\boldsymbol{r})$ or RDG, at low densities

$$s(\boldsymbol{r}) = \frac{1}{C_s}\frac{|\nabla\rho(\boldsymbol{r})|}{\rho(\boldsymbol{r})^{4/3}}. \tag{5.2}$$

Here, $C_s = 2(3\pi^2)^{1/3}$ and the $4/3$ exponent of the density ensures that $s(\boldsymbol{r})$ is a dimensionless quantity.

Two main interpretations can be given to this quantity, which are exposed in the coming subsections.

#### 5.3.1.1 Kinetic Energy Densities

It may be straightforwardly shown that $s(\mathbf{r})$ is the kernel of $t_{bose}(\mathbf{r})$ (see Eq. 4.138)[1]:

---

[1] From the definition of the Weiszäcker (Eq. 4.41) and Thomas–Fermi (Eq. 4.9) kinetic energy densities

$$t_w(\mathbf{r}) = \frac{1}{8}\frac{|\nabla\rho(\mathbf{r})|^2}{\rho(\mathbf{r})} \tag{5.3}$$

$$t_h(\mathbf{r}) = \frac{3}{10}(3\pi^2)^{2/3}\rho(\mathbf{r})^{5/3)} \tag{5.4}$$

We obtain the bosonic kinetic energy ratio defined in Eq. 4.138

$$t_{bose} = \frac{t_w(\boldsymbol{r})}{t_h(\boldsymbol{r})} = \frac{5}{12(3\pi^2)^{2/3}}\frac{|\nabla\rho\mathbf{r}|^2}{\rho(\mathbf{r})}. \tag{5.5}$$

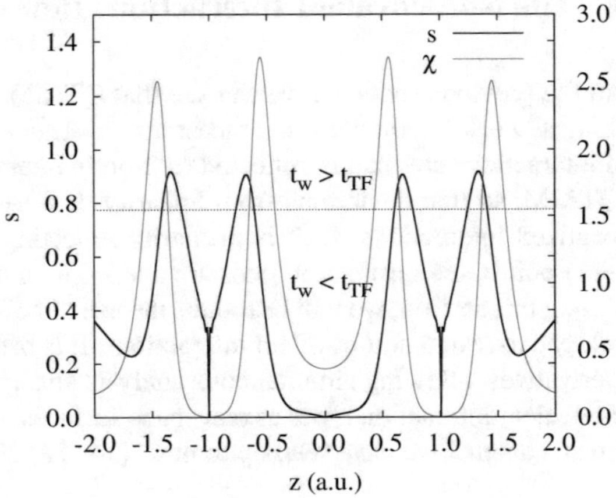


**Fig. 5.3** $s(z)$ (red line) along the internuclear axis of $N_2$. ELF core, $\chi$, has been added for comparison (green line). The regions of different kinetic energy density ratios have been marked with a dashed black line

$$s(\mathbf{r}) = \left(\frac{3}{10}\right)^{1/2} \frac{1}{2(C_F)^{1/2}} \frac{|\nabla\rho(\mathbf{r})|}{\rho(\mathbf{r})^{4/3}} \tag{5.6}$$

$$t_{bose}(\mathbf{r}) = \frac{t_w(\mathbf{r})}{t_h(\mathbf{r})} = \frac{5}{3}s(\mathbf{r})^2 \tag{5.7}$$

In order to understand Eq. 5.7, let's recall that $t_W$ is a good approximation of the real kinetic energy density whenever the region of space is described by a unique orbital, i.e. where electrons are localized. Going back to Fig. 4.5, we can see that as regions become better localized, $t$ becomes closer to $t_W$ (e.g. they are indistinguishable in the nuclei). Moreover, with Eq. 5.7 in mind, we see that for $s < \sqrt{\frac{3}{5}}$, we have $t_w < t_h$. In other words, electrons are slower than in the corresponding uniform electron gas. Indeed, in these regions, electrons are paired and the gradient of the wavefunction, given by a unique orbital, goes to zero. Regions of $s >> 1$ reflect the opposite behavior. Electrons are faster than what we would expect in the homogeneous electron gas (Fig. 5.3).

Finally, note that taking into account $\hat{t}(\mathbf{r}) = \hat{p}^2/2m$, we can also express $s$ (instead of $s^2$) in terms of the corresponding momenta

$$s(\mathbf{r}) = \left(\frac{3}{5}\right)^{1/2} \frac{p_W(\mathbf{r})}{p_h(\mathbf{r})} \tag{5.8}$$

#### 5.3.1.2 Inhomogeneities

Historically, the reduced gradient arises in the early days of density functional theory. Properties of $s(\boldsymbol{r})$ have been investigated in depth in the process of developing increasingly accurate functionals [46, 47, 49, 54, 57]. The origin of $s(\boldsymbol{r})$ can be traced back to the generalized gradient contribution to the GGA exchange energy $E_x^{GGA}$ [47]

$$E_x^{GGA} - E_x^{LDA} = -\int F(s)\rho^{4/3}(\boldsymbol{r})d\boldsymbol{r}, \tag{5.9}$$

where $F(s)$ is a function of $s(\boldsymbol{r})$ for a given spin.

Largely inspired by the Thomas–Fermi model seen in Sect. 4.3.1.1, it was used to describe the deviation from a homogeneous electron distribution.

In order to understand its relationship to inhomogeneity, let's go back to Eq. 4.6, where we see that the Seitz radius provides the size of a sphere whose volume is equal to the volume *per* free electron. Differentiating Eq. 4.6, we can easily see that the gradient of $r_s$ is directly related to $s$

$$|\nabla r_s(\mathbf{r})| = -\frac{1}{4\pi}\frac{|\nabla\rho(\mathbf{r})|}{\rho^{4/3}(\mathbf{r})} = \frac{1}{3}(2\pi)^{1/3}s(\mathbf{r}) \tag{5.10}$$

Therefore, $s$ is the rate of spatial change of the Thomas–Fermi distance metric: if we imagine our system density described by a grid where the electron density is nearly constant at each grid point, we are looking at how much the density changes from one bin to the other. Therefore, $s$ leads to a measure of local homogeneity in the electron density. There is a caveat: Since the TF model is an approximation, the inhomogeneity accounted for by $s$ is also an approximation. Hence, we expect, that it to be suitable for those molecular regions that are smooth enough, just like in the application of the LDA functional to real systems.

## 5.3.2 General Shape of *s*

### 5.3.2.1 Atoms

In order to analyze the shape of $s$ in atoms, we can assume that the spherically symmetric density of an atom $A$ is well described by a sum of exponentials, one per shell

$$\rho_A(\mathbf{r}) = \sum_{N_A} c_i^A e^{-\zeta_i^A r} \tag{5.11}$$

where $N_A$ is the number of shells of atom $A$ and $c_i^A$ and $\zeta_i^A$ are characteristic constants of each of these shells.

These Slater-type functions can be, for example, obtained from a fit to closely reproduce spherically-averaged, density functional atomic densities. When this is done, the parameters $\zeta_i^A$ are usually sufficiently different, so that each shell is described by a unique exponential (not by a sum) in most parts of space (see the approximation in blue in Fig. 5.4a). Hence, at a given shell described by $c$ and $\zeta$ each of these shells, $s$ will be given by

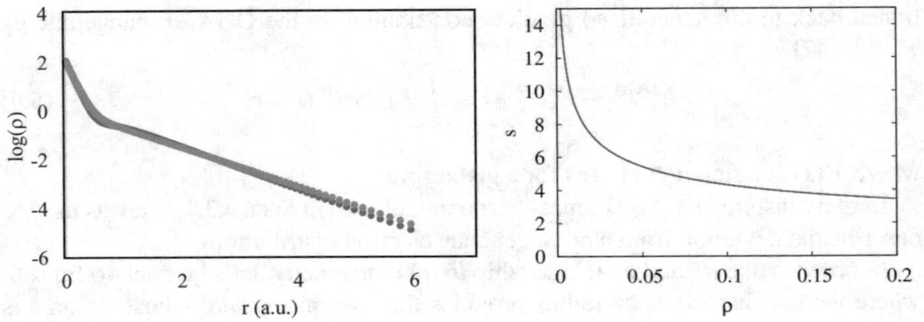


**Fig. 5.4** **a** $log(\rho)$ versus $r$ for the carbon atom: calculation in red and promolecular approximation in blue. **b** Promolecular $s(r)$ versus $\rho$ for a H atom

$$s(\mathbf{r}) = \frac{\zeta}{2(3\pi^2)^{1/3}}\rho^{-1/3} \tag{5.12}$$

This is illustrated in Fig. 5.4b, In the following, we will use promolecular densities to analyze the different trends in the atomic regions.

**At Infinity**

Starting from Eq. 5.12, it is easy to demonstrate that

$$s\,(\mathbf{r} \to \infty) = \frac{\zeta_N}{2(3\pi^2)^{1/3}}\rho^{-1/3}\,(\mathbf{r} \to \infty) \to \infty \tag{5.13}$$

This is shown in Fig. 5.4a for the hydrogen atom: at $r \to \infty$ (i.e. $\rho \to 0$), $s$ diverges.

**At the Nucleus**

The electron density at the nucleus can be approximated by a single $s$-type Slater $1s$ orbital [53]. If we take $n_{1s}$ to be the occupation of the 1s orbital, we have

$$\rho_{1s} = \frac{n_{1s}}{\pi}\left(\frac{Z_\alpha}{a_0}\right)^3 e^{-2Zr/a_0} \tag{5.14}$$

Substituting into Eq. 5.12 reveals that the reduced gradient at the nucleus has a constant value[2]:

$$s(\mathbf{R}_A) = (3\pi n_{1s})^{-1/3} \tag{5.15}$$

Table 5.3 compares the cusp values numerically obtained with those of the analytical bound of Eq. 5.15 for the first 54 elements of the Periodic Table (numerical Hartree–Fock level). Overall, we can see that a very good agreement is obtained.

[2] $s(\mathbf{R}_A) = \frac{2Z}{a_0 2(3\pi^2)^{1/3}}\left(\frac{n_{1s}}{\pi}\right)^{-1/3}\left(\frac{Z}{a_0}\right)^{-1} = (3\pi n_{1s})^{-1/3}$.

**Table 5.3** Numerical values of the lower bound of the reduced gradient for neutral atoms

| $s(0)$ | $Z = 1$ | $1 < Z \leq 54$ | $\sigma$ |
|---|---|---|---|
| Analytical (Eq. 5.15) | 0.4734 | 0.3758 | – |
| Numerical (Ref. [4]) | 0.4742 | 0.3770 | 0.0093 |
| $\Delta\%$ | 0.17% | 0.32% | |

### 5.3.2.2 Weak Interactions

We have seen that NCI reflects inhomogeneities. Let's see how this can help us in understanding weak interactions. We have used exponentials to approximate atoms, now we are going to use a sum of exponentials to approximate molecules. This was pinpointed already in Sect. 1.2. The resulting molecular density, also known as promolecular density (or independent atom model in the crystallographic community), $\rho^{pro}$, is then given by

$$\rho^{pro}(\boldsymbol{r}) = \sum_A \rho_A(\boldsymbol{r}), \tag{5.16}$$

Each atomic density, $\rho_A(\boldsymbol{r})$, is described again by a sum of exponentials.

The pitfalls of this approximation for covalent interactions were introduced in Sect. 1.2. However, a promolecular density obtained from simple exponential atomic pieces is able to predict low-density, low-reduced-gradient regions with very good accuracy due to the small relaxation in these areas: see in Fig. 5.5a how the red (SCF) and green curves (promolecular) resemble each other at $r = 0$ when H–H distances correspond to long distances (distance taken from H–H interactions in solid state).

**The 2D Plot**

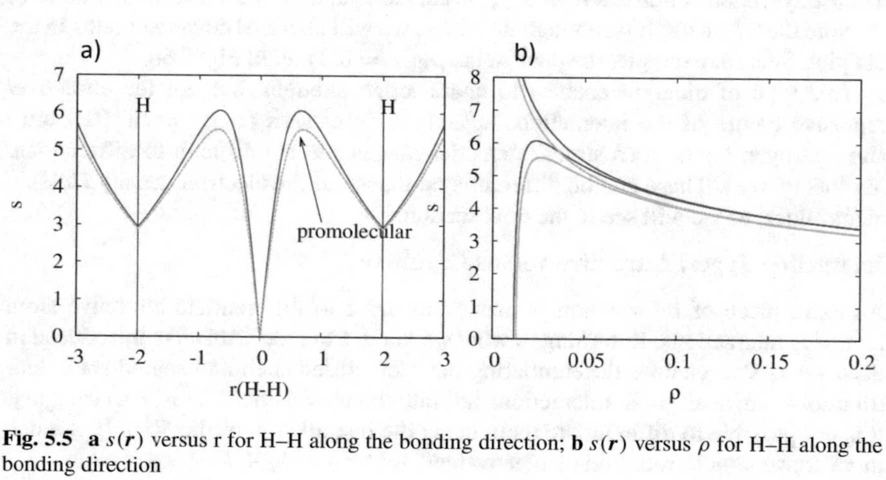


**Fig. 5.5** **a** $s(\boldsymbol{r})$ versus r for H–H along the bonding direction; **b** $s(\boldsymbol{r})$ versus $\rho$ for H–H along the bonding direction

The effect of bonding on the reduced density gradient is easy to visualize in Fig. 5.5. $s$ has a differential behavior depending on the chemical region of the molecule.

$s(\rho)$ for the tails—far away from interactions—still adopt the $\rho^{-1/3}$ behavior. This is seen in Fig. 5.5b.

However, when there is overlap between atomic orbitals, a spike in the $s(\rho)$ diagram appears. This can be seen in the middle of the H-H molecule $s(\boldsymbol{r})$ plot in Fig. 5.5a and in the low density region in of the $s(\rho)$ plot in Fig. 5.5b.

The appearance of this spike can be easily demonstrated if we calculate a model $H_2$ promolecular density made of two hydrogen atomic densities separated by a distance $R$

$$\rho(x, y, z) = \rho_A + \rho_B = \left(c_H e^{-\zeta_H \sqrt{x^2+y^2+(z-R/2)^2}}\right)^2 + \left(c_H e^{-\zeta_H \sqrt{x^2+y^2+(z+R/2)^2}}\right)^2 \quad (5.17)$$

We can make a Taylor expansion of Eq. 5.17 around $z \rightarrow 0$ to show that

$$lim_{z\rightarrow 0}\rho(x, y, z) = lim_{z\rightarrow 0}\left(e^{R/3}r + O(r^3)\right) = 0 \quad (5.18)$$

where $R$ is the distance between the atoms and $r$ is the interatomic axis. Hence, in the regions of interactions between atoms, $s$ tends to zero instead of tending to $\infty$. This extremely differential behavior ($s$ goes to 0 instead of $\infty$ when tails interact) facilitates the identification of non-covalent interactions of any nature, from hydrogen bonds to steric repulsions, van der Waals, etc.

This is illustrated for the phenol dimer in Fig. 5.6. This is a hydrogen-bonded complex, but it also exhibits non-bonding interactions within each benzene ring and a stacking interaction between the benzene rings. We thus have the three main types of non-covalent interactions. Characteristic densities of van der Waals interactions are much smaller than densities at which hydrogen bonds appear: ($\rho_{BCP}$ = ca. 0.03 a.u. for hydrogen bonds whereas $\rho_{BCP}$ = ca. 0.005 a.u. for the stacking interaction).

Note that if we use high enough densities, we will also find covalent bonds in the 2D plot. See, for example, the peak at ca. $\rho_{BCP} = 0.2$ a.u. in Fig. 5.6b.

This type of diagram enables to characterize strength, but not the attractive/ repulsive nature of the interaction. Indeed, steric clashes ($\rho_{BCP}$ = ca. 0.07 a.u.) and hydrogen bonds span similar densities ranges and are difficult to differentiate in plots of $s(\rho)$. These can be differentiated thanks to the electron density Hessian eigenvalues, as we will see in the next section.

### Interaction Types: Attractive Versus Repulsive

An extra piece of information is needed in order to differentiate attractive from repulsive interactions. Returning to what we know from the Laplacian introduced in Sect. 4.4.1, $\nabla^2\rho$ enables differentiating between closed shell and shared shell contributions. Since all weak interactions fall into the closed-shell ($\nabla^2\rho < 0$) category, it is not possible to differentiate them using the overall sign of the $\nabla^2\rho$. It is much more convenient to focus on its eigenvalues, $\lambda_i$: $\nabla^2\rho = \lambda_1 + \lambda_2 + \lambda_3$.

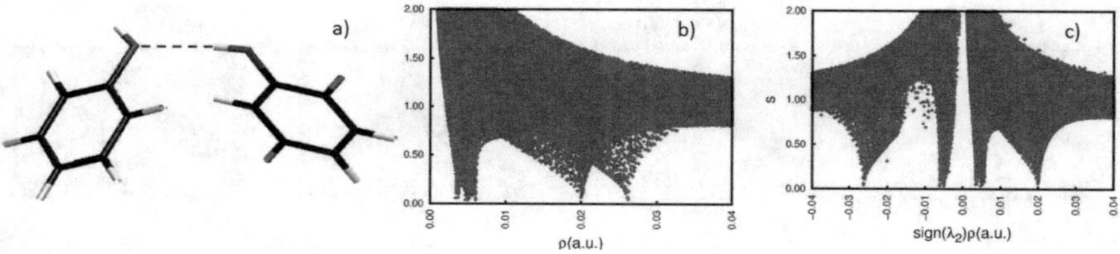


**Fig. 5.6** NCI analysis for the phenol dimer. **a** Phenol dimer (hydrogen bond is indicated with a discontinuous line) **b** $s(\boldsymbol{r})$ versus $\rho$ diagram. **c** $s(\boldsymbol{r})$ versus $\text{sign}(\lambda_2)\rho$

At the nuclei, all the eigenvalues are negative, while at the center of cages or holes (minima of $\rho$), all the eigenvalues are positive. In the remaining points of space $\lambda_3 > 0$ and $\lambda_1 < 0$, but $\lambda_2$ can be either positive or negative. In terms of $t(\boldsymbol{r})$ and $\mathcal{V}(\boldsymbol{r})$ (see Sect. 4.4.1), $\lambda_3 > 0$, $\lambda_1 < 0$ represent directions dominated by the kinetic and potential energies, respectively, whereas along the direction $\lambda_2$, there is a balance between both terms. This eigenvalue can be used to discern between attractive and repulsive contributions. Van der Waals interactions and hydrogen bonds show negative values of $\lambda_2$. Instead, in the cases where $\lambda_2$ is positive (as in rings or cages), usually several atoms interact but are not bonded, which corresponds to steric crowding according to classical chemistry.

Putting all this together, the analysis of the sign of $\lambda_2$ helps to discern the different types of weak interactions, whereas the density itself provides information about their strength. Both pieces of information can be combined in the value of $\text{sign}(\lambda_2)\rho$. This is illustrated in Fig. 5.6c, which shows a modification of the $s(\boldsymbol{r})$ plot on the left, such that the ordinate is now $\text{sign}(\lambda_2)\rho$. When the sign of $\lambda_2$ is considered, and the different nature of the phenol dimer interactions is made clear: the benzene intra-ring interactions remain at positive values, whereas the hydrogen bond now lies at negative values, i.e. in the attractive regime. The two NCIv spikes close to zero correspond to weakly-attractive dispersive interactions between the phenol rings. Therefore, the value of the $\text{sign}(\lambda_2)\rho$ at the position of the peaks in s($\rho$) plots may be used as the signature of the non-covalent interaction type. We will use this in the next section.

**The 3D Plots**

The 2D NCI plots can be used as inputs to construct 3D NCI plots, consisting of isosurfaces of $s$ enabling the visualization of non-covalent interactions. In a nutshell, it is convenient to choose a cutoff value for $s$ close to zero so as to visualize the isosurfaces corresponding to those peaks. Typically, $s$ in the range 0.3–0.5 provides an appropriate representation.

In order to discriminate between different interaction types, they are colored using the $\text{sign}(\lambda_2)\rho$ magnitude we used in the 2D plot. An RGB (red-green-blue) coloring scheme is usually chosen, where red is used for destabilizing interactions, blue for stabilizing interactions and green for delocalized weak interactions. The shading of these colors is associated with the interaction strength.

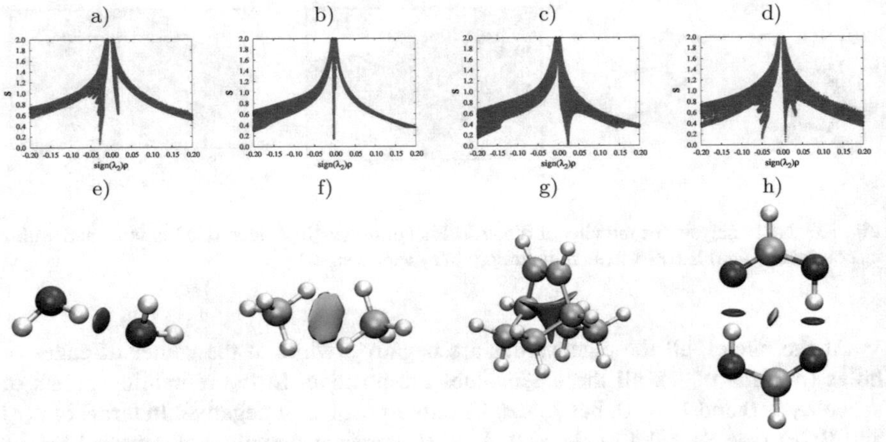


**Fig. 5.7** 2D (top) and 3D (bottom) NCI plots of test molecules (computed at the B3LYP/6-311G level of theory). From left to right: water dimer, methane dimer, bicyclooctene and formic acid dimer. The gradient isosurfaces ($s = 0.5$ a.u.) are colored on a BGR scale according to the $\text{sign}(\lambda_2)\rho$ over the range $-0.03$ to 0.03 a.u. Reprinted with permission from Ref. [13]

Figure 5.7 illustrates how this works through a set of representative molecular systems:

- a hydrogen bond (water dimer)
- a dispersive interaction ($CH_4$ dimer)
- a steric clash (bicyclo[2.2.2]octene, from now on bicyclooctene)
- an ensemble of interactions of different strength: hydrogen bonds and dispersion (formic acid dimer)

The hydrogen bond in water dimer shows a negative value of $\lambda_2$ at the critical point (Fig. 5.7a, e). Instead, $\lambda_2 \simeq 0$ (both positive and negative) for van der Waals interactions in methane dimer (Fig. 5.7b, f). Non-bonding interactions in bicyclooctene result in density depletion, so that $\lambda_2 > 0$ (Fig. 5.7c, g). It can be observed that both hydrogen bonds and steric clashes (Fig. 5.7a and c) appear at greater densities (although different $\lambda_2$ sign) than van der Waals (see Fig. 5.7b). Both strong and weak interactions are present in formic acid dimer (Fig. 5.7d and h).

Turning back to the phenol dimer example, Fig. 5.8b highlights 3D NCI surfaces. Figure 5.8c shows the BGR coloring in the $s(\boldsymbol{r})$ versus $\text{sign}(\lambda_2)\rho(\boldsymbol{r})$ to ease the interpretation of the 3D graph in Fig. 5.8d. The hydrogen bond appears in between alcohols in blue, the steric repulsion in red and the van der Waals interaction in green. The latter appears at both positive and negative $\lambda_2$ but since $\rho \simeq 0$, the sign does not play an important role.

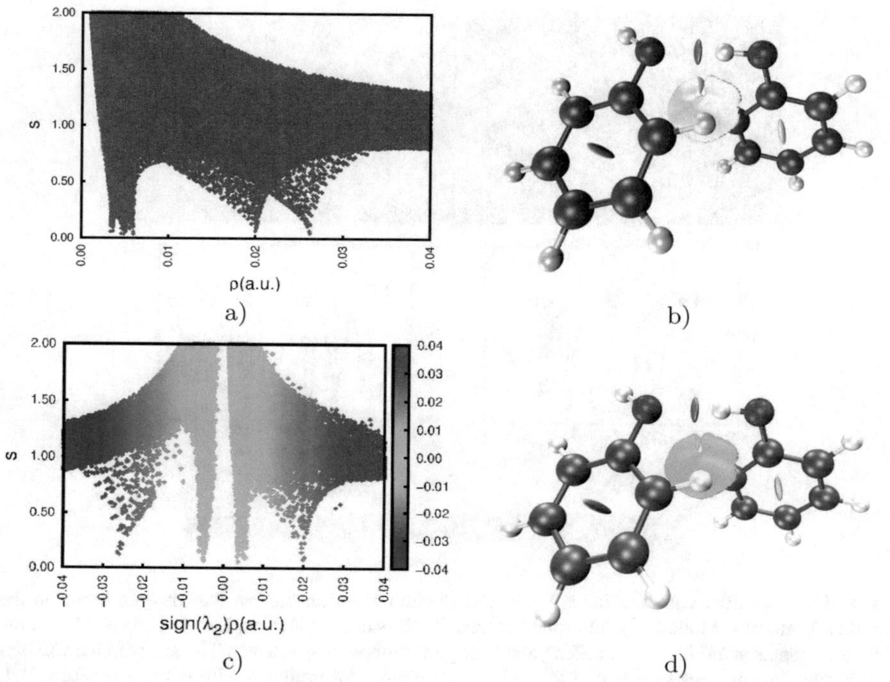


**Fig. 5.8** NCI for phenol dimer. **a** $s(\rho)$ diagram. **b** $s = 0.5$ isosurface. **c** $s$ versus sign$(\lambda_2)\rho$ diagram. **d** Colored $s = 0.5$ isosurface with a BGR code from sign$(\lambda_2)\rho$ $-0.04$(blue) to 0.04(red)

#### 5.3.2.3 Atomic and Molecular Structure

We have been focusing in the previous section on the low density region. Let's see now what the shape of NCI is at higher densities.

Recalling that $s^2 \propto t_w/t_h$ (Eq. 5.7) and looking back at Fig. 4.5, where the blue line represents $t_w/t_h$, we can see that NCI at high densities reveals electron localization.

A couple of examples are shown in Fig. 5.9. Figure 5.9a shows the result for $N_2$ as a representative of a purely covalent homonuclear molecule. Core, lone pairs and interatomic bonding regions appear as minima of $s$. The bonding region in $N_2$ corresponds to only one minimum which expands over a large region of space (both along the bonding and in the region perpendicular to it), indicating that there is a big density reconstruction upon formation of the molecule.

If we now look at a polar bond such as the one in CO (Fig. 5.9b), we see that $s$ is not concentrated at the center of the internuclear line. Instead, two minima appear: the deepest one, closer to carbon, corresponds to the QTAIM BCP, whereas the minimum closer to oxygen (which has no QTAIM analogous) corresponds to the oxygen valence shell concentration.

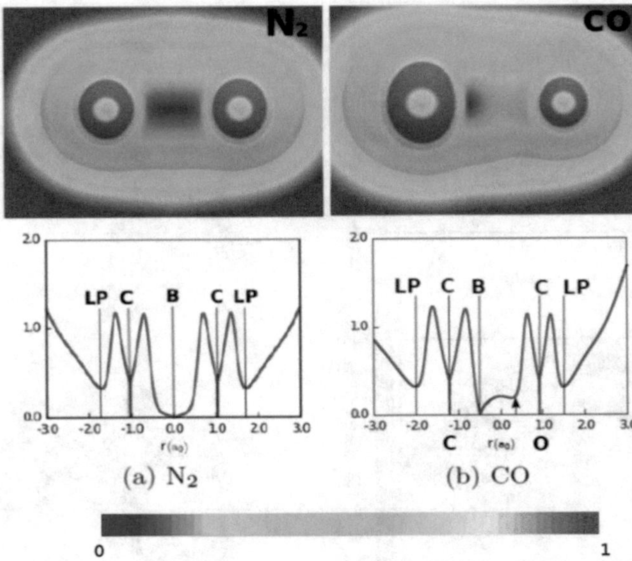


**Fig. 5.9** Top: $s(\boldsymbol{r})$ contours for **a** $N_2$, **b** CO (carbon atom on the left and oxygen atom on the right). Bottom: $s(\boldsymbol{r})$ along the internuclear axis for the same molecules. Labels C, B and LP in the bottom panels stand for core, bonding and lone pair minima, respectively. The arrow in the CO plot indicates the additional minimum in the bonding region. Reproduced with permission from Ref. [8]

It is interesting to note that the CO case is known to provide too high charges within the QTAIM approach, which has been attributed to the displacement of the BCP toward the carbon atom, yielding a very big (for QTAIM detractors, too big) oxygen basin. Within the $s$ approach, the extra minimum shows that a certain contribution from the oxygen to the bonding region is also important to understand the bonding in CO.

Finally, if we look at higher Z atoms, atomic shells can also be distinguished [11]. Vela and coworkers analyzed the behavior of $s$ in atoms across the periodic table (Li through Xe) showing that the number of maxima is in general an indicator of the row to which the atom belongs. Some exceptions in the last rows of the periodic table are nonetheless found (e.g. Cu, Pd, Ag). Results are shown for the alkali atoms in Fig. 5.10.

### 5.3.3 When to Use NCI

On many occasions, a QTAIM analysis will suffice for the analysis of weak interactions, this is especially the case for localized interactions such as strong hydrogen bonds. However, in cases where densities are very flat, QTAIM analysis can become

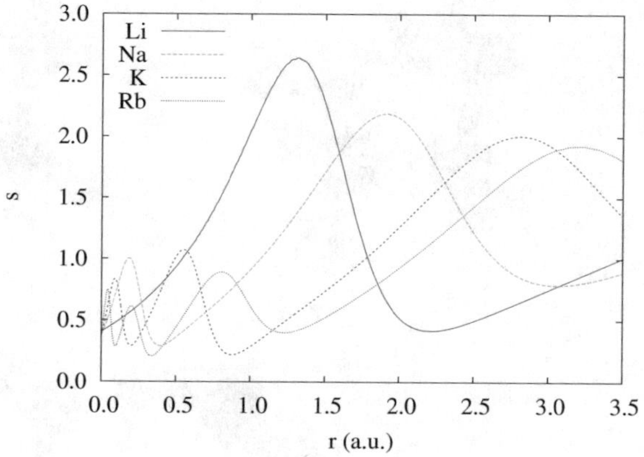


**Fig. 5.10** $s(\mathbf{r})$ versus $r$ for the alkali series

cumbersome, or even not appropriate at all. Secondly, we will also see that NCI is more stable, so it is also more appropriate when using promolecular densities.

#### 5.3.3.1 Weak Intramolecular Interactions

Experimental data suggest that the strength of the intramolecular hydrogen bonding interactions along the 1,n-alkanediols series becomes stronger as the alkane chain length increases and the OH$\cdots$O angle becomes more linear. For example, the OH-stretching vibrational mode of the hydroxyl group involved in hydrogen bonding becomes progressively more red shifted in the fundamental and overtone regions, with a corresponding increase in intensity in the fundamental region and a decrease in intensity in the first overtone region. However, when AIM theory is applied to this 1, $n$-alkanediol series, not all molecules present a hydrogen bond signature.

This is shown in Fig. 5.11 for 1, 2-ethanediol (Et2OH), 1, 3-propanediol (Pr2OH) and 1, 4-butanediol (Bu2OH). Intermolecular critical points are shown as small spheres (BCPs in purple and RCPs in yellow). Whereas a BCP and a RCP are found in Pr2OH and Bu2OH, no BCP is found for the hydrogen bond in Et2OH due to the geometrical constraints imposed by the intramolecular nature of the interaction. This image becomes more coherent if we consider the NCI surfaces. The image for Bu2OH corresponds to a typical strong hydrogen bond within the NCI framework: blue in color and very disk-shaped (see Sect. 5.3.4.6), that is, a very localized interaction, which expands from the BCP. This interaction becomes weaker in Pr2OH, and even more in Et2OH. In the latter case, the BCP is not present anymore, whereas the NCI signal is. We will see in Sect. 5.3.4.3 that this is related to a type of NCI critical points known as non-QTAIM-CPs.

In other words, the absence of a BCP should not necessarily be considered evidence as to the absence of an interaction. In constrained structures, NCI can help retrieve weak interactions where QTAIM critical points are absent.

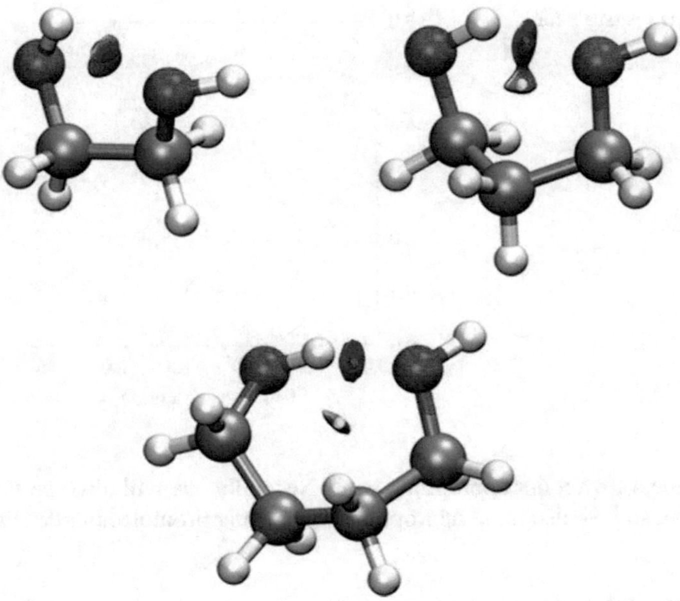

**Fig. 5.11** NCI isosurfaces for Et2OH (top left), Pr2OH (top right), and Bu2OH (bottom): s = 0.5 and a blue-green-red color scale from $-0.02 < \text{sign}(\lambda_2)\rho(\boldsymbol{r}) < +0.02$ a.u. Small spheres represent critical points: BCPs in purple and RCPs in yellow. A continuous change is observed in the isosurfaces from Et2OH to Bu2OH. Reprinted with permission from [37]. Copyright 2013

#### 5.3.3.2 Delocalized Interactions

Dispersion interactions are inherently non-local in nature, as they correlate the overall charge density distributions of individual molecular moieties to each other. However, they leave an easy to identify trace on the electron density map. Van der Waals dominated compounds show very flat densities, with very low values of $\rho$ and positive Laplacian. The flatness of the electron density profile leads in these cases to changes in the electron density topology upon small changes in geometry or calculation method. This is easily avoided by the use of NCI. From the electron density point of view, the different critical points (bonds, rings are cages) are extremely interrelated and for typical $s$ values they are all embedded into the same (in many occasions large) NCI isosurface (see Fig. 5.12). From a topological point of view, this is related to the fact that all the density critical points in the NCI surface have a very low persistence, i.e. independently of their signature, they all have very similar densities. From a chemical viewpoint, this means that all these density critical points correspond to the same van der Waals interaction, i.e. the interaction takes place between monomers and not atom pairs.

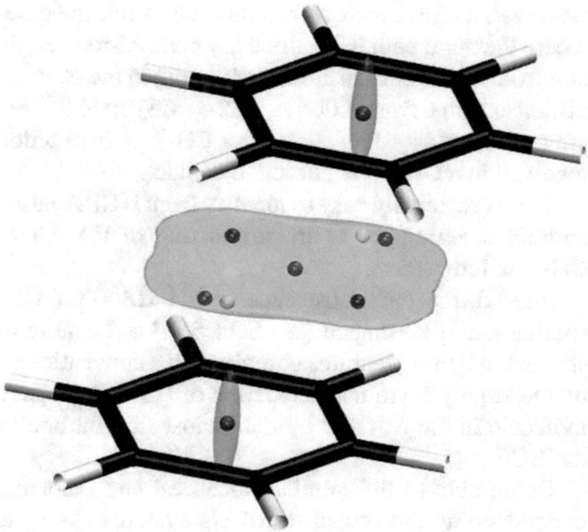

**Fig. 5.12** NCI plot of benzene dimer. The gradient isosurface ($s = 0.6$) are colored on a BGR scale according to the sign$(\lambda_2)\rho$ over the range $-0.04$ to 0.04 a.u. Critical points of the density have been included for comparison. Reprinted with permission from Ref. [15]. Copyright 2011

The presence of different critical points yields an oscillating sign$(\lambda_2)\rho$, sometimes positive, sometimes negative, which should not be taken as an indication of repulsive due to the fact that it oscillates very close to zero.

The difference between localized and delocalized interactions is even clearer in the benzene crystal, where both conventional and non conventional hydrogen bonds appear. These results have been obtained from experimental densities and a modified color scale has been used from $-0.03$ (red) to 0.035 a.u. (violet) [50].

Figure 5.13b shows the QTAIM result for a pair of benzene molecules extracted from the crystal. Two BCPs link **C3** and **C1** with **H2** and **H3** atoms, respectively. From the QTAIM perspective, this should indicate two well-defined CH$\cdots$C interactions.

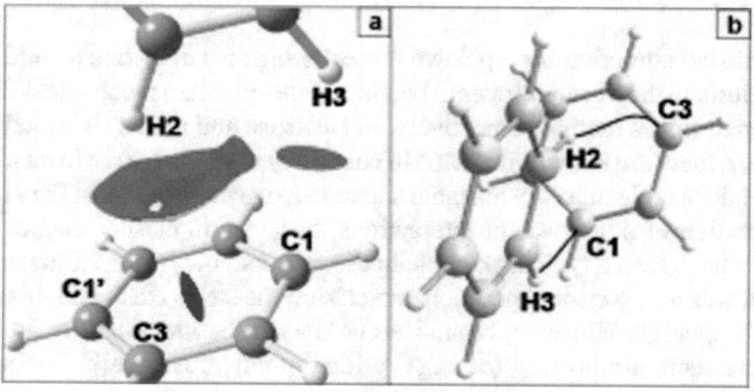


**Fig. 5.13** **a** NCI isosurfaces for a molecular pair extracted from the benzene crystal. **b** Intermolecular bond paths for the same molecular pair. Reproduced with permission from Ref. [50]

However, a closer look shows that this is not the case for the **H2**$\cdots$**C3** interaction, where the bond path is significantly bent. Moreover, the **H2** atom is roughly equidistant from all the carbon atoms belonging to the other molecule of the pair: the H$\cdots$C distances range from 3.065 Å (**H2**$\cdots$**C3**) up to 3.109 Å (**H2**$\cdots$**C1**). These features are typical of a system showing a CH$\cdots\pi$ interaction [45], and suggest that **H2** is involved in a C-H$\cdots\pi$ attractive contact.

This is extremely easy to identify from NCI. A large isosurface (Fig. 5.13a) almost entirely covering the hydrocarbon ring of the other molecule is obtained for the CH$\cdots\pi$ interaction.

Instead, the NCI isosurface of the **H3**$\cdots$**C1** CH$\cdots$C interaction looks much smaller and disk-shaped (see Sect. 5.3.4.6 for more details on how to analyze NCI shapes). All these features comply with a conventional, very weak HB, as anticipated by the largely asymmetric location of **H3** with respect to the ring atoms of the other molecule in the pair and by the almost straight bond path and density properties at the BCP.

Being able to differentiate localized and delocalized interactions is crucial for understanding the properties of big systems. As an example, the benzene T-shape conformation is more stable than its parallel-displaced conformation. However, as the number of benzene rings increases (e.g. naphthalene and anthracene dimers), the parallel-displaced conformation becomes the global minimum. Delocalized $\pi\cdots\pi$ stacking is found to be stronger than the localized T-shape interactions in larger dimers. This is related to the fact that localized interactions are rather additive (in terms of the number of atom pairs interacting), whereas delocalized interactions grow with the full system size (see Exercise 9.5.2.3).

These results suggest that systems like graphene bilayer, a really interesting material for nanotechnology, would be stabilized over saturated systems such as the graphane bilayer. This shows that getting insight into this kind of interaction is crucial to perform an accurate rational design of novel materials.

#### 5.3.3.3 Repulsive Interactions

There are no net attractive nor repulsive forces acting on a field-free quantum system at equilibrium. In this case, the overall balance of the quantum mechanical Eherenfest and Feynman forces (acting respectively on electrons and nuclei) is exactly zero.

However, there are known situations in chemistry where there are local hindrances that can render a molecule very unstable and reactive to perturbations. For an accurate characterization of non-covalent interactions, these steric clashes cannot be disregarded. We have seen a typical example in Fig. 5.7 with bicyclo[2,2,2]octene. In this case, there was a ring critical point characterizing the steric clash and this molecule is prone to opening. However, repulsive clashes can be identified by NCI even in cases where there are no ring (or cage) critical points, just like in the case of the alkanediols.

Figure 5.14a shows the $s(\rho)$ diagram for the neopentane molecule, $C(CH_3)_4$, another text-book example of steric clash. In this case, there is no critical point

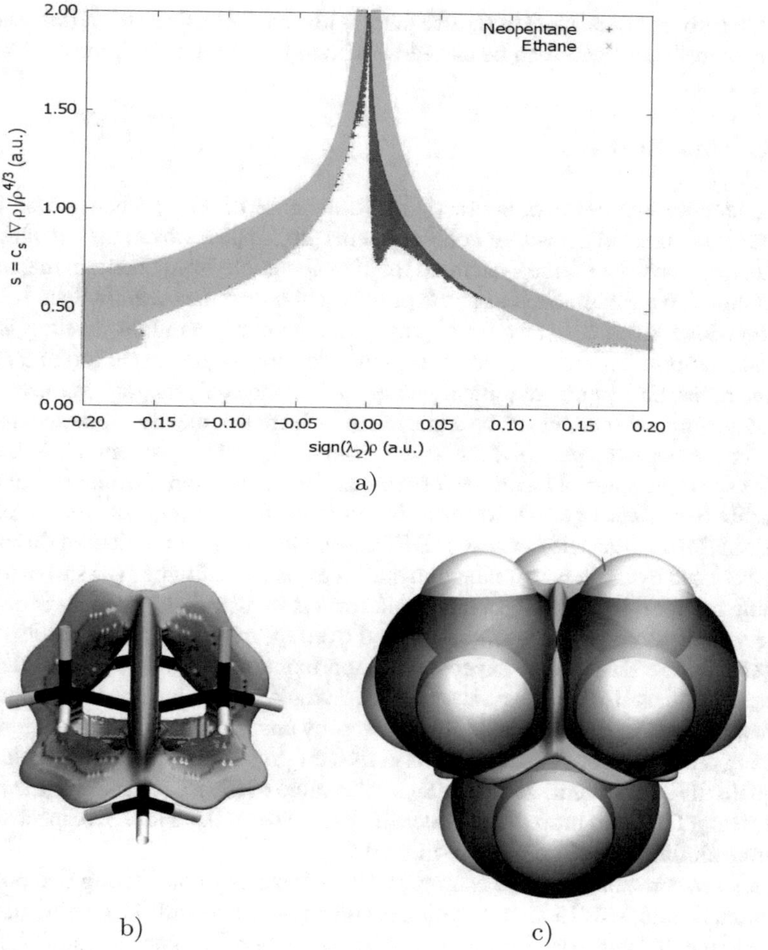


**Fig. 5.14** Plot of $s(\rho)$ (**a**) and NCI isosurfaces (**b** and **c**) for neopentane, $C(CH_3)_4$. Isosurfaces $s = 1.2$ were generated for the region below the lower edge of the ethane curve. A scale $-0.03 < \text{sign}(\lambda_2)\rho < 0.03$ a.u. was used to color the isosurfaces. Van der Waals spheres have been added to pannel c to highlight the steric nature of the interaction. Reprinted with permission from Ref. [15]. Copyright 2011

characterizing the steric clash in this molecule. However, a deviation from the $\rho^{-1/3}$ decay appears at low density and low $s$, on the positive side of the diagram (i.e. the non-bonding regime). When we find the points giving rise to the inhomogeneity, we can see that they appear between the methyl groups and correspond to the expected regions of steric hindrance (Fig. 5.14b). In Fig. 5.14c, van der Waals spheres have been added to highlight the fact that NCI surfaces appear in the region of clash between carbon atoms. Note that since there is no density critical point, this textbook intramolecular hindrance can be casted with NCI, but not with orthodox QTAIM.

#### 5.3.3.4 Biosystems

Understanding non-covalent interactions is crucial for the comprehension of the 3D structure and, thus, of the activity of biosystems [20, 24]. However, the calculation of the electron density in these systems is totally unbearable from quantum mechanical calculations. We can thus resort to the promolecular densities seen in Sect. 1.2. This has the added advantage of only requiring the geometry as input, leading to very fast calculations. Moreover, since these densities are approximate, looking at NCI regions rather than punctual critical points renders the analysis more stable.

But how precise can results be expected to be? Approximately promolecular densities for methane and phenol dimers are shown in Fig. 5.15. Resultant plots for these species show the same 2D features for relaxed (DFT) and promolecular calculations. Also, 3D isosurfaces generated from the promolecular density are very similar to those obtained at the self-consistent DFT level. Only slight quantitative differences are introduced by the density relaxation that, as expected, shift the $s(\rho)$ spikes to more bonding regimes. Specifically, a large shift toward smaller density values is obtained in the spikes corresponding to non-bonded overlap, introducing less repulsion and greater stability. This is to be expected: the approximate promolecular densities used to generate these isosurfaces have not been adjusted to alleviate Pauli repulsion. However, once the shift is taken into account (by changing the density range we are focusing on—see Sect. 5.3.4.2), results at the self-consistent and promolecular level are qualitative equivalent. As an example, the value of the isosurfaces for the phenol dimer using DFT and promolecular densities were set at 0.5 and 0.3, respectively, to generate similar 3D NCI plots (see Fig. 5.15).

Let's see some bio-inspired examples. We will start by considering two polypeptide models: one with 15 alanine residues and another one with 17 glycine residues. Whereas the first one leads to an $\alpha$-helix, the second one forms an antiparallel $\beta$-sheet (Fig. 5.16a,b).

In both cases, the blue isosurfaces reflect the presence of hydrogen bonds, though different in location. The hydrogen bonds in the $\alpha$-helix appear every $n$ and $n+4$ residues, stabilizing the helix. Instead, they are located in between the sheets in the $\beta$-sheet, stabilizing the layer arrangement. In both cases, green isosurfaces appear in between the glycine-$CH_2$ pairs, however, their role is clearly different. Whereas in the $\beta$ structure, they appear in between the sheets, they appear along the helix in the $\alpha$-helix conformation.

Figure 5.16c,d illustrates the case of ADN at two different scales. Figure 5.16d highlights the hydrogen bonds that keep the ADN basis together. We have adenine-thymine (A-T, Fig. 5.16d-top) with two hydrogen bonds and cytosine-guanine (C-G, Fig. 5.16d-bottom) with three of them (as expected from textbooks). These hydrogen bonds keep the two strands together. Figure 5.16c highlights the stabilization of the helical shape by means of $\pi$-stacking between base steps.

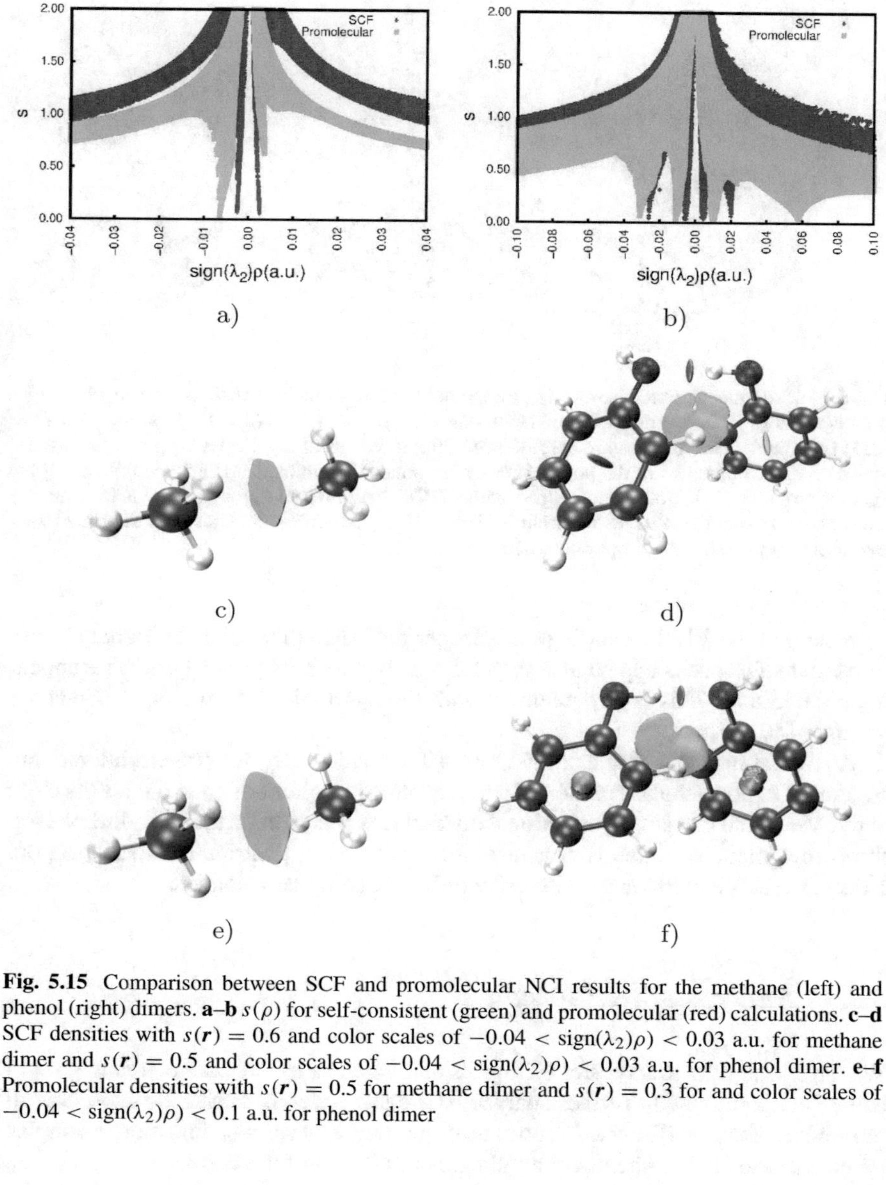


**Fig. 5.15** Comparison between SCF and promolecular NCI results for the methane (left) and phenol (right) dimers. **a–b** $s(\rho)$ for self-consistent (green) and promolecular (red) calculations. **c–d** SCF densities with $s(\boldsymbol{r}) = 0.6$ and color scales of $-0.04 < \text{sign}(\lambda_2)\rho) < 0.03$ a.u. for methane dimer and $s(\boldsymbol{r}) = 0.5$ and color scales of $-0.04 < \text{sign}(\lambda_2)\rho) < 0.03$ a.u. for phenol dimer. **e–f** Promolecular densities with $s(\boldsymbol{r}) = 0.5$ for methane dimer and $s(\boldsymbol{r}) = 0.3$ for and color scales of $-0.04 < \text{sign}(\lambda_2)\rho) < 0.1$ a.u. for phenol dimer

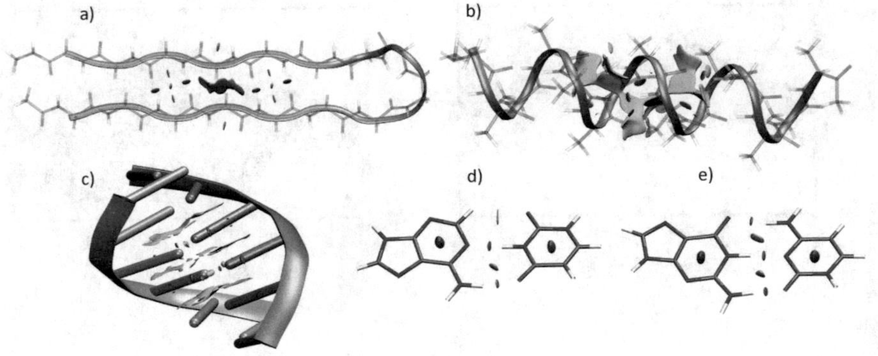


**Fig. 5.16** Promolecular NCI isosurfaces for cuboid sections of the **a** $\beta$-sheet ($s = 0.35$); **b** $\alpha$-helix polypeptides ($s = 0.35$); **c** B-form of DNA ($s = 0.25$), and **d** A-T ($s = 0.25$) and **e** C-G ($s = 0.25$) base pairs. The surfaces are colored according to values of sign$(\lambda_2)\rho$, ranging from $-0.06$ to $+0.05$ a.u. Geometries of the polypeptides were obtained with the MMFF force field using the Spartan program [17]. Both were capped with $COCH_3$ and $NHCH_3$ groups. The DNA structure was obtained using the X3DNA program [40] with ideal geometry parameters [1]. Reprinted with permission from Ref. [29]. Copyright 2010

Note that the ADN isosurfaces are deeper blue than the protein secondary structures (the ADN peaks appear at $\rho \simeq 0.065$ a.u., whereas the polypeptide ones appear at ca 0.035 a.u.) This is in agreement with the fact that ADN hydrogen bonds are stronger [30, 55, 56].

As a final biomolecular example, we will consider a ligand–protein interaction. Figure 5.17 shows the interaction of a tetracycline inhibitor bound to the tetR protein active site. Note that strong directional interactions enable to fix the ligand. However, the complementary shape is crucial in leading to a big contribution to dispersion. Hence, weak interactions are also crucial for designing new ligands.

## 5.3.4 *Analysis of NCI Results*

We have seen how to interpret NCI plots and when to use them. Let's now go into some details on how to further analyze NCI data: which isosurface to plot, how to interpret its shape and how to quantify its properties. You can also find more examples on how to use NCI in chemistry applications in Part II of the book.

### 5.3.4.1 Isosurface Value

On the one hand, we have seen with the neopentane example (Fig. 5.14) that all NCI features do not strictly have a spike that goes down to $s = 0$ so that too low a value

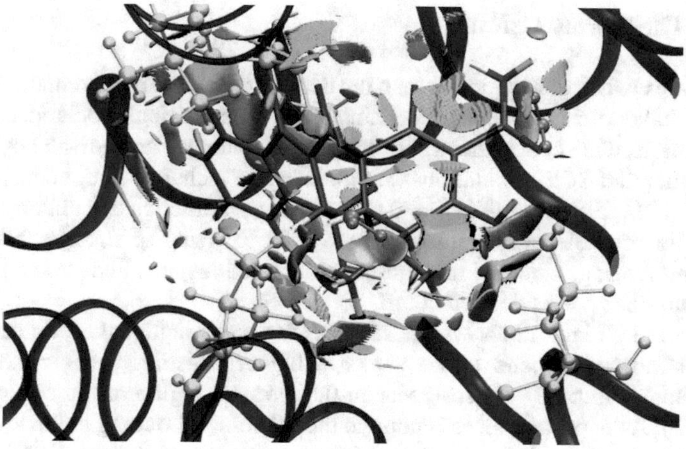

**Fig. 5.17** Promolecular NCI ($s = 0.35$) for the interaction between the tetR protein and tetracycline inhibitor. Surfaces colored in the sign$(\lambda_2)\rho$ range from $-0.06$ to $+0.05$ a.u. Reprinted with permission from Ref. [29]. Copyright 2010

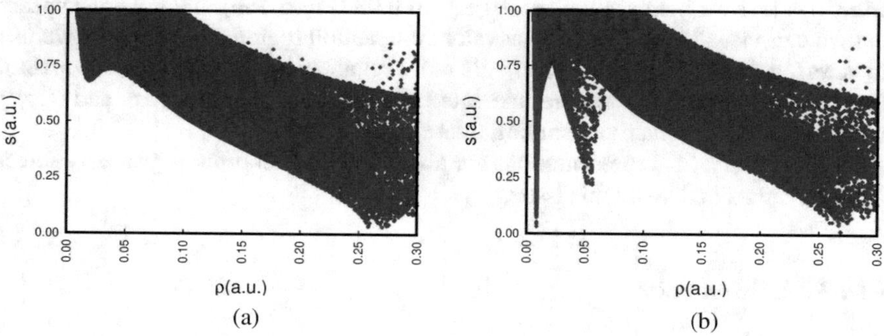


**Fig. 5.18** $s(\mathbf{r})$ against $\rho(\mathbf{r})$ plots for **a** the acetic acid molecule and **b** the acetic acid dimer as in Fig. 5.19. Reproduced with permission from [38]

might miss some of the interactions of interest. On the other hand, too high a value would disclose atomic tails of the density. If we want to make sure that all weak interactions are present, it is thus convenient to check in the 2D plot whether high peaks are present (typically if we are interested in intramolecular interactions). The isovalue is, therefore, chosen from the 2D plot so that all spikes, but only spikes (not tails), are captured.

#### 5.3.4.2 The Density Cutoff

Only features with a density lower than the density cutoff are plotted in the NCI plot. This cutoff takes different values depending on the type of calculation. Indeed, we saw in the promolecular section that a shift is needed to achieve an equivalency between promolecular and SCF calculations. This is done by a change in the density cutoff. Within `NCIPLOT`, this cutoff is $\rho = 0.05$ a.u. for wavefunction calculations and $\rho =$ 0.07 a.u. for promolecular estimates. If we want to visualize stronger interactions (e.g. strong hydrogen bonds, metallorganic interactions, covalent bonds), we will have to introduce a user-defined cutoff.

For example, Fig. 5.18 shows the 2D plots for acetic acid and acetic acid dimer. covalent bonds correspond to the peaks at higher densities, $\rho(\mathbf{r}) \approx 0.27$ (rightmost peaks in Fig. 5.18). NCIs appear in the leftmost region of the dimer 2D plot (Fig. 5.18b). As expected, there is none in the monomer. Looking at the low density peaks in the dimer, the thin peak at $\rho = 0.09$ corresponds to the van der Waals interaction in the center of the dimer, whereas the strong hydrogen bonds correspond to the peak with $\rho \approx 0.05$. Note that, in this case, the default density cutoff of 0.05 falls in the middle of the peak. When this happens, a hole inside the surface appears (Fig. 5.19a). If we increase the cutoff, e.g. to 0.08 (Fig. 5.18b), the hole disappears and we can visualize the full non-covalent interaction regions. Note that if we use no cutoff, covalent interactions and tails will also appear in the picture (Fig. 5.18c). This can be interesting if we are interested simultaneously in the weak and strong interactions (e.g. in reactivity studies, Sect. 5.3.4.4).

As a corollary, 2D representations are always advised in order to properly select a density cutoff and avoid this issue.

#### 5.3.4.3 Critical Points

We have seen in previous chapters that the analysis of critical points can enable us to condensate information about a function and understand its general shape (at least for high persistances). Let's then see where the critical points of NCI are located and how we can use them to extract information about interaction changes with the tools we have already analyzed (e.g. bifurcation trees, integrals, etc.).

**Critical Point Location**

Since the reduced density gradient contains an absolute value, it is more convenient to recall the relationship of its square with the kinetic energy densities ratio, $t_{bose}$ (Eq. 5.7). Differentiating Eq. 5.7, we can notice that the critical points of $s$ that do not coincide with those of the QTAIM match those of $t_{bose}(\mathbf{r})$ [7]

$$\nabla t_{bose}(\mathbf{r}) = \frac{10}{3} s(\mathbf{r}) \nabla s(\mathbf{r}). \tag{5.19}$$

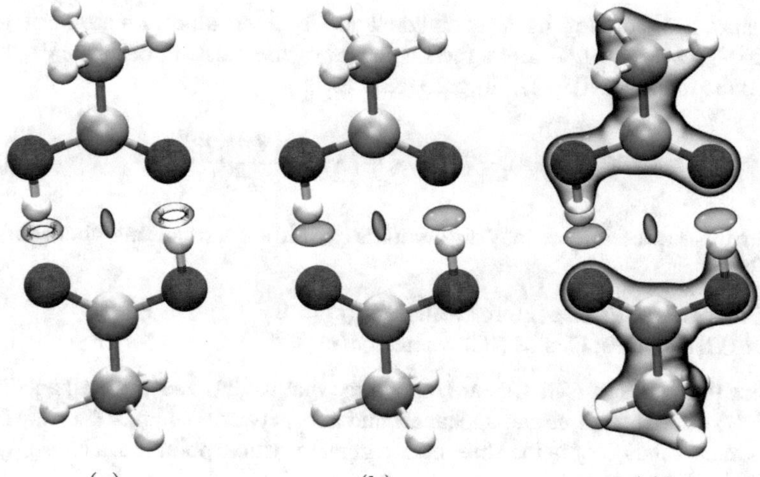


**Fig. 5.19** $s(\mathbf{r}) = 0.5$ isosurfaces on the acetic acid dimer. **a** with a $\rho(\mathbf{r}) = 0.05$ cutoff **b** with a $\rho(\mathbf{r}) = 0.08$ cutoff **c** without any cutoff. Reproduced with permission from [38]

Moreover, since $s(\mathbf{r})$ is positive semidefinite, we can relate their Hessians

$$\nabla \otimes \nabla t_{bose}(\mathbf{r}) = \frac{10}{3}\Big(\nabla s(\mathbf{r}) \otimes \nabla s(\mathbf{r}) + s(\mathbf{r})\nabla \otimes \nabla s(\mathbf{r})\Big). \tag{5.20}$$

Since at the non-QTAIM critical points $\nabla t_{bose}(\mathbf{r_{cp}}) = \nabla s(\mathbf{r_{cp}}) = 0$, and

$$\nabla \otimes \nabla t_{bose}(\mathbf{r_{cp}}) = \frac{10}{3}\Big(s(\mathbf{r_{cp}})\nabla \otimes \nabla s(\mathbf{r_{cp}})\Big), \tag{5.21}$$

then the Hessian of $t_{bose}$ at any point is equal to that of $s$ scaled by a positive quantity ($s(\boldsymbol{r})$), so that after diagonalizing it, the critical points of $s(\mathbf{r})$ and $t_{bose}(\mathbf{r})$ become identical both in location and in nature (i.e. they are homeomorphic), so we can analyze the CPs of the latter. Substituting Eq. 5.2 back into Eq. 5.19, we have

$$\nabla t_{bose}(\mathbf{r}) = \frac{X}{4c_F\rho^{8/3}(\boldsymbol{r})}\left[\mathbb{H}(\boldsymbol{r}) - \frac{4}{3}\frac{|\nabla\rho(\boldsymbol{r})|^2}{\rho(\boldsymbol{r})}\mathbb{I}\right]\nabla^t\rho(\boldsymbol{r}), \tag{5.22}$$

where $\mathbb{H}(\boldsymbol{r})$ is the electron density Hessian matrix and $I$ is the identity matrix of order 3.

Equation 5.22 is zero if and only if the expression in brackets cancels

$$\left[\mathbb{H}(\boldsymbol{r}) - \frac{4}{3}\frac{|\nabla\rho(\boldsymbol{r})|^2}{\rho(\boldsymbol{r})}\mathbb{I}\right] = 0. \tag{5.23}$$

Equation 5.23 implies that $\mathbb{H}$ is diagonal and its eigenvalues, $\lambda_i$, are positive and equal to $\frac{4}{3}|\nabla\rho(\boldsymbol{r})|^2/\rho(\boldsymbol{r})$. Since the Laplacian of the electron density, $\nabla^2\rho(\boldsymbol{r})$, is equal to the trace of $\mathbb{H}(\rho(\boldsymbol{r}))$, Eq. 5.23 leads to

$$\nabla^2\rho(\boldsymbol{r}) = \lambda_1 + \lambda_2 + \lambda_3 = 4\frac{|\nabla\rho(\mathbf{r})|^2}{\rho(\mathbf{r})}. \tag{5.24}$$

From this expression, we may differentiate two different situations where $\nabla t_{bose}(\mathbf{r})$ cancels:

1. QTAIM-CPs: CPs of $\rho(\mathbf{r})$, for which $\nabla\rho(\mathbf{r}) = \mathbf{0}$.
2. Non-QTAIM-CPs: CPs of NCI where $\nabla\rho(\mathbf{r}) \neq \mathbf{0}$.

These two types of NCI CPs are related to what we have seen for alkanediols in Sect. 5.3.3.1: NCI features can appear around a density critical point (i.e. a QTAIM-CP), as in butanediol; or in the absence of a density critical point (i.e. a Non-QTAIM-CP), as in ethanediol.

Since we have devoted Chap. 3 to the characterization of QTAIM-CPs, let's now characterize the occurrence of Non-QTAIM-CPs.

If we diagonalize $\mathbb{H}$, it can be shown that one of the non-QTAIM-CPs solutions is given by

$$\frac{\nabla^2\rho(\boldsymbol{r})}{\rho(\boldsymbol{r})} - 4\frac{|\nabla\rho(\boldsymbol{r})|^2}{\rho(\boldsymbol{r})^2} = 0 \tag{5.25}$$

This can happen in two cases that we have already covered in a phenomenological way:

1. Lewis entities (Sect. 5.3.2.3)
2. Very weak (generally intramolecular) interactions (Sect. 5.3.3.1).

Now that we know the equation to which they respond, we can go deeper into their nature.

**Lewis Entities**

The variables in Eq. 5.25 for the non-QTAIM-CPs are also involved in the one-electron potential (OEP) [28]

$$OEP = \frac{1}{4}\left[\frac{\nabla^2\rho(\boldsymbol{r})}{\rho(\boldsymbol{r})} - \frac{1}{2}\frac{|\nabla\rho(\boldsymbol{r})|^2}{\rho(\boldsymbol{r})^2}\right] \tag{5.26}$$

The regions where the OEP is negative are classically allowed: the kinetic energy takes positive values. Instead, those regions where the OEP is positive are associated with negative kinetic energy densities,[3] and therefore, are considered as

[3] Note that we are using here the $K$ definition of kinetic energy (Eq. 3.45).

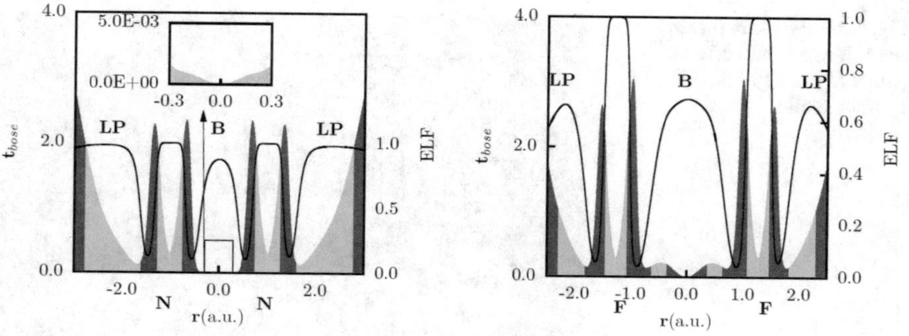


**Fig. 5.20** $t_{bose}(\boldsymbol{r})$ along with ELF values (solid black line) for $N_2$(left) and $F_2$(right). Negative (classically allowed) and positive (classically forbidden) regions of OEP are displayed as cyan and red-colored areas, respectively. Labels B and LP stand for bond and lone pair ELF basins, respectively. Reproduced with permission from Ref. [7]

classically forbidden. This has been translated chemically into Lewis regions, with atomic shells and covalent bonds typically showing negative OEP values. If we now analyze the transition from allowed to forbidden regions, we see that the ratio between $\nabla^2\rho(\boldsymbol{r})/\rho(\boldsymbol{r})$ and $(\nabla\rho(\boldsymbol{r})/\rho(\boldsymbol{r}))^2$ is equal to 1/2 for OEP $= 0$. This can be compared to Eq. 5.25, where this same ratio became 4/3 for non-QTAIM-CPs. In other words, a non-QTAIM CP of $t_{bose}(\boldsymbol{r})$ is anticipating a transition from an allowed to a forbidden region.

Figure 5.20 displays $t_{bose}(\boldsymbol{r})$ as well as ELF along the internuclear axis of $F_2$ and $N_2$. ELF is shown as a black line and colors below the $t_{bose}(\boldsymbol{r})$ line have been added to highlight OEP positive (red) and negative (cyan) regions. It can be seen how the connection between OEP and NCI enables to differentiate the core, lone pairs and interatomic bonding regions as minima separated by maxima. All CPs of $t_{bose}$ are anticipated by roots of OEP (though the opposite is not necessarily true). In line with the interpretation of $t_{bose}(\boldsymbol{r})$, these minima are regions of electron pair localization as revealed also by the ELF maxima.

Note that nuclear and bond critical points of $\rho(\boldsymbol{r})$ (QTAIM-CPs) are identified as positions where $t_{bose}(\boldsymbol{r})$ takes a value of zero. Conversely, lone pairs are not revealed by critical points of $\rho(\boldsymbol{r})$, but by critical points of the Laplacian of the electron density (Sect. 4.4.1).

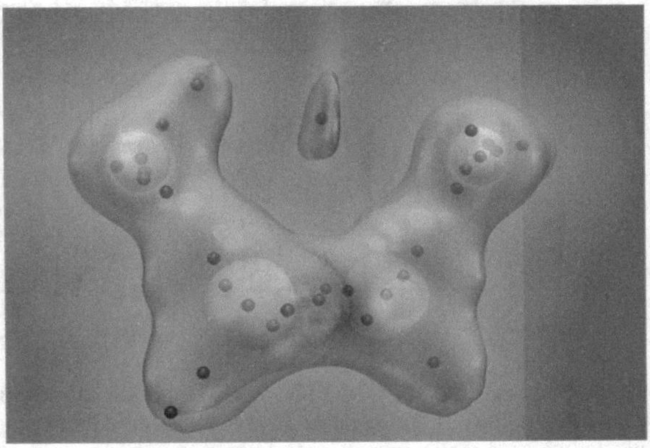

**Fig. 5.21** Maxima (breen) and minima (blue) critical points of $t_{bose}(\mathbf{r})$ for 1, 2-ethanediol

**Weak Interactions**

It is straightforward to show from Eq. 5.25 that non-QTAIM-CPs occur at $\nabla^2\rho(\boldsymbol{r}) > 0$

$$\nabla^2\rho(\boldsymbol{r}) = 4\frac{|\nabla\rho(\boldsymbol{r})|^2}{\rho(\boldsymbol{r})} \tag{5.27}$$

Indeed, non-QTAIM-CPs in the valence region are associated with closed-shell interactions not showing a BCP. This is generally the case of intramolecular weak interactions such as the Et2OH in Sect. 5.3.3.1.

Critical points for Et2OH have been highlighted in Fig. 5.21. Blue and green spheres represent minima and maxima, respectively. It is easy to see that a minimum of $s$ is found inside the NCI surface corresponding to the weak hydrogen interaction. Note that critical points also appear at the core–valence interface, which are less intuitive than ELF ones, except for the case of the C–H bond, where a minimum of $s$ can be associated with the bond.

### 5.3.4.4 Bifurcation Trees

We have just focused on the minima of NCI. But, just like in the case of ELF, it can be interesting to analyze first order saddle points and construct bifurcation trees in order to follow the evolution of interactions (Sect. 4.3.3.1). This can be done visually by changing the values of the $f$-domains and checking when surfaces merge. Note that within the NCI approach, interactions are associated with minima, i.e. we are dealing with repulsion basins or $\alpha$-limit partitions (see Sect. 2.2.2). Thus, as the value of $f$ decreases, successive splitting of the initial domains take place until all of them contain one and only one repulsor (note, that the inverse behavior is observed in $\Omega$-limit partitions, such as the one we saw for ELF, where we increase the function value in order to obtain regions with one attractor only). Here, the higher the value at the turning point, the weaker the inter-domain interaction is.

Let us see illustrate this principle with the reaction LiH+$NH_3$ → $LiNH_2$+$H_2$. We will further analyze this reaction in Part II (Sect. 6.2.3). In general, reactants and products show well differentiated fragments (Fig. 5.22), with their respective units related to the reactants (LiH and $NH_3$—Step I) or the products ($LiNH_2$ and $H_2$—Step V) at high $s$ values. Intermediate structures, instead, show a continuous NCI surface around the valence even at low $s$ values. Hence, it can be interesting to analyze the $f$-domains around the TS. Figure 5.22-Step III shows how the TS is divided into the reactants at only $s = 0.6$ (when the Li atom separates from the rest) and it is not until $s = 0.47$ that the bonds being formed and broken become irreducible domains. This homogeneity in the $s$ value characterizes the bond formation and rupture process.

#### 5.3.4.5 The Volume

In order to account for the 3D properties within NCI surfaces, density properties can be integrated to obtain the volume ($V_{NCI}$) or the electron population ($N_{NCI}$) enclosed by a volume $\Omega_{NCI}$

$$V_{NCI} = \int_{\Omega_{NCI}} d\boldsymbol{r}, \tag{5.28}$$

$$N_{NCI} = \int_{\Omega_{NCI}} \rho(\boldsymbol{r}) d\boldsymbol{r}. \tag{5.29}$$

Furthemore, attractive ($att$) and repulsive ($rep$) contributions (in the NCI method parlance) may be defined if $sign(\lambda_2)$ is considered

$$V_{NCI}^{rep} = \int_{\Omega_{NCI}} d\boldsymbol{r} \quad \forall \boldsymbol{r} \mid \lambda_2(\boldsymbol{r}) < 0, \tag{5.30}$$

$$V_{NCI}^{att} = \int_{\Omega_{NCI}} d\boldsymbol{r} \quad \forall \boldsymbol{r} \mid \lambda_2(\boldsymbol{r}) > 0, \tag{5.31}$$

Looking at the volume enables us to overcome the problems coming from symmetry imposed BCPs. The most common example of such case is the stretching of a molecule to infinity. Even though at very large distances, molecules will not interact with one another, symmetry imposes the appearance of at least one critical point between them. The use of surfaces and volumes instead of critical points to identify interactions naturally overcomes this limitation. If we look at the NCI pictures for the water dimer at very large distances ($d = 3.5$ Å), we see that NCI volume (Fig. 5.23b) basically reduces to the BCP. Since we can associate this volume (see Sect. 5.3.4.5) with the interaction energy, we easily identify that we facing the non-interacting limit. This is reflected in the $s(\rho)$ diagram (Fig. 5.23a) by an extremely sharp peak.

From the mathematical point of view, we can see by looking at Eq. 5.29 that performing the integration requires defining the NCI region ($\Omega_{NCI}$) beyond the use

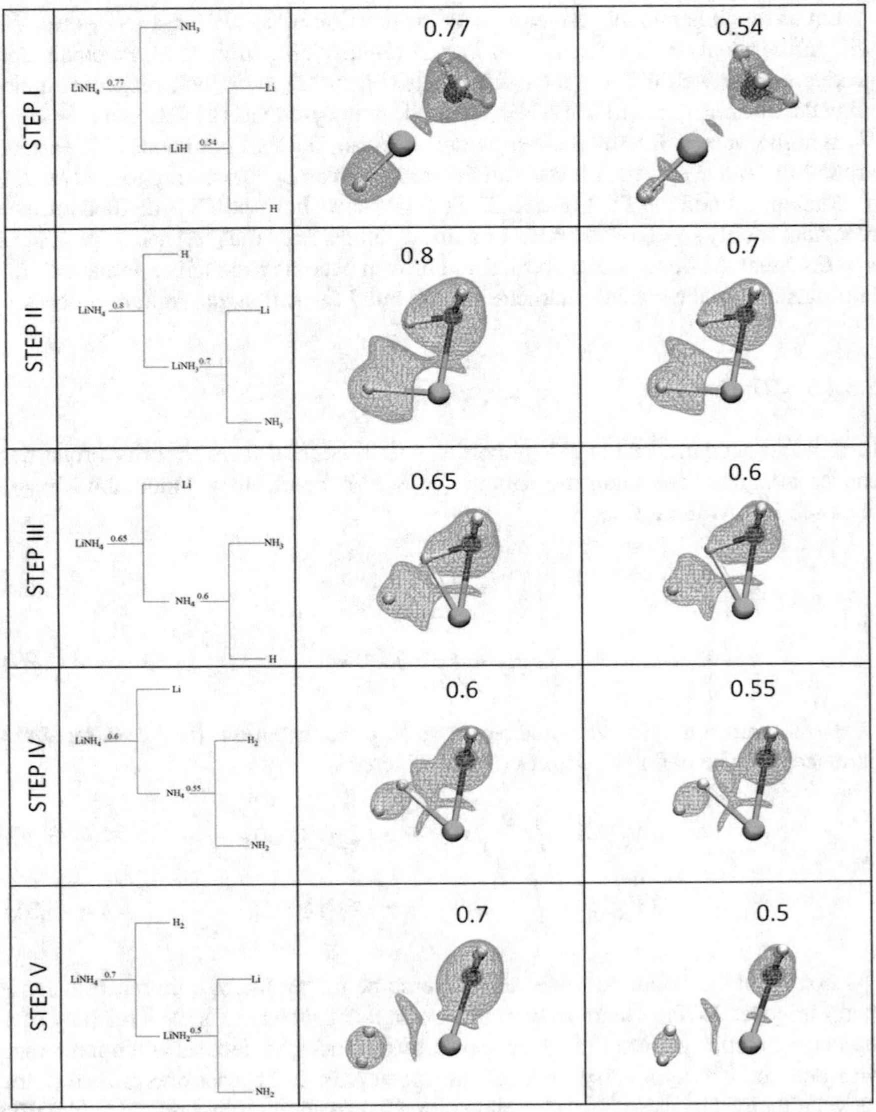


**Fig. 5.22** Bifurcation diagrams corresponding to the NCI analysis. Reprinted with permission from Ref. [2]. Copyright 2014

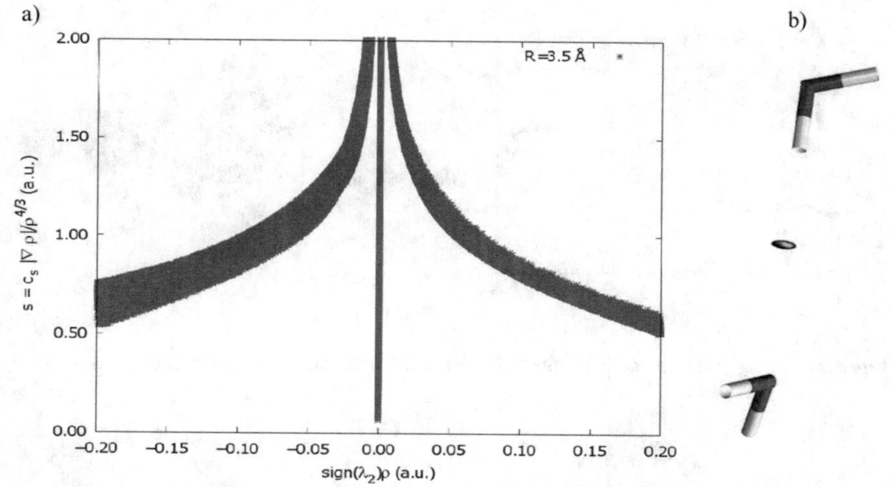


**Fig. 5.23** **a** $s(\rho)$ diagram and **b** NCI isosurface in the non-interacting regime ($d = 3.5$ Å). The isosurface was generated for $s = 0.6$ a.u. and $-0.04 < \text{sign}(\lambda_2)\rho < 0.04$ a.u

of isosurfaces. There are two main options depending on which quantity is used to establish the non-interacting reference.

**Option 1: Using $s(\rho)$ a Reference**

Since peaks appear when the system interacts, we can compare the $s(\rho)$ diagram before and after the interaction. This requires defining a reference system $(ref)$ and analyzing at each point the inequality

$$s(\boldsymbol{r}) - s^{ref}(\boldsymbol{r}) < 0. \tag{5.32}$$

Hence, both the reference and the interacting system must be computed and compared. The lower edge of $s(\boldsymbol{r})^{ref}$ is splined and all points of the interacting system $s(\boldsymbol{r})$ laying below the splined curve are localized in real space. Figure 5.24 shows the extraction of $\Omega_{NCI}$ with this method for the methane dimer as well as the 3D representation of this volume.

It is important to note that NCI volumes are obtained by integration within the region delimited by the interaction, defined as the difference with respect to a reference (the monomers). Thus, they are not isosurfaces, but the region corresponding to the peaks that appear upon the interaction of the two reference moieties [15]. In order to highlight this, Fig. 5.25 shows the integrated volumes, to be compared with isosurfaces in Fig. 5.28 for pnictogen bonds.

**Option 2: $\rho$ Reference**

Option 1 has the great disadvantage of requiring the calculation of the reference(s) on top of the full system. This can be cumbersome, or even impossible in the case

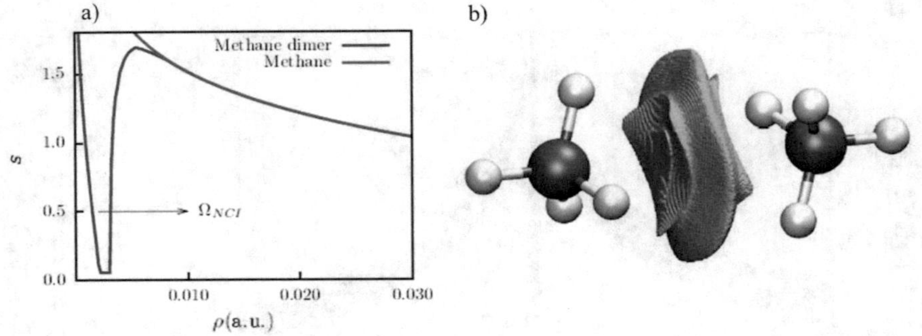


**Fig. 5.24** **a** $\Omega_{NCI}$ for the methane dimer. **b** $s(\boldsymbol{r}) = 2.0$ isosurface containing $\Omega_{NCI}$

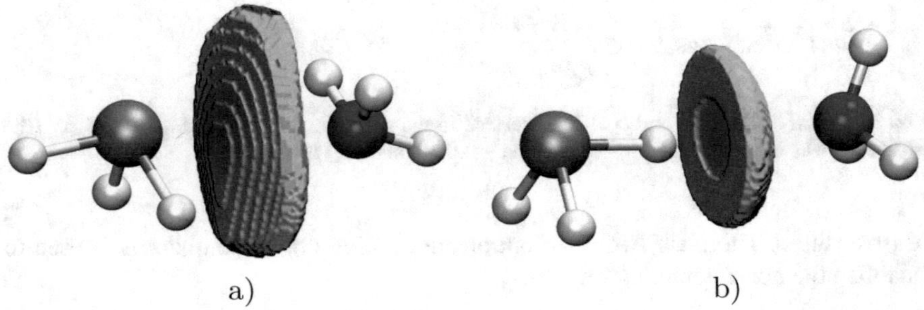


**Fig. 5.25** Comparison of interacting volumes in the $PH_3 \cdots NH_3$ complexes for **a** the pnictogen and **b** the hydrogen bond

of intramolecular interactions. To avoid these problems, and grow toward bigger systems, a second option based on the electron density has been developed.

Based on the promolecular density introduced in Sect. 5.3.3.4, the interaction region in a complex $AB$ can be determined as those parts of space where the electron density shows important contributions from monomers $A$ and $B$. In other words, we look at points that while having a low $s$ value, do not have an electron density that comes primarily from only one of the monomers

$$\mathbf{r}_i \in \Omega_{NCI} \begin{cases} s(\mathbf{r}_i) < s^{ref} \\ \rho^A(\mathbf{r}_i) < \gamma^{ref} \rho(\mathbf{r}_i) \\ \rho^B(\mathbf{r}_i) < \gamma^{ref} \rho(\mathbf{r}_i), \end{cases} \tag{5.33}$$

where $\gamma^{ref}$ is a user-defined threshold value, typically close to 0.95. Note that in this case, $s^{ref}$ is not local, but a constant value throughout molecular space. Typically, $s = 1.0$ is used (see Fig. 5.26). This might seem surprising at first, since $s = 1.0$ can spread to atomic tail regions. However, since the intramolecular regions are being deleted by the $\gamma^{ref}$ threshold, atomic tails do not interfere in the picture. This is shown

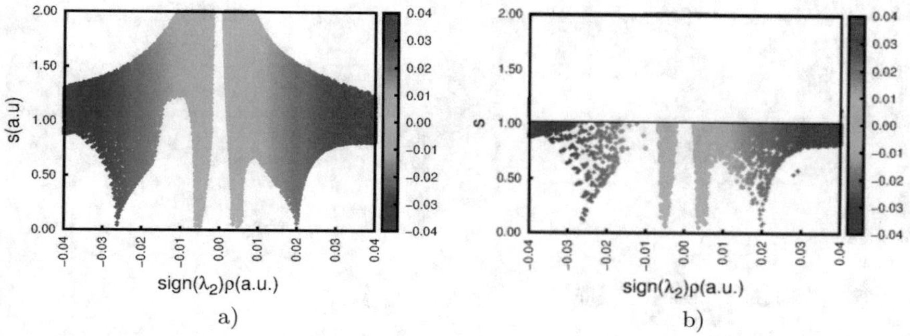


**Fig. 5.26** $\Omega_{NCI}$ obtained from the integration approach. **a** Full result. **b** $\Omega_{NCI}$ defined from the density reconstruction and constant $s$ (Eq. 5.33). Parameters used: $\gamma_j^{ref} = 0.95$ for $j$ = A, B (this can be seen from the smaller concentration of points); and $s^{ref} = 1.0$ (black line). Reprinted with permission from Ref. [6]. Copyright 2020

in Fig. 5.26: the picture after applying the intramolecular criterion (Fig. 5.26-right) shows a depletion of points in the tails with respect to the original (Fig. 5.26-left).

In order to test this approach, tests have been carried out on small molecules. The S66 dataset [48] is a widely used benchmarking reference for interaction energies in small non-covalent dimers. Following the definition of NCI volumes, $\Omega_{NCI}$ given by Eq. 5.33, we have computed the integrals of electron density at different values of $n$

$$I_n = \int_{\Omega_{NCI}} \rho^n(\mathbf{x})dx, \qquad n = 0,\ 1,\ \frac{4}{3},\ 1.5,\ \frac{5}{3},\ 2,\ 2.5,\ 3, \tag{5.34}$$

Analyzing the correlation with the dataset reference energies , we find that the highest correlation is obtained for $n = 2 - 2.5$ and $s^{ref} = 1.0$. Figure 5.27a shows the fit for $n = 2.5$ for which the correlation coefficient $R$ is 0.94.[4]

Since intermolecular interactions can span a wide range of distances, in a second step, we have carried out similar analyses for the S66x8 set [10] leading also to a good description of the involved energetics (see Fig. 5.27b). This is especially interesting since in many occasions, the main covalent framework imposes non-equilibrium distances among non-covalently bonded parts of the biomolecule(s). All in all, this approach allows the estimation of the interaction energy of the S66 and s66x8 sets from atomic densities at a trivial cost. At the time of writing this book, more tests still need to be carried out.

[4] $R = 0.93$ for $n = 2.0$.

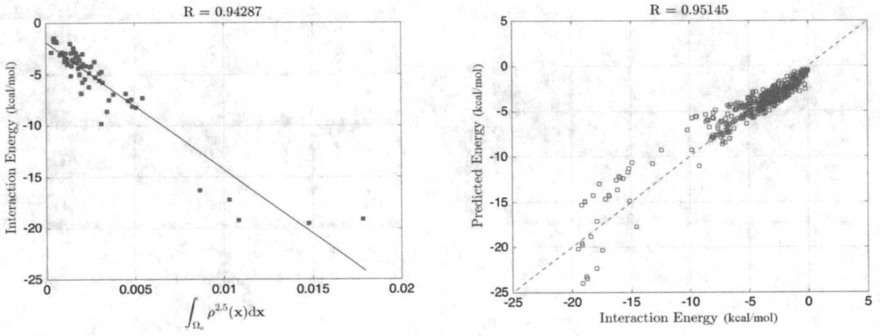


**Fig. 5.27** **a** Data fitting of the integral $I_{2.5}$ and the interaction energy for S66 dimers [48]; **b** The predicted interaction energy and the exact interaction energy for all S66x8 dimers [10]. Reprinted with permission from Ref. [6]. Copyright 2020

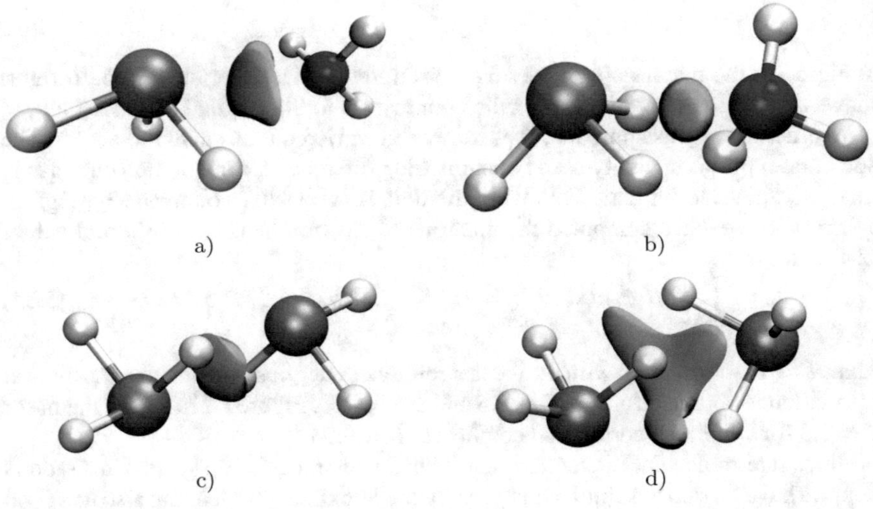


**Fig. 5.28** NCI isosurfaces of the $PH_3$-$NH_3$(top) and $PH_3$-$PH_3$ (bottom) complexes. **a, c** Pnictogen bonded and **b, d** hydrogen bonded configurations. 3D isosurfaces were produced using cutoff values of $s(\boldsymbol{r}) = 0.5$ and $\rho(\boldsymbol{r}) \leq 0.05$

### 5.3.4.6 Shape

It emerges that a one-to-one inverse correlation seems to exist between the directionality (and the strength) of specific non-covalent interaction and the surface/volume ratio of the corresponding NCI isosurface. In particular, the stronger the interaction is, the smaller and more disk-shaped the $s$ surface appears in real space (and the more negative the $\rho$ sign($\lambda_2$) values are). For example, hydrogen bonds appear as localized interactions, whereas dispersive ones appear as delocalized spread surfaces.

**Table 5.4** Classification of CPs of $\rho(\boldsymbol{r})$ according to the relative magnitude of the Hessian eigenvalues $|\lambda_1| < |\lambda_2| < |\lambda_3|$, along with the local shape of $s(\boldsymbol{r})$ around the CP

| Curvature relations | Shape | Interaction |
|---|---|---|
| $\lambda_1 \approx 0, \lambda_2 \approx 0$ | Planar | Delocalized (e.g. van der Waals) |
| $\lambda_1 \approx \lambda_2 << \lambda_3$ | Disk | Localized (e.g. hydrogen bond) |
| $\lambda_1 << \lambda_2 \approx \lambda_3$ | Cylindrical | Covalent interaction |

This effect is highlighted in Fig. 5.28 for pnictogen bonds. Complexes Fig. 5.28a, c and d are characterized by large, shapeless NCI isosurfaces. They highlight the multiatomic nature of dispersion interactions. Instead, the complex in Fig. 5.28b shows a localized round shaped NCI isosurface characteristic of a highly directional interaction, alike hydrogen bonds. This is the strong PH$\cdots$N pnictogen bond.

There are two options for quantifying this "shape" effect: local and global. One can resort to an approximate local measure at the BCP. As highlighted by Bohorquez et al. [5], the density eigenvalues at the BCP give an idea of the shapes. In Table 5.4, the local shapes of typical bonds are summarized. A diffuse region, such as the ones expected in dispersive interactions, will be characterized by a small $|\lambda_i|$ values. This is due to the fact that the electron density is very flat in this region. The effect will be specially visible in the direction perperdicular to the bond, so that both $|\lambda_1|$ and $|\lambda_2|$ should be small. Instead, a localized interaction gives rise to NCI thick disk shapes. These two characteristics are also reflected in the eigenvalues. Thick surfaces are given by higher $\lambda_3$ values at the BCP, and disk shapes (i.e. an isotropic behavior in the direction perpendicular to the bond) are characterized by $\delta = \lambda_1/\lambda_2 \simeq 1$. As the multiatomic contribution increases, $\delta$ will depart from unity.

Table 5.5 reveals this agreement between eigenvalues and the shape of the isosurface. The interaction Fig. 5.28b, a localized hydrogen bond, has the biggest eigenvalues and $\delta$ closest to one. Instead, the interaction Fig. 5.28d has the smallest eigenvalues and $\delta$ departs noticeably from unity.

These measures are local (centered at a point), and have limitations. For example, the dispersive contribution in pnictogen bonds Fig. 5.28c and d is very similar, and this is not captured by the local measures. It can then become interesting to look at global measures, such as the volume (Sect. 5.3.4.5). Table 5.5 reveals the agreement between the dispersion energy and the volume of the NCI region, $V_{NCI}$. We can see the bigger volume associated with the dispersive pnictogen interaction in comparison with the hydrogen bonded one in the $PH_3 \cdots NH_3$ complexes, in agreement with their dispersive contributions, $E_{disp}$.

**Table 5.5** NCI volumes ($V_{NCI}$, a.u.), interaction ($E_{int}$, kcal/mol), dispersion energies ($E_{disp}$, kcal/mol), $\lambda_3$, $\lambda_2$ and $\delta = \lambda_1/\lambda_2$ of the pnicogen ($PH_3 \cdots NH_3$, $H_3P \cdots PH_3$) and hydrogen bonded ($H_3P \cdots NH_3$, $PH_3 \cdots PH_3$) complexes

| System | Bond | $V_{NCI}$ | $E_{int}$ | $E_{disp}$ | $\lambda_3$ | $\lambda_2$ | $\lambda_1/\lambda_2$ |
|---|---|---|---|---|---|---|---|
| (a) $H_3P \cdots NH_3$ | Pnictogen | 26.69 | −1.60 | −3.70 | 0.0407 | −0.0058 | 0.98 |
| (b) $PH_3 \cdots NH_3$ | HB | 13.77 | −0.81 | −2.97 | 0.0473 | −0.0096 | 0.997 |
| (c) $H_3P \cdots PH_3$ | Pnictogen 1 | 41.21 | −1.10 | −4.83 | 0.0287 | −0.0042 | 0.98 |
| (d) $PH_3 \cdots PH_3$ | Pnictogen 2 | 44.34 | −0.70 | −4.83 | 0.0189 | −0.0029 | 0.739 |

## 5.4 Other Descriptors

We have already seen in Sect. 5.3.2.3 that NCI is able to simultaneously reveal pairing features and weak interactions. Other descriptors for revealing non-covalent interactions have been proposed in the literature, with a more or less pronounced focus on non-covalent interactions but similar visualization abilities. In this section, we will review the most prominent ones and their physical basis.

### 5.4.1 LED

The Localized Electron Detector (LED), $\tilde{P}(\boldsymbol{r})$ [3, 5]:

$$\tilde{P}(\boldsymbol{r}) = \frac{-\hbar}{2} \frac{\nabla\rho(\boldsymbol{r})}{\rho(\boldsymbol{r})} \tag{5.35}$$

was introduced as a measure of the bosonic character of the electron density which identifies the presence of paired electrons in atoms and molecules. As can be seen from Eq. 5.35, LED is a vector field that runs antiparallel to the gradient of the electron density $\nabla\rho(\boldsymbol{r})$, and therefore, depicts the same electron density gradient paths studied in QTAIM.

The magnitude of LED, $|\tilde{P}(\boldsymbol{r})|$ is bounded

- Just like $s(\boldsymbol{r})$, $|\tilde{P}(\boldsymbol{r})| = 0$ at the critical points of the electron density
- $|\tilde{P}(\boldsymbol{r})|$ is maximal in the neighborhood of the heaviest nucleus within the molecule (located at $\boldsymbol{R_A}$ and with atomic number $Z_A$), $lim_{\boldsymbol{r}\to\boldsymbol{R_A}}|\tilde{P}(\boldsymbol{r})| = p_0 Z_A$ (Kato's equation), where $p_0$ is the atomic unit of momentum
- At long distances, valence electron kinetic energies are limited by the ionization energy, yielding $lim_{|\boldsymbol{r}|\to\infty}|\tilde{P}(\boldsymbol{r})| = \sqrt{2m_e I}$, where $I$ is the ionization energy.

### 5.4.2 DORI

Another option for revealing weak interactions is the Density Overlap Regions Indicator (DORI), which was tailored to reveal the regions of space, where the total electron density results from a strong overlap of shell, atomic, or molecular densities [52]

$$\theta(\boldsymbol{r}) = \frac{\left(\nabla\left(\frac{\nabla\rho(\boldsymbol{r})}{\rho(\boldsymbol{r})}\right)^2\right)^2}{\left(\frac{\nabla\rho(\boldsymbol{r})}{\rho(\boldsymbol{r})}\right)^6} \tag{5.36}$$

Considering that $\theta(\boldsymbol{r})$ is unbound from above, it is not very convenient from the point of view of its visualization, so it is mapped in the [0, 1] range

$$\text{DORI}(\boldsymbol{r}) = \frac{\theta(\boldsymbol{r})}{1+\theta(\boldsymbol{r})} \tag{5.37}$$

### 5.4.3 Comparison

In order to understand the connection between NCI, LED and DORI, we can resort to the local-wave vector, $\boldsymbol{k}(\boldsymbol{r}) = -\nabla\rho(\boldsymbol{r})/\rho(\boldsymbol{r})$ [33, 44]. It has been shown that the local-wave vector is the gradient of the per particle Shannon information (as well as the square of the per particle Fisher information) [43]. Since we have

$$t_w(\boldsymbol{r}) = \frac{\rho(\boldsymbol{r})\boldsymbol{k}^2(\boldsymbol{r})}{8}, \tag{5.38}$$

descriptors based on $t_w(\boldsymbol{r})$ can also be directly linked to information theory.

More specifically, LED is given by

$$\tilde{P}(\boldsymbol{r}) = -\frac{\boldsymbol{k}(\boldsymbol{r})}{2}. \tag{5.39}$$

Comparing $|\tilde{P}(\mathbf{r})|$ with that of a suitable reference system such as the homogeneous electron gas, enables to obtain NCI's kernel, $s(\boldsymbol{r})$

$$s(\boldsymbol{r}) = \frac{|\tilde{P}(\mathbf{r})|}{p_{TF}(\boldsymbol{r})}, \tag{5.40}$$

$$= \frac{|\boldsymbol{k}(\boldsymbol{r})|}{2p_h(\boldsymbol{r})} \tag{5.41}$$

where $p_h(\boldsymbol{r})$ is the Fermi momentum $p_h(\boldsymbol{r}) = (3\pi^2\rho(\boldsymbol{r}))^{1/3}$ (see Sect. 4.3.1.1).

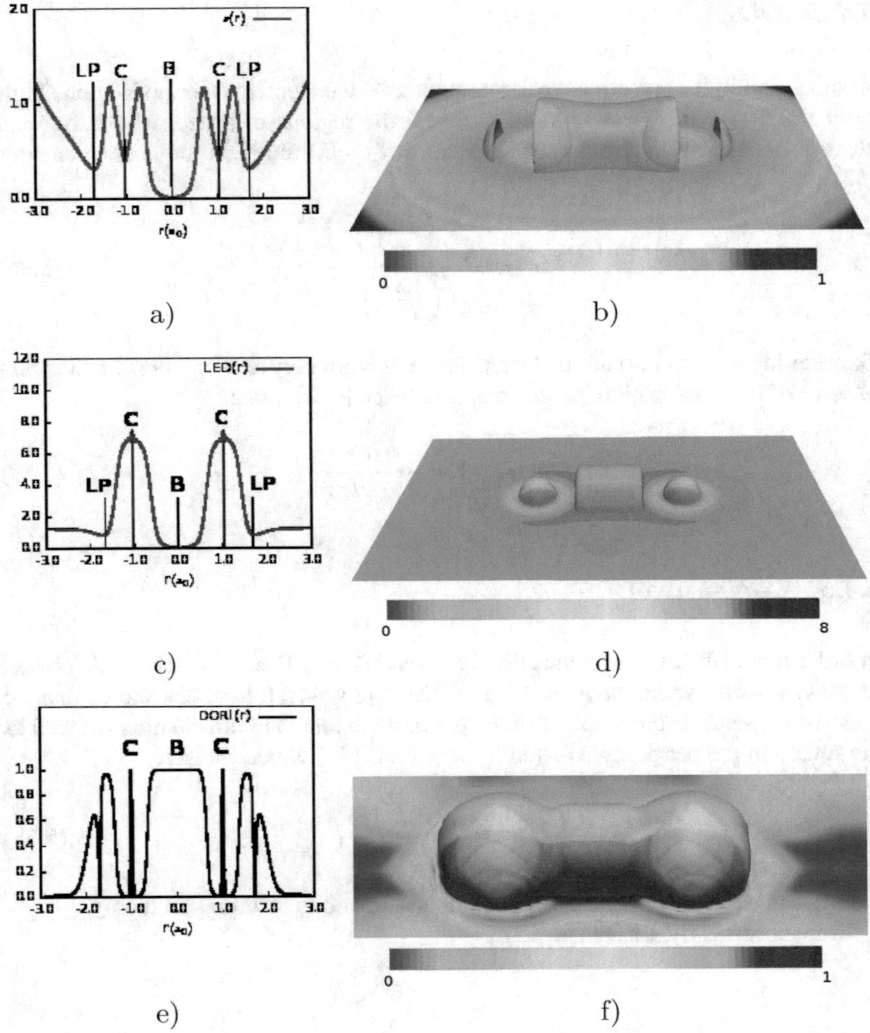


**Fig. 5.29** $s(\boldsymbol{r})$ (top), LED($\boldsymbol{r}$) (middle) and DORI($\boldsymbol{r}$) (bottom) for $N_2$. (Left) Representation along internuclear axis. (Right) $s(\boldsymbol{r}) = 0.33$, LED($\boldsymbol{r}$) $= 0.6$ (cyan) and LED($\boldsymbol{r}$) $= 6.0$ (yellow), and DORI ($\boldsymbol{r}$) $= 0.9$ isosurfaces

Conversely, a more elaborated scaling is used by the density overlap regions indicator (DORI) [52]

$$\mathrm{DORI}(\boldsymbol{r}) = \frac{\theta(\boldsymbol{r})}{1+\theta(\boldsymbol{r})} \quad \theta(\boldsymbol{r}) = \frac{(\nabla \boldsymbol{k}^2(\boldsymbol{r}))^2}{(\nabla \boldsymbol{k}^2(\boldsymbol{r}))^3}. \tag{5.42}$$

Since these three functions use the same basic variable $\boldsymbol{k}(\boldsymbol{r})$, it is not surprising that all of them carry a very similar chemical content, though with different topologies in some cases.

Let's see it through the $N_2$ example displayed in Fig. 5.29 As we have seen, core, lone pairs and interatomic bonding regions may be identified as minima of $s(\boldsymbol{r})$. LED provides a similar picture to that obtained by $s(\boldsymbol{r})$. Due to the different exponent of $\rho(\boldsymbol{r})$ in the denominator, 4/3 for $s(\boldsymbol{r})$ and 1 for LED, the difference between maxima and minima are much more highlighted in $s(\boldsymbol{r})$ than in LED. Indeed, the exponent in $s(\boldsymbol{r})$ was chosen to boost the visualization of weak interactions. On the other hand, this makes of LED a more suitable tool at higher densities. One of the strong points of LED is that it is a bounded function and different chemical entities may be visualized by choosing different LED values. In Fig. 5.29, core regions are colored in yellow and they were obtained by a LED = 6.0 isosurface, while the cyan isosurface identifying the bonding region was obtained by setting LED = 0.6. Conversely, DORI reveals a more complex picture. Core regions are characterized by minima of DORI, while it shows wide maxima in binding regions where the atomic density overlapping is much more important. In contrast with $s(\boldsymbol{r})$ and LED, where core and valence regions are separated by a maximum, DORI reveals a series of maxima and minima associated with the remaining atomic shell structure present in the molecule.

## References


1. http://rutchem.rutgers.edu/olson/tsukuba/
2. Andrés, J., Berski, S., Contreras-García, J., González-Navarrete, P.: Following the molecular mechanism for the $NH_3$ + LiH → $LiNH_2$ + $H_2$ chemical reaction: a study based on the joint use of the quantum theory of atoms in molecules (QTAIM) and noncovalent interaction (NCI) index. J. Phys. Chem. A **118**, 1663 (2014)
3. Bohórquez, H.J.: Local quantum chemistry. Ph.D. Thesis, Dalhousie University (2012)
4. Bohórquez, H.J., Boyd, R.J.: On the local representation of the electronic momentum operator in atomic systems. J. Chem. Phys. **129**(2), 024,110 (2008)
5. Bohórquez, H.J., Boyd, R.J.: A localized electrons detector for atomic and molecular systems. Theor. Chem. Acc. **127**, 393 (2010)
6. Boto, R.A., Peccati, F., Laplaza, R., Quan, C., Carbone, A., Piquemal, J.P., Maday, Y., Contreras-García, J.: Nciplot4: Fast, robust, and quantitative analysis of noncovalent interactions. J. Chem. Theory Comput. **16**, 4150–4158 (2020)
7. Boto, R.A., Contreras-García, J., Tierny, J., Piquemal, J.P.: Interpretation of the reduced density gradient. Mol. Phys. **114**, 1406 (2016)
8. Boto, R.A., Piquemal, J., Contreras-García, J.: Revealing strong interactions with the reduced density gradient: a benchmark for covalent, ionic and charge-shift bonds. Theor. Chem. Acc. **136**, 139 (2017)

9. Brandl, M., A., M.S.W., Jabs, Sühnel, J., Hilgenfeld, R.: C-h-interactions in proteins. J. Mol. Biol. **307**, 357 (2001)
10. Brauer, B., Kesharwani, M.K., Kozuch, S., Martin, J.M.L.: The s66x8 benchmark for noncovalent interactions revisited: explicitly correlated ab initio methods and density functional theory. Phys. Chem. Chem. Phys. **18**, 20,905 (2017)
11. del Campo, J.M., Gázquez, J.L., Alvarez-Mendez, R.J., Vela, A.: The reduced density gradient in atoms. Int. J. Quantum Chem. **112**, 3594–3598 (2012)
12. Cerniý, J., Hobza, P.: Non-covalent interactions in biomacromolecules. Phys. Chem. Chem. Phys. **9**, 5291 (2007)
13. Contreras-García, J., Boto, R.A., Izquierdo-Ruiz, F., Reva, I., Woller, T., Alonso, M.: A benchmark for the non-covalent interaction (nci) index or...is it really all in the geometry? Theor. Chem. Acc. **135**, 242 (2016)
14. Contreras-García, J., Johnson, E.R., Keinan, S., Chaudret, R., Piquemal, J.P., Beratan, D.N., Yang, W.: Nciplot: a program for plotting noncovalent interaction regions. J. Chem. Theory Comput. **7**, 625 (2011)
15. Contreras-García, J., Yang, W., Johnson, E.R.: Analysis of hydrogen-bond interaction potentials from the electron density: integration of noncovalent interaction regions. J. Phys. Chem. A **115**, 12,983 (2011)
16. Day, G.M., et al.: Significant progress in predicting the crystal structures of small organic molecules - a report on the fourth blind test. Acta Crystallogr. B **65**, 107 (2009)
17. Deppmeier, B.J., Driessen, A.J., Hehre, W., Johnson, J.A., Klunzinger, P.E., Watanabe, M.: Spartan ES 1.0.2. Wavefunction Inc., Irvine (2002)
18. Desiraju, G.R.: Crystal Engineering. The Design of Organic Solids. Elsevier, Amsterdam (1989)
19. DiLabio, G.A., Piva, P.G., Kruse, P., Wolkow, R.A.: Dispersion interactions enable the self-directed growth of linear alkane nanostructures covalently bound to silicon. J. Amer. Chem. Soc. **126**, 16,048 (2004)
20. Dill, K.A.: Dominant forces in protein folding. Biochemistry **29**, 7133 (1990)
21. Dutta, A., Jana, A.D., Gangopadhyay, S., Das, K.K., Marek, J., Marek, R., Brus, J., Ali, M.: Unprecedented $\pi \dots \pi$ interaction between an aromatic ring and a pseudo-aromatic ring formed through intramolecular h-bonding in a bidentate schiff baseligand: crystal structure and dft calculations. Phys. Chem. Chem. Phys. **13**, 15,845 (2011)
22. Espinosa, E., Alkorta, I., Elguero, J., , Molins, E.: From weak to strong interactions: a comprehensive analysis of the topological and energetic properties of the electron density distribution involving x–h..f–y systems. J. Chem. Phys. **117**, 5529 (2002)
23. Fenniri, H., Packiarajan, M., Vidale, K.L., Sherman, D.M., Hallenga, K., Wood, K.V., Stowell, J.G.: Helical rosette nanotubes: design, self-assembly, and characterization. J. Amer. Chem. Soc. **123**, 3854 (2001)
24. Fiedler, S., Broecker, J., Keller, S.: Protein folding in membranes. Cell. Mol. Life Sci. **67**, 1779 (2010)
25. Fuster, F., Silvi, B.: Does the topological approach characterize the hydrogen bond? Theor. Chem. Acc. **104**, 13 (2000)
26. Gavezzotti, A.: Molecular Aggregation. Structure Analysis and Molecular Simulation of Crystals and Liquids. Oxford University Press, Oxford (2007)
27. Gourlaouen, C., Gérard, H., Piquemal, J.P., Parisel, O.: Understanding lead chemistry from topological insights: The transition between holo- and hemidirected structures within the $[Pb(CO)n]^{2+}$ model series. Chem. Eur. J. **14**, 2730 (2008)
28. Hunter, G.: The exact one-electron model of molecular structure. Int. J. Quant. Chem. **29**, 197 (1986)
29. Johnson, E.R., Keinan, S., Mori-Sánchez, P., Contreras-García, J., Cohen, A.J., Yang, W.: Revealing noncovalent interactions. J. Amer. Chem. Soc. **132**, 6498 (2010)
30. Jurecka, P., Hobza, P.: True stabilization energies for the optimal planar hydrogen-bonded and stacked structures of guanine$\cdots$cytosine, adenine$\cdots$thymine, and their 9- and 1-methyl derivatives: complete basis set calculations at the MP2 and CCSD(T) levels and comparison with experiment. J. Amer. Chem. Soc. **125**, 15,608 (2003)

31. Keinan, S., Ratner, M.A., Marks, T.J.: Molecular zippers - designing a supramolecular system. J. Chem. Phys. Lett. **392**, 291 (2004)
32. Koch, U., Popelier, P.L.A.: Characterization of c-h-o hydrogen bonds on the basis of the charge density. J. Phys. Chem. **99**, 9747 (1995)
33. Kohout, M., Savin, A., Preuss, H.: Contribution to the electron distribution analysis. i. shell structure of atoms. J. Chem. Phys. **95**, 1928 (1991)
34. Kollman, P.A.: Noncovalent interactions. Chem. Rev. **10**, 365 (1977)
35. Krishnamoorthy, N., Yacoub, M.H., Yaliraki, S.N.: A computational modeling approach for enhancing self-assembly and biofunctionalisation of collagen biomimetic peptides. Biomaterials **32**, 7275 (2011)
36. Kruse, P., Johnson, E.R., DiLabio, G.A., Wolkow, R.A.: Patterning of vinylferrocene on h-si(100) via self-directed growth of molecular lines and stm-induced decomposition. Nano Lett. **2**, 807 (2002)
37. Lane, J.R., Contreras-García, J., Piquemal, J.P., Miller, B.J., Kjaergaard, H.G.: Are bond critical points really critical for hydrogen bonding? J. Chem. Theory Comput. **9**(8), 3263–3266 (2013)
38. Laplaza, R., Peccati, F., A. Boto, R., Quan, C., Carbone, A., Piquemal, J.P., Maday, Y., Contreras-García, J.: Nciplot and the analysis of noncovalent interactions using the reduced density gradient. WIREs Comput. Mol. Sci. **11**(2), e1497 (2021)
39. Lehninger, A.L., Nelson, D.L., Cox, M.M.: Principles of Biochemistry, 2nd edn. Worth Publishers, Inc. (1993)
40. Lu, X.J., Olson., W.K.: 3dna: a software package for the analysis, rebuilding and visualization of three-dimensional nucleic acid structures. Nucleic Acids Res. **31**, 5108 (2003)
41. McGaughey, B.G., Gagné, M., Rappé, A.K.: $\pi$-stacking interactions alive and well in proteins. J. Biol. Chem. **273**, 15,458 (1998)
42. Metrangolo, P., Meyer, F., Pilati, T., Resnati, G., Terraneo, G.: Halogen bonding in supramolecular chemistry. Angew. Chem. Int. Ed. **47**, 6114 (2008)
43. Nagy, A., Liu, S.: Local wave-vector, shannon and fisher information. Phys. Lett. A **372**, 1654 (2008)
44. Nagy, A., March, N.H.: Ratio of density gradient to electron density as a local wavenumber to characterize the ground state of spherical atoms. Mol. Phys. **90**, 271 (1997)
45. Nishio, M.: The CH/$\pi$ hydrogen bond in chemistry. Conformation, supramolecules, optical resolution and interactions involving carbohydrates. Phys. Chem. Chem. Phys. **13**, 13,873 (2011)
46. Pearson, E.W., Gordon, R.G.: Local asymptotic gradient corrections to the energy functional of an electron gas. J. Chem. Phys. **82**, 881 (1985)
47. Perdew, J.P., Burke, K., Ernzerhof, M.: Generalized gradient approximation made simple. Phys. Rev. Lett. **77**, 3865 (1996)
48. Rezác, J., Riley, K.E., Hobza, P.: S66: a well-balanced database of benchmark interaction energies relevant to biomolecular structures. J. Chem. Theory Comput. **7**, 2427 (2011)
49. Sahni, V., Gruenebaum, J., Perdew, J.: Study of the density-gradient expansion for the exchange energy. Phys. Rev. B **26**, 4371 (1982)
50. Saleh, G., Gatti, C., Presti, L.L., Contreras-Garcia, J.: Revealing non-covalent interactions in molecular crystals through their experimental electron densities. Chem. Eur. J. **18**, 15,523 (2012)
51. Sheiko, S.S., Sun, F.C., Randall, A., Shirvanyants, D., Rubinstein, M., Lee, H., Matyjaszewski, K.: Adsorption-induced scission of carbon-carbon bonds. Nature **440**, 191 (2006)
52. Silva, P.D., Corminboeuf, C.: Simultaneous visualization of covalent and noncovalent interactions using regions of density overlap. J. Chem. Theory Comput. **10**, 3745 (2014)
53. Simas, A.M., Sagar, R.P., Ku, A.C., Smith, V.H., Jr.: The radial charge distribution and the shell structure of atoms and ions. Can. J. Chem. **66**(8), 1923–1930 (1988)
54. Tognetti, V., Cortona, P., Adamo, C.: A new parameter-free correlation functional based on an average atomic reduced density gradient analysis. J. Chem. Phys. **128**, 034,101 (2008)
55. Viswanathan, R., Asensio, A., Dannenberg, J.J.: Cooperative hydrogen-bonding in models of antiparallel $\beta$-sheets. J. Phys. Chem. A **108**, 9205 (2004)

56. Wieczorek, R., Dannenberg, J.J.: Comparison of fully optimized $\alpha$- and $3_{10}$-helices with extended -strands. an oniom density functional theory study. J. Am. Chem. Soc. **126**, 14,198 (2004)
57. Zupan, A., Burke, K., Ernzerhof, M., Perdew, J.P.: Distributions and averages of electron density parameters: explaining the effects of gradient corrections. J. Chem. Phys. **106**, 10,184 (1997)

# Part II
# Applications

# Chapter 6
# Molecules

**Abstract** In this chapter, we will review some applications of the topological tools in the understanding of the structure, bonding and properties of molecules. We will pay special attention to controversial cases, where topology was able to provide an answer difficult to obtain from a mere inspection of the molecular geometry. At the second stage, the real space tools are applied to a selected set of problems in chemical reactivity. A brief summary of the mathematical tools used to understand topological changes, Thom's catastrophe theory, is highlighted. These insights are used to explore a few prototype cases using the QTAIM, the ELF, and the NCI methodologies.

## 6.1 Structure

In this section, we will not only see how the tools introduced in Part I are able to recover classical concepts of molecular structure, but also pushing the limits of classical chemistry and providing answers to cases where phenomenological approaches had previously failed.

### *6.1.1 QTAIM*

QTAIM has been extensively applied in all the chemical fields, particularly (but not only) in Organic Chemistry [39]. Since the number of distinct bonding situations in the chemistry of carbon compounds are limited, a very exhaustive knowledge of how the electron density responds to chemical changes in this realm has been achieved. The situation is less satisfying when we expand the possible set of intervening atoms as it is necessary for transition metal chemistry, where concepts beyond the Lewis pair, particularly multicenter bonding, have been developed to rationalize vast categories of phenomena. After forty years of QTAIM, it has become more and more clear that, in some cases, the molecular orbital description provides an easier-to-manipulate set of indices than those based on the use of the electron density and



Á. Martín Pendás and J. Contreras-García, *Topological Approaches to the Chemical Bond*, Theoretical Chemistry and Computational Modelling,
https://doi.org/10.1007/978-3-031-13666-5_6

its derivatives at CPs. After all, one may examine many orbitals but only one density. Thus, in the last two decades, an effort to recast the orbital invariant descriptors based on reduced density matrices into an orbital language has been done. Simultaneously, the use of further order reduced density matrices (like the pair density) beyond the plain electron density has allowed the expansion of the topological methods to describe phenomena that were considered out of the QTAIM arena (see Sect. 3.5.4). In the next sections, we will see several examples on how to analyze multicenter bond order from QTAIM grounds.

#### 6.1.1.1 Multicenter Bonds

The $H_3{}^+$cation, together with the diborane molecule have become the *de facto* testbed to introduce 3c–2e bonds in computational chemistry. In $H_3{}^+$, the molecule orbital picture assigns two electrons to a noodeless totally symmetric a1g orbital, which in a mininal basis set representation is given by a linear combination of the 1sa + 1sb + 1sc functions. A B3LYP/6-311G** calculation provides three equivalent H atomic basins, each with a symmetry-fixed population of $2/3\ e^-$.

It is a simple but illuminating exercise to show that with a single determinant, all the multicenter bonding descriptors are again fixed by symmetry. In this case, the $\alpha$ and $\beta$ electrons are independent, and all cumulants are the sum of an all-$\alpha$ and all-$\beta$ part which is constructed as a power of the spin-densities. In the second-order case, for instance (from Eq. 3.183)

$$\rho_2^{c,\alpha\alpha}(\boldsymbol{r}_1, \boldsymbol{r}_2) = \rho^\alpha(\boldsymbol{r}_1)\rho^\alpha(\boldsymbol{r}_2). \tag{6.1}$$

Since the electron density integral over atom $a$ is given by symmetry as $\int_a \rho d\boldsymbol{r} = 1/3$, the DI over atoms $a$ and $b$ reduces to[1]:

$$\delta^{ab} = 2\int_a\int_b \rho_2^c(\boldsymbol{r}_1, \boldsymbol{r}_2)d\boldsymbol{r}_1 d\boldsymbol{r}_2 = 4/9 \approx 0.444 \tag{6.5}$$

Similarly, the total 3-center bond index is $\delta^{abc} = 3!N_{abc} = 6 \times 2 \times 1/27 = 4/9$ (see Eq. 3.186).

Whenever a 3-center bond index is present, there are mutual electron population fluctuations among three different centers. This implies that there are also non-zero

---

[1] Resorting to Eq. 6.1, we have

$$\delta^{ab} = 2\int_a \rho_2^c(\boldsymbol{r}_1, \boldsymbol{r}_2) = 2\int_a\int_b (\rho_2^{c,\alpha\alpha}(\boldsymbol{r}_1, \boldsymbol{r}_2) + \rho_2^{c,\beta\beta}(\boldsymbol{r}_1, \boldsymbol{r}_2))d\boldsymbol{r}_1 d\boldsymbol{r}_2 = \tag{6.2}$$

$$= 2\int_a\int_b \rho^\alpha(\boldsymbol{r}_1)\rho^\alpha(\boldsymbol{r}_2) + \rho^\beta(\boldsymbol{r}_1)\rho^\beta(\boldsymbol{r}_2)d\boldsymbol{r}_1 d\boldsymbol{r}_2 = \tag{6.3}$$

$$= 2\left[\frac{1}{3}\left(\int_b (\rho^\alpha(\boldsymbol{r}_2) + \rho^\beta(\boldsymbol{r}_2)\right) d\boldsymbol{r}_2\right] = \frac{2}{3}\frac{2}{3} = \frac{4}{9}. \tag{6.4}$$

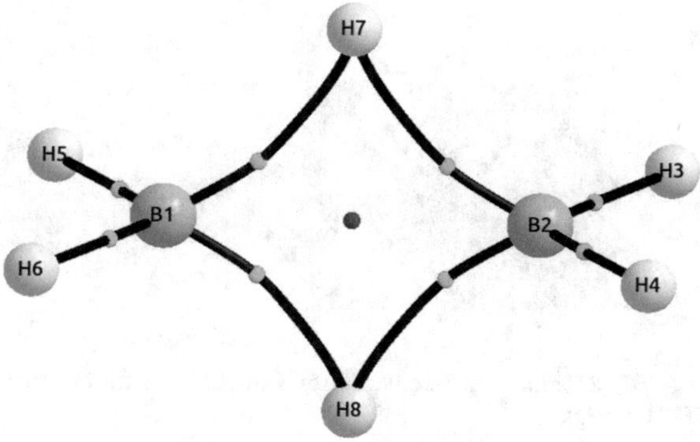


**Fig. 6.1** Structure, atomic numbering, and bond critical points (green balls) in $B_2H_6$

fluctuations between any of the three possible pairs of centers in the trio, i.e. there are non-zero 2-center bond indices as well. This shows that the 2-center index is subordinated to the 3-center one and that the existence of 2-center electron delocalization in this system is a by-product of the 3-center one. This result is obviously generalizable to a larger number of centers.

Diborane provides more interesting insights. Figure 6.1 shows its B3LYP/6-311**G geometry together with its numbering. Recall from Chap. 3 that there is no bond critical point between the borons. Since we will use a more complex label for atoms in the rest of the Section, we will simplify the notation and use DI for the delocalization index in the following. We will single out the bridging H atoms, H7 and H8, as $H_b$ and leave the H label for the rest. The QTAIM image is clearly that of a hydride, with $Q(H) = N(H) - Z_H = -0.669$ and $Q(H_b) = -0.665$. The 2-center bond orders are typical of a polar interaction, with $\delta^{B,H} = DI(B, H) = 0.516$. Notice that, for an ideal covalent model, the DI of a single bond is one, and that of a pure ionic 2c–2e interaction falls to zero, with polar bonds in between. However, $DI(B, H_b) = 0.288$ is directly telling us that the $H_b$ electrons are being shared with the two boron atoms simultaneously. The B-B delocalization is significantly smaller, $DI(B, B) = 0.063$, justifying the absence of a $(3, -1)$ CP between the two atoms and, interestingly, $DI(H, H) = 0.127$ for the vicinal H atoms and $DI(H_b, H_b)$ scales to 0.209. There is a non-negligible electron sharing between the pairs of neighboring H's, particularly for the bridging ones. These secondary interactions draw a considerably more complex image of bonding in diborane than the textbook one.

Now we turn to the expected 3-center interactions. If we take the $BH_2$ group, $DI(B, H, H) = 0.053$, a number that increases to 0.069 for $DI(B, H_b, B)$. This simple analysis shows clear indications of 3-center interactions. As they involve atoms of different electronegativities, the non-symmetrical delocalizations lead to a decrease in the DI, as it occurs in standard 2-center indices. The bridge 3-center index is, however, only slightly larger than that in the $BH_2$ moiety. Thanks to the large direct H, H exchange in the $BH_2$ group, we should also consider 3-center contributions here.

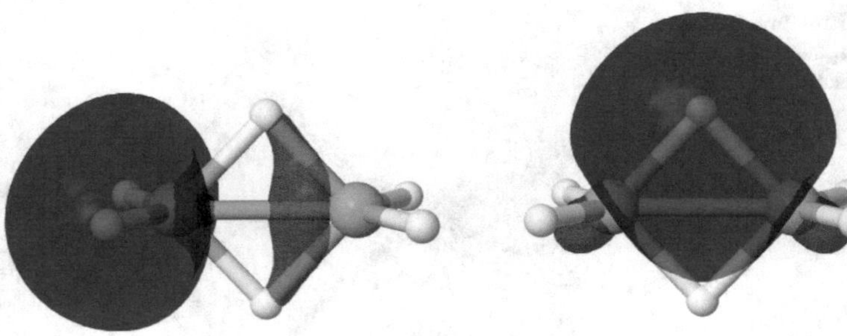

**Fig. 6.2** Dominating 3-center NAdOs in diborane for the non-bridging (left) and bridging (right) HBH groups

The 3-center DIs can be decomposed into one-electron contributions as sketched in Sect. 3.5.4. Figure 6.2 depicts the dominant 3-center NAdOs when we consider the non-bridging and bridging HBH groups. The symmetric 3-center structure of the orbitals is salient. They contribute to the total index with 0.050 and 0.068, respectively. In other words, 94 and 99% the total 3-center index is due to a single orbital contribution.

For completeness, we can also point out several IQA contributions in this case. For instance, the exchange–correlation energy between the geminal non-bridging H's is −13.1 kcal/mol, while that between the bridging ones is much larger, −26.3 kcal/mol. In contrast, the B-B contribution is just −8.1 kcal/mol. All in all, the real space picture points toward paying more attention to the interaction between the bridging H than between the B atoms to rationalize this system.

#### 6.1.1.2 The Energetics of Halogen Bonding

As the interest in non-covalent interactions blossomed, so did the zoo of new bonding types, including beryllium, tetrel, pnictogen or chalcogen bonds. Partly because of simplicity and the background of the researchers originally involved in uncovering these interactions, electrostatics was targeted as the fundamental driving force in many of these bonds, and the concept of $\sigma$-holes became essential in their understanding, particularly in the halogen bond case. In these systems, it is widely believed that the charge density anisotropy around the bonded halogen atom, which can be clearly visualized from $\nabla^2\rho$, the ELF function, and many other scalar fields leads to a torus-like density accumulation region surrounding the halogen X and to a density depletion one along the A-X internuclear region. This is already evident in Fig. 4.27, where the green region pointing toward the observer can be identified with this $\sigma$ hole. Electron rich/poor groups may then engage in electrostatic matching with these depletion/accumulation regions, respectively. This has led to emphasize the role of the electrostatic potential (ESP) in the rationalization of the geometries displayed by many of such complexes [41]. However, this traditional view has been challenged

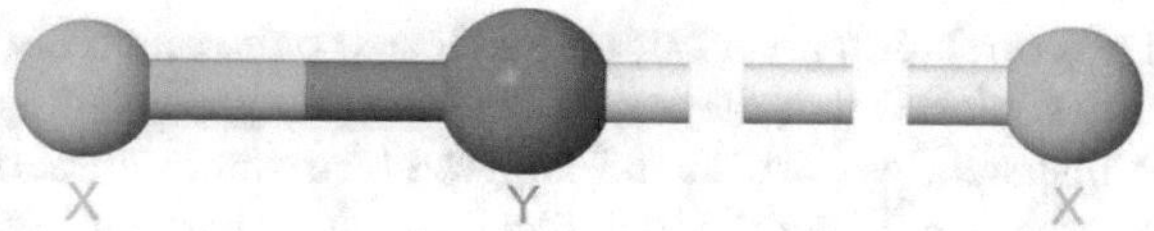


**Fig. 6.3** Asymmetric geometry of the $X_1$-$Y_2\cdots X_3^-$ complexes (X, Y = Cl, Br, I) mimicking a halogen bond

**Table 6.1** IQA formation energies together with the interaction energy between the $X_1$-$Y_2$ and $X_3$ fragments divided into classical and covalent parts. All data in kcal mol$^{-1}$

| Complex | $E^{IQA}_{form}$ | $E_{int}$ | $E_{xc}$ | $E_{class}$ |
|---|---|---|---|---|
| Cl-Cl$\cdots$ Cl$^-$ | −21.61 | −59.92 | −66.06 | 6.14 |
| Cl-Br$\cdots$ Cl$^-$ | −32.64 | −79.10 | −82.41 | 3.31 |
| Cl-I$\cdots$ Cl$^-$ | −35.47 | −90.33 | −79.89 | −10.44 |
| Br-Cl$\cdots$ Br$^-$ | −22.14 | −58.80 | −70.75 | 11.95 |
| Br-Br$\cdots$ Br$^-$ | −32.53 | −76.05 | −87.47 | 11.42 |
| Br-I$\cdots$ Br$^-$ | −29.77 | −75.41 | −75.99 | 0.59 |
| I-Cl$\cdots$ I$^-$ | −16.53 | −46.99 | −61.05 | 14.06 |
| I-Br$\cdots$ I$^-$ | −23.59 | −59.88 | −73.62 | 13.74 |
| I-I$\cdots$ I$^-$ | −27.16 | −66.43 | −75.45 | 9.02 |

with the use of other techniques. Since IQA is able to sense the strength of electrostatic and non-electrostatic interactions on the same footing, studying the role of these terms in model systems is enlightening.

To decrease the complexity of halogen bonding as much as possible, one can examine the energies of asymmetric constrained $X_1$-$Y_2\cdots X_3^-$ geometries that mimic the halogen arrangements found in crystals, as in Fig. 6.3, where $d_{12}$ is fixed to the $X_1$-$Y_1$ experimental distance and $d_{23}$ is further optimized. This was done in Ref. [28] at the M06-2X/x2c-TZVPPall level.

Although arguments based on matching blue and red regions of ESP isosurfaces are easy to grasp, one should never forget that it is always the total electrostatic interaction between the approaching entities that have to be taken into account, and that the lock-and-key ESP isosurface match may actually hide a globally destabilizing electrostatic contribution. The IQA $E^{AB}_{int}$, and its classical $V_{cl}$ and covalent-like $V_{xc}$ components provide directly the grand total interaction between two quantum atoms. These can be envisioned as what one would obtain by adding together all the ESP images at all possible different isovalues, like when reconstructing an onion from its peels. In other words, since a positive ESP value on the van der Waals envelope surface may well display a negative ESP value on another isosurface, the total electrostatic interaction between two moieties is by no means obvious from the exam of the first surface.

Data shown in Table 6.1 confirms the above suspicions. Surprisingly, all but one of the complexes studied exhibit an overall destabilizing electrostatic interaction, and it is the exchange–correlation contribution that leads to the relatively large negative interaction energies. Notice that the trends are correctly predicted by ESP docking arguments. For instance, we expect the largest electrostatic attraction between a diatomic with a large $\sigma$-hole at the I end, like ClI, with a hard $Cl^-$ anion. However, without the electron-exchange contribution provided by $V_{xc}$ none of the complexes would display stabilizing interaction energies.

## 6.1.2 ELF

For comparison with the previous QTAIM section (Sect. 6.1.1.1), we will first review multicenter bonds. Then, due to their relevance in the development of the VSEPR, we will jump into the concepts of hyper and hypovalence to show how topology can help in understanding old rules (and exceptions) of chemistry.

### 6.1.2.1 Multicenter Bonding and Agostic Interactions

Since, in the QTAIM, a bond path almost always links two nuclei, in the absence of a multi-basin delocalization index analysis, bond paths refer to 2c–2e cases.

Let's illustrate the use of synapticity in the understanding of multicenter bonding through two representative examples: diborane will allow comparison with QTAIM, and agostic interactions will allow introducing metallic interactions.

Starting with the diborane example from the previous section, the ELF image in Fig. 6.4a clearly confirms the 3c–2e bond interpretation in the shape of two trisynaptic basins, V(B, H, B) in yellow and blue.

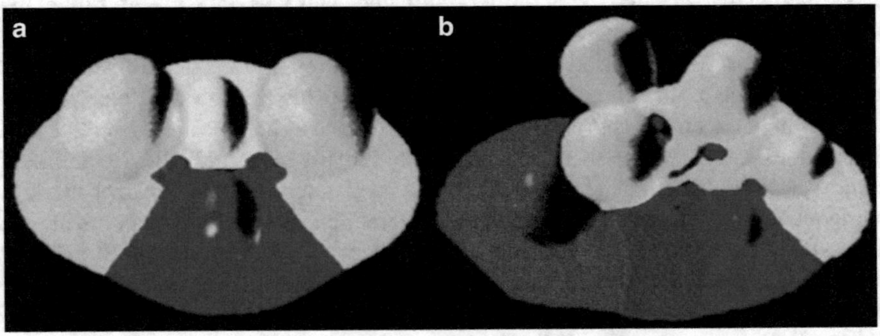


**Fig. 6.4** Localization domains of **a** $B_2H_6$. The trisynaptic domains have different colors in order to be distinguished. **b** $RuClCH_2CH_3(PH_3)_2$. The core domains are represented in magenta, the disynaptic V(A, B) and trisynaptic V(A, B, C) in green, the protonated disynaptic domains V(A, H) in light blue, the protonated trisynaptic V(A, H, B) in blue and the monosynaptic ones in redbrick. Reprinted from Ref. [46], Copyright 2002, with permission from Elsevier

Another example of multicenter bonding is given by agostic interactions [9]. These consist of an interaction between a CH group (which can be several bonds away) and a metal. Figure 6.4b shows the ELF picture for $RuClCH_2CH_3(PH_3)_2$. Whereas QTAIM identifies the multicenter nature of the interaction as a distorted bond path linking the metal with H atoms of the in-plane $CH_3$ group, the ELF picture does it by the presence of a trisynaptic basin, V(C, H, Ru). This approach has also been used in reactivity to describe dehydrogenation processes, whereby the evolution along the reaction can be viewed as the formation and rupture of an agostic bond [10].

It should be noted that, within the ELF approach, metallic bonding (Sects. 4.3.2 and 7.2.2.1) can be summarized as a multicenter bonding, usually holding small charges.

Finally, in the Chapter on weak interactions (Chap. 5), we saw that multicenter bonding can also appear among monosynaptic basins. In those cases, we will see that the atoms involved, A, and B, present monosynaptic basins, V(A) and V(B) and share a common separatrix, in which most of the spin density is shared. This leads to a big V(A)-V(B) covariance matrix element and positive (fluctuation) and a positive Core Valence Bifurcation (CVB) (Sect. 5.2.2.1).

#### 6.1.2.2 Hyper- and Hypovalency

Lewis theory is built around the pairing of electrons and the rule of eight [35]. Two main exceptions were detected at the time of developing these rules: those moieties having more than 8 electrons in their valence shell, known as hypervalent molecules (typically $PCl_5$ and $SF_6$), and those holding less than 8 electrons, which were named hypovalent (exemplified by $BF_3$).

Theoretical chemistry spent many years trying to find explanations for these exceptions. However, with the evolution of synthetic chemistry and structural determination, it became obvious by the mid of the twentieth century that hyper- and hypovalent molecules were by no means exceptions, and that the number of electron pairs around a central atom was in many occasions 5, 6, 7 and even 8. Moreover, it was also by this time that Gillespie showed that the number of electron pairs around the central atom, including lone pairs, was crucial to determine the geometry of the compounds. This observation leads to the well-known Valence Shell Electron Pair Repulsion (VSEPR) theory.

Since ELF is able to characterize bonding and localize electron pairs, it is the perfect tool to analyze these "exceptions" and see whether there is something peculiar to them. It showed not to be the case. The analysis of the bonds around the central atom as provided by ELF has confirmed that the bond in hyper- and hypovalent molecules is exactly the same as in those fulfilling the octet rule (so they do not correspond to the 3c–4e case as some had proposed) [25]. The first row molecules usually show 4 electron pairs just due to their smaller size, so that as the central atom radius increases, it can accommodate a larger number of ligands. Moreover, the analysis of bond orders has allowed to uncover cases where a first row atom becomes hypervalent. As an example: $CH_2N_2$ has been shown to hold 5 electron pairs around the nitrogen (as its higher row equivalent, $PF_5$), with the difference that two of these pairs show up as a multiple bond order, which is coherent with the smaller size of nitrogen [24].

Having access to quantitative information on electron pairs can thus help to understand the basis of classical chemistry rules, as well as their limitations.

## 6.1.3 NCI

In this section, we will see how the use of the 2D and 3D plots of NCI can allow us to understand bonding features even beyond weak and covalent bonding (Chap. 5). Indeed, we have seen that NCI leads to signals around QTAIM critical points, hence we should expect NCI isosurfaces for any type of bonding.

### 6.1.3.1 Ionic Interactions

We will start with an ionic example, NaF. Figure 6.5 top-left shows the two separate ions along with a thin but with a large interaction region. This is due to the fact that the density accumulation in ionic interactions is rather small. If we compare

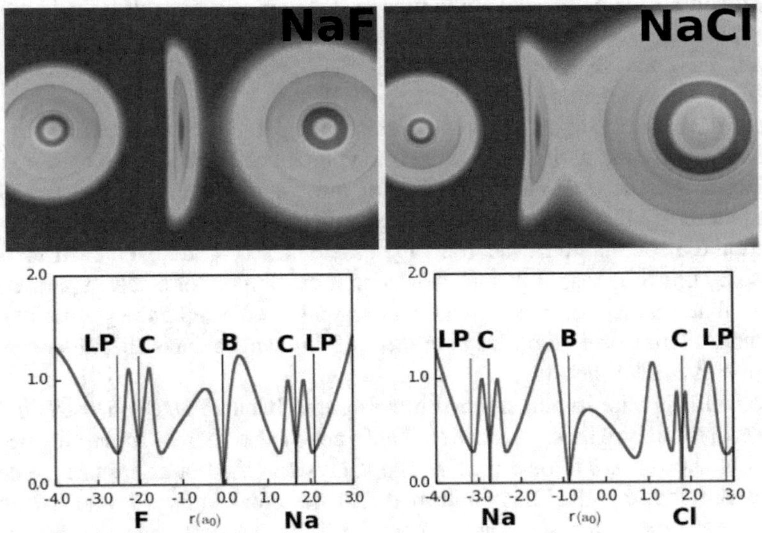


(a) NaF (b) NaCl

**Fig. 6.5** Top panels. $s(\mathbf{r})$ contours for **a** NaF, **b** NaCl. Atoms in the same order as in the notation on the top-left corner. Bottom panels. $s(\mathbf{r})$ along the internuclear axis for the same molecules. Labels C, B and LP in the bottom panels stand for core, bonding and lone pair minima, respectively. Reproduced with permission from Ref. [7]

NaF and NaCl $s$ profiles (Fig. 6.5 bottom), we see that the $s$ curve is higher for the fluorine ion than for the chlorine. This is due to the fact that the height of the tails reveals the hardness, with harder ions showing higher $s$ profiles [16]. As a result, the profile close to the cation (which is hard) is steeper than close to the anions. This leads to the convex surfaces pointing toward the anions. Moreover, the difference in slope/hardness also leads in the case of Cl to chlorine surfaces that meet at relatively low values of $s$ (Fig. 6.5 top-right).

#### 6.1.3.2 Organometallic Interactions

One of the main advantages that will be exploited in the chapter on reactivity (Sect. 6.2) is that NCI allows to follow both the non-covalent and covalent bonds as they evolve along a reaction. Let's thus see what kind of image we should expect in organocatalysis.

$(CH_3)_2Mg$ is shown in Fig. 6.6a. The density range at which the organometallic bonds appear to correspond to intermediate strength between covalent and weak interactions ($\rho \simeq 0.05$ a.u., Fig. 6.6a, i.e. cutoff default values need to be changed!).

**Fig. 6.6** $(CH_3)_2Mg$. (Top) 0.25 RDG isosurface and (bottom) $s(\mathbf{r})$ along the C-Mg-C axis. Label B in the $s(\mathbf{r})$ plot along the internuclear axis stands for bonding minimum. Arrows indicate additional minima in the bonding region

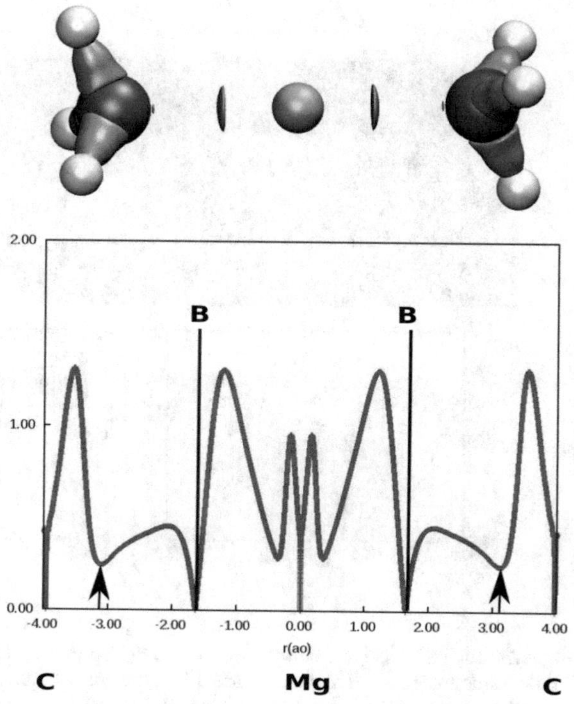


(a) $(CH_3)_2Mg$

As far as the isosurfaces are concerned, shapes are similar to ionic cases, though at higher densities and slightly larger elongations in the bonding direction. All in all, the 3D signature of these bonds is intermediate, which explains their role in catalysis (directional bond, but easy to break).

#### 6.1.3.3 Protocovalent Interactions

Protocovalent bonds, also known as charge shift bonds have led to intense studies and controversies [36, 45]. According to valence bond theory, charge shift bonds arise from the resonance between ionic forms. Typically observed in $F_2$, this would mean that there is a resonance between $F^+ - F^-$ and $F^- - F^+$. We have plotted $F_2$ and $O_2$ (also protocovalent) in Fig. 6.7a and b. We can see that the bonding region along the internuclear line is narrower, whereas an expansion in the perpendicular direction is observed. This is related to the appearance of extra non-QTAIM CPs close to the atoms. Like in the polar case, this corresponds to the existence of valence shell concentrations (in this case, on both atoms).

As can be seen for the cationic systems $t$Bu-$NH_3^+$ and $t$Bu-$OH_2^+$ in Fig. 6.8, this leads to 3D pictures similar to the ones seen in ELF (Sect. 4.3.2). NCI is able to differentiate between charge shift bonds C–N and C–O, and covalent C–C bonds.

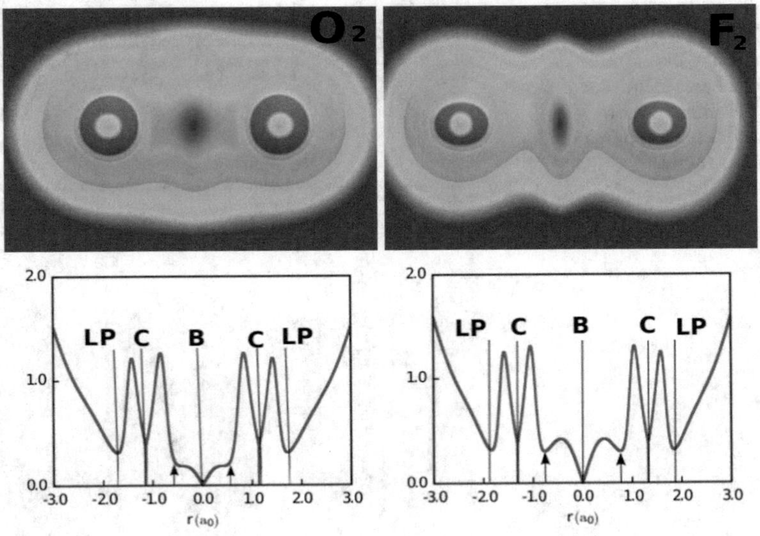

(a) $O_2$ (b) $F_2$

**Fig. 6.7** Top panels: $s(\mathbf{r})$ contours for **a** $O_2$, **b** $F_2$. Bottom panels: $s(\mathbf{r})$ along the internuclear axis for the same molecules. Labels C, B and LP in the bottom panels stand for core, bonding and lone pair minima, respectively. Arrows on $O_2$ and $F_2$ plots indicate additional minima in the bonding region. Reproduced with permission from Ref. [7]

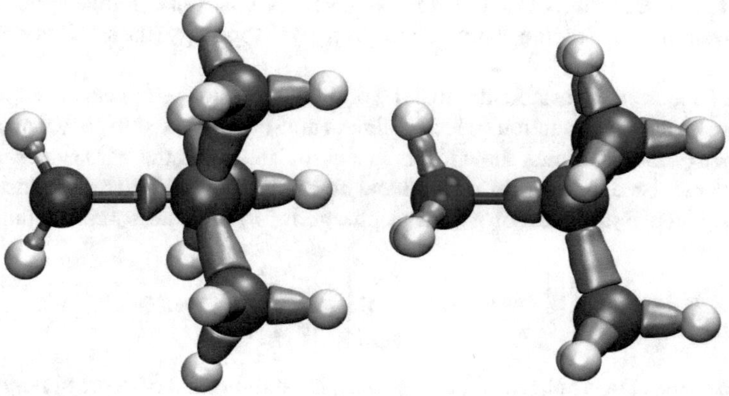

(a) $tBu\text{-}OH_2^+$ (b) $tBu\text{-}NH_3^+$

**Fig. 6.8** RDG=0.25 isosurfaces for **a** $t$Bu-$OH_2^+$ and **b** $t$Bu-$NH_3^+$. Reproduced with permission from Ref. [7]

C–N and C–O protocovalent bonds are characterized by a narrow isosurface along the interatomic axis and a stretch along the perpendicular directions, similar to that observed for $F_2$ and $O_2$. C–C bonds instead extend along the interatomic region. This difference arises from the emergence of the valence shell concentration close to the central carbon atom as seen in ELF.

## 6.2 Reactivity

### *6.2.1 Catastrophe Theory*

Some bonds require understanding how they break. For example, the separation between dative and covalent bond cannot be traced with the naked eye (Fig. 4.10a) but the differences arise at the time of bond breakage. The covalent bond leads to changes in the overall topology, whereas the dative bond does not. We will see how to analyze this process in this section. Indeed, the geometrical changes that take place along a reaction lead to electronic structure rearrangements that can be followed through the changes in the topology of the different fields we have seen in part I. In order to follow these changes, it is most common to resort to catastrophe theory [1].

We have seen in Chap. 2 that non-degenerate critical points enable to characterize the qualitative behavior of a dynamical system. Following the same principle, the stability of the dynamical system upon changes in the parameters of its potential function can be characterized by the appearance of degenerate critical points. Small changes in the potential around the control parameters, $c_\alpha^*$, for which the determinant of the potential Hessian, $V(y_k; c_\alpha)$, annihilates (i.e. catastrophe points, $(y; c^*_\alpha)$) will take the system from a region characterized by a given topology (the stability domain) to another.

Thom's theorem [1] establishes that the potential around $(y, c^*_\alpha)$ can be expressed as a sum of a quadratic function (which defines a non-degenerate subspace) and a non-Morse part (which contains the coordinates associated with the zero eigenvalues). The latter can be expressed in a canonical form, $V(y; c_\alpha)$, known as catastrophe function, which is the sum of a catastrophe germ, $u$, and the so-called universal unfolding

$$V(y; c_\alpha) = u(y_1, \ldots, y_l; c_\alpha) + \sum_{i=l+1}^{n} \lambda_i(c_\alpha) y_i^2. \tag{6.6}$$

The germ of the catastrophe, $u(y_1, \ldots, y_l; c_\alpha)$, is a polynomial of degree greater than 2 (this depends on the number of variables whose eigenvalues should become zero). The universal folding provides the degeneration of the critical point and hence the (de)stability to the potential.

As an example, we will consider in more detail the easiest catastrophe, known as the fold catastrophe. This corresponds to a universal folding $F(x) = x^3 + vx$ with germ $f(x) = x^3$ (Fig. 6.9). The manifold of the catastrophe, $M$, is the parabola defined by the projection of the inflexion points of $F$ on the $x$, $v$ plane. At the vertex of the parabola, a degenerate critical point appears which characterizes the catastrophe ($v = 0$). This point separates the space in two domains: one in which there is a maximum and a minimum ($v < 0$) and another one in which there are no extrema ($v > 0$). We will see several examples of the Fold catastrophe in Sect. 6.2.3.

Catastrophe theory has been commonly used in the description of reactivity for dynamical systems induced by the gradient of the functions we have seen in Part I. Within the QTAIM, numerous reactions (isomerizations, ring breaks) have been analyzed. However, its use has been probably more widely spread for ELF, due to its ability to characterize electron pair redistributions [32]. In solid state, catastrophe theory has allowed to classify phase diagrams and justifies the characterization of phase transition properties with a unique parameter [15].

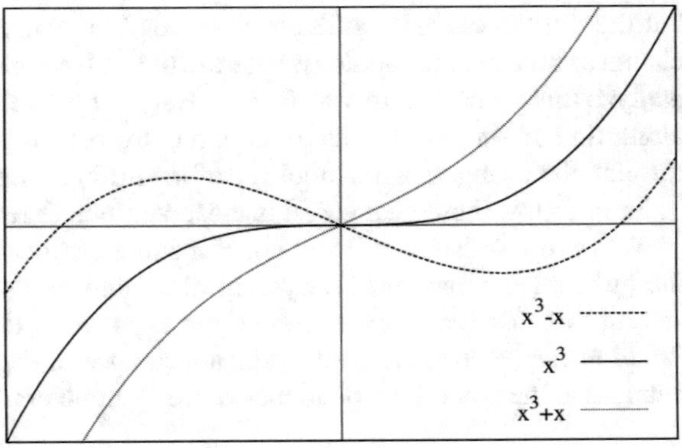


**Fig. 6.9** Appearance of instability for $F(x) = x^3 + ux$ when $u = 0$

It is important to note that although within stability domains, the changes in the quality of the wavefunction do not affect its topological description, this is not the case in the limits within domains. Hence, greater care is advised in the choice of simulation levels if catastrophe points are going to be crossed.

### *6.2.2 QTAIM*

#### 6.2.2.1 Formation of the Water Molecule

Early applications of Catastrophe theory to reactivity within the QTAIM were already introduced by Bader and coworkers, and a comprehensive presentation can be found in Bader's book [4]. In order to take profit from the previous consideration on the fold catastrophe, we will briefly present the approach of an oxygen atom O toward a hydrogen molecule that breaks the dihydrogen bond to form the open $H_2O$ structure. Notice that due to spin preservation, the attacking oxygen is in its $^1D$ excited state and that, for the sake of simplicity, $C_{2v}$ symmetry is preserved. In this sense, we envision an oxygen atom moving in the $z$ direction along a midpoint axis that bisects the dissociating dihydrogen. The distance from this atom to the H–H midpont is called $r$. At intermediate $r$ values, a ringed structure in which all the three atoms are connected and a (3,+1) CP appears is found. The evolution of $\rho$ as $r$ decreases describes a fold catastrophe. At $r > 2.4$ a.u., approximately, $\rho$ is cubic in $z$ with a minimum at the ring point, and a maximum at the H–H BCP. As $r$ decreases, the position of the maximum and the minimum approach each other, until at $r \approx 2.4$ a.u. they coalesce in a saddle point and disappear. There is a single control parameter leading to a singularity in $\rho$ that determines the opening of the ring.

The topological description of this fold catastrophe is given by $\rho(\epsilon) \approx a\epsilon^3 + b\epsilon$, where $\epsilon = z - z_c$ and $z_c$ is $z$ at the ring critical point. Notice that the approach of an O atom to a hydrogen molecule occurs in a 2D plane, and thus it requires two control parameters. The general description of the evolution of a three-membered ring is described by a 2-parameter unfolding, the elliptic umbilic catastrophe, which is more general than the one-parameter fold one that we are describing, since we are fixing the symmetry of the system to $C_{2v}$ [5].

Besides describing and characterizing the steps through which a given chemical process goes, a recurrent theme in the literature has been the possible relation between the energetics of a reaction and its topological behavior. Despite initial hopes, it is now clear that the topology of the energy landscape is not univocally related to that of the electron density. An intrinsic reaction coordinate (IRC) can be decomposed into different topological regions by means of a general procedure, based on the analysis of the derivative of the energy with respect to the IRC reaction coordinate, called the reaction force [49]. The stability regions of the reaction force do not necessarily coincide with those of the electron density, although in many cases a fruitful comparison is possible.

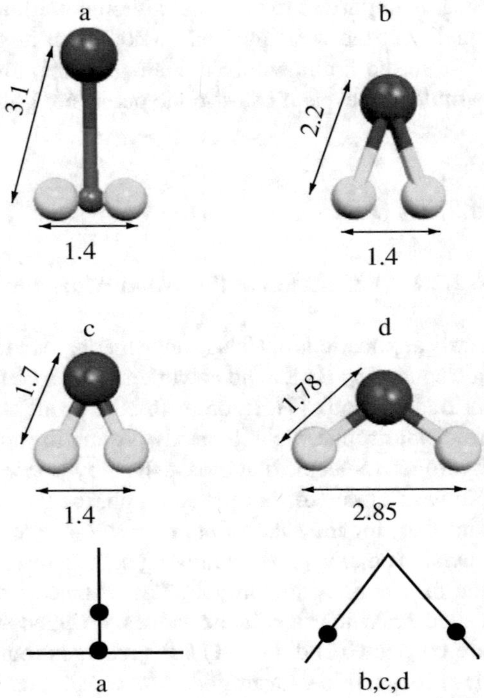


**Fig. 6.10** $C_{2v}$ approach of a $^1D$ oxygen atom to a $H_2$ molecule. Four situations labeled $a, b, c, d$ are depicted, together with their topological structures. All distances in a.u

Although energy-density isomorphism does not exist, the study of their co-evolution along a given chemical process is usually of utmost importance. Staying with the fold catastrophe, we now compare a CAS[6,5]//6-311G** IQA calculation to the density topology in the $C_{2v}$ approach of O to $H_2O$. Figure 6.10 and Table 6.2 contain a sketch of the $a$, $b$, $c$, $d$ structures analyzed and the IQA relevant interaction energies. Most of the chemistry of the transition is contained in the table. At large O-H distances, when the catastrophe has not yet been met, isolated O and $H_2$ moieties can be distinguished. The charge transfer from dihydrogen to O is small (although non-negligible), and the exchange–correlation energy between the two H atoms is much larger than that between O and H. The H–H interaction energy is largely stabilizing. The fold catastrophe, accompanied by the unfolding of two O–H BCPs, occurs close to the $b$ structure. Now the charge transfer has increased considerably, the value of $V_{xc}^{\mathrm{OH}}$ approaches that of a covalent bond and, most interestingly, the total H–H interaction energy changes sign, driven by the classical repulsion between the heavily charged H atoms. As we approach the equilibrium structure of water at $d$, all the descriptors converge to those in $H_2O$.

The changes just described are rather typical: it is in the vicinity of the density topology changes that the stronger energy variations usually take place: although there is no exact energy-density isomorphism, the simultaneous study of their varia-

**Table 6.2** Relevant IQA quantities at the *a*, *b*, *c*, *d* $H_2O$ geometries sketched in Fig. 6.10. All data in a.u

| Property | a | b | c | d |
|---|---|---|---|---|
| $Q_H$ | 0.088 | 0.312 | 0.546 | 0.560 |
| $E_{int}^{OH}$ | −0.059 | −0.227 | −0.518 | −0.490 |
| $V_{cl}^{OH}$ | 0.001 | −0.052 | −0.304 | −0.292 |
| $V_{xc}^{OH}$ | −0.060 | −0.175 | −0.214 | −0.198 |
| $E_{int}^{HH}$ | −0.138 | 0.016 | 0.252 | 0.124 |
| $V_{cl}^{HH}$ | 0.050 | 0.117 | 0.278 | 0.130 |
| $V_{xc}^{HH}$ | −0.188 | −0.098 | −0.026 | −0.006 |
| $E_{H_2}$ | −1.018 | −0.722 | −0.345 | −0.485 |
| $E_O$ | −74.731 | −74.783 | −74.577 | −74.633 |

tion along a chemical process is desirable and fruitful, since it allows a local understanding of the chemical changes in the high energy region.

**Energetics and the Meaning of Bond Critical Points**

The intimate relation between reactivity, energetics and, ultimately, the interpretation of topological chemical descriptors can be exemplified by examining briefly a still active controversy on the meaning of bond critical points. According to the orthodox QTAIM interpretation, the presence of a bond path between two atoms is a necessary and sufficient condition for the atoms to be bonded to one another if there are no forces acting on the nuclei [4]. However, in many cases, it has been found that not always a BCP is located where chemists would find bonds [17]. There are several well-known examples and we have already reviewed some of them in Sect. 5.3.3. One is the presence of BCPs in sterical clashes, a controversy that started with Cioslowski and Mixon [14] in *ortho*-substituted biphenyls and which now includes many other examples such as the ubiquitous anion–anion BPs found in many inorganic solids [37]—which have been interpreted by some authors as clear cases of BPs in nonbonding or repulsive interactions. Reservations increased when BCPs in otherwise sterically interpreted endohedral inclusion compounds were found [26].

A possible solution to these objections was found by examining the IQA contributions between different atomic pairs along several simple chemical reactions [38]. In summary, bond critical points appear between those atomic pairs for which the exchange–correlation energy component, $V_{xc}^{AB}$, dominates over all other possibilities. These are called privileged exchange channels. When only two atoms compete for electron delocalization, there is only one possible channel, and a BCP between them always appears. On the contrary, when atom *A* competes with a series of other atoms, let us say *B* and *C* for the sake of clarity, the BCP appears between the pair giving rise to the largest $V_{xc}$ contribution. Tognetti and Joubert have elaborated this idea even further [48].

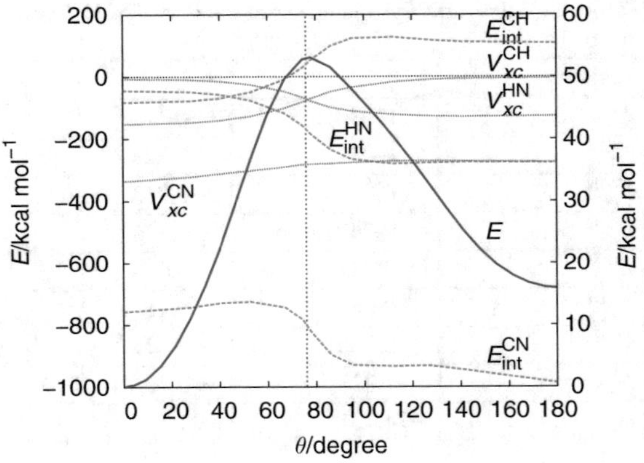


**Fig. 6.11** Evolution of relevant IQA interaction components in the HCN (left) to CNH isomerization as described in the text. The evolution coordinate is the ∠H*cm*C $\theta$ angle, where *cm* stands for the center of mass of the CN subsystem. Topological changes are shown with vertical dotted lines

A very simple example allows to clarify this idea: the isomerization of HCN to CNH, which is close to a fold catastrophe. This reaction has been studied many times from the QTAIM point of view, since the transition state lies very close to the topological change [47]. As the H atom traverses from the C- to the N- end of the molecule, the original HC BCP changes to the final H–N one after passing through a very narrow region of stability of a ringed QTAIM structure. A valence CAS/6-311G++(d, p) calculation shows that the greater the stability of the HCN isomer is, the larger the covalency of the HC *versus* the HN interaction. Figure 6.11 shows nicely how the change in the position of the BCP coincides with the crossing of the HC and HN exchange–correlation contributions. Notice how in the CNH energetic region the CH interaction is basically electrostatic, whereas the HN and CN interactions display both large electrostatic and exchange–correlation components. At the HCN end, the situation has changed dramatically, and CN is now the only pair with a relevant ionic interaction term. $V_{xc}^{HC}$ and $V_{xc}^{HN}$ cross at almost exactly the catastrophe point. The total HC and HN interaction energies do also cross, but at much lower angles.

### 6.2.3 QTAIM+ELF

We have seen that QTAIM, ELF and NCI provide complementary views of reactivity. Hence, when different types of bonds are being formed and broken, it can be interesting to carry out a simultaneous assessment of the three of them. Let's see one example, the molecular mechanism of the $NH_3$+LiH → $LiNH_2$+$H_2$ reaction, which has a potential interest in hydrogen storage. This reaction comprehends the evolution of three prototypical chemical interactions: the H–H covalent bond, the $Li^+$ $H^-$ ionic bond and the N$\cdots$LiH interaction.

The reaction $NH_3$+LiH → $LiNH_2$+$H_2$ begins with the formation of the intermolecular complex $H_3N\cdots Li^+H^-$ ($C_{3v}$) stabilized by the lithium bond $N\cdots Li^+H^-$

and ends with the complex $[H_2N]^-Li^+ \cdots H_2$ ($C_{2v}$) stabilized by two lithium bonds $N^-Li^+ \cdots H$. The transition state lies 14.12 kcal/mol above the $H_3N \cdots Li^+H^-$ complex. The studied reaction is endothermic with an activation barrier of 3.99 kcal/mol including $\Delta$ZPVE [30].

Remember that we already used this example to introduce NCI bifurcation trees in Sect. 5.3.4.4, so we will focus here on QTAIM and ELF analyses.

#### 6.2.3.1 QTAIM

In this section, we will resort to CP properties ($\rho_{bcp}$ and $\nabla^2\rho_{bcp}$) to identify bonding types at different points of the IRC, and illustrate the use of catastrophe theory to characterize the transformations.

From inspection of Fig. 6.12, we can see that the initial structure shows five BCPs: three correspond to the N-H bonds in ammonia and the other two reflect the $N \cdots Li^+$ and $Li^+H^-$ interactions. The nature of the bonds can be identified from a quick inspection at Fig. 6.13b: whereas $\nabla^2\rho(\boldsymbol{r}) > 0$ for $N \cdots Li^+$ and $Li^+H^-$ for closed-shell interactions, $\nabla^2\rho(\boldsymbol{r}) < 0$ in the case of N-H bonds, a indicating shared electrons. The final complex $H_2N\text{-}Li^+ \cdots H_2$ shows an ionic nitrogen—lithium interaction ($\nabla^2\rho(\boldsymbol{r}) > 0$) and a covalent H-H bond ($\nabla^2\rho(\boldsymbol{r}) < 0$).

Let us now see how to resort to catastrophe theory to understand these changes. In total, 5 different CPs distributions are encountered (Fig. 6.12), which leads to 5 stability domains.

In the first stability domain, the $Li^+H_b^-$ bond bends toward the N–$H_a$ bond. In the first catastrophe (from step I to step II in Fig. 6.12), two new CPs, a BCP and an RCP, appear approximately in the region of interaction between the $H_a$ atom from ammonia and the $H_b$ atom from lithium hydride ($H_a \cdots H_b$). This can be classified as a fold catastrophe.

During step II, there is a "shift" of the RCP to the neighborhood of the BCP of the ionic interaction between $Li^+$ and $H_b^-$ ions. An analysis of $\rho(\boldsymbol{r})$ and $\nabla^2\rho(\boldsymbol{r})$ evolution shows that the electron density value for the N-$H_a$ bond decreases and that of $H_a \cdots H_b$ interaction increases rapidly.

The second catastrophe, also of the fold type, happens when the RCPs and BCPs involved in the $Li^+H_b^-$ interaction are annihilated. Within QTAIM, this means that the ionic lithium–hydrogen bond is broken so that the reacting system is now formed by the $H_2N \cdots Li$ and $H_a \cdots H_b$ subsystems. In step III, the character of the $H_a \cdots H_b$ interaction changes qualitatively. The $H_a \cdots H_b$ distance shortens along this step, the value of electron density subsequently rises and the Laplacian switches from positive to negative. Indeed, a negative sign of $\nabla^2\rho(\boldsymbol{r})$—typical for covalent interactions—is to be expected for the covalent $H_a$-$H_b$ bond at the end of the reaction. The opposite behavior is observed for the $N \cdots H_a$ interaction. It should be noted that the TS is located along these dramatic changes.

The third catastrophe is also a fold and leads to the formation of two new CPs RCP and BCP approximately between the Li and $H_b$ atoms. During step IV, the $H_a \cdots H_b$ distance becomes increasingly shorter and the (3, +1) CP moves toward

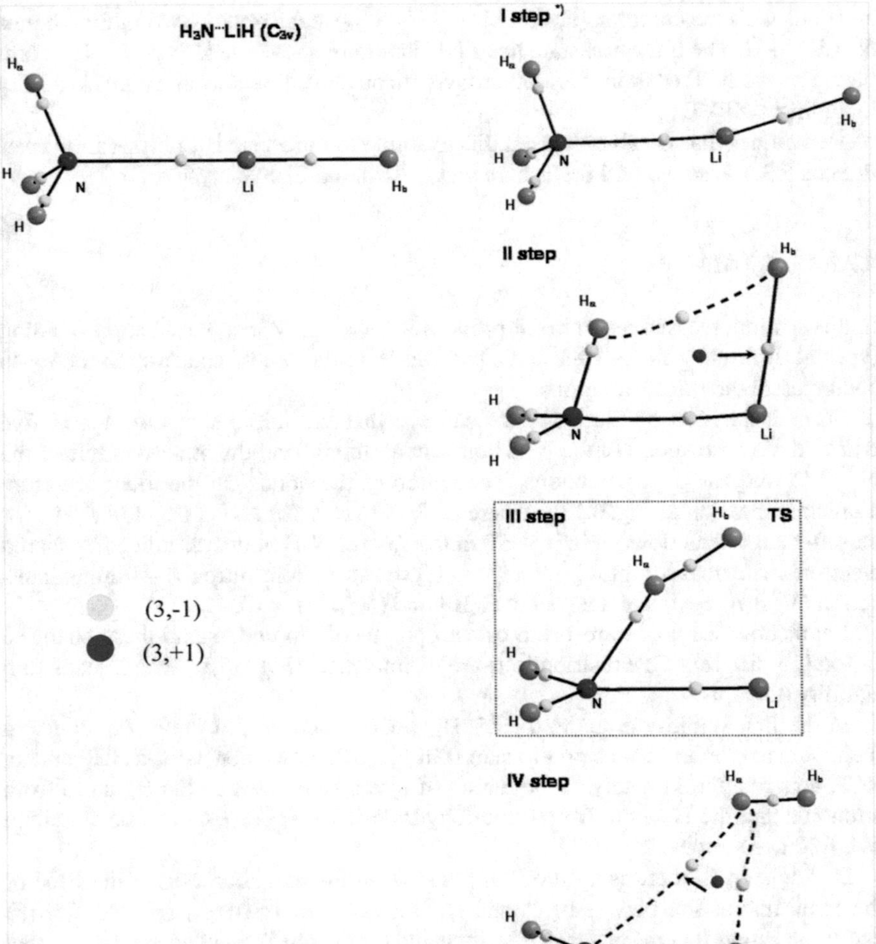

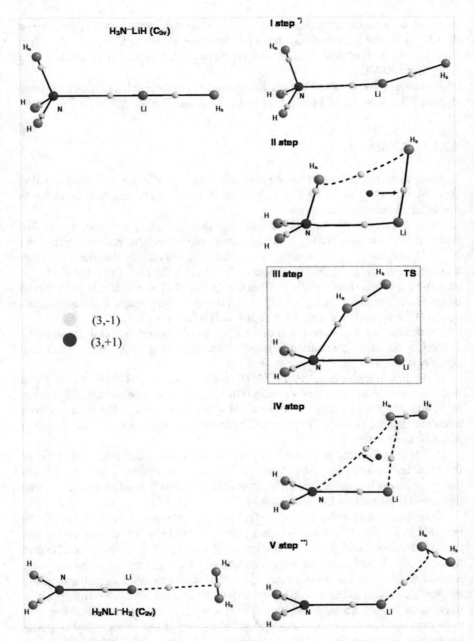

**Fig. 6.12** Evolution of the CPs for the electron density field. The initial and final points on the IRC path are shown on the left (top and bottom, respectively). Reprinted with permission from Ref. [2]. Copyright 2014

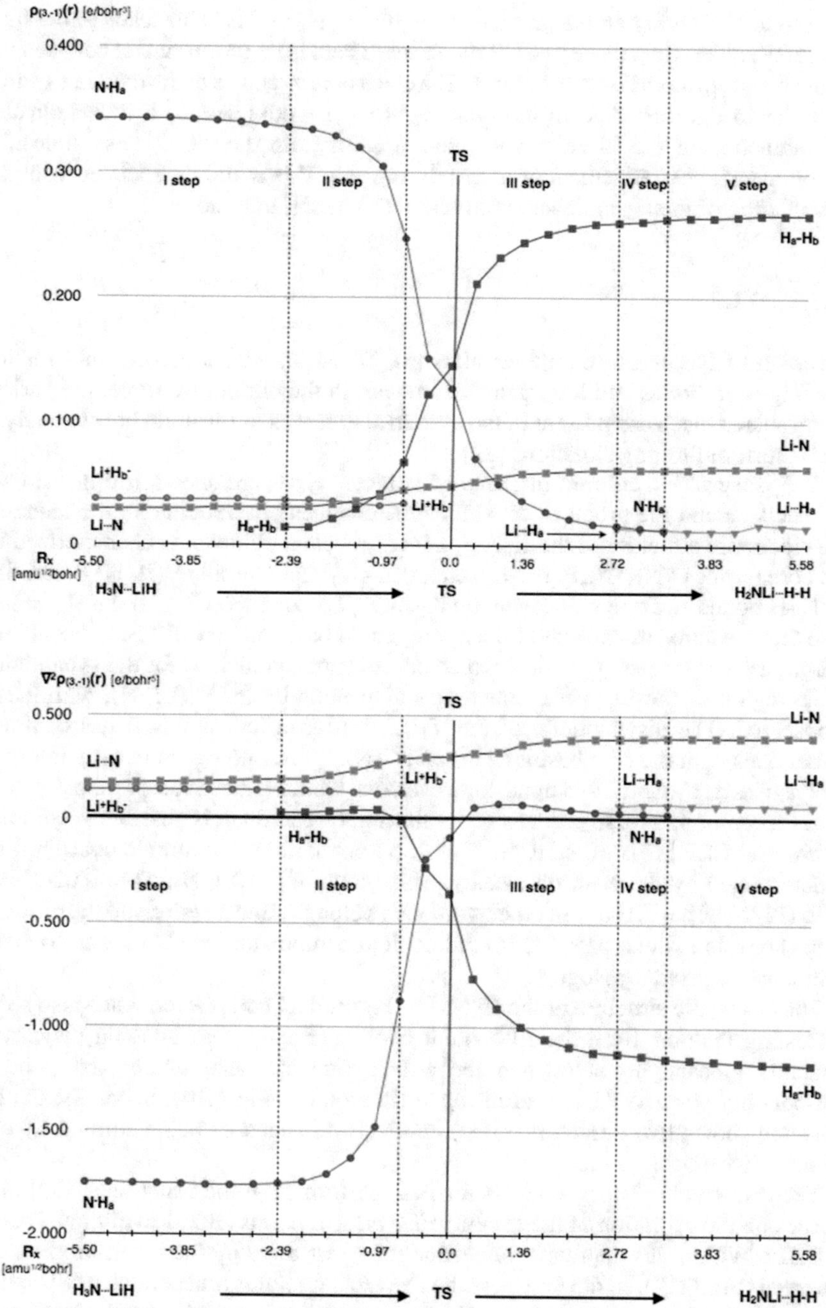


**Fig. 6.13** Evolution of the $\rho(\boldsymbol{r})$ and $\nabla^2\rho(\boldsymbol{r})$ fields along the IRC path at some specific $(3, -1)$ CPs. Reprinted with permission from Ref. [2]. Copyright 2014

the second BCP lying on the gradient path joining $H_a \cdots H_b$. This leads to the final fold catastrophe when these two CPs coalesce. There is no more contact between the nitrogen and $H_a$ atoms from ammonia. The electronic structure of the reacting system is similar to that observed in the final $H_2NLi \cdots H_2$ complex. The BCP indicates the interaction of the Li atom with the $H_b$ atom from the dihydrogen molecule. The geometric rearrangement of atoms during step V is mainly associated with the reorientation of the $H_2$ molecule with respect to the N–Li bond.

#### 6.2.3.2 ELF

The use of ELF is usually complementary to QTAIM since it can provide information on charges on bonds and lone pairs. Moreover, in the current example, the variety of non shared electrons present in the system also serves to illustrate how to analyze ELF results in limiting situations [29].

ELF isosurfaces, colored in terms of attractor types are shown in Fig. 6.14 for the reactants and the products as well as the changes observed along the reaction. The electronic structure of the $H_3N \cdots Li^+H_b^-$ reactant complex is characterized by two core basins C(N), C(Li), three protonated disynaptic basins V(H, N) describing the N–H bonds and one monosynaptic basin $V(H_b)$ corresponding to the $H_b^-$ anion. Note that the latter, as expected for a hydrogenoid basin, is very diffuse. Overall, the charges are very close to the ones expected. It is not so much so for the synapticity. In the region of the $Li \cdots N$ interaction a disynaptic basin, $V_1$(Li, N), with $2.2e^-$ is observed. The disynapticity of the $V_1$(Li, N) basin means that it has common separatrices with the core basins C(Li) and C(N). Thus, one would usually interpret this basin as a covalent Li–N bond. However, one would rather expect an electrostatic donor-acceptor nature. In order to verify this interpretation, it is possible to combine both $\rho$ and ELF [43]: we split the $V_1$(Li, N) population into atomic contributions as determined by the electron density. This yields $N[V_1(\text{Li, N})|\text{Li}] = 0.02e^-$ and $N[V_1(\text{Li, N})|\text{N}] = 2.18e^-$. In other words, $V_1$ belongs almost entirely to the nitrogen atom (i.e. it is its lone pair). This kind of field combination is also a way to avoid symmetry imposed topologies.

The electronic structure of the $H_2NLi \cdots H_2$ product complex contains also some interesting features from the ELF point of view. Firstly, two bonding disynaptic attractors appear lying below and above the symmetry plane which correspond to non-bonding electron density of nitrogen (in violet in Fig. 6.14). Secondly, the $H_2$ molecule shows two attractors $V(H_a)$ and $V(H_b)$ instead of being formed by one attractor (with $\simeq 2e^-$).

The catastrophe theory analysis for ELF performed at the usual level [32] only focuses on the evolution of the (3, −3) ELF critical points (Bond Evolution Theory or BET). Within this approach, only four steps separated by three catastrophes are observed (Fig. 6.14). In the first step, the reactant structures are maintained. The first catastrophe is observed when the $V(H_1$, Li, N) attractor is annihilated (the N–H bond is broken) and two new attractors, $V(H_1)$ and V(N), along with a (3, −1) CP appear. This catastrophe is known as a cusp. Since ELF enables us to follow localization

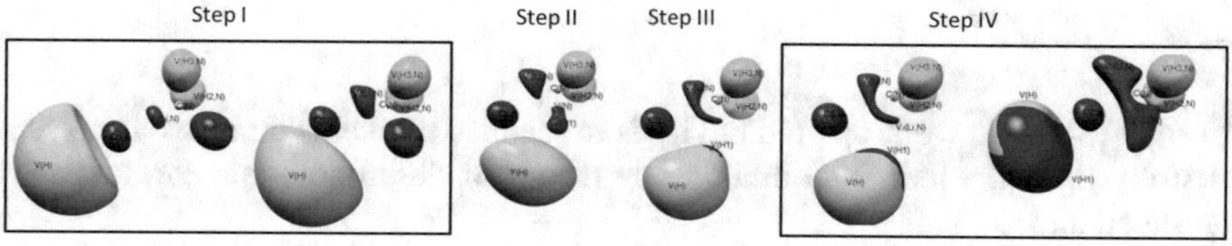


**Fig. 6.14** Evolution of ELF basins (ELF = 0.88). Deformations along steps I and IV after and prior to the reactants and products, respectively, are shown for the sake of clarity. Figure taken from Ref. [29]

regions, we can also seek to understand how the charge is being distributed among them: the N lone pair keeps most of the charge, $1.21e^-$, whereas the hydrogen atom H1 only holds $0.65e^-$. In the second step, the electrons flow from the region of the destroyed N–H1 bond to the Li–N region until the V(N) basin is annihilated in a fold catastrophe. Finally, the saturation of the Li–N bond (like the protonation rules seen at the beginning of the chapter) leads to a qualitative change: a new V(Li, N) basin appears in a fold catastrophe.

It should be noted that the use of only the attractors leads to limitations when the electron localization does not follow common patterns. Firstly, the lithium hydride is bound by electrostatic forces ($Li^+H^-$) and topological analysis of ELF does not reveal any bonding disynaptic attractor in the interatomic region. Hence, finding the moment when such $Li^+H^-$ bond is broken is problematic. Second, the dihydrogen molecule ($H_2$) topological analysis of ELF reveals only a large area comprising two protons with an almost constant value equal to one. Therefore, numerical problems may occur related to the localization of the attractor V(H, H). To overcome these difficulties, the analysis of ELF maxima evolution can be expanded to include the meaningful BCPs ($Li^+H^-$ and H-$H_1$) [29] or NCI.

## 6.3 Trying to Go Predictive

One of the most spread criticisms regarding Quantum Chemical Topology is its lack of predictivity, as opposed to the alleged ability of molecular orbital pictures to provide chemical insight with ease. It is thus important to show that, at least formally, topological approaches can also be used to forge predictive models. In the next sections, we will provide some examples of predictive and inverse design QCT approaches.

### 6.3.1 QTAIM

A possible way to work on predictivity is to identify problems that have been successfully tackled within MO thinking by means of descriptors that can be easily translated into QCT.

For instance, a particularly successful model in chemical reactivity that uses energy decompositions is the so-called distortion/interaction or activation strain model (ASM) [6, 20]. It is based on taking into account the usually forgotten role of geometrical deformation (strain, distortion) in building the activation barrier in a bimolecular chemical reaction. The barrier is not only a result of how the reactants interact with each other, but also of the energy cost that they have to invest to deform until they adopt the geometry at the top of the barrier. Despite its simplicity, this model has been successfully applied to rationalize/predict many observations in substitutions, eliminations, or pericyclic processes, among many others.

Usually, the ASM is used together with a Ziegler–Rauk-like energy decomposition method [21, 51], but nothing prevents it from being used with a QCT-based one, like IQA. All that is needed is a decomposition of the energy change of reactants along a reaction coordinate, $\xi$: $\Delta E(\xi) = \Delta E_{\text{strain}}(\xi) + \Delta E_{\text{int}}(\xi)$, where $\Delta E_{\text{strain}}(\xi)$ is the distortion cost, the energy needed to take the reactants to the geometry found at the transition state (TS), and $\Delta E_{\text{int}}$ is the stabilization energy released when the constrained moieties are allowed to interact. A truly simple, but insightful corollary of this partition is that if two reactions proceed with the same $\Delta E_{\text{strain}}$, the one with more stabilizing $\Delta E_{\text{int}}(\xi)$ will display an earlier TS with a smaller barrier, $\Delta E^{\ddagger}$.

Note that EDA decompositions use non-stationary intermediate states that violate antisymmetry and that have, thus, no physical meaning. On the contrary, no artificial state is used in IQA, and if we start with two real space fragments, $A$, $B$, we can exactly decompose $\Delta E = E_{\text{int}}^{AB} + E_{\text{def}}^{AB}$ at each $\xi$. Now the deformation energy of any fragment $X$ can be further partitioned into a geometric and an electronic component, $E_{\text{def}}^{X} = E_{\text{def,geom}}^{X} + E_{\text{def,el}}^{X}$, so that the sum of the geometric deformations of $A$ and $B$ equals $\Delta E_{\text{strain}}$.

This method has been applied, for instance, to understand the *endo* rule in Diels–Alder reactions [12]. In the case of the cycloaddition of *cis*-1, 3-butadiene reacts with maleic anhydride, def2-QZVPPD/B3LYP calculations show that the endo/exo reactions have activation barriers of 17.50/18.65 kcal/mol, thus favoring a kinetically controlled endo pathway by 1.15 kcal/mol. Details are found in Fig. 6.15.

The analysis of the deformation energy, in agreement with what was noticed by Fernández and Bickelhaupt, [20], shows how $\Delta E_{\text{strain}}$ favors the endo pathway. However, the electronic component of the deformation is slightly larger for the endo route in the earlier stages of the reaction, to become almost degenerate to that of the exo path at about $R(\text{C-C}) \approx 2.8$ Å. This is in agreement with a larger Pauli repulsion for the endo approach pathway that is added to the $\Delta E_{int}$ in the standard EDA decomposition. As regards the IQA $E_{int}$, the electrostatic component is almost negligible until past the barrier, in disagreement with models that use these interactions to rationalize the endo rule. $E_{int}$ is dominated by $V_{xc}$. The sum of $V_{xc}$ plus $E_{def}$,

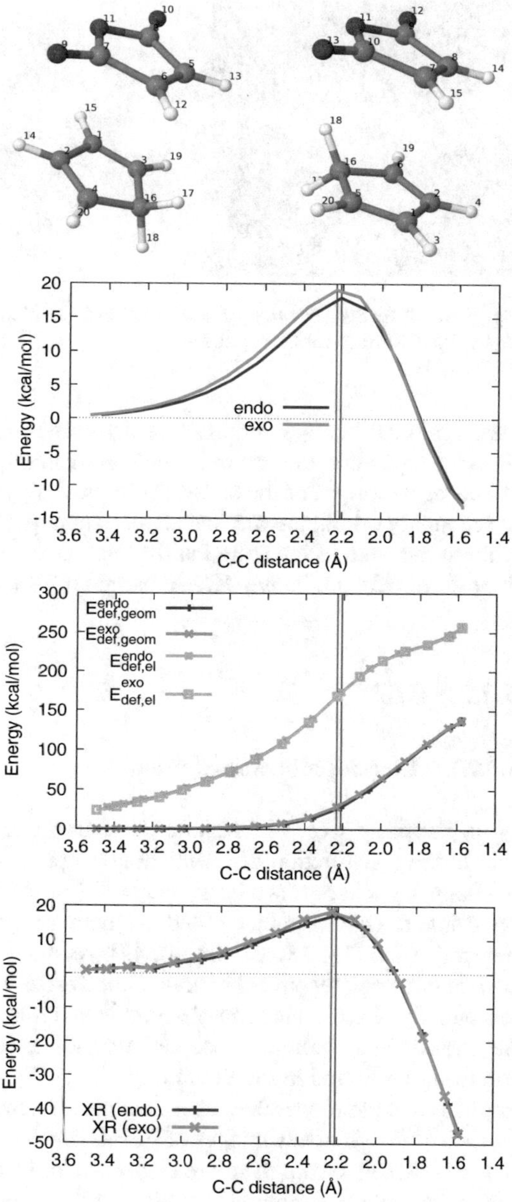


**Fig. 6.15** Top: endo (top-left) and exo (top-right) pathways in the cyclopentadiene + maleic anhydride cycloaddition. The energy is projected onto the distance of the C–C bond that is being formed, and the transition state geometry is highlighted by vertical lines. Middle: geometric and electronic deformation energies along the IRC. Bottom: exchange–repulsion (XR) energy

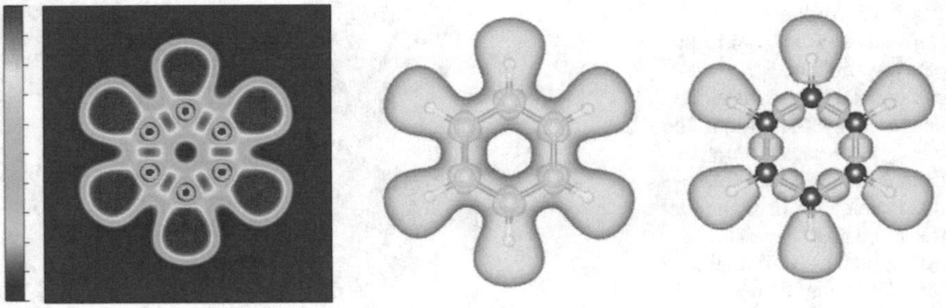

**Fig. 6.16** ELF representation of benzene. Left: 2D plot. Center: 3D isosurface for ELF = 0.5, Right: 3D isosurface for ELF = 0.7

the so-called exchange–repulsion term, mimics quite well the binding energy curve, showing again that the origin of the barrier is $E_{def}$. IQA thus rules out electrostatic effects as the origin of the endo rule in these simple cases, allowing for a small role of secondary orbital interactions. It also supports the basic claims of the ASM model without the need of invoking Pauli-violating states. Since the ASM has been shown to be extremely predictive, IQA-dressed QCT enjoys the same predictability.

## *6.3.2 ELF*

### 6.3.2.1 Electrophilic Substitution

From the early works of Kekulé [31], aromatic compounds have played a major role in the construction of chemical concepts. Aromaticity is a ubiquitous concept in chemistry that covers a large number of properties including energy stabilization, bond-length equalization, exalted magnetic properties and electron conjugation and delocalization [13, 18, 19, 33, 40, 42]. As many of the chemical concepts we have seen in this book (chemical bonding, bond order, hardness, electron population, etc.), aromaticity is not a quantum observable and it lacks a mathematical definition. This fact results in a plethora of aromaticity indices (energetic, structural, magnetic and electronic) reported in the literature [13, 19, 33, 40], which have been successfully applied to explain countless chemical phenomena.

One of the great advantages of ELF is that it allows to identify both localized and delocalized units from the mere inspection of its bifurcation diagram (Sect. 4.3.3.1), which thus allows to predict the electrophilic reactivity of various aromatic sites upon substitution [22].

As shown in Fig. 6.16, all ELF domains are united for ELF $< 0.5$. As we increase this value (ELF $= 0.58$), a reduction takes place leading to six protonated domains, $V(C_i, H_i)_{i=1,6}$ with a toroidal volume which includes all carbon cores and C–C bonds. This is usually called the “aromatic domain”. For ELF $= 0.65$, this volume splits into six C–C bonds. This last value, is the one usually used in the analysis of substituted aromatic compounds within the ELF framework. Negatively charged centers lead to a greater localization, hence their corresponding localization domains bifurcate at lower values. This allows to identify them as electrophilic sites.

This analysis can be enriched if we focus on the $ELF^{\pi}$ (Sect. 4.3.1.7) bifurcation tree, allowing us to distinguish between aromatic and antiaromatic compounds. We have seen that the separation between $\pi$ superbasins (Fig. 6.16) takes place in the upper zone of the diagram. An extensive study of cases has shown that this separation usually takes place in the range 0.64–0.91 in aromatic compounds, as opposed to the range 0.11–0.35 in antiaromatic ones.

Note that the lack of a single aromaticity scale is one of the most limiting factors in the "aromaticity" debate. The currently available scales, whether based on energy and structural behavior or on electronic delocalization, were constructed to agree with the image in traditional organic aromatic compounds, so their applicability to other types of compounds, such as metal clusters, is not clear. Calibration of aromaticity by means of the $ELF_{\pi}$ not only makes it possible to discriminate aromatic vs antiaromatic compounds, but it can also be used to quantify their resonance and to establish a scale of aromaticity.

**Protonation Sites**

An extremely interesting application of ELF has been the prediction of protonation sites. The prediction is based on following a set of rules based on energetic principles:

- the description of the complex as the initial step of the proton transfer reaction according to the rules of Rice and Teller [44]. This means that the smallest displacement is favored since they imply a decrease in the energetic barrier of an elementary reaction (minimization of the number of broken bonds along the process). For example, the favored geometry is that where the atoms move toward the position they will occupy in the product.
- the prediction of the final geometry (i.e. the angular conformation of the hydrogen compound) is given by Legon and Millen rules [34]. The direction is determined so as to favor the direct interaction with the electronic clouds (lone pairs, $\pi$ electrons), etc.

Since this means that we need to be able to localize the electronic clouds, electron localization sources of information are especially suited to apply these rules. Carroll et al. [11] used the Laplacian of the electron density to accurately predict the structure of several complexes. Resorting to ELF, it has been shown that it is even possible to discriminate between several potential protonation sites following a minimal topological change rule:

1. Protonation takes place on the most populated accessible valence basin for which the addition of the proton does not lead to a topological change.
2. The population of the basin V(H, B) of the protonated base should not largely exceed $2e^{-}$. In these cases, the population is smaller in the protonated form than in the neutral. The population decrease can be achieved by participation of the original basin, or even better, by charge transfer to neighboring basins.

3. If there is no monosynaptic basin accessible, the protonation can take place on a disynaptic basin.
4. A protonated disynaptic basin can be protonated if it belongs to an electropositive center.

Let's see an example worked out by Fuster and Silvi [23]. The CO molecule has 2.57e at V(C), whereas the population of V(O) is larger than $4e^-$. Hence, the preferential protonation site is V(C) (rules 1 and 2), leading to $H^+CO$. The formation of a hypothetical $COH^+$ would lead to a huge charge transfer of $2.3e^-$ to the V(C, O) bond (plus a topological change!) in order to alleviate the steric repulsion within V(O). Instead, the protonation on V(C) leads to only $0.6e^-$ being transferred to the CO bond.

### *6.3.3 NCI*

Reactivity is on many occasions controlled by weak interactions at the transition state. In this section, we will show a couple of examples where NCI is the preferred tool to identify, characterize and design reactivity in terms of weak interactions.

Let's start with an example where the mere inspection of NCI enables to understand the outcome of a reaction upon substitution, the nickel(0)-catalyzed denitrogenative trasannulation of benzotriazinones with alkynes [50]. The outcome and experimental conditions of the reaction are determined by the substituent on the nitrogen atom of the benzotriazinone. Whereas the phenyl-substituted substrate proceeds at room temperature, the methyl-substituted one requires higher temperatures. The hydrogen-substituted benzotriazinone do not react even upon heating at 100°C (Fig. 6.17).

In all cases, the alkyne insertion reaction into the Ni–C bond constitutes the rate-determining step (see the transition state (TS), indicated by a ‡ in Fig. 6.17). The activation barrier follows the expected trend: 22.0 kcal/mol (R = Ph) < 25.2 kcal/mol (R = Me) < 28.5 kcal/mol (R = H). These numbers can be greatly explained by dispersion. For R = Ph, important $\pi \cdots \pi$ stacking interactions appear at the TS (Fig. 6.18b), which are replaced by weaker CH$\cdots\pi$ interactions for R = Me not possible in the hydrogen- substituted counterpart (Fig. 6.18a). In this case, just by looking at the interactions in the TS, we are able to predict which substituent better favors the trasannulation.

These type of weak interactions can be further manipulated in order to control selectivity, regioselectivity and even enantioselectivity. Let's see one example. One of the first successes of computational reactivity studies was the rationalization of the mechanism involved in proline-catalyzed intra- and intermolecular aldol condensation reactions by Houk and List in 2003 [27]. The authors suggested a mechanism with one-proline where the TS was associated with the stereogenic carbon–carbon bond formation (see Fig. 6.19). This organocatalysis cornerstone study is still used

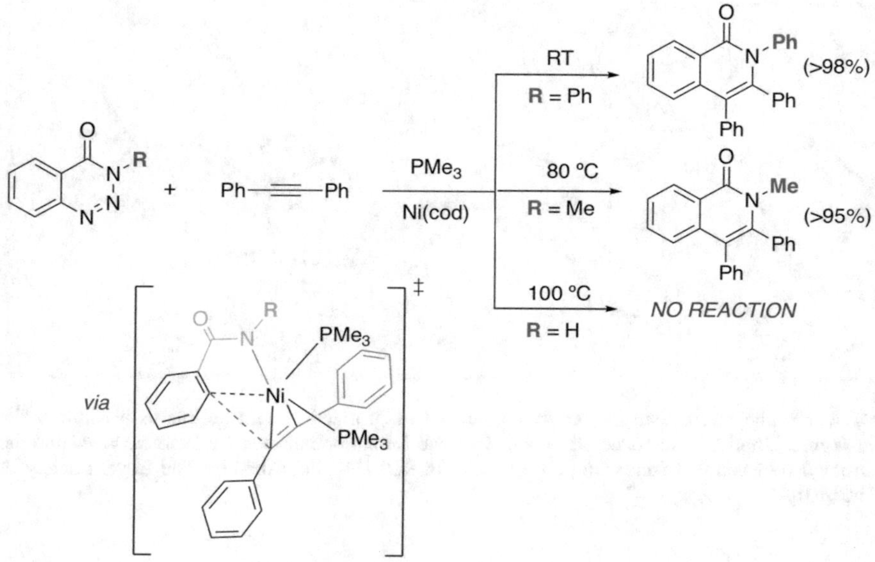


**Fig. 6.17** Nickel(0)-catalyzed denitrogenative trasannulation of benzotriazinones with alkynes. Taken with permission from Ref. [8]

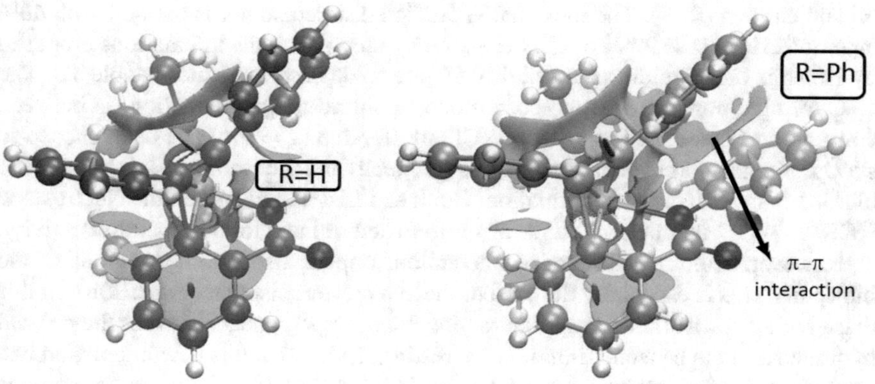


**Fig. 6.18** NCI analysis of the key transition states involved in the nickel(0)-catalyzed denitrogenative trasannulation reaction of benzotriazinones with alkynes. Taken with permission from Ref. [8]

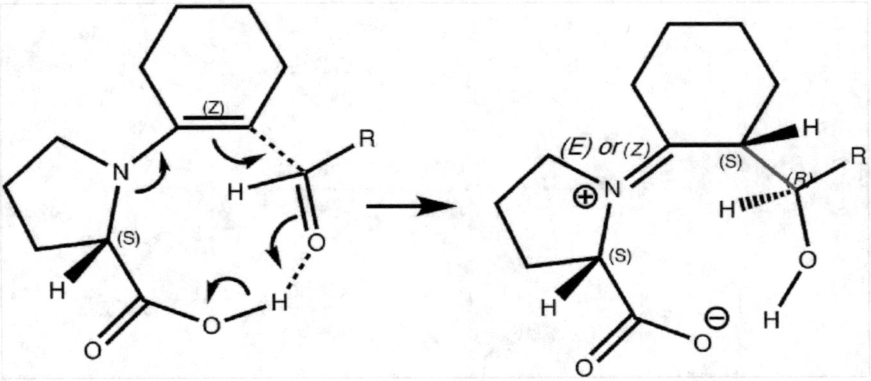


**Fig. 6.19** The mechanism for the intermolecular aldol reaction with proline as a chiral auxiliary (green), and key stereocenters created by bond formation between the enamine C=C and the carbonyl (red bond). Taken with permission from Ref. [3]—Published by The Royal Society of Chemistry

to understand other proline-mediated reactions, spanning from aldol or Mannich reactions to Michael additions.

Four different diastereoisomers can be formed: anti (S, R), syn (S, S), ent-anti (R, R) and ent-syn (R, S). The most stable predicted stereoisomer is the anti with 60% enantiomeric excess for R = Ph. Let's start by identifying the interactions proposed by Houk as those guiding the stability. Figure 6.20a shows the most stable TS. The C–C forming bond (red arrow) corresponds to a mid-range interaction (in between covalent and non-covalent), so the NCI cutoff value ($s = 0.1$) has been chosen to show both the non-covalent interactions (in green) and the forming C–C bond (deep blue). Along with the C–C formation, Houk and List had identified the electrostatic $NCH^{\delta+} \cdots O^{\delta-}$ interaction (purple arrow) as a determinant for the enantioselectivity.

However, important dispersive interactions appear also in other regions of the molecule. This is especially the case of the syn diastereoisomer (Fig. 6.20b). In this diastereoisomer, a tilted T-shape interaction (blue arrow) appears between the $\pi$ facial hydrogen and the benzene ring. It is interesting to note that this new interaction had not been identified before by mere geometric inspection [3]. However, its presence allows us to explain the fact that the syn conformers are the ones with the greatest dispersion correction.

Thus, a combination of both $NCH^{\delta}+ \cdots O^{\delta}-$ electrostatics and dispersion determine the outcome of the reaction, with only anti and syn as observable diastereoisomers.

Understanding which interactions stabilize a given TS is crucial for the proposal of new transformations where the outcome is modified. In this case, the identification of the NCI interactions in the syn TS allowed us to propose that the proline-mediated reaction between acetone and pyrrole-2-carboxaldehyde should reverse the anti/syn ratio, which was indeed the result of the corresponding simulation. This highlights the

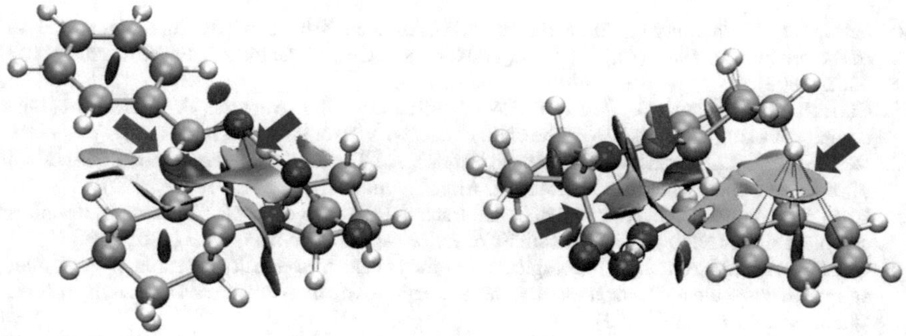

**Fig. 6.20** $s = 0.5$ isosurfaces for selected transition states for the aldol reaction between cyclohexanone and benzaldehyde catalyzed by proline. Left: the most stable TS, aka ent. Right: TS with the most important vdW interactions (measured by the D3 correction), aka syn. Isosurfaces are colored on a BGR scale according to the sign$(\lambda_2)\rho$ over the range −0.03–0.03 a.u. Arrows highlight important interactions: in red, the forming C-C bond, in purple the N$\cdots$O-C interaction and in blue the T-shape interaction

usefulness of the 3D analysis of weak interactions to understand subtle differences in reactivity.

## References


1. http://home.swipnet.se/w48087/faglar/materialmapp/teorimapp/ekt1.htm
2. Andrés, J., Berski, S., Contreras-García, J., González-Navarrete, P.: Following the molecular mechanism for the $NH_3 + LiH \rightarrow LiNH_2 + H_2$ chemical reaction: a study based on the joint use of the quantum theory of atoms in molecules (QTAIM) and noncovalent interaction (NCI) index. J. Phys. Chem. A **118**, 1663 (2014)
3. Armstrong, A., Boto, R.A., Dingwall, P., Contreras-García, J., Harvey, M.J., Mason, N.J., Rzepa, H.S.: The houk-list transition states for organocatalytic mechanisms revisited. Chem. Sci. **5**, 2057 (2014)
4. Bader, R.F.W.: Atoms in Molecules. Oxford University Press, Oxford (1990)
5. Bader, R.F.W., Nguyen-Dang, T.T., Tal, Y.: A topological theory of molecular structure. Rep. Prog. Phys. **44**(8), 893 (1981)
6. Bickelhaupt, F.M., Houk, K.N.: Analyzing reaction rates with the distortion/interaction-activation strain model. Angew. Chem. Int. Ed. **56**(34), 10070–10086 (2017)
7. Boto, R.A., Piquemal, J., Contreras-Garcia, J.: Revealing strong interactions with the reduced density gradient: a benchmark for covalent, ionic and charge-shift bonds. Theor. Chem. Acc. **136**, 139 (2017)
8. Boto, R.A., Woller, T., Contreras-García, J., Fernández, I.: Analysis of reactivity from the noncovalent interactions perspective. In: Mahmudov, K.T., Gurbanov, A.V., da Silva, M.F.C.G., Pombeiro, A.J.L. (eds.) Non-covalent Interactions in Catalysis, pp. 628–643. Royal Society of Chemistry (2019)
9. Brookhart, C.M., Green, M., Wong, L.: Carbon-hydrogen-transition metal bonds. Prog. Inorg. Chem. **36**, 1 (1988)

10. del Carmen Michelini, M., Sicilia, E., Russo, N., Alikhani, M.E., Silvi, B.: Topological analysis of the reaction of $Mn^+$ ($^7$s,$^5$s) with $H_2O$, $NH_3$, and $CH_4$ molecules. J. Phys. Chem. A **107**, 4862 (2003)
11. Carroll, M.T., Chang, C., Bader, R.F.W.: Prediction of the structures of hydrogen-bonded complexes using the laplacian of the charge density. Molec. Phys. **63**, 387 (1988)
12. Casals-Sainz, J.L., Francisco, E., Martín Pendás, Á.: The activation strain model in the light of real space energy partitions. Z. Anorg. Allg. Chem. **646**(14), 1062–1072 (2020)
13. Chen, Z., Wannere, C.S., Corminboeuf, C., Puchta, R., Schleyer, P.V.R.: Nucleus-independent chemical shifts (NICS) as an aromaticity criterion. Chem. Rev. **105**, 3842 (2005)
14. Cioslowski, J., Mixon, S.T.: Topological properties of electron-density in search of steric interactions in molecules – electronic-structure calculations on ortho-substituted biphenyls. J. Amer. Chem. Soc. **114**, 4382 (1992)
15. Coleman, A.J., O'Shea, D.: Local characterization of phase diagrams. Phys. Rev. B **22**, 3428 (1980)
16. Contreras-García, J., Calatayud, M., Piquemal, J.P., Recio, J.M.: Ionic interactions: comparative topological approach. Comput. Theor. Chem. **998**, 193 (2012)
17. Farrugia, L.J., Evans, C., Tegel, M.: Chemical bonds without "chemical bonding"? a combined experimental and theoretical charge density study on an iron trimethylenemethane complex. J. Phys. Chem. A **110**, 7952 (2006)
18. Feixas, F., Matito, E., Poater, J., Solà, M.: Metalloaromaticity. WIRes Comput. Mol. Sci. **3**, 105 (2013)
19. Feixas, F., Matito, E., Poater, J., Solà, M.: Quantifying aromaticity with electron delocalisation measures. Chem. Soc. Rev. **44**, 6434 (2015)
20. Fernández, I., Bickelhaupt, F.M.: The activation strain model and molecular orbital theory: understanding and designing chemical reactions. Chem. Soc. Rev. **43**(14), 4953–4967 (2014)
21. Fernández, I., Frenking, G.: The diels-alder reaction from the EDA-NOCV perspective: a re-examination of the frontier molecular orbital model. Eur. J. Org. Chem. **2019**(2–3), 478–485 (2018)
22. Fuster, F., Sevin, A., Silvi, B.: Topological analysis of the electron localization function (ELF) applied to the electrophilic aromatic substitution. J. Phys. Chem. A **104**, 852 (2000)
23. Fuster, F., Silvi, B.: Does the topological approach characterize the hydrogen bond? Theor. Chem. Acc. **104**, 13 (2000)
24. Gerratt, J., Cooper, D.L., Karakov, P.B., Raimondi, M.: Modern valence bond theory. Chem. Soc. Rev. **26**, 87 (1997)
25. Gillespie, R.J., Silvi, B.: The octet rule and hypervalence: two misunderstood concepts. Coord. Chem. Rev. **53**, 233 (2002)
26. Haaland, A., Shorokhov, D.J., Tverdova, N.V.: Topological analysis of electron densities: is the presence of an atomic interaction line in an equilibrium geometry a sufficient condition for the existence of a chemical bond? Chem. Eur. J. **10**, 4416 (2004)
27. Hoang, L., Bahmanyar, S., Houk, K.N., List, B.: Kinetic and stereochemical evidence for the involvement of only one proline molecule in the transition states of proline-catalyzed intra- and intermolecular aldol reactions. J. Amer. Chem. Soc. **125**, 16 (2003)
28. Jiménez-Grávalos, F., Gallegos, M., Martín Pendás, Á., Novikov, A.S.: Challenging the electrostatic $\sigma$-hole picture of halogen bonding using minimal models and the interacting quantum atoms approach. J. Comput. Chem. **42**(10), 676–687 (2021)
29. Kalinowski, J., Berski, S., Latajka, Z.: AIM and BET approach for ionic and covalent bond evolution in reaction of hydrogen elimination from ammonia and lithium hydride. Chem. Phys. Lett. **501**, 587 (2011)
30. Kar, T., Scheiner, S., Li, L.: Theoretical investigation on the mechanism of $LiH + NH_3 \rightarrow LiNH_2 + H_2$ reaction. J. Mol. Struct. **857**, 111 (2008)
31. Kekulé, A.: iuntersuehungen fiber aromatische verbindungen. Annalen der Chemie **137**, 129 (1865)
32. Kruger, M.B., Nguyen, J.H., Li, Y.M., Caldwell, W.A., Manghnami, M.H., Jeanloz, R.: Equation of state of $\alpha$-$si_3n_4$. Phys. Rev. B **55**, 3456 (1997)

33. Krygowski, T.M., Szatylowicz, H., Stasyuk, O.A., Dominikowska, J., Palusiak, M.: Aromaticity from the viewpoint of molecular geometry: application to planar systems. Chem. Rev. **114**, 6383 (2014)
34. Legon, A.C., Millen, D.J.: Determination of properties of hydrogen-bonded dimers by rotational spectroscopy and a classification of dimer geometries. Faraday Discuss. **73**, 71 (1982)
35. Lewis, G.N.: The atom and the molecule. J. Amer. Chem. Soc. **38**, 762 (1916)
36. Llusar, R., Beltrán, A., Andrés, J., Noury, S., Silvi, B.: Topological analysis of electron density in depleted homopolar chemical bonds. J. Comput. Chem. **20**(14), 1517 (1999)
37. Martín Pendás, A., Costales, A., Luaña, V.: Ions in crystals: the topology of the electron-density in ionic materials. III. geometry and ionic-radii. J. Phys. Chem. B **102**, 6937 (1998)
38. Martín Pendás, A., Francisco, E., Blanco, M.A., Gatti, C.: bond paths as privileged exchange channels. Chem. Eur. J. **13**, 9362 (2007)
39. Matta, C.F., Boyd (eds.): The Quantum Theory of Atoms in Molecules: From Solid State to DNA and Drug Design. Wiley-VC (2007)
40. Poater, J., Duran, M., Solà, M., Silvi, B.: Theoretical evaluation of electron delocalization in aromatic molecules by means of atoms in molecules (aim) and electron localization function (elf) topological approaches. Chem. Rev. **105**, 3911 (2005)
41. Politzer, P., Murray, J.S.: Electrostatics and polarization in $\sigma$- and $\pi$-hole noncovalent interactions: an overview. Chem. Phys. Chem. **21**(7), 579–588 (2020)
42. Schleyer, V.R.P., Jiao, H.: What is aromaticity? Pure Appl. Chem. **68**, 209 (1996)
43. Raub, S., Jansen, G.: A quantitative measure of bond polarity from the electron localization function and the theory of atoms in molecules. Theor. Chem. Acc. **106**, 223 (2001)
44. Rice, F.O., Teller, E.: The role of free radicals in elementary organic reactions. J. Chem. Phys. **6**, 489 (1938)
45. Shaik, S., Danovich, D., Silvi, B., Lauvergnat, D.L., Hiberty, P.C.: Charge-shift bonding-a class of electron-pair bonds that emerges from valence bond theory and is supported by the electron localization function approach. Chem. Eur. J. **11**(21), 6358 (2005)
46. Silvi, B.: The synaptic order: a key concept to understand multicenter bonding. J. Molec. Struct. **614**, 3 (2002)
47. Tal, Y., Bader, R.F.W., Nguyen-Dang, T.T., Ojha, M., Anderson, S.G.: Quantum topology. IV. Relation between the topological and energetic stabilities of molecular structures. J. Chem. Phys. **74**, 5162 (1981)
48. Tognetti, V., Joubert, L.: On the physical role of exchange in the formation of an intramolecular bond path between two electronegative atoms. J. Chem. Phys. **138**(2), 024,102 (2013)
49. Toro-Labbé, A., Gutiérrez-Oliva, S., Murray, J.S., Politzer, P.: The reaction force and the transition region of a reaction. J. Mol. Model. **15**(6), 707 (2008)
50. Wang, N., Zhend, S., Zhang, L.L., Guo, Z., Liu, X.: Nickel(0)-catalyzed denitrogenative transannulation of benzotriazinones with alkynes: mechanistic insights of chemical reactivity and regio- and enantioselectivity from density functional theory and experiment. ACS Catal. **6**, 3496 (2016)
51. Zhao, L., von Hopffgarten, M., Andrada, D.M., Frenking, G.: Energy decomposition analysis. Wiley Interdiscip. Rev. Comput. Mol. Sci. **8**(3), e1345 (2017)

# Chapter 7
# Solid State

**Abstract** We will examine in this section some of the special features of the application of the topology to condensed phase systems, in particular to periodic crystals. We have already presented what type of information can be extracted from the topological analysis in isolated molecules. The conclusions obtained thereby are basically transferable to the study of molecular crystals, as long as we restrict ourselves to the molecules that constitute the condensed phase system. Much more interesting is the study of the **intermolecular behavior** in molecular crystals, or the **interatomic** one in non-molecular systems. As we will see, the application of the theory to the latter regenerates the geometrization of the chemical physics of solids so common in the the first half of the last century. The study of intermolecular interactions using topology is relatively novel, partly due to the difficulty with which such systems are handled by the available *ab initio* methods, and partly because of the lack of existing chemical intuition as far as their characterization and bond properties are concerned. We will start by examining some essential characteristics of the topological method in periodic domains. Following the same approach as in the previous chapter, we will then analyze structure and reactivity in crystals, characterizing prototype examples and then bonding and coordination changes during phase transitions.

## 7.1 Periodicity and Topology

In this section, we will extract the main consequences of the existence of point and translational symmetry on the topology of electronic density in periodic systems and their properties. We will be using the main concepts of periodic systems. Readers who are not acquainted with solid-state fundamentals are referred for example to Ref. [30].



Á. Martín Pendás and J. Contreras-García, *Topological Approaches to the Chemical Bond*, Theoretical Chemistry and Computational Modelling,
https://doi.org/10.1007/978-3-031-13666-5_7

### 7.1.1 Topological Features

We will first focus on the domain over which the fields and scalar densities that we have been studying are defined in condensed phases. Since the imposition of Born–von Karman boundary conditions is general, it will only be necessary to analyze a single unit cell of the crystal, made up of the points

$$\begin{aligned} \boldsymbol{r} &= x\boldsymbol{a} + y\boldsymbol{b} + z\boldsymbol{c}, \\ (x, y, z) &\in \mathbb{B}^3, \end{aligned} \tag{7.1}$$

where $(\boldsymbol{a}, \boldsymbol{b}, \boldsymbol{c})$ are the primitive displacements (generators of the translation group) and $\mathbb{B}^3$ is the unit cube, $(0 \leq x, y, z \leq 1)$. Periodic boundary conditions identify the opposite faces of the unit cube, $x(0) = x(1), y(0) = y(1), z(0) = z(1)$. Thus, the domain of our functions is homomorphic to a three-dimensional torus, $\mathbb{T}^3$ (i.e. every turn on the torus for a given coordinate is equivalent, just as passing to the following cell on $x, y, z$).

This has several important consequences from the topological point of view. Firstly, the Morse sum that holds for the full set of critical points of a system is 0 (remember it was equal to 1 in molecules):

$$n_{(3,-3)} - n_{(3,-1)} + n_{(3,+1)} - n_{(3,+3)} = 0 \tag{7.2}$$

where $n_{(3,-3)}$ represents the number of $(3, -3)$ critical points and so on. The theoretical proof can be found in Sect. 2.3.

Secondly, periodicity enforces, in one way or another, that all basins be finite and, in this sense, it is perfectly feasible to assign a well-defined volume to all the units induced by the gradient function in the crystal. Moreover, this means that the sum of all these volumes recovers the total volume of the unit cell.

Finally, important differences appear with respect to critical points. An essential difference with respect to isolated molecular systems is that the existence of $(3, +3)$ CPs is enforced by Weierstrass' theorem. This is easily understood through a 1D analog. Imagine a Gaussian-like function in $\mathbb{R}$. This has just one maximum, so that the 1D Morse relation $n_{(1,-1)} - n_{(1,+1)} = 1$ is clearly satisfied. If we now close the 1D line to form a $\mathbb{T}$ ring, we must necessarily add a minimum, so that $n_{(1,-1)} - n_{(1,+1)} = 0$ now. In three dimensions, the Morse relations (Eq. 2.3) force the existence of all types of critical points (maxima, minima and both types of saddle points). This is relevant from the computational point of view since the great number and variety of critical points generally present in a unit cell leads to important algorithmic differences with respect to the molecular realm [41, 42]. As an example, it suffices to form all the dyads, triads and quartets of atoms to find all critical points of the density in a molecule (a description of the algorithms can be found in Ref. [5]). This approach usually fails in solids, where there is an infinite number of critical points, and different techniques to cope with the problem have been devised (see Chap. 8 for more details).

The specific point group to which the system under study belongs provides additional rules to locate the position of the CPs of a scalar function. Let us consider, for example, a pure rotation axis $C_n$, which, without loss of generality, we may assume along the $z$-axis. The behavior of a scalar function $f$ under a rotation $\hat{C}_n$ belonging to the point symmetry group is such that $f(\boldsymbol{h}) = f(\mathbb{R}\boldsymbol{h})$, where $\boldsymbol{h}$ is a general point (with an arbitrary origin with respect to the rotation axis), and

$$\mathbb{R} = \begin{pmatrix} \cos\phi & \sin\phi & 0 \\ -\sin\phi & \cos\phi & 0 \\ 0 & 0 & 1 \end{pmatrix} \tag{7.3}$$

is the matrix corresponding to the $\hat{C}_n$ rotation operation, where $\phi = 2\pi/n$ is the angle of rotation. For a small enough value of $\boldsymbol{h}$, a first-order Taylor expansion allows us to reach the following expression:

$$f(\mathbf{0}) + \nabla f(\mathbf{0}) \cdot \boldsymbol{h} = f(\mathbb{R}\mathbf{0}) + \nabla f(\mathbb{R}\mathbf{0}) \cdot \mathbb{R}\boldsymbol{h}. \tag{7.4}$$

Since $\mathbb{R}\mathbf{0} = \mathbf{0}$, we can remove the point at which we evaluate the gradient from our notation ($\mathbf{0}$ in all cases), so that the expression simplifies to

$$\nabla f \cdot \boldsymbol{h} = \nabla f \cdot \mathbb{R}\boldsymbol{h}. \tag{7.5}$$

Decomposing $\boldsymbol{h}$ into components perpendicular and parallel to the axis of symmetry, $\boldsymbol{h}_\parallel + \boldsymbol{h}_\perp$, we obtain

$$\nabla f \cdot \left(\boldsymbol{h}_\parallel + \boldsymbol{h}_\perp\right) = \nabla f \cdot \left(\mathbb{R}\boldsymbol{h}_\parallel + \mathbb{R}\boldsymbol{h}_\perp\right). \tag{7.6}$$

Since $\boldsymbol{h}_\parallel$ is on the axis of symmetry, $\mathbb{R}\boldsymbol{h}_\parallel = \boldsymbol{h}_\parallel$, so

$$\nabla f \cdot \boldsymbol{h}_\perp = \nabla f \cdot \mathbb{R}\boldsymbol{h}_\perp. \tag{7.7}$$

If we also decompose $\nabla f$ into perpendicular and parallel components, $\nabla f_\parallel + \nabla f_\perp$, only the component perpendicular to the axis in the first scalar product survives ($\nabla f_\parallel \cdot \boldsymbol{h}_\perp = 0$). Since $\mathbb{R}\boldsymbol{h}_\perp$ must remain perpendicular to the axis of symmetry, the same happens to the second scalar product:

$$\nabla f_\perp \cdot \boldsymbol{h}_\perp = \nabla f_\perp \cdot \mathbb{R}\boldsymbol{h}_\perp. \tag{7.8}$$

Rearranging,

$$\nabla f_\perp \cdot (\boldsymbol{h}_\perp - \mathbb{R}\boldsymbol{h}_\perp) = 0. \tag{7.9}$$

$\boldsymbol{h}_\perp - \mathbb{R}\boldsymbol{h}_\perp$ is a vector perpendicular to the axis of symmetry, which only vanishes if $\cos\phi = 1 \Rightarrow \phi = 2k\pi$, a situation that corresponds to the identity. Since $\boldsymbol{h}$ is arbitrary, this means that $\nabla f_\perp = \mathbf{0}$: symmetry annihilates the derivatives in directions

**Table 7.1** Symmetry of the fixed positions that ensure the presence of a critical point. Adapted with permission from Ref. [41]

| System | | | | | |
|---|---|---|---|---|---|
| Triclinic | $C_i(\bar{1})$ | | | | |
| Monoclinic | $C_{2h}(2/m)$ | | | | |
| Orthorhombic | $D_2(222)$ | $D_{2h}(mmm)$ | | | |
| Tetragonal | $C_{4h}(4/m)$ | $D_4(422)$ | $D_{2d}(\bar{4}2m)$ | $D_{4h}(4/mmm)$ | |
| Trigonal | $C_{3i}(\bar{3})$ | $D_3(32)$ | $D_{3d}(\bar{3}m)$ | | |
| Hexagonal | $C_{3h}(\bar{6})$ | $C_{6h}(6/m)$ | $D_6(622)$ | $D_{3h}(62m)$ | $D_{6h}(6/mmm)$ |
| Cubic | $T(23)$ | $T_h(m3)$ | $O(432)$ | $T_d(\bar{4}3m)$ | $O_h(m3m)$ |

perpendicular to the axes of symmetry when evaluated at a point that belongs to that axis. Likewise, very simple arguments can also be applied to other elements of the symmetry of the crystallographic point group.

This means that the simultaneous coexistence of certain combinations of symmetry elements at a given point will ensure that the gradient of a scalar function vanishes at that point that will necessarily be a critical point. For instance, an inversion center must be a critical point of any scalar field in a crystal. Table 7.1 shows the symmetry of these special positions for all the crystalline systems. Their specific location can be obtained very easily from the tables of Wyckoff positions [22] for each spatial group: they are all the positions with three fixed coordinates. It is important to notice that all the other special positions of the space group, although not ensuring the existence of a critical point, do limit strongly their location. Thus, those special positions with one or two fixed parameters must present null gradients for specific values of these parameters, although in many cases these points will correspond to positions of greater symmetry, already determined previously. Following this scheme, a good number of critical points given by the crystal symmetry can be located by mere inspection of the crystallographic tables.

Besides introducing restrictions on the number of critical points of each type that can coexist in a unit cell, periodicity also has important consequences on the topology induced by a gradient field. There is a partition of space in three-dimensional regions surrounded by separatrices that is thinner than the partition into basins we have seen up to now. We define [41] a **primary bundle** as the set of trajectories of the gradient field with common $\alpha-$ and $\omega-$ limits or, in other words, as the bundle of trajectories that begin at a given minimum and end at a given maximum. The limiting surface of a primary bundle is a separatrix. If a gradient trajectory would intersect it, we would have found a gradient line that would be born at a minimum or would die at a maximum that does not belong to the primary bundle.

All CPs must be located on the surface of a primary bundle. The general structure of a non-degenerate bundle is thus rather simple. It consists of a maximum, a minimum and, say, $n$ ring and $n$ bond points linked together in a particular way: at the maximum, a trajectory appears linking it with each of the bond points.

Without loss of generality, let us see how it works for the electron density. Gradient lines flow from each BCP linking it with the ring point(s) and, finally, each ring point is linked to the minimum, which also receives field lines from the bond point(s).

This scheme introduces an isomorphism between a primary bundle and a convex polyhedron. At every critical point, we associate a vertex of the polyhedron and an edge to each path on the surface of the bundle that connects maxima with bonds, bonds with rings and rings with cages. The faces of the polyhedron are determined by the attraction surfaces of a bond (or the repulsion surfaces of a ring), formed by the lines that start at a BCP and die at the minimum in the first case (or by those that start at the maximum and die at the ring). It can be proven that these polyhedra have $2n+2$ vertices, $2n$ faces and $4n$ edges and that, therefore, fulfill Euler's relation: faces + vertices = edges + 2 [10].

Hence, the basic topological structure of the crystal is that of its different primary bundles and their interconnections. It is easy to verify that, considering only non-degenerate primary bundles, the minimum number of BCPs (and of RCPs) is $n = 2$. To understand this, let us look at Fig. 7.1a. In this plot, with $n = 1$, all trajectories have appropriate beginnings and endings, and no CP is apparently degenerate. However, this alleged two-sided polyhedron is quite peculiar: on each of its faces, there exist a couple of edges that are made to coincide with each other, thus introducing a degeneracy in the bundle. If this were possible, the connection between the BCP and the RCP would be a line, so that the polyhedron would collapse onto a two-dimensional object. Instead, Fig. 7.1b shows a non-degenerate polyhedron: each BCP has an attraction basin, curved, but with edges and vertices different from each other. The same situation appears when considering the repulsion basins of the RCPs.

The partition into bundles is not commonly used to describe the topology of the electron density. Actually, the bundles are usually grouped together to introduce thicker divisions of space, in which we take regions limited by larger separatrices as primary objects of study. The usual practice has been to identify these basic objects as the union of all the primary bundles that share the same maximum. The interior of such an object is nothing else but the attraction basin of the maximum. All the other CPs in the set of primary bundles that define an atomic basin lie on its surface, defining another correspondence with sets of polyhedra, the **attraction polyhedra**. In an attraction polyhedron, we assign a vertex to each cage, a face to each BCP point (which constitutes its two-dimensional attraction basin) and an edge to each ring point (corresponding to its 1D attraction basin). Thus, an attraction polyhedron with $m$ vertices is composed of $m$ primary bundles.

The recognition of bundles as the intrinsic topological units that form the crystal allows us to group them in other ways, giving rise to alternative fruitful perspectives of the same reality. An immediately affordable grouping recipe is found by associating all the bundles that share the same minimum to form other basic topological objects. This prescription is symmetric to the previous one regarding the exchange of attraction basins by repulsion basins: the set of points in the space that share the same $\alpha$- limit. Thus, a **repulsion polyhedron** has a cage point inside. On its surface, the attractors define vertices, the ring points and their 2D repulsion basins define faces

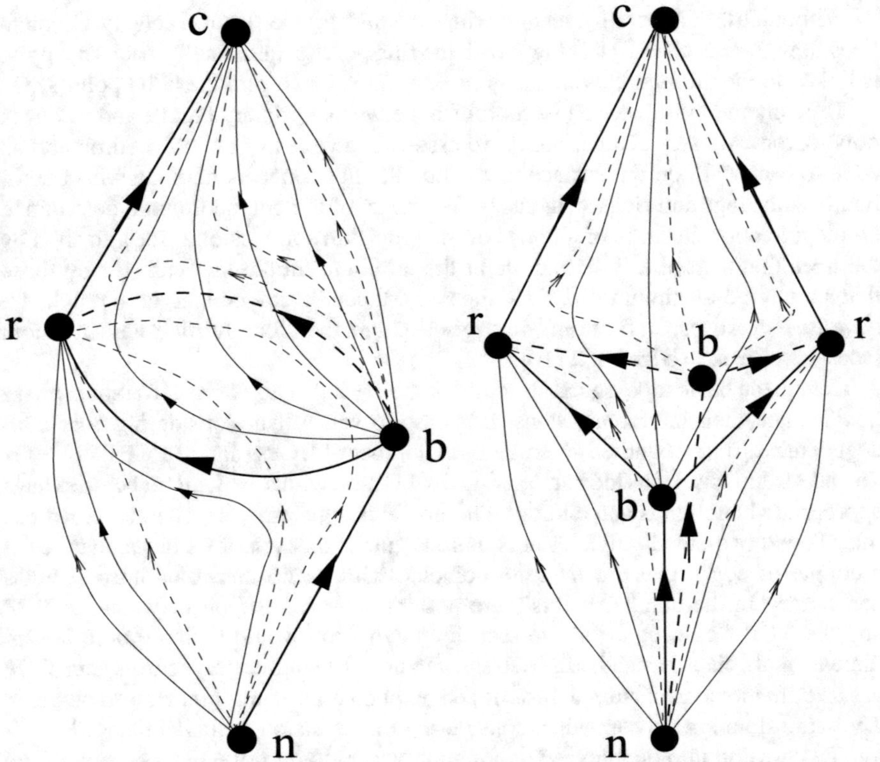


(a) Primary bundle with $n = 1$. (b) Primary bundle with $n = 2$.

**Fig. 7.1** Primary bundles with one and two bonds

and the bond points and their one-dimensional repulsion basins define edges. Please note that this prescription is applicable to any of the scalar fields we have introduced in part I.

## *7.1.2 Crystal Properties*

We have seen that the cell volume, $V$, can be recovered from the sum of the different attraction basins volumes, $V_{\Omega_i}$:

$$V = \sum_i V_{\Omega_i} = \sum_i V_i. \tag{7.10}$$

This additivity allows to decompose the static thermodynamic properties (at T = 0 K) into basin contributions. Let us briefly examine the basics of these ideas.

Global extensive properties can be easily divided into a sum of basin contributions extended to a unit cell, this being a direct consequence of the additivity of basin volumes. Thus, the expected value of an extensive observable $\hat{X}$ can be expressed as a sum of basin averages $X_{\Omega_A}$, obtained by integrating the corresponding operator densities over the volume expanded by the basins. In this way, we can decompose the molar energy of a crystal, as shown with the IQA method (Sect. 3.5.2). More interesting is to show that the same can be done with intensive properties (not dependent on the size).

Let us take, for instance, the isothermal compressibility of a crystal, which is defined, via the pressure $p$ and the volume $V$, as follows:

$$\kappa = -\frac{1}{V}\left(\frac{\partial V}{\partial p}\right)_T . \tag{7.11}$$

This volume additivity in Eq. 7.10 is the key piece that allows us to connect the thermodynamic properties of the crystals with different microscopic quantities. Substituting 7.10 in 7.11 provides

$$\kappa = -\frac{1}{V}\left(\frac{\partial(\sum_i V_i)}{\partial p}\right)_T = \frac{1}{V}\sum_i\left(\frac{\partial V_i}{\partial p}\right)_T , \tag{7.12}$$

and introducing $-1/V$ into the summation and multiplying and dividing by $V_i$, the previous expression is transformed into

$$\kappa = \sum_i\left(\frac{V_i}{V}\right)\left(-\frac{1}{V_i}\left(\frac{\partial V_i}{\partial p}\right)_T\right). \tag{7.13}$$

By analogy with the definition of the macroscopic isothermal compressibility of the crystal (Eq. 7.11), we can assign to each basin of the crystal a compressibility given by

$$\kappa_i = -\frac{1}{V_i}\left(\frac{\partial V_i}{\partial p}\right)_T , \tag{7.14}$$

which results in a measure of the variation per unit volume suffered by the volume of the $i$-th basin as external pressure is applied. Replacing this definition in 7.13, we obtain an expression for the compressibility of the crystal as a function of the basin compressibilities:

$$\kappa = \sum_i \frac{V_i}{V}\kappa_i . \tag{7.15}$$

This relationship indicates that the compressibility of a crystal can be decomposed into basin contributions whose average recovers the total compressibility. The weight of each of the microscopic magnitudes is given by the relative volume occupied by each basin in the volume of the cell. Similar procedures can be used to break down other intensive properties of crystals, such as the pressure. The interested reader can find more information in Ref. [40].

## 7.2 Structure

### *7.2.1 QTAIM*

#### 7.2.1.1 Coordination and Shape

The introduction of attraction and repulsion polyhedra in the previous section allows us to establish unambiguously concepts so important in the physics and chemistry of solid state as the **coordination index** of an atom and its **associated coordination polyhedron**. It suffices to use the concepts of attraction polyhedra. To each atomic species in the crystal, we associate an atomic or attraction polyhedron with as many faces as different bonds the atom establishes with its neighbors, resulting in a continuous-to-discrete correspondence between the set of all possible nuclear configurations and the set of topological polyhedra. Imagine a simple cubic atomic lattice, with only one non-equivalent atom at (0, 0, 0) that establishes BCPs with its six nearest octahedral neighbors. By symmetry, the BCPs lie at the midpoint between any two nearest neighbors. Their 2D attraction basins are necessarily planar, and it is a very simple exercise to check that the attraction basin of any atom is a cube with the atom in its center and a BCP at the center of each of its six faces. This is its attraction polyhedron.

Repulsion polyhedra are directly associated with the network of bonds of the structure or **molecular graph** of the crystal, showing edges along the bond paths. Both visions thus complement each other. Following the recipe, the definition of the number of bonds that are connected to an atom acquires an objective solid foundation.

Primary bundles can be grouped to form the topological equivalent of the Wigner–Seitz cell of the crystal, already introduced by Zou and Bader [62] as the smallest connected region of the space limited by a zero-flux surface that exhibits both the translational and local point group invariance of the crystal.

The homomorphism between bundles or atomic basins and convex polyhedra can be pushed a little further. For this, it is necessary to remember the concept of **Voronoi polyhedron**. Voronoi polyhedra are the Wigner–Seitz cells of a hypothetical Bravais lattice that contains a node at each nuclear position [3, 30]. Its importance lies in that they exhibit the full local point group symmetry, sharing faces with their neighbors and filling the space.

It is easy to prove that, in the case where all the atoms of the lattice are equivalent, the proximity polyhedra coincide with the QTAIM atomic basins. This is the case in simple metals, for instance. As soon as we consider different atomic species, the sizes and shapes of the proximity polyhedra adapt themselves to the coordination and to the characteristic sizes of each of the constituents. Using the property that atomic surfaces are orthogonal to the main direction of positive curvature at the BCPs, we can define the **proximity-weighted Voronoi polyhedra**, taking the BCP as the point where the planes that form the faces of the polyhedra are located, instead of taking the midpoint of the line that joins the two atoms, as in the naïve plain Voronoi case.

These polyhedra have proved to be fairly accurate approximations to the shape of the atomic basins. This similarity is based on the fact that their faces turn out to be a linear approximation to the attraction basins of the BCPs, which actually curl as a result of an expansion induced by the larger atoms at the expense of the smaller ones. According to this idea, the faces of the weighted proximity polyhedra will distort, losing their planarity until they define the atomic basin (see Fig. 7.4 for an example).

#### 7.2.1.2 Prototype Crystals

We will examine only one prototype ionic system, the LiI crystal in its NaCl (B1) phase (ambient pressure phase) which belongs to the space group $Fm\bar{3}m$. The theoretical equilibrium network parameter is $a = 4.259$ Å. In this space group, the special positions are those presented in Table 7.2. We should note that there are four fixed points of different symmetry. Two of them are found at the Lithium atom, which occupies position 4$a$, and at the iodine nucleus, which is found at position 4$b$. According to what has previously been shown, positions 8$c$ and 24$d$, which have also three fixed coordinates, should also be critical points.

From Table 7.2, we observe that all the important critical points in the crystal lie on two different crystal planes, the [100] and the [110] ones (as well as in its symmetry

**Table 7.2** Wyckoff special positions for the $Fm\bar{3}m$ space group

| Point group | Label | Multiplicity | Position |
|---|---|---|---|
| $C_1$ | $l$ | 192 | $(x, y, z)$ |
| $C_s$ | $k$ | 96 | $(x, x, z)$ |
| $C_s$ | $j$ | 96 | $(0, y, z)$ |
| $C_{2v}$ | $i$ | 48 | $(\frac{1}{2}, y, y)$ |
| $C_{2v}$ | $h$ | 48 | $(0, y, y)$ |
| $C_{2v}$ | $g$ | 48 | $(x, \frac{1}{4}, \frac{1}{4})$ |
| $C_{3v}$ | $f$ | 32 | $(x, x, x)$ |
| $C_{4v}$ | $e$ | 24 | $(x, 0, 0)$ |
| $D_{2h}$ | $d$ | 24 | $(0, \frac{1}{4}, \frac{1}{4})$ |
| $T_d$ | c | 8 | $(\frac{1}{4}, \frac{1}{4}, \frac{1}{4})$ |
| $O_h$ | $b$ | 4 | $(\frac{1}{2}, \frac{1}{2}, \frac{1}{2})$ |
| $O_h$ | $a$ | 4 | $(0, 0, 0)$ |

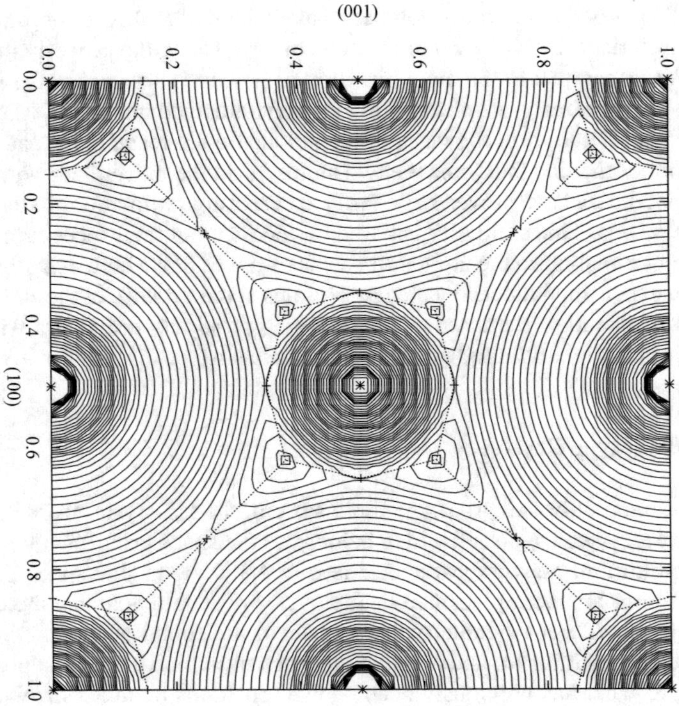


**Fig. 7.2** Representation of isodensity lines in the [100] plane for the LiI crystal. The isolines scale is logarithmic, and 50 values have been taken from the minimum to the absolute maximum. Critical points as well as the intersections of the zero-flux surfaces have been superimposed with the following criteria: points for surface boundaries, asterisks at nuclear positions, crosses for BCPs, squares for ring points and, finally, triangles for cage points. The axes labels correspond to crystallographic directions, and crystallographic units are used throughout. Adapted with permission from Ref. [41]

equivalent ones). The curves of electron isodensity in both planes are presented in Figs. 7.2 and 7.3, respectively.

The lithium and iodide ions turn out to be essentially spherical. If we take into account that non-nuclear critical points necessarily appear on the interatomic surfaces, we see that the distance from the critical points neighboring the Li nucleus appears at a relatively constant distance, irrespective of direction. This distance defines an ionic **directional radius**. The iodide anion shows small but significant deviations from sphericity, mainly along the direction of anion–anion contacts. From Figs. 7.2 and 7.3, it is clear that iodide ions form a compact cubic packing. This allows us to define an ionic radius for the $I^-$ from the lattice parameter $a$ only, as suggested repeatedly in the literature.

Secondly, the separatrices show very slight deviations of linearity or planarity, and the atomic surfaces are faithfully approximated by the weighted proximity polyhedra. Actually, the structure of the basins in the region close to the Li–I BCP is very similar

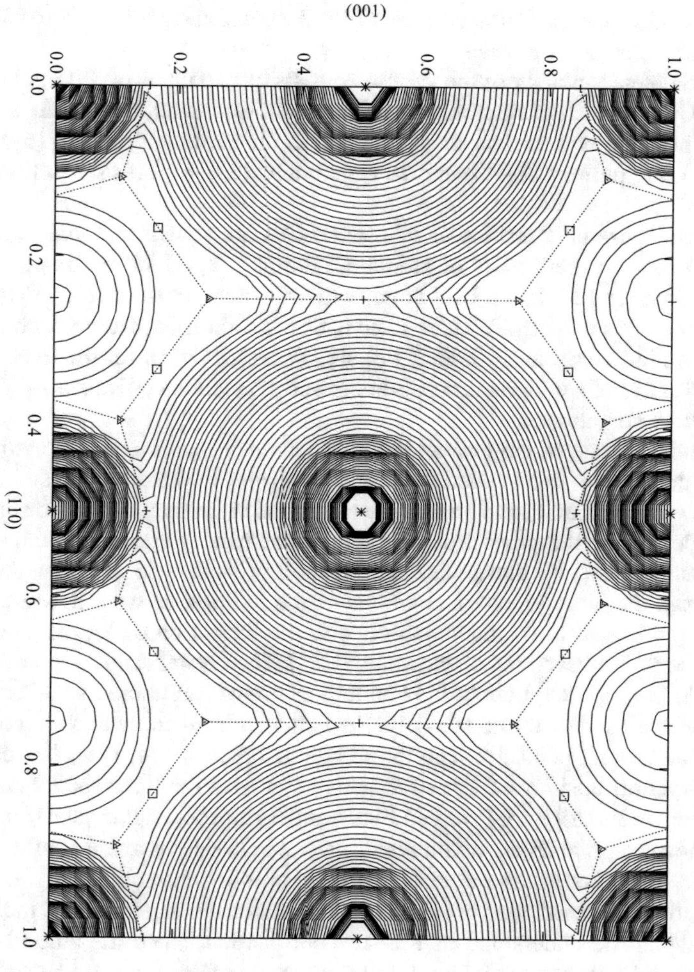


**Fig. 7.3** Representation of isodensity lines in the [110] plane for the LiI crystal. The isolines scale is logarithmic, and 50 values have been taken from the minimum to the absolute maximum. Critical points as well as the intersections of the zero-flux surfaces have been superimposed with the following criteria: points for surface boundaries, asterisks at nuclear positions, crosses for BCPs, squares for ring points and, finally, triangles for cage points. The axes labels correspond to crystallographic directions, and crystallographic units are used throughout. Adapted with permission from Ref. [41]

to that found in the free molecule (see Fig. 3.2), and the lithium cation is convex against a concave iodide anion.

The complete topology of the system consists of two types of different and independent CPs for each of the four possible categories: nuclei, bonds, rings and cage points. There are two nuclear critical points in the primitive cell, 12 $(6 + 6)$ bond points, 20 ring points $(12 + 8)$ and 10 cage points $(2 + 8)$. This set fulfills Morse's relation.

The positions and densities of this set of CPs are shown in Table 7.3. The two BCPs connect different pairs of atoms. CP number 3, which is found occupying Wyckoff position (24*e*), is a Li–I bond, while CP number 4, fixed by symmetry at position (24*d*), is a I–I BCP. There is no cation–cation BCP, from which it follows that lithium has a coordination index of six, whereas it is 18 for the iodide moiety, $6 + 12$: 6 I–Li links with its first neighbors and 12 second equivalent I–I links with its 12 second neighbors.

We find here one of the problems that we already mentioned: a bond point that does not correspond to a traditional link. In this case, the electron density evaluated at the two different BCPs is virtually the same, and there seems to be no clear reason to exclude one or the other from the list of chemically significant links. One of the reasonings used by those who are not fond of such type of anion–anion links comes from the large topological charge of the iodines, as we will see later. This large charge is said to be incompatible with the formation of a *bond* between the iodine anions. However, the electrostatic field in the crystal stabilizes the anions, so we should have to doubt about the bonding nature of such links. Another common argument against the anion–anion interaction is related to symmetry constraints. Indeed, the critical point between the iodide anions, $(0, 1/4, 1/4)$, lies at the 24*d* special Wyckoff position of Table 7.2. This point fixes the three spatial coordinates and is thus necessarily a CP forced by symmetry in rock-salt crystals. It has been shown, however, that anion–anion bonds persist in lattices without symmetry, being practically ubiquitous in the electron densities of inorganic solids.

Accepting the description provided by the electron density topology, the LiI structure of should be understood as a 6–18 coordination, since the anion has 6 first neighbors and 12 second neighbors, links, and not as 6–6. It should be stressed that position 24*d*, which as shown is necessarily a critical point in alkaline halides in the B1 phase, needs not be always a BCP. In many alkali halides, it corresponds to a ring critical point. In these cases, there are no anion–anion BCPs.

Using Table 7.3, and taking into account the value of the lattice parameter $a$, we can obtain without difficulty the directional ionic radii in the crystal: 0.9377 Å for lithium along the cation–anion direction, and 2.1448 or 2.1797 Å for iodide, along the I–I link or Li–I bond directions, respectively. This proves once again the considerable spherical deformation suffered by the anions. These values are in good agreement with Shannon and Prewitt radios, 0.9 and 2.06 Å [55].

Table 7.4 contains the properties of $\mathbb{H}\rho$ at all the non-nuclear CPs. Both BCPs are of the closed-shell type. If we analyze the geometrical disposition of the CPs, it turns out, for instance, that the basin of critical point number 3, the Li–I bond, is parallel

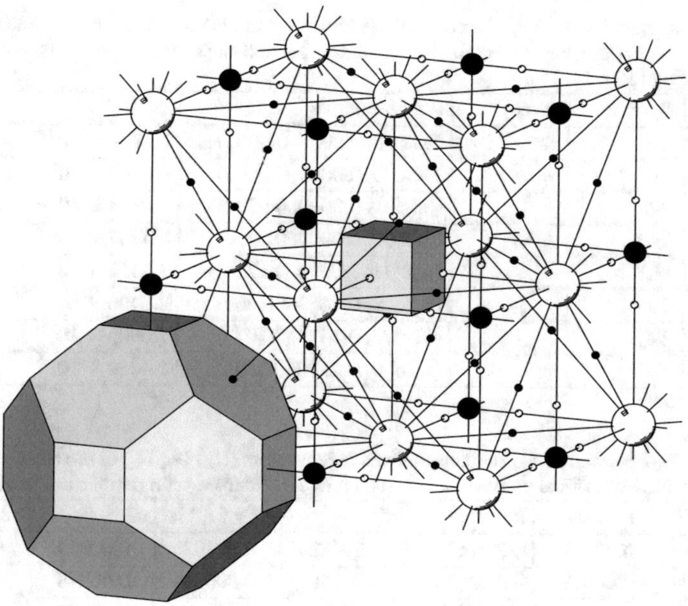

**Fig. 7.4** Weighted proximity polyhedra for the LiI structure, represented on the unit cell. The lines correspond to the I–I and Li–I bonds. The filled circles to the I–I bond points, and the hollow circles to the Li–I CPs. The relative position of cations and anions has been reversed with respect to Figs. 7.2 and 7.3, to allow a better view. The white spheres are iodide ions, and the black spheres lithium cations. Reproduced with permission from Ref. [41]

to the $yz$-plane in the proximity of the CP. The bond path is simply the $x-$ axis, and the bond angle is 180°.

Following an analogous reasoning for the six bonds around the lithium atom (see the central atom in Fig. 7.2), we can use the chemical intuition to imagine its proximity-weighted polyhedron, which will have the shape of a regular cube, slightly bulged to form the basin. It can be readily verified that the cage point ($c7$) forms the eight vertices of the cube, and that one ring ($r5$) is located in the middle of the twelve edges of the cube. The edges, seen from the center of the cube, are straight lines parallel to Cartesian axes (this can be checked by looking at the eigenvectors associated to the negative eigenvalue of point $r5$ in Table 7.4), which curve themselves slightly inwards in the vicinity of the vertices. We can convince ourselves of the veracity of the latter statement by considering lithium (0, 0, 0) at the bottom left of Fig. 7.2. The nearest ring point, at (0.1221, 0, 0.1221), is at the center of an edge along the $y$-axis. This edge ends at two cage points, located above the plane of the Figure at (0.10731, 0.10731, 0.10731) and (0.10731, −0.10731, 0.10731). The cage points are actually closer to the $yz$-plane than the rings.

Similar arguments can be applied to the iodine atom. In this case, the square faces are associated with each of the six Li–I bonds along the (100) lines, and the

**Table 7.3** Independent CPs in LiI. Positions in crystallographic units, and all other magnitudes in a.u. The symbols *n*, *b*, *r* and *c* refer to nuclear, bond, ring and cage CPs, respectively. Adapted with permission from Ref. [41]

| Number | Symm. | Type | Representative | $\rho$ |
|---|---|---|---|---|
| 1 | 4*a* | *n* | (0.0000, 0.0000, 0.0000) | $1.3722 \times 10^2$ |
| 2 | 4*b* | *n* | (0.5000, 0.0000, 0.0000) | $1.0503 \times 10^5$ |
| 3 | 24*e* | *b* | (0.1521, 0.0000, 0.0000) | $4.4858 \times 10^{-3}$ |
| 4 | 24*d* | *b* | (0.2500, 0.0000, 0.2500) | $5.5567 \times 10^{-3}$ |
| 5 | 48*h* | *r* | (0.1221, 0.0000, 0.1221) | $2.3112 \times 10^{-3}$ |
| 6 | 32*f* | *r* | (0.1669, 0.1669, 0.1669) | $2.1400 \times 10^{-3}$ |
| 7 | 32*f* | *c* | (0.1073, 0.1073, 0.1073) | $1.6903 \times 10^{-3}$ |
| 8 | 8*c* | *c* | (0.2500, 0.2500, 0.2500) | $1.4466 \times 10^{-3}$ |

**Table 7.4** $\nabla^2\rho$, curvatures ($\lambda_i$) and normalized eigenvectors ($U_x$, $U_y$, $U_z$) of the Hessian matrix of $\rho$ at the non-nuclear critical points of LiI. All data in a.u. Adapted with permission from Ref. [41]

| Number | *b*3 | *b*4 | *r*5 | *r*6 | *c*7 | *c*8 |
|---|---|---|---|---|---|---|
| $\nabla^2\rho$ | 0.03277 | 0.01485 | 0.04472 | 0.00803 | 0.00969 | 0.00632 |
| $\lambda_1$ | −0.00473 | −0.02695 | −0.00158 | −0.00098 | 0.00278 | 0.00211 |
| $U_x$ | 0.00000 | 0.00000 | 1.00000 | −0.57735 | 0.40825 | 0.57735 |
| $U_y$ | 0.70711 | 0.70711 | 0.00000 | −.57735 | 0.40825 | 0.57735 |
| $U_z$ | 0.70711 | −0.70711 | 0.00000 | 0.57735 | 0.81650 | 0.57735 |
| $\lambda_2$ | −0.00473 | −0.00269 | 0.00557 | 0.00451 | 0.00278 | 0.00211 |
| $U_x$ | 0.00000 | 1.00000 | 0.00000 | 0.40825 | −0.70711 | 0.70711 |
| $U_y$ | −0.70711 | 0.00000 | 0.70711 | 0.40825 | 0.70711 | −0.70711 |
| $U_z$ | 0.70711 | 0.00000 | 0.70711 | 0.81650 | 0.00000 | 0.00000 |
| $\lambda_3$ | 0.04223 | 0.01479 | 0.01479 | 0.00451 | 0.00412 | 0.00211 |
| $U_x$ | 1.00000 | 0.00000 | 0.00000 | −0.70711 | −0.57735 | 0.40825 |
| $U_y$ | 0.00000 | 0.70711 | −0.70711 | 0.70711 | −0.57735 | 0.40825 |
| $U_z$ | 0.00000 | 0.70711 | 0.70711 | 0.00000 | 0.57735 | −0.81650 |

hexagonal faces, which turn out to be flat due to symmetry constraints, to the twelve I–I bonds in the ⟨100⟩ directions. In Fig. 7.4, the weighted proximity polyhedra of both atoms display their mutual relative positions and the space-filling property. Figure 7.5 presents the atomic basins obtained from the crystalline wave function. Notice the great size difference between the iodine and lithium atoms. Comparing both figures, we check that proximity-weighted polyhedra are, in this case, a very good approximation to the atomic basins.

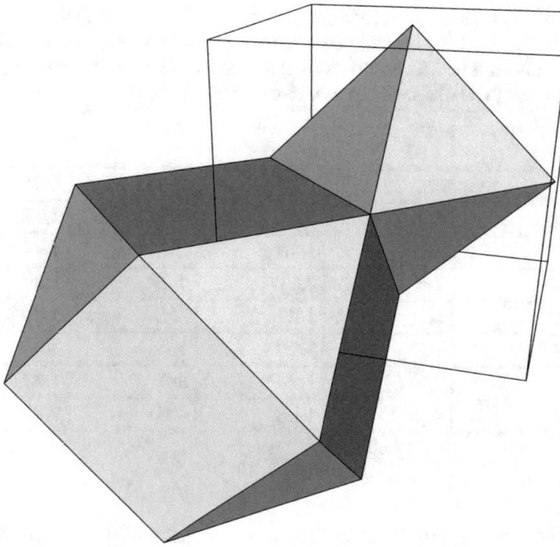

**Fig. 7.5** Coordination polyhedra of the LiI structure inside the cubic unit cell. The atomic positions are equivalent to those of the previous figures. Note that here, the polyhedra are in contact with each other. Reproduced with permission from Ref. [41]

The duals of these attraction polyhedra[1] (the coordination polyhedra) can easily be found and are represented in Fig. 7.5. The 18-fold coordination of the iodides gives rise to the truncated cube shown, while the six bonds of lithium ions generate a regular octahedron.

If we choose to focus on the repulsion polyhedra to analyze the critical points of the crystal, clear spatial relationships appear among them. Based on the data shown in Table 7.3, we see that there are only two independent cage points, $c7$ and $c8$. We can use Fig. 7.4 to visualize the situation. The repulsion polyhedron of critical point $c7$ is, for instance, the triangular pyramid formed by the central lithium atom and the three anions located at $(1, \frac{1}{2}, \frac{1}{2})$, $(\frac{1}{2}, 1, \frac{1}{2})$ and $(\frac{1}{2}, \frac{1}{2}, 1)$. The point $c7$ is located on the axis of this pyramid, at 0.0594 crystallographic units from the base. It is easily verifiable that the four faces of the pyramid are really triangular rings of bonds, and that all edges (or bond paths) are perfectly straight (they are symmetry lines). There is no stress due to curved bonds whatsoever. The base of the pyramid, on the contrary, is slightly bulged. Ring point $r6$ has a free geometrical parameter along the diagonal of the cubic unit cell. If we would ignore this lack of planarity, this point would be found at $(\frac{2}{3}, \frac{2}{3}, \frac{2}{3})$. Its actual position is (0.6669.0.6669.0.6669), at only $2.5 \times 10^{-4}$ crystallographic units from its ideal position. Ring stress is thus negligible for this crystal. The other cage, $c8$ (a fixed point with symmetry $T_d$ at the 8$c$ Wyckoff position), is at the center of the regular tetrahedron formed by the four anions of any of the eight small cubes that make up the cubic unit cell. Its four

[1] In geometry, the dual polyhedron is a figure where the vertices of one correspond to the faces of the other, and the edges between pairs of vertices of one correspond to the edges between pairs of faces of the other. Hence, the dual of a tetrahedron is another tetrahedron (inverted with respect to the first one); the dual polyhedron of a cube is an octahedron (and vice versa), etc.

**Table 7.5** Shell structure (up to third neighbors) of the CPs around each non-equivalent critical point in the LiI crystal. The *nn*, *nnn* and *nnnn* tags are abbreviations for first, second and third neighbors. The first columns indicate the number and type of neighboring CPs. The second columns indicate their distances in a.u. to the reference CP

| Critical point | *nn* | | *nnn* | | *nnnn* | |
|---|---|---|---|---|---|---|
| $n1$ | $6b3$ | 1.77 | $12r5$ | 2.01 | $8c7$ | 2.17 |
| $n2$ | $6b3$ | 4.05 | $12b4$ | 4.12 | $24r5$ | 4.63 |
| $b3$ | $4r5$ | 1.46 | $1n1$ | 1.77 | $4c7$ | 1.84 |
| $b4$ | $2r5$ | 2.11 | $4r6$ | 2.38 | $4c7$ | 2.91 |
| $r5$ | $2c7$ | 1.27 | $2b3$ | 1.46 | $1n1$ | 2.01 |
| $r6$ | $1c7$ | 1.20 | $1c8$ | 1.68 | $3r5$ | 2.08 |
| $c7$ | $1r6$ | 1.20 | $3r5$ | 1.27 | $3b3$ | 1.84 |
| $c8$ | $4r6$ | 1.68 | $4c7$ | 2.88 | $6b4$ | 2.91 |

faces are equivalent equilateral triangles of anion–anion links. The network of links in this crystal is the union of three-member rings of bonds surrounding tetrahedral or pyramidal cages. Notice that the cage points of a crystal can be correlated with the concept of **holes** in a structure: those points inside the crystal where it is more likely to find excess ions or impurities. The existence of tetrahedral holes in the sodium chloride structure is, in fact, one of the classical examples in structural chemistry.

It is possible to obtain additional information by examining the distribution of critical points around a given one. When this analysis is applied to nuclear positions, we generate the shell structure of the crystal, the radial distribution function of neighbors used in solid-state physics. It is easy to visualize the geometry of critical points around a given one using Fig. 7.4. For example, if we take point $c8$ (the tetrahedral hole), its first neighbors are constituted by the four centers of the faces of its repulsion tetrahedron, followed by the four cages located tetrahedrally around it in the inside of the four pyramids and by the six I–I bonds at the centers of the edges of its repulsion tetrahedron.

We must also emphasize the high density of CPs in a crystal, and their mutual proximity. In the case we are studying, the minimum distance between two different CPs is 0.63 Å, between the points $r6$ and $c7$. Even more illustrative of the proximity between CPs is the fact that in the neighbors list of BCP $b3$ (the Li–I bond), the iodine nucleus is not present among the twenty-first shells of neighbors around it. These results are summarized in Table 7.5.

Let's finally present some atomic properties in this system. Formal atomic charges are $Q_{\text{Li}} - Q_{\text{I}} = 0.970$ electrons. For the lithium atom, we have a volume $V_{\text{Li}} = 29.619\ \text{bohr}^3$, that is, 7.496% of the total volume of the cell. The iodide volume is $V_{\text{I}} = 365.521\ \text{bohr}^3$, or 92.5% of its volume. This analysis places LiI as a highly ionic compound, with very small lithium cations and the iodide anion occupying most of the cell volume.

**Table 7.6** Total ($B$) and local ($B_i$) bulk moduli, in GPa, along with the occupation factors $f_i = V_i/V$ for some spinels

| System | $f(A)$ | $f(B)$ | $f(\mathrm{O})$ | $B$ | $B_A$ | $B_B$ | $B_\mathrm{O}$ |
|---|---|---|---|---|---|---|---|
| $MgAl_2O_4$ | 0.08319 | 0.09259 | 0.76278 | 215.2 | 282.1 | 331.9 | 201.6 |
| $MgGa_2O_4$ | 0.08761 | 0.16375 | 0.74864 | 211.2 | 261.2 | 283.9 | 196.1 |
| $ZnAl_2O_4$ | 0.13543 | 0.15755 | 0.70702 | 214.8 | 246.0 | 335.2 | 203.2 |
| $ZnGa_2O_4$ | 0.13558 | 0.09537 | 0.76904 | 213.3 | 241.1 | 308.6 | 195.7 |

#### 7.2.1.3 Properties: Atomic Compressibility

Let's see an example of property partition within QTAIM in spinel oxides of stoichiometry $AB_2O_4$, where $A$ and $B$ are divalent and trivalent cations, respectively. It is a well-known experimental fact that bulk moduli of most of these compounds are practically constant and equal to 200 GPa. The QTAIM partition transparently explains the origin of this constancy.

Table 7.6 shows the results for several spinels. On the one hand, we observe that the bulk moduli of trivalent cations are huge, considerably larger than those of their divalent counterparts, and that the latter, in turn, are also quite larger than those of the oxide anion. This fact agrees with chemical intuition, which assigns inversely proportional polarizabilities to the formal charge of the cations. We can also verify that the oxide bulk modulus is practically constant, an example of **transferability** in the inorganic world. Finally, the origin of the constancy of the bulk moduli in these crystals has to be sought in the enormous proportion of the volume occupied by the oxygen anions, a consequence of both stoichiometry and the size of the oxide basin.

In other words, the compressibility of a system is determined by that of the most deformable and abundant component. Since the earth's crust is dominated by metal oxides with relatively high oxidation state cations, its compressibility is dominated by that of the oxide ions.

### 7.2.2 *ELF*

One of the main differences of applying ELF to the solid state is that now, even more than for QTAIM, the number of critical points is usually very high. However, this can also become an advantage. As we have already seen for QTAIM, one of the main uses of topology in the solid state is to understand coordination and phase organization. In the next sections, we will see how ELF can also provide information in these directions.

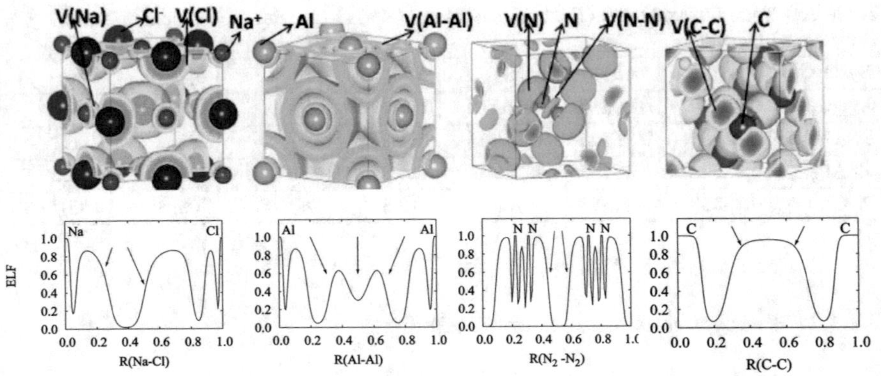


**Fig. 7.6** Bonding patterns in solids. From left to right: NaCl, Al, $N_2$ and diamond. Top: 3D isosurfaces. Bottom: Representative 1D profiles

#### 7.2.2.1 Prototype Crystals

One of the big advantages of ELF in the solid state is its ability to reveal the nature of bonding from a mere inspection of the ELF isosurfaces.

A covalent solid is characterized by clear bonding basins. For instance, diamond displays a continuous 3D network of tetrahedrally coordinated carbons, bound together by means of single-bond basins, B(C–C) (Fig. 7.6-right). In this case, perfect electron pairs (nearly 2 electrons per bonding basin) are found, as expected from the perfect covalency of the compound. Polar compounds lead to a deformation of the bond basin (see BN in Fig. 7.11 below).

Just like for QTAIM, we can further analyze covalent bonding in diamond by looking at the complete list of critical points. Table 7.7 shows the full list of critical points in the conventional unit cell, which can be used to construct the bifurcation tree in Fig. 7.7. The first branching is related to the separation of the solid into cores and a common valence, formed by the bonds, at ELF $= 0.08$ (C(C)–V(C,C)). The valence domain is separated into irreducible bond basins at 0.59 (V(C,C)–V(C,C)). All basins are highly localized, as can be deduced from the ELF values in Table 7.7: ELF $> 0.9$ for the maxima (cores and bonds) and ca. 0.08 for the valence separation.

Closed-shell interactions, such ionic or van der Waals atomic solids, yield spherical isosurfaces around the core positions (Fig. 7.6).

In this case, the bifurcation trees can be used to analyze core deformation due to polarization, and hence, coordination. The fluorine valence becomes separated at ELF $= 0.855$ into 6 pieces (Fig. 7.7), oriented toward the counterions (Fig. 7.8). As expected, the bigger atoms/ions show greater deformations. This can be seen by comparing LiF and BeO: the valence shell of BeO, more polar than LiF, loses to a greater extent its spherical shape.

As the polarizability of the anion increases, more complex patterns appear for the valence topology. New sets of valence attractors in the bigger ion appear toward

**Table 7.7** Set of critical points found for a covalent crystal (diamond C, a = 3.102Å). Positions $(x, y, z)$, multiplicities (M), ELF value at the CP ($\eta$) and chemical meaning (CHM) are collected

C(diamond)

| Type | $x$ | $y$ | $z$ | M | $\eta$ | CHM |
|---|---|---|---|---|---|---|
| (3, −3) | 0.1250 | 0.1250 | 0.1250 | 8 | 0.9999 | C(C) |
| (3, −3) | 0.0000 | 0.2500 | 0.2500 | 16 | 0.9352 | V(C,C) |
| (3, −1) | 0.0745 | 0.3750 | 0.3750 | 48 | 0.5954 | C(C)–V(C,C) |
| (3, −1) | 0.0710 | 0.1790 | 0.1783 | 32 | 0.0833 | V(C,C)–V(C,C) |
| (3, +1) | 0.2403 | 0.2403 | 0.2405 | 32 | 0.4675 | Ring |
| (3, +1) | 0.0293 | 0.1250 | 0.1250 | 48 | 0.0736 | Ring |
| (3, +1) | 0.2500 | 0.2500 | 0.5000 | 16 | 0.0427 | Ring |
| (3, +3) | 0.3750 | 0.3750 | 0.3750 | 8 | 0.0236 | Cage |
| (3, +3) | 0.1829 | 0.1829 | 0.1779 | 32 | 0.0670 | Cage |
| Morse | | | | 0 | | |

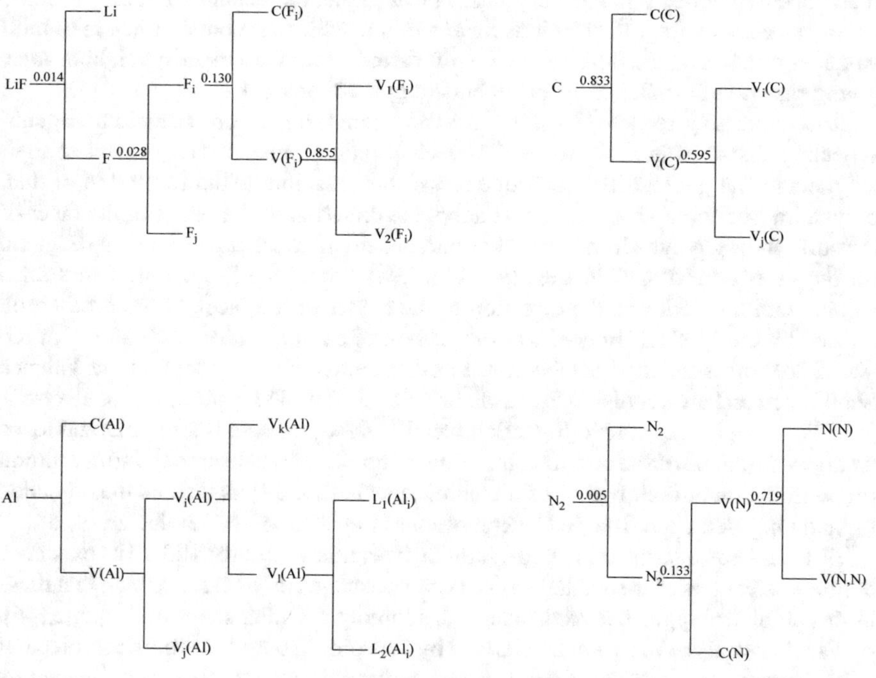


**Fig. 7.7** Bifurcation trees for LiF, diamond, Al and $N_2$

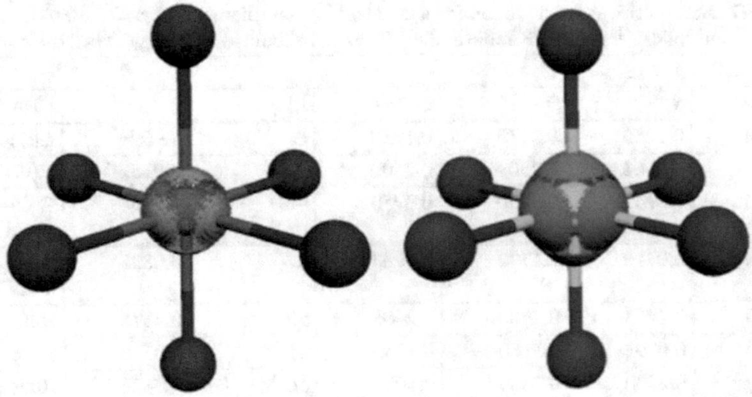

**Fig. 7.8** Isosurfaces of the valence ELF at the (3, −1) ELF value for the B1 structure of **a** LiF and **b** BeO. Adapted with permission from Ref. [15]. Copyright 2008 American Chemical Society

their second neighbors in such crystals. For example, the octahedra of $Cl^-$ become cubooctahedra in $Br^-$. If both ions have similar radii, the repulsion between their valences predominates. Both valences shift further from their nearest neighbor ions, giving rise to cubic valence arrangements (e.g. both ions in KF).

It is interesting to note that for third row atoms, inner core attractors are subsequently disposed in order to minimize electronic repulsion with respect to their neighbors. The geometrical consequence of this principle is the formation of dual polyhedra (see footnote 1): attractors adopt the direction of the center of the faces of the surrounding polyhedron, e.g. cubes and octahedra. Such a pattern is highlighted for the outer core of $Cl^-$ in LiCl (see Fig. 7.9a), where the L-shell attractors form a cube with the vertex in the direction to the center of the faces of the octahedron formed by the M shell. In contrast, the L shell of a cubic cation such as $K^+$ in KF (see 7.9b) forms the dual octahedron. In other words, we see here that the Valence Shell Electron Pair Repulsion Principle or VSEPR [23] also applies to core electrons [17]. For even bigger atoms (transition metals), the outer core is still polarizable, so it adopts the shape of the coordination. These simple rules subsume the structuration found in the outer-core region of transition metals [24, 32] as well as the so-called Ligand Opposed Core Charge Concentrations (LOCCC) of the Laplacian [9, 38].

ELF has been also useful in understanding vacancies in ionic solids. The existence of non-nuclear basins associated with oxygen vacancies in MgO has shown that these basins emulate oxygen basins themselves, adopting a similar shape and charge [46].

Van der Waals solids are characterized by the voids in between atoms or molecular units. Figure 7.6 shows the case of the $N_2$ molecular crystal. $N_2$ molecules can be easily identified. The value of ELF at the intermolecular points is crucial in order to study the relationship between molecular units. As reflected in the bifurcation tree (Fig. 7.7), the van der Waals forces that stabilize the crystal give rise to a first reduction into $N_2$ molecular units at a very low ELF value (ELF = 0.005), whereas the

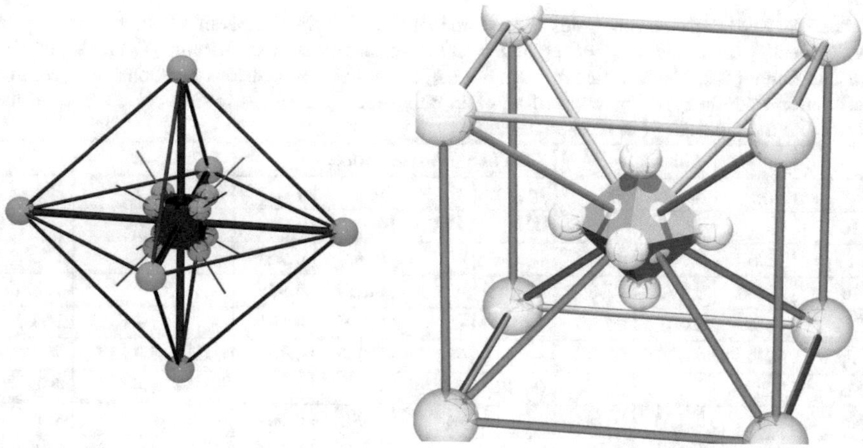

**Fig. 7.9** ELF maxima organization in **a** $Cl^-$ in LiCl (L shell is cubic, M shell is octahedral) and **b** $K^+$ in KF (L shell is octahedral, M shell is cubic). Reprinted with permission from Ref. [17]. Copyright 2011

covalency within molecules is revealed by the common valence they share (formed by bonds, V(N, N), and lone pairs V(N)) at ELF = 0.7187.

Metals are characterized by a big valence delocalization. The standard definition of bonding in metals is based on band theory, so that the closing of the energetic *gap* between the conduction and valence bands is one of the necessary conditions for metallicity. This definition in reciprocal space has complicated its inclusion in the theory of chemical bonding. Thus, Lewis avoided it in his book [36] and Pauling described it as a partial covalent bond between neighbors. This argument has been put forward [1, 2] to eliminate metal bonding from chemical language. However, the delocalization and lack of directionality that characterize the metal bond make it difficult to conjugate it with the typical image of a covalent bond.

The study carried out by Silvi and Gatti [56] is revealing in this sense. It proposes that metals should show a very flat ELF profile (already introduced in Sect. 4.3.2). We can define a delocalization window to quantify this "flatness" [56]:

$$\Delta \mathrm{ELF} = \mathrm{ELF}^{val}_{(3,-3)} - \mathrm{ELF}^{val}_{(3,-1)}. \tag{7.16}$$

For the aluminum crystal in Fig. 7.6, $\Delta$ELF = 0.043. This image would therefore be in accordance with that provided by Pauling, in which the metallic bond could be identified with a weak covalent bond, which is also generally multicentric.

**Table 7.8** Zero pressure properties of bulk and ELF basins of the crystalline phases for most of the elements with atomic numbers up to $Z = 20$. E stands for element. Volume ($V_0$) in Å$^3$, crystal bulk modulus ($B_0$), average electron density ($\rho_i$) in number of electrons per bohr$^3$ and volume occupation fraction ($f_i$). The controlling basin bulk modulus ($B_{0,val}$) in GPa corresponds to the basin with the greatest $f_i$, as marked in bold

| | | Core | | Valence | | Bond | | Lone pair | | |
|---|---|---|---|---|---|---|---|---|---|---|
| E | $B_0$ | $\rho_i$ | $f_i$ | $\rho_i$ | $f_i$ | $\rho_i$ | $f_i$ | $\rho_i$ | $f_i$ | $B_{0,val}$ |
| He | 0 | – | – | 0.085 | 1.000 | – | – | – | – | – |
| Li | 13.6 | 0.302 | 0.049 | – | – | 0.007 | **0.951** | – | – | 13.5 |
| Be | 121.0 | 0.477 | 0.081 | – | – | 0.039 | **0.919** | – | – | 108.4 |
| C | 418.5 | 2.637 | 0.020 | – | – | 0.105 | **0.979** | – | – | 364.2 |
| $N_2$ | 62.4 | 4.634 | 0.002 | – | – | 0.066 | 0.142 | 0.020 | **0.855** | 54.8 |
| Ne | 3.1 | 18.819 | 0.001 | 0.048 | **0.999** | – | – | – | – | 4.9 |
| Na | 7.0 | 0.470 | 0.083 | – | – | 0.005 | **0.917** | – | – | 8.5 |
| Mg | 36.1 | 0.684 | 0.094 | – | – | 0.015 | **0.905** | – | – | 32.6 |
| Al | 57.5 | 0.955 | 0.073 | – | – | 0.026 | **0.927** | – | – | 52.7 |
| Si | 74.2 | 1.471 | 0.069 | – | – | 0.042 | **0.931** | – | – | 69.4 |
| P | 88.8 | 2.061 | 0.049 | – | – | 0.051 | **0.951** | – | – | 89.6 |
| $Cl_2$ | 24.5 | 4.519 | 0.029 | – | – | 0.125 | 0.075 | 0.032 | **0.897** | 26.6 |
| Ar | 0.7 | 6.033 | 0.006 | 0.030 | **0.994** | – | – | – | – | 0.3 |
| K | 4.7 | 0.150 | 0.238 | – | – | 0.002 | **0.762** | – | – | 4.5 |
| Ca | 17.6 | 0.272 | 0.232 | – | – | 0.008 | **0.768** | – | – | 17.1 |

#### 7.2.2.2 Properties: Valence Compressibility

As we have seen for the QTAIM, in most cases the sum over the basin contributions in Eq. 7.15 is reduced to just one atom, that with the greatest compressibility. In general, this is a good strategy applicable to other topologies: starting by identifying which is the most sensitive basin to pressure effects (i.e. the *controlling* basin). In the case of ELF, this means the valence, since it is the basin with the greatest compressibility (less localized) and usually also the one occupying the biggest volume. This is illustrated in Table 7.8. Not only the bulk modulus of each crystal and that of the *controlling* basin (the softest valence) follow the same periodic trend, but also their absolute values are very similar.

We can identify two main classes of compounds as far as the effect of pressure is concerned, which agree with the main classification provided by the Laplacian of the electron density. On the one hand, metal and covalent crystals (i.e. "shared electron systems") and on the other hand molecular and rare-gas solids (i.e. "closed-shell systems"). In all cases, the contribution from the K-shell cores of these elements is almost negligible.

Being related to the bond type, the changes are periodic. Along the first row, and as we go from group I to group VIII, the valence is first associated with a metallic bond basin ($B_{0,val}$ in Li is 13.5 GPa and the calculated crystal value is 13.6 GPa), then if forms a covalent bond basin ($B_{0,val}$ in diamond (carbon) is 364.2 GPa and $B_0$ is 418.5 GPa), after that we see the formation of lone pair basins ($B_{0,val}$ in $N_2$ is 54.8 GPa and in the crystal $B_0$ is 62.4 GPa), and the series ends with a closed-shell valence basin in Ne, where $B_{0,val}$ is 4.9 GPa and the bulk value is 3.1 GPa. Similar results are obtained for the calculated elements of the second and third periods.

It is interesting to note that this sequence follows the same ordering as the volume occupied by the respective valences. Moreover, this corresponds to a well-known volume ordering:

$$\text{Lone pair} > \text{Multiple bond} > \text{Single bond} > \text{Core},$$

used within the VSEPR model. Since the behavior under pressure is to a large extent dictated by the volume (i.e. the $PV$ term), we can also expect the reactivity under pressure to follow the same sequence. In other words, most commonly if there are lone pairs in a molecular solid, they will transform under pressure, leading to the formation of smaller units, e.g. new bonds (i.e. to polymerization). This can be used to **predict** the response to pressure.

One way to exploit this predictive behavior is to use an energetic model that includes chemical quantities explicitly. We have already seen one such model in Sect. 4.3.4.2, the Bond Charge Model (BCM). But in order to apply it, we first need to extend it to the solid state. For the sake of clarity, we will restrain ourselves to a given spatial group. The bond energy of an AX zinc-blende lattice in which each lattice atom is tetrahedrally coordinated via four bonds to their nearest neighbors can be expressed within the BCM framework as follows:

$$E_B = -\frac{Mq_B^2}{R_B} + \frac{4D'q_B}{R_B^2}, \tag{7.17}$$

$M$ is an average Madelung constant for these lattices with a $-q_B$ charge at each bond position.[2] The energy of an AX unit in the solid is then given by $E = E_A + E_X + 4E_B$, where $E_A$ and $E_X$ account for the energy of the corresponding core electrons which are not involved in the bonding and, in practice, are independent of geometrical parameters.

If we reasonably assume that $(\partial E_B/\partial R_B) \simeq 0$ at equilibrium, we have that

$$E_B = -\frac{4D'q_B}{R_B^2} = -\frac{Mq_B^2}{2R_B}. \tag{7.18}$$

As we saw in Sect. 4.3.4.2, we can obtain the bond energy in Eq. 7.18 from ELF quantities. Data for the BCM as derived from ELF topology ($q_B$, $R_B$) and Madelung constants, $M$, are collected in Table 7.9.

[2] Note that since the position of the bond charge will vary, this constant will not be the same for all zinc-blende solids in this BCM approach.

**Table 7.9** $q_B$ and $R_B$ parameters of ELF-BCM for a broad group of zinc-blende-type solids (i.e. diamond-type for homonuclear) from the IV, III–V and IIb–VI groups [14]. $M$ and $E_B$ stand for the Madelung constant and the bonding energy of the crystal, respectively. Bulk moduli at zero pressure, $B_0$, were taken from Ref. [11] and optical band gaps from Ref. [39]. Lengths are in Å, charges in electrons, $E_B$ in atomic units, $B_0$ is GPa and $E_{gap}$ in eV. Reproduced with permission from Ref. [12]

| AX | $R_B$ | $q_B$ | $M$ | $E_B$ | $B_0$ | $E_{gap}$ |
|---|---|---|---|---|---|---|
| **Group IV** | | | | | | |
| C | 0.938 | 1.950 | 10.856 | −11.644 | 442 | 5.2 |
| Si | 1.132 | 1.950 | 8.577 | −7.623 | 98 | 1.12 |
| Ge | 1.003 | 2.100 | 7.414 | −8.625 | 77.2 | 0.66 |
| SiC | 0.985 | 1.950 | 9.356 | −9.556 | 211 | 2.86 |
| **Group III–V** | | | | | | |
| BN | 0.936 | 1.975 | 11.006 | −12.136 | 397 | 5.0 |
| BP | 1.050 | 2.000 | 9.662 | −9.739 | 166 | 4.2 |
| BAs | 1.023 | 2.050 | 8.711 | −9.468 | 144 | 3.0 |
| BSb | 0.999 | 1.988 | 8.554 | −8.954 | 108 | 2.6 |
| AlP | 1.122 | 1.975 | 8.925 | −8.206 | 86 | 3.1 |
| AlAs | 1.080 | 2.025 | 8.028 | −8.068 | 77 | 2.16 |
| AlSb | 1.093 | 1.963 | 7.304 | −6.810 | 58 | 1.52 |
| GaAs | 0.992 | 2.100 | 7.662 | −9.008 | 75 | 1.35 |
| GaSb | 0.986 | 2.038 | 6.764 | −7.538 | 57 | 0.71 |
| InP | 1.012 | 1.970 | 8.494 | −8.615 | 71 | 1.25 |
| InAs | 0.958 | 2.020 | 7.268 | −8.187 | 60 | 0.35 |
| **Group II–VI** | | | | | | |
| ZnS | 1.001 | 2.025 | 10.130 | −10.976 | 77 | 3.6 |
| ZnSe | 0.970 | 2.075 | 8.631 | −10.134 | 62 | 2.6 |
| ZnTe | 0.971 | 2.038 | 7.415 | −8.389 | 51 | 2.2 |
| CdS | 0.976 | 1.975 | 10.016 | −10.512 | 62 | 2.5 |
| CdSe | 0.938 | 2.025 | 8.591 | −9.697 | 53 | 1.7 |

As expected, the valence is distributed in the bonds, so that the compressibility of the solids can be approximated to a good extent by the compressibility of the bonds ($B_{0,val}$), whereas cores are assumed to remain untouched: $B_0 \simeq B_{0,val}$ [16, 17]. The bulk modulus of the solid is then given by

$$B_{0,B} = \left[ V_B \left( \frac{\partial^2 E}{\partial V_B^2} \right) \left( \frac{\partial V_B}{\partial V} \right) \right]_0 = \left[ \frac{M q_B^2}{32\sqrt{3} R_B R^3} \right]_0 . \tag{7.19}$$

This simple electrostatic model that takes the bonding pattern into account enables to predict solid-state macroscopic properties with rather good accuracy (see Fig. 7.10) and provides a way to unveil the basins *controlling* the property under scrutiny.

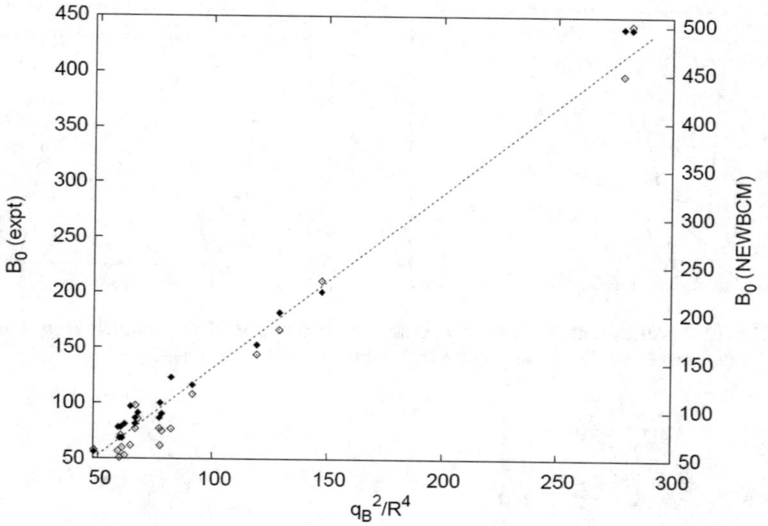


**Fig. 7.10** Experimental (left axis) and estimation from Eq. 7.19 (aka, NEWBCM, right axis) bulk modulus. Bulk moduli in GPa and $R$ in a.u. Reproduced with permission from Ref. [14]

Casting Table 7.9, it is easily observed that the hardest materials (diamond and BN) not only have the smallest $R_B$, but also the highest $\nu = R_B/R$ ratio. A characteristic comes also into play from $M$. One of the parameters determining the value of $M$ is the distance from each core to the bond, i.e. $R(A - B)$ and $R(X - B)$. The best combination is when these two distances are equal. By definition, this will be the case of homonuclear solids (see Fig. 7.11 top). But it can also occur in heteronuclear solids. This is the case of BN. In this system, a good combination of atomic sizes and polarity compensate to provide $r_1/r_2 = 1.03$. This is shown in Fig. 7.11 bottom: although the N core is smaller, the position of the bond is shifted toward Boron, so that the core–bond distances are similar for both atoms. This provides non-intuitive guide for the **design of superhard materials**.

### 7.2.3 NCI

We have seen in Sect. 5.3.3.2 one example of NCI in a molecular crystal, benzene, where both CH$\cdots$C and CH$\cdots\pi$ interactions are present. Let's now turn to a controversial case, the dihydrogen bond. These interactions can become especially important in solid-state and surface chemistry. We will see one example of each.

The term dihydrogen bond [19] was coined to describe an interaction of the type D-H$\cdots$H-E, where D is a typical hydrogen donor such as N or O. The interesting thing about this type of bond is that the acceptor atom is also a hydrogen. Thus,

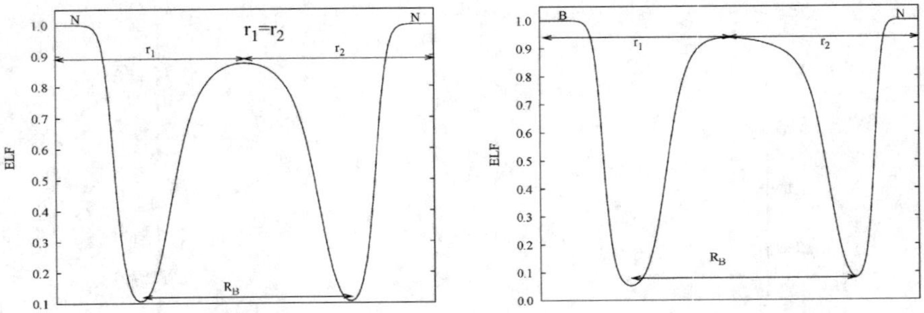


**Fig. 7.11** ELF profile along C–C bond in diamond (left) and B–N in cubic BN (right). Arrows highlight the core–bond distances. Reproduced with permission from Ref. [14]

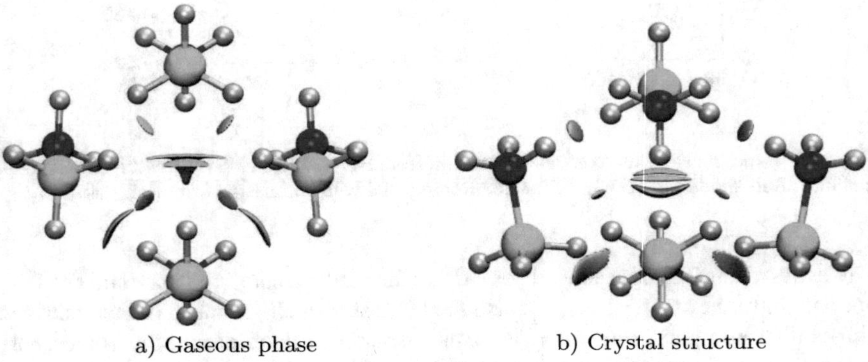


**Fig. 7.12** Dihydrogen interactions in a $BH_3NH_3$ tetramer in **a** the fully optimized gas-phase geometry and **b** the solid-state geometry. NCI surfaces correspond to $s = 0.4$ a.u. with a color scale of $-0.03 < \text{sign}(\lambda_2)\rho < +0.03$ a.u. Reprinted with permission from Ref. [13]. Copyright 2011 American Chemical Society

the accepting hydrogen atom must be negatively charged and E has to be an atom capable of accommodating a hydridic hydrogen. Transition metals and boron are some known examples of atoms occurring at position E.

$BH_3NH_3$ constitutes a typical example of dihydrogen complex [44, 47, 50]. Figure 7.12 shows the NCI results for the tetramer $(BH_3NH_3)_4$, whose geometry has been derived from the solid state. Each $BH_3NH_3$ molecule interacts with its neighbors by one dihydrogen bond, which shows the typical small round shape of localized interactions.

NCI enables us to see that dihydrogen bonds look like any other hydrogen bond, but are they really so relevant to the solid state? We will resort to self-assembly monolayer (SAM) models to highlight their relevance. SAMs form ordered two-dimensional layers, mainly caused by dispersion forces among neighboring molecules. We have created one-dimensional octylamine polymers in order to separately analyze the interactions along each direction. Figure 7.13 shows the 1D-

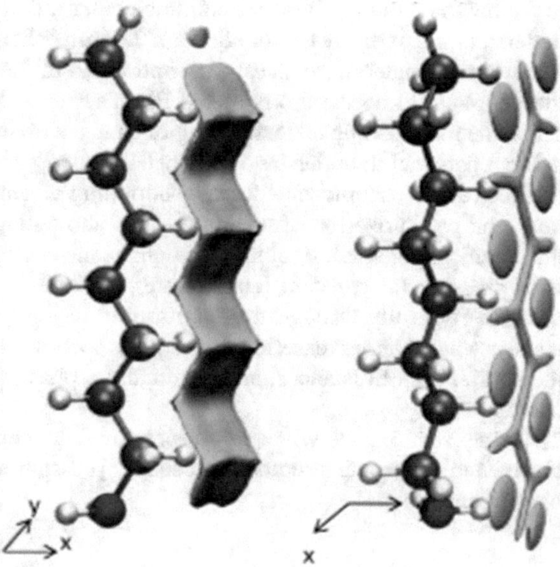


**Fig. 7.13** Octylamine molecules along $x$-(left) and $y$-(right) directions in a square 2D model superstructure. Top: structure with interacting hydrogen atoms are connected by lines. Bottom: NCI isocountours

polymers and their NCI surfaces along the $x$-direction (left) and $y$-direction (right). Along the $y$-direction, octylamine molecules are oriented face to face, so that direct H–H contacts appear. This is not the case in the $x$-direction, where interactions are less directional. These 1D-polymers enable us to illustrate the difference between H–H and van der Waals contacts within the NCI approach (Fig. 7.13-bottom). Whereas NCI isosurfaces are delocalized along the $x$-direction, they are very localized (small round shape) for the H–H contacts in the $y$-direction. Interestingly, these localized interactions lead to more stable polymers than the delocalized ones, pointing at the stabilizing nature of the H–H interactions. This is an important issue, since H–H interactions have been classically considered as repulsive, whereas topological analysis usually characterizes them as attractive. This simple system enables us to show that H–H interactions can indeed contribute to the stabilization of the molecular packing.

## 7.3 Phase Transitions

Materials properties are the result of the interactions between their atomic and/or molecular constituents. We have seen how the properties of a material are determined by its crystal structure and chemical composition. When the material is subjected to an external field such as high pressure, its physical and chemical structure, and thence their properties may change. Indeed, high-pressure studies are widely used to obtain new compounds and materials, allowing to find structure–property relationships and to follow the coupling between intramolecular and intermolecular interactions. An understanding of this behavior allows to control, enhance and optimize the properties of devices in material sciences, (e.g. optics and multiferroics) as well as to interpret the geophysical observations of the Earth's interior (e.g. seismic discontinuities). Thus, the microscopic understanding of changes in materials' electronic structure is a direct pathway to understand their behavior.

From a microscopic viewpoint, a shortening of nearest neighbor distances occurs upon the pressurization of a sample. This shortening is accompanied by a rise in the chemical potential, $\mu$, of the solid. Since the chemical potential always increases with pressure (at constant temperature), volume is reduced to keep $\mu$ as low as possible (typically through the "controlling basins"). The process continues up to a stage where the increase in connectivity becomes energetically competitive with the shortening of distances, and eventually a phase transition to a denser structure occurs.

In this section, we will apply topological concepts in order to understand the changes in electronic structure induced by pressure in a wide variety of solid-solid phase transitions.

### *7.3.1 QTAIM*

The QTAIM has found interesting applications in the understanding of the mechanism of several types of pressure-induced phase transitions, which can be followed using the same general principles devised for standard chemical processes once a continuous periodic path (with the corresponding symmetry subgroup) connecting the initial and final structures is found. This is equivalent to the intrinsic reaction coordinate (IRC) used in molecular computational chemistry.

One of the earliest examples of phase transition mechanism is the B1–B2 (sodium chloride to cesium chloride) in simple alkali halides [43]. We can assume a collective and concerted movement of all the atoms from one phase to the other (this is known as a "martensitic approach"), so that perfect periodic lattices are found at each point of the transition coordinate. To do that, one has to decide the number of crystallographically non-equivalent ions per primitive unit cell, thus leading to a hierarchical set of possible transition mechanisms. For alkali halides, if this number is set to one so that we have one alkali and one halide atom per unit cell, then the number of free parameters defining the transition space is nine, three angles and distances defining the unit cell and three internal positions $x$, $y$, $z$ that describe where the anion lies with respect to the cation, that we can locate at the origin of the cell (0, 0, 0) without loss of generality: $\alpha, \beta, \gamma, a, b, c, x, y, z$. The B1 and B2 phases lie at particular points of this 9D space:

$$\begin{aligned} \mathrm{B1} &: \alpha = \beta = \gamma = 60^\circ,\ a = b = c,\ x = y = z = 1/2. \\ \mathrm{B2} &: \alpha = \beta = \gamma = 90^\circ,\ a = b = c,\ x = y = z = 1/2. \end{aligned} \tag{7.20}$$

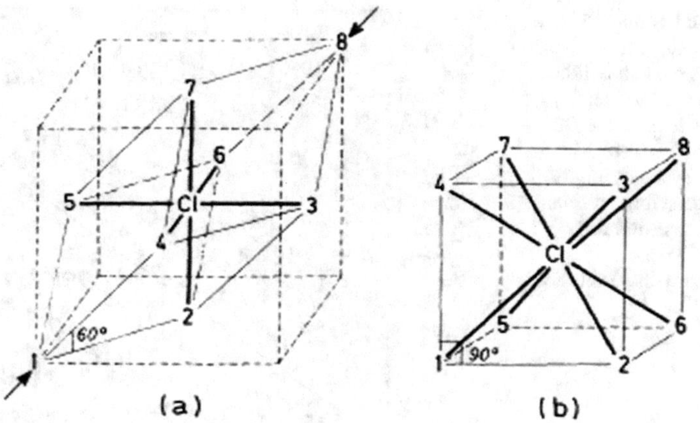


**Fig. 7.14** Primitive B1 (**a**) and B2 **b** cells. The rhombohedral B1 cell is embedded in the face-centered structure. The arrows illustrate the deformation leading from B1 to B2. Reproduced with permission from Ref. [43]

At this point, it is relevant to recall that reaction coordinates are defined as gradient-following minimum energy paths that connect reactants and products, and that the gradient of a function cannot alter its point (or space) group symmetry. For instance, if we compute the gradient of the energy for a $H_2O$ molecule in a $C_{2v}$ configuration and follow it, we will only explore $C_{2v}$ geometries. This implies that once a system starts moving onto a given symmetric reaction path, it cannot leave its symmetry until a critical point on the potential energy surface, where the gradient vanishes, is touched. By analyzing the energy Hessian at the initial B1 geometry on the 9D space, it has been found that in the alkali halides the lowest energy pathway preserves the $\alpha = \beta = \gamma$, $a = b = c$, $x = y = z = 1/2$ symmetry, which belongs to the $R\bar{3}m$ space group. This reduces the number of parameters from 9 to 2, the rhombohedral angle and the lattice parameter, so the transition may be seen as taking place as in Fig. 7.14.

The rhombohedral angle can be chosen as an appropriate reaction coordinate in these systems. As the pressure increases, the enthalpy diagram passes from a thermodynamically stable B1 phase through a double well at the transition pressure where the two phases have the same enthalpy and are separated by a barrier. Within this scenario, the stability/metastability of the B1 or B2 phases has been found to be determined by the classical Pauling ratio of cationic to anionic radii, $r_+/r_-$, although calculated with QTAIM topological bond distances, not with tabulated data. Figure 7.15 shows how Pauling's critical ratios $r_c = \sqrt{3} - 1, \sqrt{2} - 1$ for the compact filling of spheres in the sixfold or fourfold coordination separate the stability regions of the alkali halides when topological radii are used.

A full topological classification of all possible transition mechanisms [18] has shown that the different catastrophes in a solid-state transition can be classified as in molecules. Seven different sequences were found, which evolved more or less

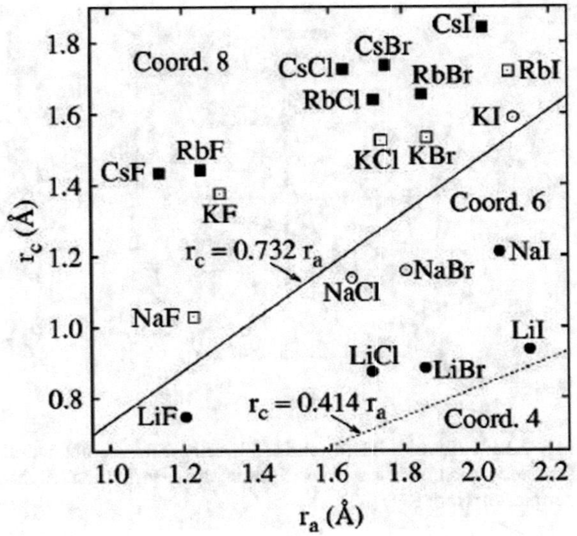


**Fig. 7.15** B1 cationic versus anionic topological radii for the 20 simple alkali halides. Circles are used to identify unstable systems in the B2 phase, empty circles degenerate systems, empty squares metastable crystals in the B2 phase and full squares stable ones. Reproduced with permission from Ref. [18]

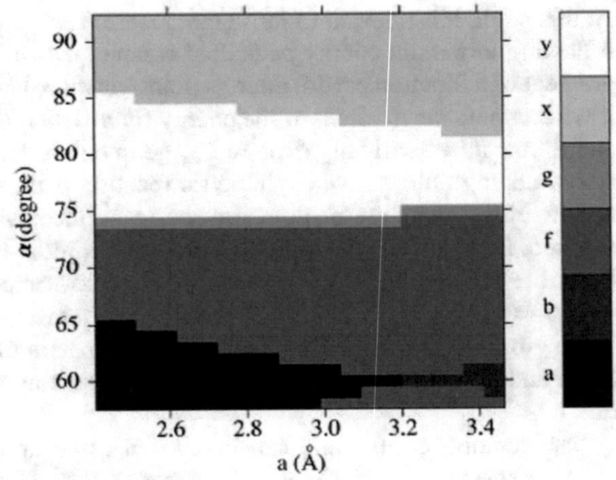


**Fig. 7.16** Topological structures of NaF over the $(a, \alpha)$-plane. $a, b, f, g, x, y$ are the topologically distinct arrangements. Reproduced with permission from Ref. [18]

smoothly with $r_+/r_-$. Figure 7.16 shows how the different topologies evolve on the $(a, \alpha)$-plane in the NaF transition. In this particular system, six different arrangements of critical points suffice to describe the transition surface.

Over the years, the use of topological descriptors in condensed matter theory has, slowly but steadily, permeated the discipline. QTAIM charges are now routinely used, although Mulliken-like decompositions (after the so-called projected density of states) have not been abandoned. A field where topological indices have proven especially valuable is that of intermetallic compounds, where traditional electron counting rules are difficult to apply. The group of Prof. Yuri Grin in Dresden has provided many fruitful examples in this regard [25].

## 7.3.2 *ELF*

ELF has allowed understanding the changes in bonding upon pressurization. One such example is the understanding of the incommensurate high-pressure phase of Sb [48]. Indeed, it was shown that the stabilization of the incommensurate phase is due to the absence of bonds between the atoms that form the host sublattice in the commensurate competing phase. ELF has also been extensively used in determining intermetallic and Zintl compounds at high temperatures and pressures [4, 35].

Building from our previous discussion where the stability of Lewis pairs under pressure was put in relationship with their volume, we will now see several examples of how electronic structure evolves with pressure leading to phase transitions, and how it is also related to volume.

Since we are mainly interested in the bond changes taking place, we will classify the examples in terms of the nature of the physical and chemical changes [8]. We will make three groups:

- Transitions with change in coordination. In other words, the main source of phase reorganization under pressure is the achievement of a more efficient atomic packing which does not involve a change in the chemical bonding type.
- Transitions with change in the chemical bonding type. These transitions are thus characterized by the appearance of new electronic characteristics in the solid. Metallization would constitute a typical example.
- Transitions with disorder. Although these transitions could fit into one of the categories above, we believe that they show some specific characteristics of their own, and their conceptual understanding is easier if considered alone.

In all cases, we will follow a martensitic approach.

### 7.3.2.1 With Changes in Coordination

Coordination counting can sometimes become a hard task. Establishing the limit of what belongs and what does not belong to the active sphere of coordination might become a matter of threshold flavor. We have seen already in Sect. 7.2.2.1 for LiF and BeO that ELF can be a very useful tool in revealing the organization of the valence maxima. Outer-core maxima (OCM) reflect in an indirect and subtle way coordination of the solid. The OCM of the soft ion (usually anions) are disposed along the nearest neighbors directions and reflect the polarization of the ion. If we analyze the outer core of a hard ion instead, the OCM are disposed so as to occupy the interstitial voids in between the bonding basins. We will review the increase in coordination of a covalent and of an ionic crystal. In both cases, we will see that the

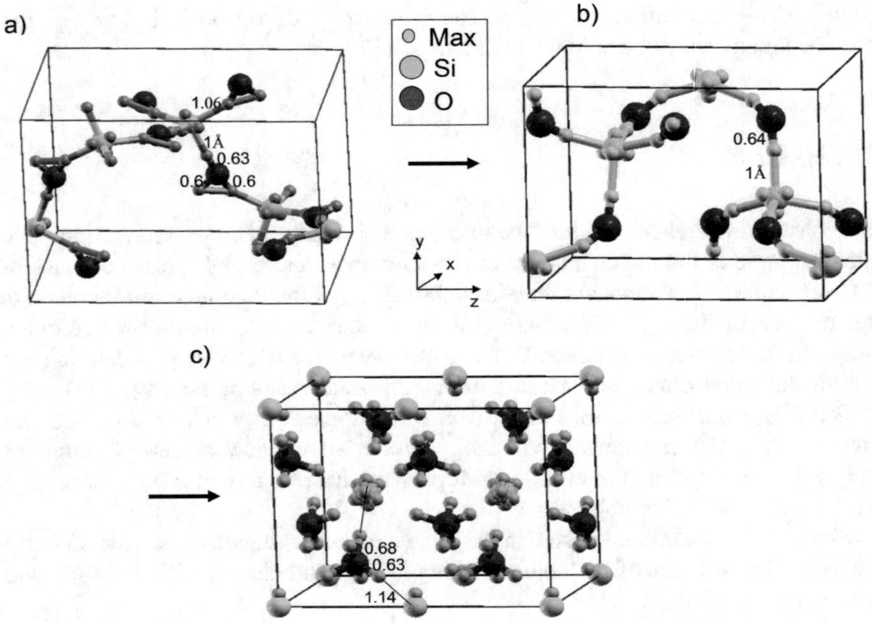


**Fig. 7.17** ELF valence attractors (Max) in the initial (**a**), intermediate (**b**) and final **c** phases of the transition. **a** $\alpha$-cristobalite→stishovite transition

coordination change can be followed by the outer-core distribution of the polarizable atoms.

The $\alpha$-critobalite→stishovite transformation in $SiO_2$ involves going from a fourfold coordinated silicon in $\alpha$-cristobalite to a sixfold one in stishovite. This change in coordination can be described with a transition path of $P4_12_12$ symmetry [57]. ELF valence attractors in the initial, intermediate and final phases of the transition are depicted in Fig. 7.17. The $\alpha$-cristobalite phase presents three maxima of ELF around the oxygen centers. One of them, at 0.6 Å from the oxygen, is a lone pair with 6.8 $\bar{e}$, while the other two, at 1.0 and 1.06 Å from oxygen, are bonding basins which lead to a sharing of 1.1 $\bar{e}$ with the neighboring silicon centers. Silicon cores are surrounded by 4 maxima (with 2.49 $\bar{e}$ each), as expected for a fourfold coordination. Note that these maxima occupy the dual tetrahedron positions with respect to the coordination.

The stishovite form presents five maxima around the oxygen atoms: two lone pair basins at 0.56 Å from the oxygen, holding 1.8 $\bar{e}$ each, and three bonding basins which share 1.1-1.6 $\bar{e}$ with the neighboring silicon atoms, generating the 4+2 coordination. Silicon nuclei cores are surrounded by 8 maxima which form a cube, typical of an octahedral coordination (remember that the cube is the dual polyhedron of the octahedron—see footnote 1).

We have seen a transition in a covalently bonded solid, let's now analyze an ionic one: the B3→ B1 phase transition in BeO which takes place at around 100 GPa [15].

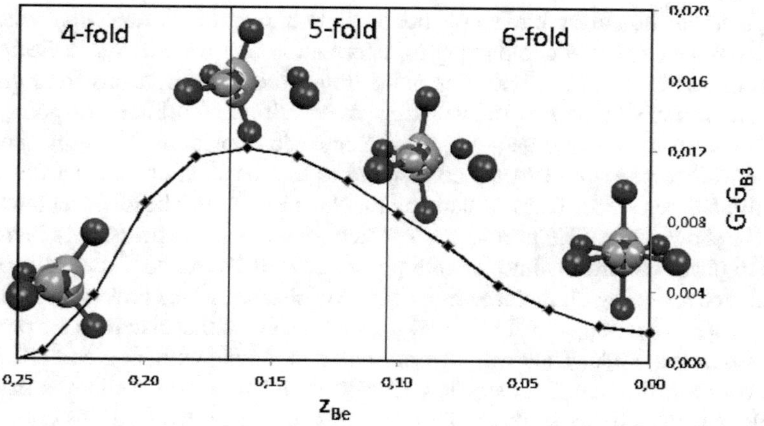


**Fig. 7.18** Volumes of core and valence basins, oxygen (K+L) and unit cell (Z = 1) along the B3→ B1 phase transition path in BeO. Appearing oxygen basins (bold squares) emerge at $z_{Be} = 0.18$ and $z_{Be} = 0.10$

Both the B3 and the B1 phases are ionic, so all electrons are distributed in atomic shells around the ions: one shell for beryllium cation (K shell) and two shells for the oxide (K and L). The transition consists of a change in coordination, from four-folded tetrahedral (B3) to six-folded octahedral (B1). We have used an orthorhombic Imm2 unit cell [45] to monitor the transition. Using $z_{Be}$ as the transition coordinate (which goes from $z_{Be} = 0.25$ in the B3 phase to $z_{Be} = 0.0$ in B1), we can follow the changes in OCM arrangements of the anion to analyze the polarization reorganization (Fig. 7.18). Three domains of structural stability are found:

- $z_{Be} > 0.18$: fourfold coordination.
- $0.18 > z_{Be} > 0.10$: fivefold coordination.
- $z_{Be} < 0.10$: sixfold coordination.

At $z_{B}e = 0.18$, a new OCM appears leading to a fivefold coordination. Note that this means that the transition goes through an intermediate coordination. Moreover, the appearance of this OCM occurs close to the transition state. This new OCM is located along the plane that bisects the angle formed by the O atom and the two approaching Be atoms. The new OCM divides into two at $z_{Be} = 0.10$, leading to the final coordination.

#### 7.3.2.2 With Changes in Chemical Bonding

High-pressure techniques are now available that enable to induce important bonding changes [20, 26]. These systems are thus a wonderful playground for bonding analysis. In general, since stable phases require the minimization of Gibbs energy, the voluminous Lewis entities transform into more compact ones: lone pairs and multiple bonds transform into single bonds. Polymerization of molecular solids is a good example of such a compacting transformation that we will see in Sect. 7.3.3. Another less intuitive transformation is the transition of alkali metals from metallic to pseudo-ionic, which leads to insulating phases. From an orbital viewpoint, these transitions are understood as s→p, s→d electronic transitions. If we use bonding techniques, we can see that they correspond to the localization of electrons in the interstitial regions [53]. This is illustrated for Na in Fig. 7.19. The low-pressure phase Na–cI16 shows a flat ELF profile, as expected for a metal. As pressure is increased, the cI16 phase transforms into the oP8 phase. Low ELF attractors start to appear in the valence revealing an increase in electron localization. This process is enhanced when the hP4 phase appears. In the hP4 phase, we can see that electrons are perfectly localized in the voids. They can be assimilated to anions without a core. The ELF value between the basins is very low (recall Eq. 7.16), meaning there is basically no delocalization between them. This explains the name given to this compound, "pseudo-ionic", as well as its properties. Indeed, the hP4 phase is a direct band gap insulator (transmission spectroscopy measures reveal a gap of 2.1 eV).

### 7.3.3 NCI

We will make the most here of the ability of NCI to show both non-covalent and covalent interactions in order to describe the changes in the cohesion of solid hydrogen, from purely long-range electrostatic and van der Waals interactions between fragments (molecular phase) to covalency.

At low pressures and temperatures, hydrogen is a molecular solid. At higher pressures ($p \sim 150$ GPa to $p > 300$), phase III appears [61], which is formed by rings of three $H_2$ molecules [49]. In order to analyze the evolution of this molecular motif in the solid, it is possible to analyze its behavior under pressure from a molecular perspective. Pressure can be simulated as a decrease in distance, $d$, between the center of the array and each $H_2$ molecule.

This allows to follow the transformations in the $H_2$ molecules. Several steps are found: firstly the H–H bond is slightly compressed. As the molecules get closer, molecules interact stronger with their neighbors and the H–H bond elongates. Finally, the inter and intramolecular interactions become equivalent, leading to a polymerization into a $H_6$ ring [33, 34].

Figure 7.20 shows the NCI description of this polymerization. At long distances ($d = 250$ pm), only van der Waals interactions are observed. The density at the H–H BCP is the same as for the isolated $H_2$ molecules ($\rho = 0.255$ a.u). As the pressure increases, van der Waals interactions become stronger at the expense of covalent bonds, which become weaker. The electron density at the BCP steadily decreases down to 0.173 a.u at $d = 89$ pm. At this stage, only one type of bond is observed, so complete polymerization has taken place. After polymerization, the system compresses uniformly.

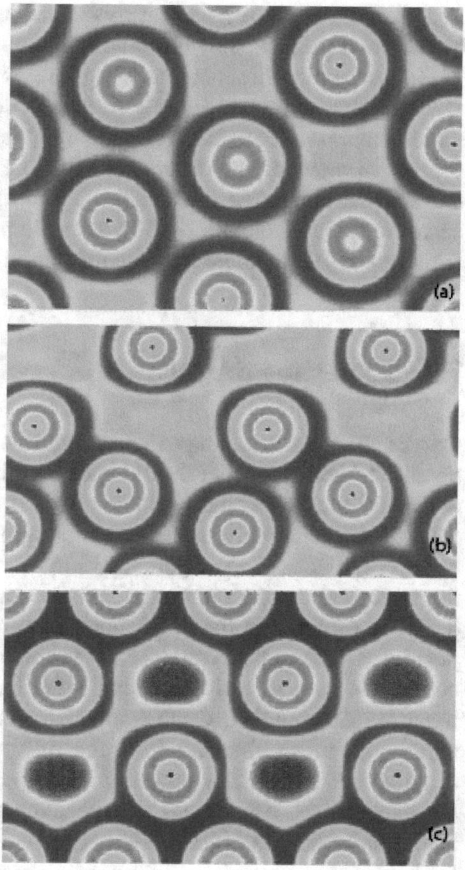


**Fig. 7.19** ELF contour plots for Na along compression: cI16 (top), op8 (middle) and hP4 (bottom)

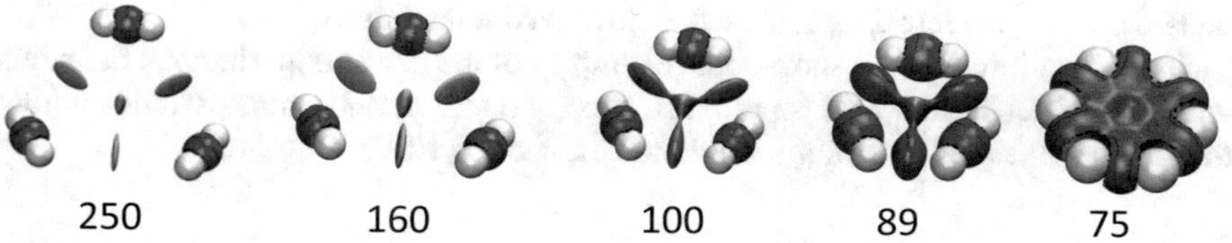


**Fig. 7.20** NCI for *d* (the internuclear distance between the H atoms in a dihydrogen molecular unit) equal to 250, 160, 100, 89 and 75 pm. Adapted from Ref. [52] with permission from the PCCP Owner Societies

This analysis allows to capture the changes in chemical bonding, but if we check the corresponding densities for the NCI peaks it also allows to discern the moments at which $H_2$ molecules do not behave like true $H_2$, and that at which they become completely polymerized [52].

It is interesting to note that this same analysis has been carried out for a number of solid phases, showing that the evolution follows the same pattern as in the "3 $H_2$", excepting the fact that jumps are observed at phase transitions. This allows to establish a correspondence between the degree of equalization of intra- and intermolecular H–H separations in the "3 $H_2$" system and the pressure $p$ at which the conformation would be observed [33]. For example, this analysis allows concluding that the pressure range at which the C2/c structure is observed ($p = 125 - 250$ GPa, corresponding to $d = 95 - 110$ pm), solid hydrogen has already begun its transition from true $H_2$ molecules to completely polymerized atomic hydrogen [52].

## 7.4 Experimental Densities

One of the great advantages of electron density-based indexes is that they can be obtained from experiment, and hence bonding analysis can be carried out for X-ray densities. Moreover, the promolecular densities we have already discussed (Sect. 5.3.2.2) are extremely used in crystallography to solve crystal structures. Known as IAM densities in the crystallography framework (for Independent Atom Model), the structural resolution starts from the superposition of spherical, neutral atomic densities centered at the (experimental or theoretical) best estimates of the atomic positions. Although this model lacks the electron density deformation due to chemical bonds, we have already seen that it is able to recover some important electronic features. Most generally, the results from experimentally derived electron densities agree with those obtained from the fully periodic wavefunction. However, some disagreements generally appear with respect to the strength and the localization of the interactions. This is not unexpected since the difference between the periodic theoretical calculation and the experimental multipolar model is clearly not constant for each point **r** and it is likely to be greater, on percentage, when small and flat densities are considered, e.g. in weak delocalized interactions.

In this section, we will show some examples of the use of experimental densities within the QTAIM and NCI topologies. Next, we will provide some details on how this can be extended to orbital-based indices, such as ELF.

### *7.4.1 QTAIM*

An example where QTAIM can provide an interesting insight into experimental structures is in the analysis of packing. For example, austdiol—a fungal metabolite (Fig. 7.21a)—crystallizes in the unusual space group $P2_12_12$ as a consequence of the coexistence of two main packing patterns in the bulk: relatively strong O–H $\cdots$ O hydrogen bonds involving an hydroxyl, the keto and the aldehyde group HBs in the $(a, b)$-plane (Fig. 7.21b) and very weak CH$\cdots$O contacts HBs interactions along $c$

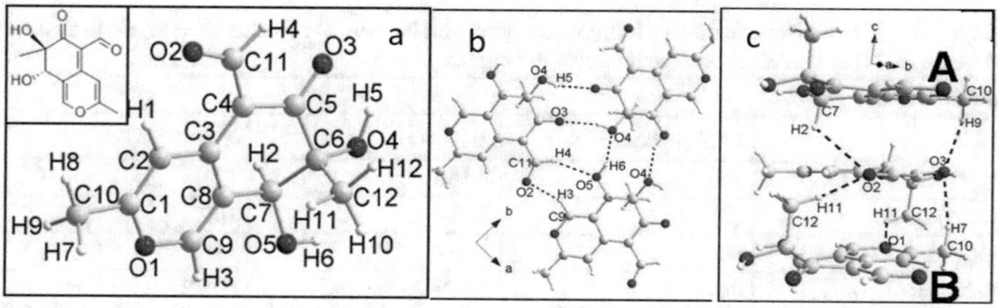


**Fig. 7.21** **a** Atom labels and chemical scheme (top left inset) of austdiol; **b** Hydrogen bond network in the (*a*, *b*)-plane; **c** Crystal packing along the c-direction. Reproduced with permission from Ref. [54]

**Table 7.10** Geometrical and electron density features of some intermolecular and intramolecular X–H ··· O (X = C, O) BCPs in austdiol. Results are experimental except otherwise stated. Electron densities in e/Å$^{-3}$, Laplacian in e/Å$^{-5}$, distances in Å and angles in degrees. Standard uncertainties are given in parentheses*

| BCPv | $\rho$(theo) | $\rho$ | $\nabla^2\rho$ | dH···O | $\widehat{XHO}$ |
|---|---|---|---|---|---|
| Intermolecular | | | | | |
| C(12)–H(11)···O(1) | 0.025 | 0.03(1) | 0.33(<1) | 2.979 | 116.9 |
| C(12)–H(11)···O(2) | 0.058 | 0.05(3) | 0.83(2) | 2.443 | 163.4 |
| C(10)–H(7)···O(3) | 0.054 | 0.04(3) | 0.41(1) | 2.544 | 166.9 |
| C(10)–H(9)···O(3) | 0.067 | 0.03(4) | 0.61(2) | 2.420 | 150.5 |
| C(9)–H(3)···O(2) | 0.116 | 0.13(3) | 1.40(4) | 2.174 | 150.7 |
| C(11)–H(4)···O(5) | 0.058 | 0.05(2) | 0.56(1) | 2.519 | 154.1 |
| O(4)–H(5)···O(3) | 0.095 | 0.09(4) | 1.11(5) | 2.114 | 158.4 |
| O(5)–H(6)···O(4) | 0.133 | 0.14(4) | 1.34(6) | 2.023 | 153.2 |
| Intramolecular | | | | | |
| C(2)–H(1)···O(2) | 0.138 | 0.11(3) | 2.00(3) | 2.145 | 121.8 |

*Note that standard uncertainties on the Laplacian are well known to be underestimated using XD2006 [59]. Standard uncertainties on the geometrical parameter cannot be correctly estimated since the hydrogen positions were not refined; however, the uncertainty on *d*H··· O should be in the order of 0.001 while the ones on the angle in the order of 0.1 (see Ref. [51])

(Fig. 7.21c). Therefore, an overall zigzag ribbon HB pattern [6] takes place along the *a*-axis.

The density has been obtained from diffraction data collected at low temperatures (70 K). Heavy atom positions were refined against X-ray structure factors, whereas positions and thermal motion of hydrogen atoms were constrained at the published values [6]. Table 7.10 reports the geometrical and topological features for inter- and intramolecular X–H ··· O (X = C, O) contacts resulting from this refinement. We will consider as true HBs only those H··· O contacts where the Koch and Popelier criteria [31] are satisfied as introduced in Sect. 5.2.1.

**Table 7.11** Geometrical features of some intramolecular X–H$\cdots$O contacts (X = C, O) in austdiol revealed by NCI. Distances in Å and angles in degrees

Intramolecular

| NCI | BCP presence | dH$\cdots$O | $\widehat{XHO}$ |
|---|---|---|---|
| C(2)–H(1) $\cdots$ O(2)* | yes | 2.145 | 121.8 |
| C(11)–H(4) $\cdots$ O(3) | no | 2.359 | 100.5 |
| O(4)–H(5) $\cdots$ O(3) | no | 2.052 | 116.9 |
| O(5)–H(6) $\cdots$ O(4) | no | 2.589 | 102.2 |
| C(12)–H(10) $\cdots$ O(5) | no | 2.667 | 92.6 |
| C(9)–H(3) $\cdots$ O(5) | no | 2.461 | 92.1 |

*Same BCP as in Table 7.10

We can see there are quite strong intermolecular H$\cdots$O HBs in the $(a, b)$-plane (e.g. C9–H3 $\cdots$ O2 and O5–H6 $\cdots$ O4 with $\rho = 0.13$ and 0.14 e/Å$^{-3}$, respectively), together with more complex HBs arising from the formation of cyclic H-bonded patterns (see the two O4–H5$\cdots$O3 interactions forming a cycle in Fig. 7.21b).

If we analyze the differences between experimental and theoretical densities, the differences in the weak densities explained above are clearly illustrated. We observe an apparent "strengthening" of the two weakest intermolecular HBs (namely H7 $\cdots$ O3 and H9 $\cdots$ O3) in passing from experiment to theory (see Table 7.10).

Recall that weak delocalized interactions are the typical case of non-QTAIM NCI critical points: competing intermolecular HBs weaken intramolecular contacts, so that they are not observed within QTAIM, but they do show an NCI signature. Hence, in molecular crystals, a better agreement can be obtained if we do not focus on BCPs (Sect. 5.3.4.3). Instead, NCI may be useful to highlight the environmental effects on the relative strength of such contacts. We will do so in the next section.

## *7.4.2 NCI*

NCI results from austdiol are shown in Table 7.11 [54]. Comparing with Table 7.10, we can see that many more interactions appear at the intramolecular level, where non-QTAIM NCI critical points are expected.

Two warning points should be raised here. Firstly, due to the change in electron density absolute values for weak interactions we have already highlighted at the beginning of the section, and the dependence of the RDG with the electron density (recall $s \propto \rho^{-1/3}$), a change in the RDG isovalue for different interaction types is sometimes required to fully match theoretical and experimental results.

We will illustrate this point in famotidine, an heteroatom-rich antiulcer drug (Fig. 7.22) which shows several sulfur interactions. BCP results are summarized in Table 7.12. They enable to showcase the different types of S interactions and their classification. Rosenfield et al. [29] systematically investigated the geometrical pref-

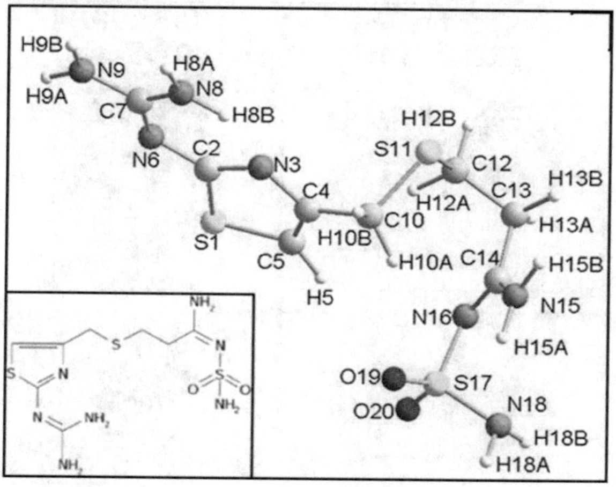


**Fig. 7.22** Atom labeling in famotidine. Inset: chemical scheme of famotidine. Reproduced with permission from Ref. [54]

**Table 7.12** Intermolecular BCPs involving sulfur atoms in famotidine. Labels as in Fig. 7.22. $\theta$ refers to the out-of-plane angle. Electron densities in e/Å$^{-3}$, Laplacian in e/Å$^{-5}$ and angles in degrees

| BCP | $\rho$ | $\nabla^2\rho$ | $\theta$ |
|---|---|---|---|
| S1 $\cdots$ H13 | 0.031(1) | 0.514(4) | 42.7 |
| H18 $\cdots$ S11 | 0.047(8) | 0.603(4) | 3.4 |
| S1 $\cdots$ S11 | 0.031(2) | 0.359(2) | 82.3–71.1 |
| H5 $\cdots$ S11 | 0.024(5) | 0.306(2) | 70.1 |
| H12 $\cdots$ S11 | 0.019(1) | 0.332(6) | 74.9 |
| S1 $\cdots$ S11 | 0.008(<1) | 0.115(<1) | 47.8/70.2 |

erences of the non-covalent contacts involving divalent S atoms at distances within the sum of the van der Waals radii. They concluded that nucleophiles tend to lie almost in the X-S-Y-plane, and preferably along the X-S- or Y-S-directions.

From the electron density, we can see that the smaller the out-of-plane angle, the higher is the electron density at the BCP, and therefore, the stronger the bond. For example, the out-of-plane angle for the H18$\cdots$S11 interaction is very close to zero, leading to a strong and directional interaction with $\rho = 0.047$ e/Å$^{-3}$. Instead, as the angle increases, like in the case of H12$\cdots$S11 ($\theta = 74.9°$), the interactions become weaker and more delocalized ($\rho = 0.019$ e/Å$^{-3}$).

On passing from weaker to stronger S–H interactions in famotidine, the shape of the NCI isosurface changes from van der Waals-like (broad in space and sometimes with zones exhibiting both $\lambda_2 > 0$ and $\lambda_2 < 0$, Fig. 7.23a) to HB-like (disk-shaped, $\lambda_2 < 0$, Fig. 7.23b).

These two interactions also enable to see how the NCI isosurface needs to be adjusted when comparing experimental and theoretical data. Typically, for recovering similar pictures, an experimental isosurface of 0.6 needs to be reduced (e.g. to 0.4).

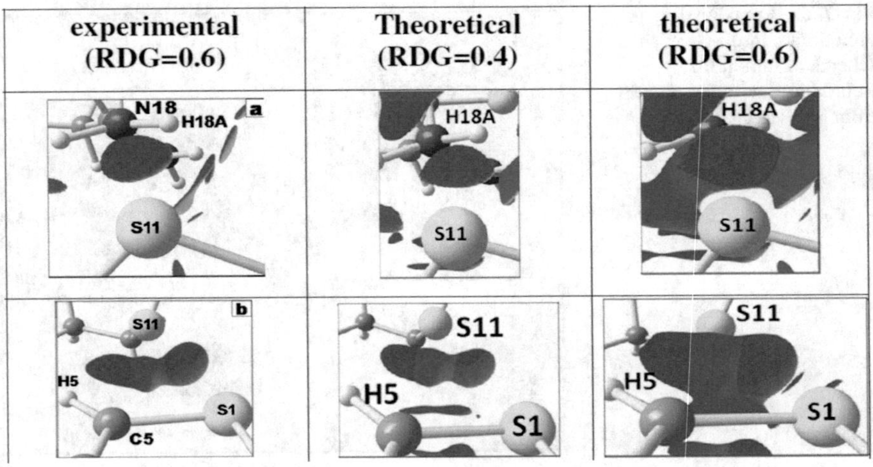


**Fig. 7.23** NCI isosurfaces for intermolecular interactions involving S atoms in famotidine from **a** experimental densities; **b** periodic wavefunction using $s(\boldsymbol{r}) = 0.4$; periodic wavefunction using $s(\boldsymbol{r}) = 0.6$. Labeling as in Fig. 7.22. Reproduced with permission from Ref. [54]

Secondly, the sign($\lambda_2$) quantity may be basically indeterminate in regions characterized by flat electron density, as it may subtly depend on the choice of the Hamiltonian and basis set (as concerns theory) or on the overall quality of the multipole electron density (as concerns the experiment). Therefore, great care should be employed in assigning an attractive or repulsive nature to interactions characterized with low $\lambda_2$.

### *7.4.3 ELF*

The use of ELF from experimental electron densities requires a little more development than QTAIM or NCI since it depends on orbital curvatures. Hence, other formulations are needed. Among them, we can highlight approximated methods (where the kinetic energy is approximated from the electron density and its derivatives) and quantum crystallography methods (where the wavefunction is obtained for the structure factors). These methods can be considered as semi-empirical applications of the ELF, since in addition to the experimental density, they adopt the same quantomechanical models used by the *ab initio* methods [27].

In the next sections, we will briefly develop both approaches and provide an example of how they work.

#### 7.4.3.1 Density-dependant Approximation of *t*

The formulation of ELF in terms of kinetic energy (Sect. 4.3.1.3) allows the development of an approximate expression of the ELF (AELF), solely as a function of the electronic density and its derivatives [58]. To do this, it is sufficient to start from Savin's expression and replace the definition of $t(\vec{r}) = (1/2)\sum_i \nabla\phi_i(\vec{r})\nabla\phi_i(\vec{r})$ dependent on orbitals with the second-order expansion used in DFT, dependent on the electron density (Kirzhnits approximation):

$$t(\vec{r}) \simeq \frac{3}{10}(3\pi^2)^{2/3}[\rho(\boldsymbol{r})]^{5/3} + \frac{1}{72}\frac{|\nabla\rho(\boldsymbol{r})|^2}{\rho(\boldsymbol{r})} + \frac{1}{6}\nabla^2\rho(\boldsymbol{r}). \quad (7.21)$$

This approach is valid for smooth (not necessarily small) changes in electron density. Thus, the expression of Pauli kinetic energy density is as follows:[3]

$$t_P(\boldsymbol{r}) \simeq \frac{3}{10}(3\pi^2)^{2/3}[\rho(\boldsymbol{r})]^{5/3} - \frac{1}{9}\frac{|\nabla\rho(\boldsymbol{r})|^2}{\rho(\boldsymbol{r})} + \frac{1}{6}\nabla^2\rho(\boldsymbol{r}). \quad (7.25)$$

Since this function depends exclusively on density and its derivatives, it is susceptible to be used with experimental densities.

#### 7.4.3.2 Hybrid Methods for Obtaining Experimental Wavefunctions

It is possible to mix experimental and theoretical approaches in the so-called "constrained wavefunction approach". It is based on solving the Schrödinger equation with a Lagrange multiplier that constrains the *ab initio* calculation to reproduce the experimental structure factors within a desired level of agreement [28, 37]. This hybrid approach has shown to provide answers to widespread quantum and crystallographic problems: anisotropic refinement of hydrogen atoms [60], consideration of relativistic effects [7] or even the description of multireference character systems [21].

Note that AELF and constrained wavefunctions might lead to differences, so they must be handled with care. The comparative analysis of AELF and ELF derived

---

3 Recall that Pauli kinetic energy density is given by

$$t_P(\mathbf{r}) = t(\mathbf{r}) - \frac{1}{8}\frac{|\nabla\rho(\boldsymbol{r})|^2}{\rho(\boldsymbol{r})}. \quad (7.22)$$

Using Eq. 7.21,

$$t_P(\boldsymbol{r}) \simeq \tfrac{3}{10}(3\pi^2)^{2/3}[\rho(\vec{r})]^{5/3} + \tfrac{1}{72}\tfrac{|\nabla\rho(\vec{r})|^2}{\rho(\vec{r})} + \tfrac{1}{6}\nabla^2\rho(\vec{r}) - \tfrac{1}{8}\tfrac{|\nabla\rho(\boldsymbol{r})|^2}{\rho(\boldsymbol{r})} \quad (7.23)$$

$$= \tfrac{3}{10}(3\pi^2)^{2/3}[\rho(\boldsymbol{r})]^{5/3} - \tfrac{1}{9}\tfrac{|\nabla\rho(\boldsymbol{r})|^2}{\rho(\boldsymbol{r})} + \tfrac{1}{6}\nabla^2\rho(\boldsymbol{r}). \quad (7.24)$$

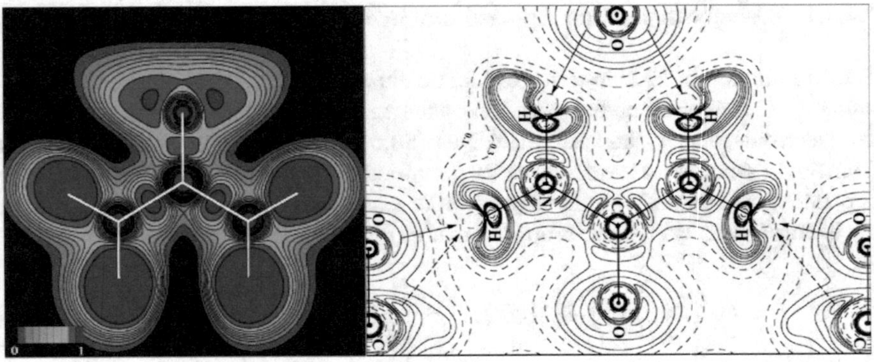

**Fig. 7.24** Semi-empirical ELF maps for urea in the plane (110) **a** From a HF constrained wavefunction; **b** AELF from multipolar analysis of R-X diffraction data. The continuous lines mark $\eta \geq 0.5$ a.u. at 0.05 a.u. intervals. Arrows mark hydrogen bonds. Reproduced with permission from Refs. [58] and [28]

from a HF constrained wavefunction in urea is shown in Fig. 7.24. AELF has a double peak in the C–N bond region instead of the single bond found in both the *ab initio* and the constrained calculations. Moreover, the AELF shows a shell structure in hydrogen. Interestingly, the latter structure, which is consistent with that described by Laplacian, would be responsible for the key-lock mechanism believed to hold the hydrogen bonds in the structure. Thus, on the one hand, the expression of Kirzhnits in the AELF seems not to be able to reproduce correctly the covalent bonds, but on the other hand, the restricted wave function is not flexible enough to reproduce subtle changes in the crystalline field. As they stand right now, the semi-empirical versions of the ELF must therefore be applied carefully and the selection of the approach should be made on the basis of the particular characteristics of the problem to solve.

## References


1. Anderson, W.P., Burdett, J.K., Czech, P.T.: What is the metallic bond? J. Am. Chem. Soc. **116**, 8808 (1994)
2. Andrés, J., Beltrán, A.: An ab initio perturbed ion study of bulk ceria. Chem. Phys. Lett. **221**, 249 (1994)
3. Ashcroft, N.W., Mermin, N.D.: Solid State Physics. Saunders College Plublishing (1988)
4. Becker, D., Beck, H.P.: Ab initio investigations of TLI-type compounds under high pressure. Z. Kristallogr. **219**, 348 (2004)
5. Biegler-König, F.W., Bader, R.F.W., Tang, T.: Calculation of the average properties of atoms in molecules. ii. J. Comput. Chem. **3**, 317 (1982)
6. Brock, C.P.: Crystal packing in vicinal diols CnHm(OH)$_2$. Acta Crystallogr.llogr. B **58**, 1025 (2002)
7. Bučinský, L., Jayatilaka, D., Grabowsky, S.: Relativistic quantum crystallography of diphenyl- and dicyanomercury. Theoretical structure factors and Hirshfeld atom refinement. Acta Crystallogr. A **75**(5), 705 (2019)

8. Buerger, J.: Phase Transformation in Solids. Wiley, New York, USA (1951)
9. Bytheway, I., Gillespie, R.J., Tang, T.H., Bader, R.F.W.: Core distortions and geometries of the difluorides and dihydrides Ca, Sr, and Ba. Inorg. Chem. **34**, 2407 (1995)
10. Carmo, M.: Differential Geometry of Curves and Surfaces. Dover Publications, Mineola, New York (2016)
11. Cohen, M.L.: Calculation of bulk moduli of diamond and zinc-blende solids. Phys. Rev. B **32**, 7988 (1985)
12. Contreras-Garcí, M.L.: Calculation of bulk moduli of diamond and zinc-blende solids. Phys. Rev. B **32**, 7988 (1985)
13. Contreras-García, J., Johnson, E.R., Keinan, S., Chaudret, R., Piquemal, J.P., Beratan, D.N., Yang, W.: NCIPLOT: A program for plotting noncovalent interaction regions. J. Chem. Theory Comput. **7**, 625 (2011)
14. Contreras-García, J., Marques, M., Menendez, J., Recio, J.: From elf to compressibility in solids. Int. J. Mol. Sci. **16**, 8151 (2015)
15. Contreras-García, J., Martín Pendás, A., Recio, J.M.: How electron localization function quantifies and pictures chemical changes in a solid: the B3-B1 pressure induced phase transition in beo. J. Phy. Chem. B **112**, 9787 (2008)
16. Contreras-García, J., Mori-Sánchez, P., Silvi, B., Recio, J.M.: A quantum chemical interpretation of compressibility in solids. J. Chem. Theory Comput. **5**, 2108 (2009)
17. Contreras-García, J., Recio, J.M.: On bonding in ionic crystals. J. Phys. Chem. C **115**, 257 (2011)
18. Costales, A., Martín Pendás, A., Luaña, V., Blanco, M.A.: Ions in crystals: the topology of the electron density in ionic materials. v. the B1$\rightleftharpoons$B2 phase transition in alkali halides. Phys. Rev. B **62**, 12,028 (1998)
19. Custelcean, R., Jackson, J.E.: Dihydrogen bonding: structures, energetics, and dynamics. Chem Rev **101**(7), 1963–1980 (2001)
20. Erements, M.L., Hemley, R.J., Mao, H., Gregoryanz, E.: Semiconducting non-molecular nitrogen up to 240GPa and its low-pressure stability. Nature **411**, 170 (2001)
21. Genoni, A.: A first-prototype multi-determinant X-ray constrained wavefunction approach: the X-ray constrained extremely localized molecular orbital–valence bond method. Acta Crystallogr. A **73**, 312 (2017)
22. Giacovazzo, C.: Fundamentals of crystallography. International Union of Crystallography Oxford University Press, Chester, England Oxford New York (1992)
23. Gillespie, R.J.: The valence-shell electron-pair repulsion (vsepr) theory of directed valency. J. Chem. Educ. **40**, 295 (1963)
24. Gillespie, R.J., Silvi, B.: The octet rule and hypervalence: two misunderstood concepts. Coord. Chem. Rev. **53**, 233 (2002)
25. Grin, Y., Fedorchuk, A., Faria, R., Wagner, F.: Atomic charges and chemical bonding in Y-Ga compounds. Crystals **8**, 99 (2018)
26. Hemley, R.J.: Effects of high pressure on molecules. Annu. Rev. Phys. Chem. **51**, 773 (2000)
27. http://www.cpfs.mpg.de/ELF
28. Jayatilaka, D., Grimwood, D.J.: Electron localization functions obtained from X-ray constrained Hartree-Fock wavefunctions for molecular crystals of ammonia, urea and alloxan. Acta Crystallogr. A **60**, 111 (2004)
29. Rosenfield, R. E. Jr., Parthasarathy, R., Dunitz, J.D.: Directional preferences of nonbonded atomic contacts with divalent sulfur. 1. Electrophiles and nucleophile. J. Am. Chem. Soc. **99**, 4860 (1977)
30. Kittel, C.: Introduction to Solid State Physics. Wiley (2004)
31. Koch, U., Popelier, P.L.A.: Characterization of C-H-O hydrogen bonds on the basis of the charge density. J. Phys. Chem. **99**, 9747 (1995)
32. Kohout, M., Wagner, F.R., Grin, Y.: Electron localization function for transition-metal compounds. Theor. Chem. Acc. **108**, 150 (2002)
33. Labet, V., Gonzalez-Morelos, P., Hoffmann, R., , Ashcroft, N.W.: A fresh look at dense hydrogen under pressure. i. an introduction to the problem, and an index probing equalization of H–H distances. J. Chem. Phys. **136**, 074,501 (2012)

34. Labet, V., Hoffmann, R., Ashcroft, N.W.: A fresh look at dense hydrogen under pressure. ii. chemical and physical models aiding our understanding of evolving H–H separations. J. Chem. Phys. **136**, 074,502 (2012)
35. Leoni, S., Carrillo-Cabrera, W., Schnelle, W., Grin, Y.: $BaAl_2Ge_2$: synthesis, crystal structure, magnetic and electronic properties, chemical bonding, and atomistic model of the $\alpha \leftrightarrow \beta$ phase transition. Solid State Sci. **5**, 139 (2003)
36. Lewis, G.N.: Valence and the structure of atoms and molecules. Dover, New York (1966)
37. Lyssenko, K.A., Antipin, M.Y., Gurskii, M.E., Bubnov, Y.N., Karionova, A.L., Boese, R.: Characterization of the $B \cdots \pi$-system interaction via topology of the experimental charge density distribution in the crystal of 3-chloro-7-$\alpha$-phenyl-3-borabicyclo[3.3.1]nonane. Chem. Phys. Lett. **384**, 40 (2004)
38. Malcolm, N.O.J., Popelier, P.L.A.: An improved algorithm to locate critical points in a 3D scalar field as implemented in the program morphy. J. Comput. Chem. **24**, 437 (2003)
39. Manca, P.: A relation between the binding energy and the band-gap energy in semiconductors of diamond or zinc-blende structure. J. Phys. Chem. Solids **20**, 268 (1961)
40. Martín Pendás, A.: Stress and pressure in the theory of atoms in molecules. J. Chem. Phys. **117**, 965 (2002)
41. Martín Pendás, A., Costales, A., Luaña, V.: Ions in crystals: The topology of the electron density in ionic materials. I. Fundamentals. Phys. Rev. B **55**, 4275 (1997)
42. Martín Pendás, A., Costales, A., Luaña, V.: Ions in crystals: The topology of the electron-density in ionic materials. III. geometry and ionic-radii. J. Phys. Chem. B **102**, 6937 (1998)
43. Martín Pendás, A., Luaña, V., Recio, J.M., Flórez, M., Francisco, E., Blanco, M.A., Kantorovich, L.N.: Pressure-induced B1–B2 phase transition in alkali halides: General aspects from first-principles calculations. Phys. Rev. B **49**, 3066 (1994)
44. Matta, C.F., Hernández-Trujillo, J., Tang, T., Bader, R.F.W.: Hydrogen-hydrogen bonding: A stabilizing interaction in molecules and crystals. Chem. Eur. J. **9**, 1940 (2003)
45. Miao, M.S., Lambrecht, W.R.L.: Universal transition state for high-pressure zinc blende to rocksalt phase transitions. Phys. Rev. Lett. **94**, 225,501 (2005)
46. Mori-Sánchez, P., Recio, J.M., Silvi, B., Sousa, C., Martín Pendás, A., Luaña V.and Illas, F.: Rigorous characterization of oxygen vacancies in ionic oxides. Phys. Rev. B **66**, 075,103 (2002)
47. Morrison, C.A., Siddick, M.M.: Dihydrogen bonds in solid $bh_3nh_3$. Angew. Chem. Int. Ed. **116**, 4780 (2004)
48. Ormeci, A., Rosner, H.: Electronic structure and bonding in antimony and its high pressure phases. Z. Kristallogr. **219**, 370 (2004)
49. Pickard, J., C., Needs, R.J.: Structure of phase III of solid hydrogen. Nat. Phys. **3**, 473 (2007)
50. Popelier, P.L.A.: Characterization of a dihydrogen bond on the basis of the electron density. J. Phys. Chem. A **102**, 1873 (1998)
51. Presti, L.L., Soave, R., Destro, R.: On the interplay between $CH \cdots O$ and $OH \cdots O$ interactions in determining crystal packing and molecular conformation: an experimental and theoretical charge density study of the fungal secondary metabolite austdiol ($C_{12}H_{12}O_5$). J. Phys. Chem. B **110**, 6405 (2006)
52. Riffet, V., Labet, V., Contreras-García, J.: A topological study of chemical bonds under pressure: solid hydrogen as a model case. Phys. Chem. Chem. Phys. **19**, 26,381 (2017)
53. Rousseau, B., Ashcroft, N.W.: Interstitial electronic localization. Phys. Rev. Lett. **101**, 046,407 (2008)
54. Saleh, G., Gatti, C., Presti, L.L., Contreras-Garcia, J.: Revealing non-covalent interactions in molecular crystals through their experimental electron densities. Chem. Eur. J. **18**, 15,523 (2012)
55. Shannon, R.D., Prewitt, C.T.: Effective ionic radii in oxides and fluorides. Acta Crystallogr. B **25**, 925 (1969)
56. Silvi, B., Gatti, C.: Direct space representation of the metallic bond. J. Phys. Chem. **104**, 947 (2000)
57. Silvi, B., Jolly, L., Darco, P.: Pseudopotential periodic hartree-fock study of the cristobalite to stishovite phase transition. Theochem. J. Mol. Struct. **92**, 1 (1992)

58. Tsirelson, V., Stash, A.: Determination of the electron localization function from electron density. Chem. Phys. Lett. **351**, 142 (2002)
59. Volkov, A., Gatti, C., Abramov, Y., Coppens, P.: Evaluation of net atomic charges and atomic and molecular electrostatic moments through topological analysis of the experimental charge density. Acta Crystallogr. A **56**(3), 252 (2000)
60. Woińska, M., Grabowsky, S., Dominiak, P.M., Woźniak, K., Jayatilaka, D.: Hydrogen atoms can be located accurately and precisely by x-ray crystallography. Sci. Adv. **2**(5) (2016)
61. Zha, C.S., Liu, Z., Hemley, R.J.: Synchrotron infrared measurements of dense hydrogen to 360 GPa. Phys. Rev. Lett. **108**, 146,402 (2012)
62. Zou, P.F., Bader, R.F.W.: A topological definition of a Wigner-Seitz cell and the atomic scattering factor. Acta Crystallogr. A **50**, 714 (1994)

# Part III
# Exercises

# Chapter 8
# Algorithms and Software

**Abstract** Once we have seen the theory behind Quantum Topology, it is important to put it into practice and understand how to compute it. In this chapter, we briefly consider the most important algorithms and software used in the topological analysis of wavefunctions. We start with the algorithms designed to search for critical points and numerically integrate densities over real space basins and end with a quick tour around the main software packages used both in the molecular and solid state realms.

## 8.1 Algorithms

The actual implementation of all the ideas and indices described in this monograph rests on the existence of computationally efficient algorithms. Topological approaches to chemical bonding theory work in two different modes: the local one, in which indices based on RDMs are obtained at the critical points (CPs) of a given field, and the global one, where integrals over the attraction or repulsion basins of the field are sought for. Thus the basic algorithms needed to implement a topological analysis of a field can be roughly divided into those of CP location and those of basin integration. No unsurmountable problems have been found in the former case, but basin integration is still a computationally intensive task that has no optimal solution. We will briefly discuss a summary of the standard methods that are used by most of the codes that will be enumerated in the next Section.

### *8.1.1 CP Location*

Locating the position of a CP of a scalar field $f(\boldsymbol{r})$ is equivalent to solving a set of three coupled equations in the $(x, y, z) \equiv \boldsymbol{r}$ variables: $\nabla f(\boldsymbol{r}) = \mathbf{0}$. These are obviously a set of non-linear equations with, in general, many solutions. Without loss of generality, we will suppose that the field to be studied is the density, $\rho$. In the molecular realm, the electron density is usually expanded in terms of a Gaussian basis set, so that once the orbital coefficients are obtained through the appropriate

 
Á. Martín Pendás and J. Contreras-García, *Topological Approaches to the Chemical Bond*, Theoretical Chemistry and Computational Modelling,
https://doi.org/10.1007/978-3-031-13666-5_8

electronic structure code, the density and its derivatives can be obtained analytically at any point by means of straightforward methods. In condensed phases systems, the density can be either expanded in Gaussians, like in the CRYSTAL code [6] or, more typically, in plane waves [11, 14]. In both cases, $\rho$ and its derivatives can be analytically reconstructed. In some cases, only a grid of values of $\rho$ is available. Several techniques can then be applied to obtain its derivatives. Usually, the density at non-grid points is interpolated by means of cubic splines, so that noise is decreased or suppressed.

Once the density and its derivatives can be obtained at any point in space, a solver is used to find CPs. Typically, the Newton–Raphson (NR) method is employed (see Ref. [19] for details on the method). It has provided good results and is considerably efficient: a search is started at point $\boldsymbol{r}_0$ that is iterated through

$$\boldsymbol{r}_{n+1} = \boldsymbol{r}_n - \mathbb{H}^{-1}(\boldsymbol{r}_n)\nabla\rho(\boldsymbol{r}_n). \tag{8.1}$$

In this expression, $\mathbb{H}$ is the Jacobian of the system. Since we solve for the gradient of the density, the Jacobian coincides with the Hessian matrix. The NR procedure is quadratically convergent, but has no solution assured, and may oscillate or diverge if several conditions on the derivatives are not met.[1] The iteration procedure is stopped when a predefined convergence (for instance in $\epsilon = |\boldsymbol{r}_{n+1} - \boldsymbol{r}_n|$, the distance to the critical points we are approaching to) has been met, or when oscillations are detected. Since we would usually like to find all the CPs in a molecular system or in the unit cell of a crystal, it is customary to start a multitude of NR searches at a properly chosen grid in the molecular space or crystalline unit cell. In molecules, searches are usually started in the middle of all atom pairs to look after BCPs, or at the center of atom triads or tetrads to locate RCPs or CCPs, respectively. After all searches converge, the appropriate Morse condition (e.g. $n + b - r + c = 1$ in molecules) is checked. To do this, each CP has to be classified. This is easily done by diagonalizing the Hessian and counting the number of positive and negative eigenvalues. If the Morse equality is not satisfied, the grid is made thinner, and the search is restarted. Notice that the fulfillment of the Morse condition does not guarantee that all the CPs have been found, so care has to be taken in this regard.

CP location is usually accompanied by the obtention of the chemical graph, i.e. the flux lines emanating from the BCPs that climb up to the nuclei. This is equivalent to solving a set of ordinary differential equations (ODEs)

$$\boldsymbol{r}(t) = \boldsymbol{r}(0) + \int_0^t \nabla\rho(\boldsymbol{r}(\tau))d\tau. \tag{8.2}$$

This is usually done with standard ODE solvers, like 4th order Runge–Kutta or several predictor–corrector techniques (see Ref. [19] for details on numerical methods to

[1] In one dimension, the NR procedure iterates the following expression: $x_{n+1} = x_n - f(x_n)/f'(x_b)$. It can be shown [19] that the method converges only if $|g'(x)| < 1$, where $g(x) = x - f(x)/f'(x)$. Similar expressions can be found in 3D.

solve differential equations). Since, in this case, $\boldsymbol{r}(0)$ is a BCP with null gradient, the flux line (the bond path) needs to be seeded by locating two points as close as necessary to (but not at) the BCP on the bond path. This is done by selecting the eigenvector of the Hessian at the BCP with $\lambda > 0$, $\boldsymbol{v}_+$, so that $\boldsymbol{r}(0) = \boldsymbol{r}_{\mathrm{BCP}} \pm \epsilon \boldsymbol{v}_+$.

Similar algorithms can be used to obtain interatomic surfaces (IASs). They are defined as the geometric locus of all flux lines that end at a given BCP, and may be obtained by integrating a set of ODEs that start on a circle that surrounds the BCP at a distance $\epsilon$ in the plane defined by the two eigenvectors corresponding to $\lambda < 0$. Some of these flux lines will end up at another CP, and some of them will go to infinity. In the latter case, a maximum distance threshold is usually enforced. After a set of points of the IAS has been obtained, the surface may be smoothed or triangularized, if necessary.

### 8.1.2 Basin Integration

A much more computationally intensive procedure is basin integration. In its simplest form, it can be stated as the problem of finding a one- or two-electron atomic observable defined as

$$\langle O \rangle_A = \int_A o(\boldsymbol{r}) d\boldsymbol{r}, \quad \text{or}$$
$$\langle G \rangle_{AB} = \int_A \int_B g(\boldsymbol{r}_1, \boldsymbol{r}_2) d\boldsymbol{r}_1 d\boldsymbol{r}_2, \tag{8.3}$$

respectively. There are two coupled but unrelated problems here. On the one hand, choosing the specific quadratures to perform the 3D (or 6D) integration. On the other hand, determining how the basin boundary is built. The second problem is far more difficult than the first one, and determines a gross classification of integration methods: explicit, if the basin boundary is explicitly constructed and implicit, if it is not. Since the basin boundary is not defined through a local property, it is difficult to escape to its explicit construction.

#### 8.1.2.1 Implicit Algorithms

The implicit algorithms that have been cleverly devised to deal with basin integration properties are based on a coordinate transformation that effectively transforms a topological basin into an infinite 3D ball. They are based on the concept of natural coordinate system, introduced by Biegler-König and coworkers [3].

Let us consider an attractor $A$ of our field which we take as the Cartesian origin of coordinates and let us construct a sphere of small radius $\beta$ centered on $A$. Any point on the surface of this sphere is characterized by the spherical angles $(\theta, \phi)$ in

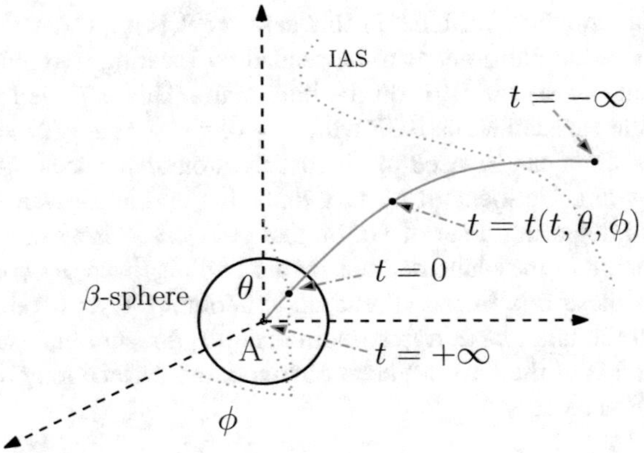


**Fig. 8.1** The natural coordinate system. A set of points on the surface of a small $\beta$-sphere surrounding a nucleus $A$ is chosen. A gradient line (in violet) is started at each of them at $t = 0$, in the ascending and descending directions. The set $(t, \theta, \phi)$, where $\theta$ and $\phi$ are the $t = 0$ spherical coordinates of each point on the sphere, maps the entire atomic basin. The $\omega$-limit of any point on the sphere is the $A$ nucleus, while the $\alpha$-set can be infinite, or at a critical point onto an interatomic surface (IAS)

a spherical polar representation of our Cartesian system. The flux line or trajectory of any point $\boldsymbol{r}$ belonging to the basin of $A$ will cross the $\beta$-sphere at a specific set of angles according to Eq. 8.2. The point $\boldsymbol{r}$ can thus be described by the set $(t, \theta, \phi)$ determined through the flux line that starts at the $(\theta, \phi)$ angles on the $\beta$-sphere at $t = 0$. This is the *natural* coordinate system, in which the basin of $A$ is covered when $t$ is allowed to go from $-\infty$ to $+\infty$. Figure 8.1 sketches this coordinate systems.

Since the mapping $h$

$$h : x, y, z \rightarrow t, \theta, \phi, \tag{8.4}$$

is one to one, a change of variables leads to

$$\langle O \rangle_A = \int_0^{2\pi} d\phi \int_0^{\pi} d\theta \int_{-\infty}^{+\infty} f(\boldsymbol{r}(t, \theta, \phi))|J(h^{-1}(t, \theta, \phi))|dt, \tag{8.5}$$

where the determinant of the Jacobian of the inverse mapping needs to be found. Since $J = (\boldsymbol{r}_t, \boldsymbol{r}_\theta, \boldsymbol{r}_\phi)$, where the subindices indicate partial differentiation, $\partial J/\partial t = \partial((\boldsymbol{r}_t, \boldsymbol{r}_\theta, \boldsymbol{r}_\phi))/\partial t$. The first component, $\partial \boldsymbol{r}_t/\partial t = \partial(\nabla\rho)/\partial t = \mathbb{H}\boldsymbol{r}_t$, with $\mathbb{H}$ the Hessian matrix. Similar expressions can be found for the other two variables, so that finally, $\partial J/\partial t = \mathbb{H}J$. We can now use a theorem from the theory of determinants, $d|A|/dt = \mathrm{Tr}A^{-1}dA/dt$, to write

$$\partial|J|/\partial t = \nabla^2\rho|J|. \tag{8.6}$$

This is a differential equation that can be solved on the fly while the integral shown in Eq. 8.5 is computed.

Despite the elegance of the natural coordinate formalism, its practical implementation is not easy, and the final accuracy of the results has to be carefully checked.

#### 8.1.2.2 Explicit Algorithms

A bunch of explicit algorithms has been proposed over the years to deal with the basin integration problem. They can be further divided into grid- or non-grid-based. A general difficulty that needs to be taken into account stems from the need to simultaneously obtain several basin observables, that may behave differently over the basin. For instance, in order to obtain an atomic population, the density itself, a well-behaved non-negative integrand, needs to be manipulated. If, at the same time, we want to examine the integrals of the Laplacian of the density, $\int_A \nabla^2 \rho d\boldsymbol{r}$, and the electron-nucleus attraction $\int_A -Z_A/r d\boldsymbol{r}$, the integration algorithm will have to be robust enough to deal with a widely oscillating function in the first case (which actually integrates to zero, so it is customarily used as a measure of the overall basin integration accuracy) and with a diverging integrand at the nucleus in the second. Since, in many cases, the interesting integrands possess nuclear divergences or at least are dominated by core regions, uniform grids at all points within the basin will not easily achieve the required precision.

This has been solved differently if the density and the fields can be obtained analytically at any point or if they are built on a grid. In the first case, the nuclear dominance leads naturally to the use of a polar spherical coordinate system. In its simplest versions, fixed quadratures are used for the angular and radial parts. An efficient angular grid is that of Lebedev [15], which is built to integrate exactly spherical harmonic expansions up to a given order. At each $\theta, \phi$ ray a radial grid is built, which goes from $r = 0$ (the nucleus) to $r = r_{\max}(\theta, \phi)$ (the ray-dependent basin boundary). If an explicit representation of the surface (a triangularized IAS, for instance) exists, this limit is obtained by intersecting the ray, usually by interpolation. If not, the ray limit can be determined by a simple bisection: after one point $r_1$ is found inside the basin (the left end of a bisecting segment) and another one $r_2$ out of it (the right end) for a given ray, the midpoint $r_{av}$ of the segment is built and a flux line is started from it. If its end is the nucleus of interest, it becomes the left end of the new segment. Otherwise, it becomes the right one. These steps are iterated up to a predefined accuracy. This is a very inefficient but robust procedure, although several cautions need to be taken since a ray may intersect the basin boundary several times. In both the interpolation or bisection cases, once $r = r_{\max}$ has been obtained, a radial quadrature is chosen. The latter can be exponentially or logarithmically mapped, so that the density of points close to the nucleus be a high as necessary. Figure 8.2a shows a triangularized IAS as well as a ray at a given set of angles $(\theta, \phi)$. Figure 8.2b sketches the bisection process.

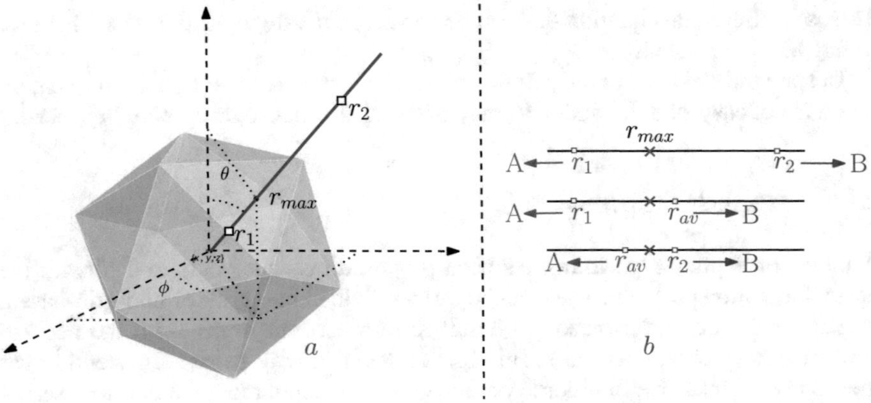


**Fig. 8.2** Triangularization of an interatomic surface by a set of vertices forming triangles (left, **a**). At a given set of polar angles $(\theta, \phi)$, a ray is also shown in blue. After two points on the ray, $r_1$ and $r_2$ are found so that a gradient line started from the first ends up on nucleus $A$ and the one started from the second ends up in another different nucleus $B$, a bisection process (right panel, **b**) finds iteratively the limit of the IAS, $r_{max}$: the average point $r_{av}$ is found and a gradient line is started from it. If it ends up onto $B$, $r_{av}$ becomes $r_2$. Otherwise, $r_{av}$ becomes $r_1$. This is repeated until $|r_1 - r_2| < \epsilon$

#### 8.1.2.3 Grid Algorithms

In many cases, the density or the field to be integrated over a basin are either too cumbersome to obtain analytically or only available on a grid, usually a fixed, uniform and cubic one. This is a rather common situation in solid state codes. Much effort has been devoted to devise algorithms that do not rest on the calculation of the basin boundary at specific points of a given quadrature. One of the best known is due to Yu and Trinkle [27]. It starts from a uniform mesh in which voxels are assigned to grid points by means of the Voronoi construction: the polyhedron formed by all points closer to the reference mesh point than to any other. The different voxels are visited sequentially and assigned to a given basin (i.e. colored). To each grid point, $i$, a weight, $w_i^A$, measuring its contribution to basin $A$ is given such that $\sum_A w_i^A = 1$. Voxels entirely belonging to basin $A$ display $w_A = 1$ and those intersected by one or several IASs values which can be fractional. With this

$$\int_A f(\mathbf{r})d\mathbf{r} \approx \sum_i V_i w_i^A f_i, \tag{8.7}$$

where $V_i$ is the volume of the $i$th voxel (usually fixed), and $f_i$ is the value of the field $f$ at grid point $i$. If the weights $w_i^A$ are defined as the fraction of points in $V_i$ whose trajectory ends in basin $A$, then the above expression converges quadratically.

To obtain the weights, a probability density of points inside a voxel, $p(\mathbf{r}, t)$ is allowed to evolve according to the gradient field that defines the basin partition.

In the case of $\rho$, the current density $\boldsymbol{j} = p(\boldsymbol{r}, t)\nabla\rho(\boldsymbol{r})$ would satisfy the continuity equation

$$\frac{\partial p(\boldsymbol{r}, t)}{\partial t} + \nabla \cdot \boldsymbol{j}(\boldsymbol{r}, t) = 0. \tag{8.8}$$

If for voxel $X$ we define $p_X(t) = \int_X p(\boldsymbol{r}, t)d\boldsymbol{r}$ and we assign a uniform initial density $p(\boldsymbol{r}, 0) = V_X^{-1}$ if $\boldsymbol{r} \in x$ and $p(\boldsymbol{r}, 0) = 0$ otherwise, then

$$\frac{dp_X(t)}{dt} = -\int_X \nabla \cdot \boldsymbol{j}\, d\boldsymbol{r} = \int_{\partial x} p(\boldsymbol{r}, t)\nabla\rho(\boldsymbol{r}) \cdot d\boldsymbol{S}. \tag{8.9}$$

Now the surface of the voxel is divided into its facets (that connect it to its neighboring Voronoi polydedra $X'$). These have a well defined area, $a_{XX'}$. Also known is the distance between the neighboring $X$'s, $l_{XX'}$. Approximating the gradient with a ramp function

$$\frac{\partial p_X(t)}{\partial t} = -p_X(t)\sum_{X'} \tau_{XX'}, \tag{8.10}$$

where $\tau$ now only depends on the densities at $X$ and $X'$ and the facet areas and inter-grid distances. This equation has the solution $p_X(t) = \exp(-t\sum_{X'} \tau_{XX'})$. With this, one can also approximate

$$J_{XX'} = \int_0^\infty dt \int_{\partial V_{XX'}} p(t)\nabla\rho(\boldsymbol{r}) \cdot d\boldsymbol{S} \approx \int_0^\infty p_X(t)\tau_{XX'}dt \approx \frac{\tau_{XX'}}{\sum_{X'} \tau_{XX'}}. \tag{8.11}$$

This flux defines the fraction of points in voxel $X$ that evolve to $X'$. Since $w_X^A$ is the fraction of $X$ points ending in basin $A$

$$w_X^A = \sum_{X'} J_{XX'} w_{X'}^A. \tag{8.12}$$

The algorithm starts by classifying all grid points from highest to lowest $\rho$. Then, sequentially, each point $X$ is either: (i) a local maximum with respect to its neighbors. The grid point is a new basin $A$, with $w_X^A = 1$; (ii) an interior point: for all $X'$ with $\rho(X') > \rho(X)$ the weights are known and equal to 1 for the same basin $A$. Then $w_X^A = 1$; (iii) a boundary point, then the weights are assigned with Eq. 8.12.

The Yu-Trinkle algorithm is very fast and stable, and guarantees that the basin partition of an observable is additive. This means that $\sum_A O_A = O$ exactly, since all weights necessarily add to one. However, it does not assure the accuracy of the reconstructed values, since numerical errors due to the usually fixed grid may be very large in the case of divergent integrands.

#### 8.1.2.4 Adaptive Algorithms

We will also briefly describe another type of algorithm that does not rest on bipartition and that can nevertheless provide accurate basin integrals at a moderate computational effort by using adaptive grids. One of them was proposed by Rafat and Popelier [20], using a finite element decomposition of the basin. The core of the algorithm lies in growing the basin from a sphere centered at the nucleus, guided by the reverse trajectories started on a mesh of points on the surface that escape from it. As the gradient paths evolve in steps, the basin grows by layers, which are divided into finite elements. However, numerous subtleties are needed to ensure smooth basins, and the method also rests on growing IASs independently.

The method starts by building a triangulation of a $\beta$-sphere. The triangles are allowed to be curved, via quadratic approximations. Gradient paths are evolved reversely for each vertex (mesh point) and stopped at a given value of $\rho$. Each of these steps is known as a growth level. The connections between the mesh on the $\beta$-sphere (level-0) and the points at level-1 are preserved, so that each triangle in the former has an equivalent one in the latter. Thus, a 3D triangular prism goes from each triangle in one level to its corresponding one in the next. When the trajectories diverge in one of these steps, stretched prisms may appear. This is avoided by adding new points in the middle of the side of a triangle whenever the length of the side becomes larger than a predefined value. This may give rise to 2, 3, or 4 triangles as sketched in Fig. 8.3.

IASs are treated similarly. A small triangulated circle around the BCP is constructed, and the full IAS is grown in 2D steps as before. The growing sphere and IASs points are connected using more complex finite elements including hexahedral or brick ones. If any side diverges, the element is again subdivided. Special techniques, that will not be discussed, are needed to deal with ring critical points. The interested reader can find more details in Ref. [20].

After the full growth has finished, the basin has been transformed into an adaptive mesh of elementary finite elements. Both 2D and 3D integrations can now be performed. We will focus on 3D integrations. Taking $\{V_i\}$ as the set of 3D finite elements, the basin observable is given by $\int_A f(\boldsymbol{r})d\boldsymbol{r} = \sum_i \int_{V_i} f(\boldsymbol{r}_i)d\boldsymbol{r}$. The integral over each finite element is obtained from the position of a set of quadrature points

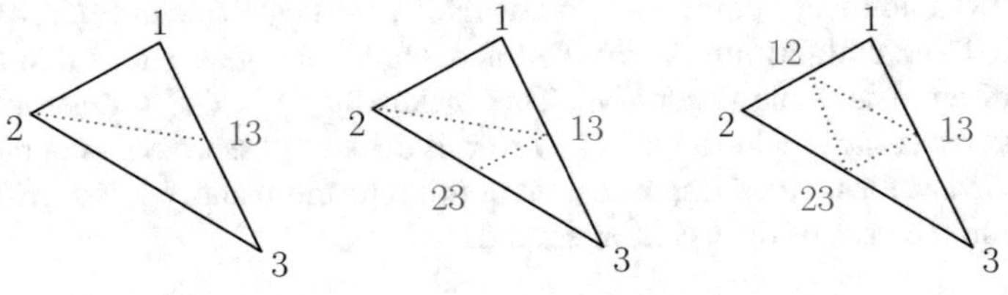


**Fig. 8.3** Adaptive division of a triangle. Three cases that lead to 2, 3, or 4 sub-triangles are shown. Each edge is subdivided depending on whether its length exceeds a predetermined value. If this is the case, a new node is added at the center of the divided edge

inside the element, which are usually described in natural coordinates $p, s, t$. Also, the so-called shape function $N$s are used to interpolate the value of the function $f$ at a quadrature point from those at the vertices of the finite element

$$f(x, y, z) = \sum_j N_j(p, s, t) f(x_j, y_j, z_j). \tag{8.13}$$

The $N_j$ are used to map local to global coordinates, since if we choose $f = x$, then $x = \sum_j N_j(p, s, t)x_j$, with similar expressions for $y$ and $z$. Now weights $w_{pst}$ are associated to each quadrature point $(p, s, t)$ in local coordinates, so that

$$f(\boldsymbol{r}_i) = \sum_{i,j,k} f(\boldsymbol{r}_{ijk})\, w_{ijk}\, |J(p_i, s_j, t_k)|. \tag{8.14}$$

In this expression, we run over all quadrature points in the finite element, and J is the Jacobian of the global to local coordinate transformation. This is obtained for each type of finite element. Usually, the weights are obtained from Gauss–Legendre product rules with a small number of points.

## 8.2 Available Software

A considerable number of codes, both freely available (or under open software licenses) or proprietary have appeared over the years, whose basic task is to analyze the topological features of scalar fields. We will only comment briefly on a few ones. The oldest suite of programs devoted to QTAIM analyses is AIMPAC [3, 13], from the laboratory of R. F. W. Bader. It introduced the Newton–Raphson searching technique as well as the PROAIM algorithm in which the IASs are approximated by triangulation. Many other codes used to compute electron densities from X-ray data, like TOPXD [26], VALTOPO [1] or WinXPRO [25], use variants of these ideas, as well as the PROMOLDEN [17] code. Biegler-König and coworkers have used the natural coordinates algorithm in AIM2000 [2], a Windows code that unfortunately has ceased to be updated. IASs have also been analytically fitted through Chebyshev polynomials in MORPHY [18]. Widely used general purpose codes with a graphical interface (GUI) are AIMAll [12] and MULTIWFN [16]. For QTAIM in the solid state, several options are also available beyond those used by X-ray crystallographers which have already been commented. TOPOND, by C. Gatti, is now implemented in the CRYSTAL code [7], and CRITIC2 [21], is able to perform topological analyses and basin integration for several fields using the output grids of a wide variety of solid state codes.

We will succinctly review some of the capabilities of AIMAll, ToPMoD, NCI-PLOT and CRITIC2 as simple examples of molecular and solid state packages since they will be further used in the exercises (Chap. 9). We will also briefly review NCIMilano as a code for experimental densities.

### 8.2.1 AIMAll

AIMAll is a cross-platform suite (available for Linux, MacOS and Windows). It provides thorough molecular QTAIM analyses from electronic structure calculations that are based on Gaussian expansions. As most of the QTAIM molecular codes, it reads the wavefunction (geometry, primitives, molecular orbitals) that is output from several electronic structure packages in the original `.wfn` format used since AIMPAC, although a more modern XML format (`.wfx`) has also been adopted. Although not necessary, AIMAll is usually run in conjunction with the GAUSSIAN suite [8] and it also accepts `.chk` files. It is an efficient and fast implementation (basically automatic), parallelized in a shared memory environment, which allows the simultaneous calculation and visualization of the topology of the electron density as well as other scalar fields. It also allows computing local (at critical points) and global (integrated both in IASs and basins) properties. AIMAll is proprietary software. A demonstration version is available that limits the number of basis functions and the total number of atoms that can be used.

The AIMAll package is a set of fully integrated, yet independent codes (see Chap. 9 for more details). It is possible to run an interactive AIMAll session by invoking the AIMstudio code. The user then uploads a wavefunction file (`.wfn`, `.wfx`) and runs the AIMQB dialog. Although reasonable default values are available, it is possible at this stage to fix user-defined conditions and integration options. For example, it is possible to define the basin integration method, the atoms to be integrated, the fineness of the integration grid, etc. It is also at this stage that the different options to run IQA analysis are selected. Note that the full IQA calculation might be rather time-consuming. AIMQB outputs several text files (`.sum`, `.sumviz`) which can then be read by AIMStudio to produce graphical output.

A limited, yet informative list of its capabilities, includes:

- Automatic topological analysis with a large number of electron density critical point and path descriptors, including electron density at all CPs, ellipticities at BCPs, energy densities, total electrostatic potential, etc.
- Analysis of the topological features of $\nabla^2\rho$, the virial field, magnetically induced current density distributions and molecular orbitals.
- Basin and surface integration (with multiple integration algorithms) of a large number of properties: atomic charges and moments, atomic kinetic and potential energies, full implementation of IQA for HF, LSDA, B3LYP and M06-2X wavefunctions. Atomic volumes and atomic surface areas. Localization and delocalization indices, atomic magnetic response properties, visualization of IASs, mapped isosurfaces including electrostatic potential properties, source functions, etc.

  It is relevant to notice that AIMAll works both with natural orbitals and their occupations when using correlated wavefunctions. This means that all quantities that depend on the pair density (e.g. the delocalization indices) are approximated and should be taken with care.

### 8.2.2 ToPMoD

The ToPMoD package enables the calculation of the ELF function on a 3D grid. Assignment of the basins and the calculation of the basin populations and of their variance is carried out with the already seen algorithm for basin integration in grids. It also uses `.wfn` files as input. The modules are written in FORTRAN 90 in order to enable dynamic memory allocation. In order to carry out a standard ELF analysis three programs have to be run in the following order:

- `grid90`: calculates ELF on a 3D grid parallel to the standard axis defined in the MO calculation (this allows to exploit the factorization of the gaussian functions)
- `bas90`: assigns the grid points to basins
- `pop90`: calculates the basin populations and variances.

Other modules are available for more specific functionalities:

- `top_sym`: exploits abelian symmetry operations
- `search90`: localizes the critical points of the ELF gradient field
- `mod_wfn`: enables to remove atoms in the case of large systems in order to focus the calculation on the region of interest
- `sym_wfn`: symmetrizes the wfn file for $^2\Pi$ states of linear molecules
- `bas_to_syn`: assigns basin types from the `bas.sbf` file
- `sbf_to_cube`: converts `sbf` files to `cube` files for visualization.

If correlated wavefunctions are used, external pair densities have to be fed. Otherwise, the code can work with natural orbital approximations. It is also able to obtain rough approximations of QTAIM topologies and basin integrated properties.

### 8.2.3 NCIPLOT

NCIPLOT is focused on the calculation of NCI on grids. A 3D grid is defined around the system of interest. $\rho$ and $RDG$ are then computed on the nodes of this grid, along with $\lambda_2$. The main piece of information is $\rho$. It can be computed from the wave function (`wfn` or `wfx` files) or from the coordinates in promolecular calculations (`.xyz` file). Since NCI analysis only requires information on the $s(\rho)$ peaks, multilevel grids are used to improve the performance. The multilevel algorithm uses several regular grid separations by progressively rejecting points with high $\rho(\mathbf{r})$ or $s(\mathbf{r})$, i.e. $n$ times higher than reference values $\rho_{ref}$ and $s_{ref}$. The procedure starts with a coarse grid (big squares in Fig. 8.4), computes the required functions in all points of the grid, then builds an $\alpha_l$-times denser grid containing only the regions of space in which both $s(\mathbf{r}_j) < \alpha_l s_{ref}$ and $\rho(\mathbf{r}_j) < \alpha_l \rho_{ref}$ (middle size squares in Fig. 8.4). After the identification of important grid points, the scalar functions are computed again, and the process proceeds until the desired local grid density is achieved (violet squares in Fig. 8.4). Only the density and RDG at the active grid points should be

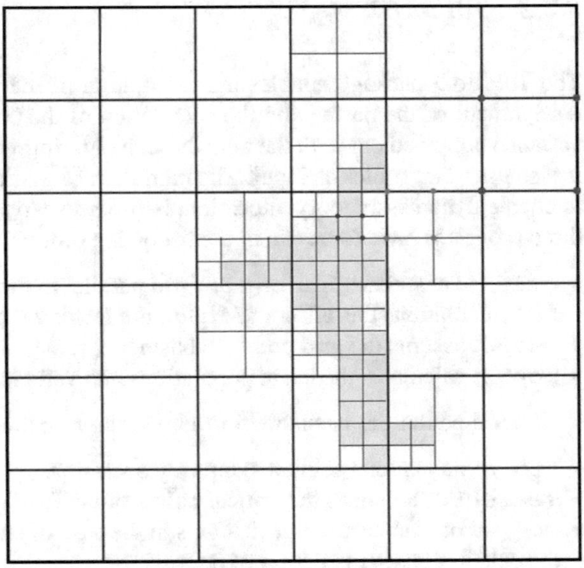

**Fig. 8.4** 2D schematic diagram of multilevel grids, from coarse to fine grids. The density and the RDG values are computed at the active points of the coarse, medium and fine grids (respectively, for instance, the red, orange and blue points). The small blue boxes are the active boxes of the finest grid, which really contribute to form the isosurface. Adapted with permission from Ref. [4]. Copyright 2020

computed, which is much less expensive than dealing with the finest grid directly. Naturally, the number of iterations and the value of $\alpha_l$ can be adjusted to ensure maximal performance for different steps.

By default, the multilevel is turned off, using only one-level grid without acceleration (keywork `CG2FG` option is set to 1 1). In practice, the multigrid is usually set with 3- or 4-level grids for acceleration (keyword `CG2FG` set typically to 3 4 2 1 or 4 8 4 2 1). It is observed that the multilevel grid method can decrease the time of computing $\rho$ and $s$, from several to tens of times. Furthermore, the percentages of active points in the finest grid are the same for one grid and multilevel grids, implying that this method does not ignore any grid point of interest.

NCI regions are not defined in terms of basins, but rather trying to keep only intermolecular interactions. From the promolecular approach exposed in Sect. 5.3.3.4, the interacting region from a complex AB can be determined as those regions of space where the electron density has important contributions from monomers A and B. In other words, most of the electron density does come neither from A, nor from B

$$\mathbf{r}_i \in \Omega_{NCI} \begin{cases} \rho^A(\mathbf{r}_i) < \gamma^{ref}\rho(\mathbf{r}_i) \\ \rho^B(\mathbf{r}_i) < \gamma^{ref}\rho(\mathbf{r}_i) \end{cases} \tag{8.15}$$

where $\gamma$ is a threshold value, typically around 0.9. Since the definition is done in terms of the electron density, it suffices to use a large constant value of $s$ to enclose the interacting volume. Typically, a value of $s^{ref} = 1.0$ encloses all relevant interaction regions: bonds and non-covalent interactions.

### 8.2.4 CRITIC2

CRITIC2 is a standalone code written in FORTRAN 90 that performs topological analyses in periodic solids. It is publicly available under the GNU General Public License. It lacks a specific graphical interface, although it can output graphical information readable by several visualization packages. Its input is highly structured and keyword oriented, making it easy for the standard user to write a new or adapt a previous input file. The code reads output from a number of electronic structure codes in the solid state, like WIEN2k [24], ELK (available at https://elk.sourceforge.net), ABINIT [10], Quantum ESPRESSO [9], VASP [14], GAUSSIAN [8] and any other which is able to write the scalar fields in a 3D grid. It can also be used to interconvert several crystal descriptions and to perform arithmetic operations on the scalar fields. The code is parallelized in a shared-memory environment.

Critical points are sought by finding the Wigner–Seitz cell, reducing it by symmetry to its irreducible part, and dividing the latter into a set of tetrahedra from which a number of Newton–Raphson searches are started. Basin integrals are obtained through bisection if the `integrals` keyword is selected, via the QTREE [22] algorithm with the `qtree` keyword or the Yu-Trinkle algorithm [27] with the `yt` keyword.

### 8.2.5 NCIMilano

NCIMilano, written in FORTRAN 90, allows to calculate NCI from single-crystal X-ray diffraction data. It reads the electron density grid or cube files provided by XD2006, GAUSSIAN09 and TOPOND (PL3D option) set of programs. XD reconstructs the electron density following the Hansen–Coppens formalism:

$$\rho(\vec{r}) = \rho_A(\vec{r}) + \Delta\rho(\vec{r}) \tag{8.16}$$

where $\rho_A(\vec{r})$ is the contribution of a given atom or group of atoms (even a molecule) to the density at point $\vec{r}$, and $\Delta\rho(\boldsymbol{r})$ is coming from the multipoles centered on the remaining atoms in the unit cell. Since multiple functions decay exponentially with the distance to a point, we can assume that the electron density at a given molecular point, is given by the first part of Eq. 8.16 as long as the first neighbors are taken into account. For a crystal where interactions are given by dimers, it suffices then to extract molecular pairs from the crystal. Accordingly, the main features of electron density in the space between a pair of nearest neighbor molecules depend almost exclusively on their (composing) pseudoatoms. This strategy implies that the contribution from multipoles centered on atoms beyond the molecular pair is ignored, their effect being only indirectly taken into account though the multipolar expansion of the molecular pair pseudoatoms.

Its operational mode is very similar to NCIPLOT: it calculates $s$ and sign($\lambda_2$)$\rho$ in the range decided by the user. The $\lambda_2$ is obtained numerically. The user may apply a real space cutoff to exclude some regions from the calculation of $s$. In the output, the two cube files ($s$ and sign($\lambda_2$)$\rho$), along with a file containing $\rho$ and $s$ for each point of the input grid, are written.

## References


1. Bianchi, R., Forni, A.: VALTOPO: a program for the determination of atomic and molecular properties from experimental electron densities. J. Appl. Crystallogr. **38**(1), 232 (2005)
2. Biegler-König, F., Schönbohm, J.: Update of the AIM2000-program for atoms in molecules. J. Comput. Chem. **23**(15), 1489 (2002)
3. Biegler-König, F.W., Nguyen-Dang, T.T., Tal, Y., Bader, R.F.W., Duke, A.J.: Calculation of the average properties of atoms in molecules. J. Phys. B **14**, 2739 (1981)
4. Boto, R.A., Peccati, F., Laplaza, R., Quan, C., Carbone, A., Piquemal, J.P., Maday, Y., Contreras-García, J.: NCIPLOT4: fast, robust, and quantitative analysis of noncovalent interactions. J. Chem. Theory Comput. **16**, 4150–4158 (2020)
5. Coppens, P.: Electron density from x-ray diffraction. Annu. Rev. Phys. Chem. **43**, 663–692 (1992)
6. Dovesi, R., Causà, M., Orlando, R., Roetti, C., Saunders, V.R.: Ab initio approach to molecular crystals: a periodic Hartree-Fock study of crystalline urea. J. Chem. Phys. **92**, 7402 (1990)
7. Erba, A., Baima, J., Bush, I., Orlando, R., Dovesi, R.: Large-scale condensed matter DFT simulations: performance and capabilities of the CRYSTAL code. J. Chem. Theory Comput. **13**(10), 5019 (2017)
8. Frisch, M.J., Trucks, G.W., Schlegel, H.B., Scuseria, G.E., Robb, M.A., Cheeseman, J.R., Zakrzewski, V.G., Montgomery, J.A., Stratmann, R.E., Burant, J.C., Dapprich, S., Millam, J.M., Daniels, A.D., Kudin, K.N., Strain, M.C., Farkas, O., Tomasi, J., Barone, V., Cossi, M., Cammi, R., Mennucci, B., Pomelli, C., Adamo, C., Clifford, S., Ochterski, J., Petersson, G.A., Ayala, P.Y., Cui, Q., Morokuma, K., Malick, D.K., Rabuck, A.D., Raghavachari, K., Foresman, J.B., Cioslowski, J., Ortiz, J.V., Baboul, A.G., Stefanov, B.B., Liu, G., Liashenko, A., Piskorz, P., Komaromi, I., Gomperts, R., Martin, R.L., Fox, D.J., Keith, T., Al-Laham, M.A., Peng, C.Y., Nanayakkara, A., Gonzalez, C., Challacombe, M., Gill, P.M.W., Johnson, B.G., Chen, W., Wong, M.W., Andres, J.L., Head-Gordon, M., Replogle, E.S., Pople, J.A.: Gaussian98, Revision A.6. Gaussian Inc., Pittsburgh, PA (1998)
9. Giannozzi, P., Baroni, S., Bonini, N., Calandra, M., Car, R., Cavazzoni, C., Ceresoli, D., Chiarotti, G.L., Cococcioni, M., Dabo, I., Corso, A.D., de Gironcoli, S., Fabris, S., Fratesi, G., Gebauer, R., Gerstmann, U., Gougoussis, C., Kokalj, A., Lazzeri, M., Martin-Samos, L., Marzari, N., Mauri, F., Mazzarello, R., Paolini, S., Pasquarello, A., Paulatto, L., Sbraccia, C., Scandolo, S., Sclauzero, G., Seitsonen, A.P., Smogunov, A., Umari, P., Wentzcovitch, R.M.: QUANTUM ESPRESSO: a modular and open-source software project for quantum simulations of materials. J. Phys.: Condens. Matter **21**(39), 395,502 (2009)
10. Gonze, X., Amadon, B., Anglade, P.M., Beuken, J.M., Bottin, F., Boulanger, P., Bruneval, F., Caliste, D., Caracas, R., Côté, M., Deutsch, T., Genovese, L., Ghosez, P., Giantomassi, M., Goedecker, S., Hamann, D., Hermet, P., Jollet, F., Jomard, G., Leroux, S., Mancini, M., Mazevet, S., Oliveira, M., Onida, G., Pouillon, Y., Rangel, T., Rignanese, G.M., Sangalli, D., Shaltaf, R., Torrent, M., Verstraete, M., Zerah, G., Zwanziger, J.: ABINIT: first-principles approach to material and nanosystem properties. Comp. Phys. Commun. **180**(12), 2582 (2009)
11. Gonze, X., Rignanese, G., Verstraete, M., Beuken, J., Pouillon, Y., Caracas, R., Jollet, F., Torrent, M., Zerah, G., Mikami, M., Ghosez, P., Veithen, M., Raty, J., Olevano, V., Bruneval, F., Reining, L., Godby, R., Onida, G., Hamann, D.R., Allan, D.C.: A brief introduction to the ABINIT software package. Zeitschrift fur Kristallographie **220**, 558 (2005)

12. Keith, T.A.: The AIMALL program (2015). The code is http://aim.tkgristmill.com
13. Keith, T.A., Laidig, K.E., Krug, P., Cheeseman, J.R., Bone, R.G.A., Biegler-König, F.W., Duke, J.A., Tang, T., Bader, R.F.W.: The aimpac95 programs (1995)
14. Kresse, G., Furthmüller, J.: Efficient iterative schemes for ab initio total-energy calculations using a plane-wave basis set. Phys. Rev. B **54**(16), 11,169 (1996)
15. Lebedev, V.I., Laikov, D.N.: A quadrature formula for the sphere of the 131st algebraic order of accuracy. Dokl. Math. **59**, 477 (1999)
16. Lu, T., Chen, F.: Multiwfn: a multifunctional wavefunction analyzer. J. Comput. Chem. **33**(5), 580 (2011)
17. Martín Pendás, A., Francisco, E.: Promolden: a QTAIM/IQA code (Available from the authors upon request by writing to ampendas@uniovi.es)
18. Popelier, P.: Integration of atoms in molecules: a critical examination. Mol. Phys. **87**(5), 1169 (1996)
19. Press, W.H., Flannery, B.P., Teukolsky, S.A., Vetterling, W.T.: Numerical recipes in FORTRAN: the art of scientific computing. Cambridge University Press, Cambridge England New York, NY, USA (1992)
20. Rafat, M., Popelier, P.L.A.: Topological atom–atom partitioning of molecular exchange energy and its multipolar convergence. In: The Quantum Theory of Atoms in Molecules, p. 121. Wiley-VCH Verlag GmbH & Co. KGaA, Weinheim, Germany (2007)
21. de-la Roza, A.O., Johnson, E.R., Luaña, V.: Critic2: a program for real-space analysis of quantum chemical interactions in solids. Comput. Phys. Commun. **185**(3), 1007 (2014)
22. de-la Roza, A.O., Luaña, V.: A fast and accurate algorithm for QTAIM integration in solids. J. Comput. Chem. **32**(2), 291 (2010)
23. Saleh, G., Gatti, C., Presti, L.L., Contreras-Garcia, J.: Revealing non-covalent interactions in molecular crystals through their experimental electron densities. Chem. Eur. J. **18**, 15,523 (2012)
24. Schwarz, K., Blaha, P., Madsen, G.: Electronic structure calculations of solids using the WIEN2k package for material sciences. Comp. Phys. Commun. **147**(1–2), 71 (2002)
25. Stash, A., Tsirelson, V.: WinXPRO: a program for calculating crystal and molecular properties using multipole parameters of the electron density. J. Appl. Crystallogr. **35**(3), 371 (2002)
26. Volkov, A., Gatti, C., Abramov, Y., Coppens, P.: Evaluation of net atomic charges and atomic and molecular electrostatic moments through topological analysis of the experimental charge density. Acta Crystallogr. A **56**(3), 252 (2000)
27. Yu, M., Trinkle, D.R.: Accurate and efficient algorithm for bader charge integration. J. Chem. Phys. **134**(6), 064,111 (2011)

# Chapter 9
# Exercises

**Abstract** This chapter contains a set of proposed exercises spanning the main topics that have been developed in the book.

## 9.1 Topology

### *9.1.1 Exercises*

Take the scalar field defined in $\mathbb{R}^2$ by $V(x, y) = x^2(x - 1)^2 + y^2$.

1. Obtain the position of the critical points.
2. Construct the Hessian to characterize their nature.
3. Find the repellors, the $\alpha$ limit sets and the separatrix.

## 9.2 Before Starting: Obtaining the Electronic Structure Files

### *9.2.1 Molecular Calculations: The* `wfn` *File*

Most topological packages read the wavefunction of a molecular system in a format known as `.wfn`. Besides this, AIMAll reads also an XML-like format `.wfx`, which allows for the use of pseudopotentials. These files are produced by several electronic structure packages like Gaussian (use the keyword `output=wfn` or `output=wfx`), GAMESS (use `aimpac=.true.`) and ORCA (use the `!AIM` directive). Other codes may write MOLDEN [4] files which may then be transformed to `.wfn` by the MOLDEN2AIM [5] utility.

Let's see an example for Gaussian. It suffices to set `output=wfn` in the Route section and give the name of the wfn file at the end of the molecule specification:



Á. Martín Pendás and J. Contreras-García, *Topological Approaches to the Chemical Bond*, Theoretical Chemistry and Computational Modelling,
https://doi.org/10.1007/978-3-031-13666-5_9

```
# P HF/6-311++G(2df,2p) opt output=wfn

tetrahedrane

0 1
C -0.524831 0.524831 0.524831
C 0.524831 -0.524831 0.524831
C -0.524831 -0.524831 -0.524831
C 0.524831 0.524831 -0.524831
H -1.134794 1.134794 1.134794
H 1.134794 -1.134794 1.134794
H -1.134794 -1.134794 -1.134794
H 1.134794 1.134794 -1.134794

tetrahedrane.wfn
```

The first word in the title (contiguous non-blank characters) is used to form the file names in several of the programs we will be illustrating. Hence, it might be dangerous to use non-alphanumerical characters. Notice that there should be a blank line after that containing the name of the `.wfn` file.

This is valid for a monodeterminantal wavefunction (Hartree–Fock or DFT). Remember that if you are using post-Hartree–Fock methods, as well as NaturalOrbitals or NaturalSpinOrbitals, you need to specify that this is the density you want to use (`density=current`):

```
#CISD/6-31G** popt pop=no density=rhoci output=wfn

H2O

0 1
H
O 1 r
H 2 r 1 a
r=1.0
a=110.0

h2o-no.wfn
```

### 9.2.2 Solid-State Files

Density and ELF grids can be calculated in solids with most packages nowadays. We will show here two options, VASP and Quantum Espresso (QE).

#### 9.2.2.1 With VASP

It suffices to set the options in the INCAR

```
CHGCAR=.True.
ELFCAR=.True.
```

ELFCAR (ELF) and CHGCAR (density) are 3D grids that can be read directly by many visualization programs.

#### 9.2.2.2 With Quantum Espresso

For QE, you have to ask to produce cube files for the charge and ELF. For example,

```
&INPUTPP
   prefix       = 'C'
   outdir       = './'
   plot_num     = 8
   filplot      = 'C.ELF3D.dat'
/

&PLOT
   iflag           = 3
   output_format = 6
   fileout         = 'C.ELF3D.cube'

   nx = 500
   ny = 500
   nz = 500
 /
```

The keywords meaning is to be interpreted as follows:

| Input | Meaning |
|---|---|
| filplot | file `filplot` contains the quantity selected |
| plot_num = 8 | quantity selected = ELF |
| iflag = 3 | We will produce a 3D plot |
| output_format = 6 | The format of the file will be that of a Gaussian cube file (3D) |
| fileout | name of the file to which the plot is written |
| nx, ny, nz | number of points in the parallelepiped |

## 9.3 AIMAll

We will perform QTAIM analysis with AIMAll (www.tkgristmill.com) for a set of molecules. AIMAll is an efficient, fast, cross-platform code, mostly automatic which enables the following:

1. Simultaneous calculation and visualization of QTAIM results.
2. Parallel implementation for multi-core machines under a shared memory model.
3. Large number of properties computed.

It has a detailed HTML manual with many worked-out examples. It should be noted that the code is proprietary with a non-commercial prize ranging from 300 to 2500 USD (2021). Nonetheless, a free limited version for 12 atoms and 400 basis functions is available. In order to download it, you can register at http://aim.tkgristmill.com/register.html and download the Win, Mac or Linux installer.

### *9.3.1 Basic Instructions*

AIMAll is composed of the following:

1. `AIMQB`: This is the basic AIMAll program. Within a Windows environment, we can execute AIMQB through the AIMStudio dialogs. It is also possible to drag-and-drop a wavefunction file (`.wfn`, `.wfx` or `fchk`) directly over the AIMStudio window.
   Alternatively, it is possible to run it through the command line: `aimqb [options] [wfnfile | wfxfile | fchkfile]`
2. `AIMEXT`: This code is used for the critical point analysis and the construction of the grid files.
   It can be invoked through the command line as follows:
   `aimext [-progress] [-wsp] [wfnfile | wfxfile] [-input...]`
3. `AIMINT`: This is the code used for the integration of scalar densities over atomic basins.
   The command line is as follows:
   `aimint [-nproc=...] [-wstat] [inpfile wfnfile | wfxfile]`
4. `AIMSUM`: This code writes a summary of the results.
5. `AIMSTUDIO`: This is the graphical interface to the above codes.
6. `AIMUTIL`: This code is used by the other programs of the suite but not directly by the user.

In order to get acquainted with the graphical interface, we will solve an exercise for the cyclopropanone molecule, computed at the HF//6-311G** level. Remember that you must start by constructing a `.wfn` file. The optimized geometries (in a.u.) are given in Table 9.1.

**Table 9.1** Cartesian coordinates (in a.u.) of the atoms in the HF//6-311G** optimized cyclopropanone

| Ethane | | | |
|---|---|---|---|
| C1 | 0.00000000 | 1.47509449 | −1.60614823 |
| C2 | −0.00000000 | −1.47509449 | −2.16578522 |
| H5 | 1.71662262 | −2.41352344 | −2.16578522 |
| H6 | −1.71662262 | −2.41352344 | −2.16578522 |
| C7 | 0.00000000 | 0.00000000 | 0.72911717 |
| O8 | 0.00000000 | 0.00000000 | 2.94527708 |

### 9.3.2 *Running AIMQB*

On opening `AIMStudio`, select the `AIMQB` in the `Run` dialog and open the cyclopropanone `.wfn` file, as shown in Fig. 9.1. There you can further select the number of processors to be used as well as many options regarding the size of integration grids and integration methods, as well as the type of properties to compute or skip. Usually, it is reasonable to leave the default values. However, if IQA quantities are needed, make sure that in the properties dialog you select all the energy components available. This will increase substantially the CPU time. Once you have selected what to compute, type OK.

When the calculation finishes, a set of files will have been written on the same folder where the `.wfn` file was located. Particularly important are the `.sum` and `.sumviz` files, which can be edited manually or appropriately filtered. They contain all the QTAIM descriptors. A new subfolder, called `cylopropanone_atomicfiles`, will also have been created. Several files, one per atom, will have been written in that subfolder. They contain information about integrations, critical points of the Laplacian or interatomic surfaces. These are needed to properly visualize the results in the AIMStudio suite (Fig. 9.2).

### 9.3.3 *Visualizing Results*

Now return to `AIMStudio` and, within the `File` dialog, select `Open in new window` and click on the cyclopropanone `.sumviz` file. The molecule will be displayed, and a set of default properties shown.

BCPs are shown in green, and RCPs in red. By navigating over the secondary dialog (Atoms, BCPs, RCPs, CCPs, NNACPs, Bond Paths, Ring Paths, IAS EV Paths, IASs, Basin Paths, etc.), you can select a wide variety of properties to be shown at CPs and also build relief maps and contours, show interatomic surfaces if previously constructed, etc. In the present case, for instance, the QTAIM atomic charge of the O atom is −1.271 a.u., and the density at the C1–C2 BCP is 0.2212 a.u.

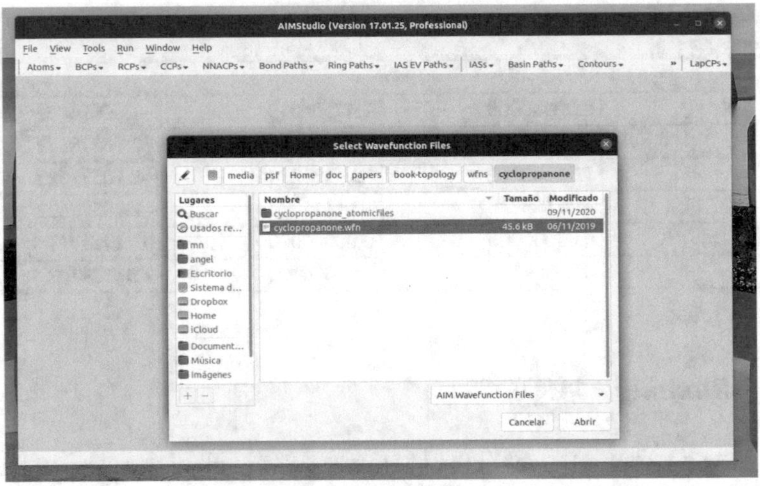


**Fig. 9.1** How to select a `.wfn` file in AIMAll to run `AIMQB`

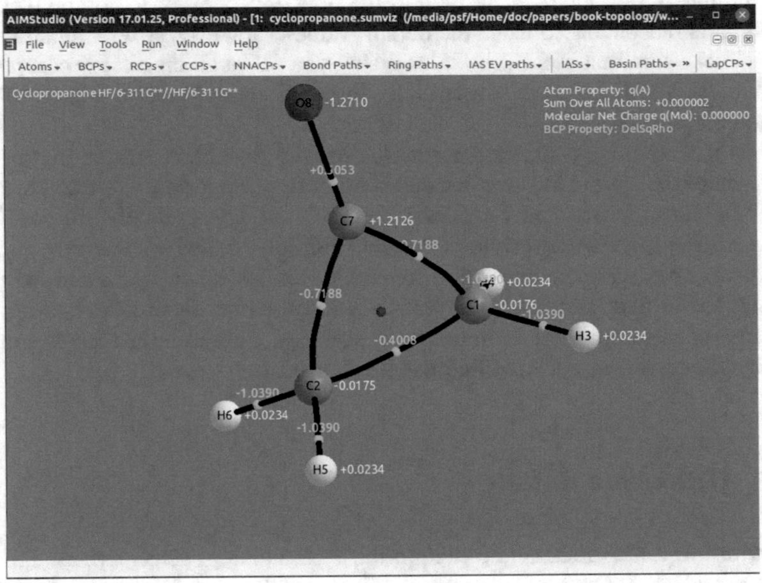


**Fig. 9.2** Densities at BCPs and QTAIM atomic charges as displayed for cyclopropanone in `AIMStudio`

**Table 9.2** A Table of BCP properties directly built from the `Table` subdialog of the BCP dialog in cyclopropanone. All data in a.u

| BCP | Atoms | Rho | DelSqRho | Ellipticity | BPL | BPL-GBL_I |
|---|---|---|---|---|---|---|
| BCP3 | C1–H3 | 0.286791 | −1.039000 | 0.027832 | 2.004117 | 0.000261 |
| BCP4 | C1–H4 | 0.286791 | −1.039000 | 0.027832 | 2.004117 | 0.000261 |
| BCP6 | C2–H5 | 0.286791 | −1.039000 | 0.027832 | 2.004117 | 0.000261 |
| BCP7 | C2–H6 | 0.286791 | −1.039000 | 0.027832 | 2.004117 | 0.000261 |
| BCP8 | C7–O8 | 0.439993 | +0.505345 | 0.033635 | 2.216152 | 0.000000 |
| BCP2 | C1–C7 | 0.275692 | −0.718781 | 0.239387 | 2.792340 | 0.030214 |
| BCP5 | C2–C7 | 0.275692 | −0.718781 | 0.239387 | 2.792340 | 0.030214 |
| BCP1 | C1–C2 | 0.221224 | −0.400827 | 0.577977 | 2.962151 | 0.011970 |

Notice how the C–C bond paths are not linear but curved, an indication of bond strain in the ring (see Sect. 3.2).

The length of a bond path can be determined on the BCP dialog under the subdialog `Properties-> Length`. There you can compare the length of the bond paths (BPL) with the Geometric distance between the nuclei `(GBL_I)`. A convenient way to examine the data is to select the `Table` subdialog. In this way, Table 9.2 is produced.

2D contours and isosurfaces are easily built with `AIMStudio`. To superimpose a 2D contour map of the Laplacian of the density in a given plane, for instance:

1. Select the `Contours` Dialog, and then `New 2D grid` (Fig. 9.3-left).
2. A dialog will pop up. Select the function you want to map. In our case, the Laplacian.

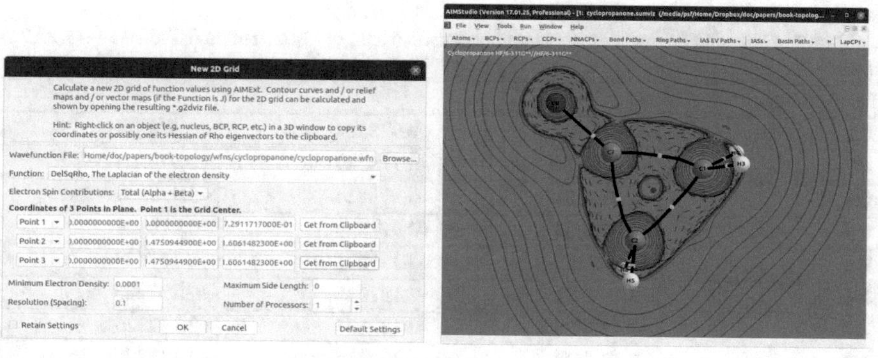


**Fig. 9.3** Dialog to build a contour 2D grid file (left). Laplacian onto the ring plane superimposed on the molecular structure. Positive isolines are rendered in solid blue, while negative ones in dashed red

3. Now choose a plane in which to compute the contours. The easiest way is to select three atoms that will define the plane. To do that, right-click on an atomic position, copy its coordinates to the Clipboard and then click on the appropriate `Get from Clipboard` button.
4. Now choose other options and click on *OK*.
5. A file called `cyclopropanone_delsqrho.g2dviz` will have been created.
6. Open this file in the current window. You should obtain a figure similar to Fig. 9.3-right.

## 9.3.4 *Exercises*

### 9.3.4.1 Tetrahedrane

1. Build the `.wfn` file for the tetrahedrane molecule. Its optimized coordinates can be found in Table 9.3. (in a.u.)
2. Locate the cage critical point. Order the CPs by increasing density values and discuss the rationale behind the numbers.

| $CP$ | $\rho$ | $\nabla^2\rho$ |
|---|---|---|
| | | |

3. Do a similar table for the BCPs containing the difference between bond path lengths and geometric lengths ($\Delta d$). Relate the behavior to bond strain.

**Table 9.3** Cartesian geometries of the tetrahedrane, in a.u., optimized at the HF/6-311++G(2d,2p) level

| Tetrahedrane | | | |
|---|---|---|---|
| C1 | +0.978824 | +0.978824 | +0.978824 |
| C2 | –0.978824 | –0.978824 | +0.978824 |
| C4 | –0.978824 | +0.978824 | –0.978824 |
| C3 | +0.978824 | –0.978824 | –0.978824 |
| H5 | +2.137225 | +2.137225 | +2.137225 |
| H6 | –2.137225 | –2.137225 | +2.137225 |
| H7 | +2.137225 | –2.137225 | –2.137225 |
| H8 | –2.137225 | +2.137225 | –2.137225 |

**Table 9.4** Cartesian geometries of the $H_2O$ molecule, in a.u., optimized at the HF/6-311++G(2d,2p) level

| Water | | | |
|---|---|---|---|
| O1 | 0.00000000 | 0.00000000 | 0.213011788 |
| H1 | 0.00000000 | 1.42132487 | –0.852047151 |
| H1 | 0.00000000 | 1.42132487 | +0.852047151 |

| $CP$ | $\rho$ | $\nabla^2\rho$ | $\Delta d$ |
|---|---|---|---|
| | | | |

4. Compare the densities at the BCP for the C–H bonds (0.2868 a.u. in cyclopropanone, 0.2910 a.u. in cubane). Do the same with the C–C bonds. Why are there two sets of C–C densities in cyclopropanone (0.2757 a.u., 0.2212 a.u.) and only one in cubane (0.2522 a.u.)?

#### 9.3.4.2 $H_2O$

1. Open and run `AIMQB` in the water molecule example, whose cartesian coordinates are shown in Table 9.4.
2. Tick the `Write Interatomic Surfaces` option in the `AIMQB properties` window. Go to `File: Open in current window`, then move to `atomicfiles` folder and select the `.iasviz` files sequentially. Play with the thickness of the paths and with the presence/absence of strips. Change color. Draw contours, vector maps and Relief surfaces: Select `New 2D grid` → `Rho`. Define the plane (to use the molecular plane, get the coordinates of the 3 atoms from the clipboard, click on one atom, right mouse-click and copy, paste). Load in current window the new `.g2dviz` file; select `options`. Now add a relief map (load the `.g2dviz`, select `Relief map` and load in the same window). Superimpose a map of the Laplacian (add a new 2D grid and load it in the same window). Draw isosurfaces: Make a 3D grid in the surface dialog. Select the scalar field `rho`. Load the `.3dgviz` file. Select the Isosurface. Map another field on the surface (the `ESP`). Locate nucleophilic and electrophilic regions.
3. Examine the CPs of the Laplacian in the $H_2O$ molecule: run `AIMQB` with the option to compute the critical points of $\nabla^2\rho$ turned on. Then discover the four types of CPs of the Laplacian. Use the `Table` feature to locate the lone pairs. Measure their distance to the O atom. Measure dihedral angles (select `CPs + Tools + Selection Tables` ) between pairs of CPs. Compare bonding and non-bonding charge concentration regions, and test VSEPR ideas.

4. Examine atomic properties in the water molecule: open the `Atoms` dialog. Run over Properties: QTAIM charges, integrated Laplacians L(A) and integrated Kinetic Energy. Compare the charge of the H atoms with Mulliken's charge ($Q(\mathrm{O}) = -0.4954$ a.u.). Explain the difference. Show the atomic dipole moments. Which atom dominates the total molecular dipole? Examine the localization and delocalization indices: get the `LIs` and `DIs`. Obtain the partition of $N_A$ into localization and delocalization contributions. Why is the localization index smaller than the atomic population? What does it mean that the O–H delocalization index is considerably smaller than one, although we have a $\sigma$ bond?

#### 9.3.4.3 Density Correlations

1. Open and run `AIMQB` on ethane, ethylene, acetylene and benzene. The geometry section of the `.wfn` files (optimized at the HF/6-311G* level) is as follows:

Ethane

```
C  1  (CENTRE 1)   0.00000000   0.00000000  -1.44224916  CHARGE =  6.0
C  2  (CENTRE 2)   0.00000000   0.00000000   1.44224916  CHARGE =  6.0
H  3  (CENTRE 3)  -1.65547465  -0.95578874   2.18212032  CHARGE =  1.0
H  4  (CENTRE 4)   1.65547465  -0.95578874   2.18212032  CHARGE =  1.0
H  5  (CENTRE 5)   0.00000000  -1.91157747  -2.18212032  CHARGE =  1.0
H  6  (CENTRE 6)   1.65547465   0.95578874  -2.18212032  CHARGE =  1.0
H  7  (CENTRE 7)  -1.65547465   0.95578874  -2.18212032  CHARGE =  1.0
H  8  (CENTRE 8)   0.00000000   1.91157747   2.18212032  CHARGE =  1.0
```

Ethylene

```
C  1  (CENTRE 1)  -1.24368052   0.00000000   0.00000000  CHARGE =  6.0
C  2  (CENTRE 2)   1.24368052   0.00000000   0.00000000  CHARGE =  6.0
H  3  (CENTRE 3)  -2.31252591  -1.73138844   0.00000000  CHARGE =  1.0
H  4  (CENTRE 4)   2.31252591  -1.73138844   0.00000000  CHARGE =  1.0
H  5  (CENTRE 5)  -2.31252591   1.73138844   0.00000000  CHARGE =  1.0
H  6  (CENTRE 6)   2.31252591   1.73138844   0.00000000  CHARGE =  1.0
```

Acetylene

```
C  1  (CENTRE 1)   0.00000000   0.00000000  -1.12396774  CHARGE =  6.0
C  2  (CENTRE 2)   0.00000000   0.00000000   1.12396774  CHARGE =  6.0
H  3  (CENTRE 3)   0.00000000   0.00000000  -3.10719699  CHARGE =  1.0
H  4  (CENTRE 4)   0.00000000   0.00000000   3.10719699  CHARGE =  1.0
```

Benzene

```
C  1  (CENTRE 1)   1.31175800   2.27203150   0.00000000  CHARGE =  6.0
C  2  (CENTRE 2)  -1.31175800   2.27203150   0.00000000  CHARGE =  6.0
C  3  (CENTRE 3)  -2.62351599   0.00000000   0.00000000  CHARGE =  6.0
C  4  (CENTRE 4)  -1.31175800  -2.27203150   0.00000000  CHARGE =  6.0
C  5  (CENTRE 5)   1.31175800  -2.27203150   0.00000000  CHARGE =  6.0
```

```
C    6   (CENTRE  6)   2.62351599  0.00000000  0.00000000  CHARGE =  6.0
H    7   (CENTRE  7)   2.32586849  4.02852239  0.00000000  CHARGE =  1.0
H    8   (CENTRE  8)  -2.32586849  4.02852239  0.00000000  CHARGE =  1.0
H    9   (CENTRE  9)  -4.65173697  0.00000000  0.00000000  CHARGE =  1.0
H   10   (CENTRE 10)  -2.32586849 -4.02852239  0.00000000  CHARGE =  1.0
H   11   (CENTRE 11)   2.32586849 -4.02852239  0.00000000  CHARGE =  1.0
H   12   (CENTRE 12)   4.65173697  0.00000000  0.00000000  CHARGE =  1.0
```

2. Open also one spreadsheet and annotate the values of the densities at the C–C BCPs.

3. Accepting that the standard bond orders of the C–C bond are 1, 2 and 3 for ethane, ethylene and acetylene, fit the densities to a straight line and compute the expected bond order for benzene from its density. Compare with the results obtained from the delocalization indices.

#### 9.3.4.4 IQA in LiH

1. Perform IQA on the LiH molecule, optimized at the M06-2X level with a cc-pVDZ basis set. Its equilibrium distance is 3.03081298 a.u. Run `AIMQB` with `IQA` enabled.

2. What is the topological charge of Li? Comment on its difference with respect to that of Mulliken's analysis (0.1 a.u.). Compute the self-energy of each atom and compare it to that of the isolated atoms. Why is the energy of the Li atom so different from the reference?

3. Compute the total interaction energy and decompose it into classical and covalent. Compute also the localization indices and the delocalization index. What does this tell us about the LiH bond?

#### 9.3.4.5 IQA in the Water Dimer

1. Perform an IQA analysis on the water dimer, optimized at the HF/6-311+G(d,p) level, considering each water molecule as a quantum fragment. The `.wfn` file geometry header is

```
                          Water dimer
-----------------------------------------------------------------------
O    1   (CENTRE  1)  -0.26401035  0.35698104  0.00000000  CHARGE =  8.0
H    2   (CENTRE  2)   1.52285627  0.28841270  0.00000000  CHARGE =  1.0
H    3   (CENTRE  3)  -0.71944864  2.07662573  0.00000000  CHARGE =  1.0
O    4   (CENTRE  4)   5.37764284 -0.02525261  0.00000000  CHARGE =  8.0
H    5   (CENTRE  5)   6.11688900 -0.77863105 -1.43524223  CHARGE =  1.0
H    6   (CENTRE  6)   6.11688900 -0.77863105  1.43524223  CHARGE =  1.0
```

2. Open `h2oh.sumviz` and build the fragments' self-energies from the IQA components. Open `h2oh2oh.sumviz` and get the deformation energy of the proton donor (PD) and proton acceptor (PA) molecules. Get the PD-PA charge transfer. Study the changes in interactions: covalent and classical, and relate them to the polarization induced by the dipolar field.
3. Compute the covalent and ionic contributions of the hydrogen bond. Why is the O–O delocalization large?

## 9.4 TopMod

### *9.4.1 Basic Instructions*

We have seen that `TopMod` is a suite of programs which enable to obtain several ELF utilities:

1. `grid09`: construct ELF on a grid.
2. `bas09`: search of ELF maxima.
3. `pop09`: integrals (e.g. volume and charge).
4. `sbf_to_cube`: 3D visualization.

If run interactively, the program prompts for each piece of input needed. The coming sections specify the required input.

A detailed running example will be given for the ethane molecule whose coordinates are given in Table 9.5. Remember you should start by constructing the `.wfn` file (see Sect. 9.2.1). At the end of the section, more exercises will be presented to be solved by the student.

**Table 9.5** Cartesian coordinates (in a.u.) of the ethane molecule

| Ethane | | | |
|---|---|---|---|
| C | 0.195959235 | 0.019217850 | 0.050821215 |
| C | –0.761484257 | –0.859999137 | –0.760789410 |
| H | –0.666128834 | –0.666832301 | –1.834968938 |
| H | 0.100871815 | –0.173749767 | 1.125221663 |
| H | 1.239474582 | –0.165905540 | –0.226657702 |
| H | –0.005020453 | 1.083598946 | –0.112558319 |
| H | –1.804979024 | –0.674723689 | –0.483217157 |
| H | –0.560188237 | –1.924437623 | –0.597379576 |

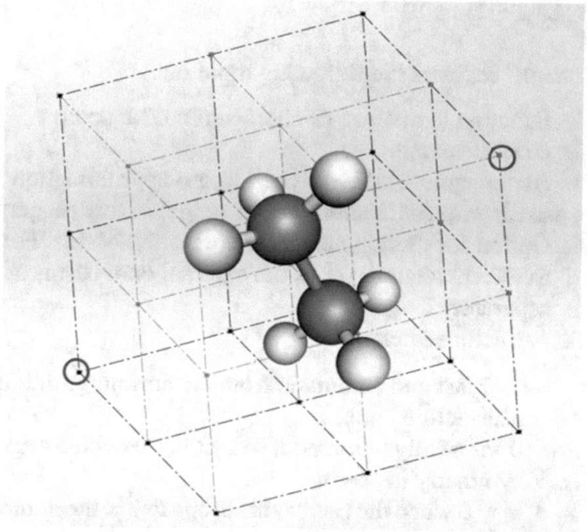

**Fig. 9.4** Example of 2×2×2 cube around the ethane molecule. The opposite corners used to define it in the input have been highlighted in red

#### 9.4.1.1 `grid09`

1. `wfn` file name (maximum 40 characters).
2. Origin as x,y,z coordinates (3 numbers): coordinates of the origin of the box defining the grid. For the whole molecule, it suffices to just copy–paste the edges suggested by the code upon running it.
3. Edge as x,y,z coordinates (3 numbers): length of the edges of the box in Bohr. For the whole molecule, it suffices to just copy–paste the edges suggested by the code upon running it. Figure 9.4 illustrates the position of the origin and edge coordinates for the ethane molecule with red circles.
4. Number of intervals along x,y,z (3 numbers) along each direction to determine the grid (in Fig. 9.4, it is 2x2x2). Usually, 100x100x100 is a good number for small molecules.

The final input file should look as follows:

```
ethane.wfn
-8.410916   -8.636660   -8.467588
15.753183   15.684365   15.593949
100 100 100
```

Upon running `grid09`, the files `title_elf.sbf`, `title_rho.sbf` and `title_lap.sbf` files will be generated, in which `title` is the first non-blank characters of the title in the `wfn` file.

#### 9.4.1.2 bas09

bas09 requires the following input lines:

1. Function to analyze (3 characters): elf or rho.
2. wfn file name.
3. Accuracy level. If we type a 0, no approximation in the derivative evaluation is used, whereas 1 and 2 use of progressively bigger cutoffs.
4. Option for merging external core–shell basins. For y, the full core is considered together, whereas for n, the external core basins as in in Fig. 4.11 are considered separately.
5. Attractor search mode:

   - $< 0$ fast and automatic from the current grid. It only finds attractors within the user-defined box.
   - 0 for a fully automatic, search in the whole molecular space (recommended if symmetry is used).
   - $n > 0$ when the user wants to specify $n$, the number of expected non-protonated valence basins.

6. Assign grid points: y assigns the grid points to basins.

The program writes one output file: title_ebas.sbf or title_ras.sbf according to the chosen function to be used by pop09 and one input file for bas09 named temp.bas.

The temp.bas file corresponds to the input attractor mode, and it can be generated with a different wavefunction (for example with a smaller basis set) which nevertheless yields the same overall topology and without grid point assignment. In this case, it is not necessary to run grid09 before. This procedure is recommended for large systems. Do not forget to update the wfn file name in temp.bas if another wfn file is used.

Input example:

```
elf
ethane.wfn
1
y
0
y
```

The output looks like this:

```
1   0.370    0.036    0.096   1.000  C(C1)           209    209    0.81  5
2  -1.439   -1.625   -1.438   1.000  C(C2)           209    209    0.81  5
3  -1.252   -1.247   -3.527   1.000  V(H1,C2)      47923  22663   87.32  2
4   0.183   -0.341    2.185   1.000  V(H2,C1)      47934  22669   87.34  2
5   2.399   -0.326   -0.446   1.000  V(H3,C1)      47949  22704   87.48  2
6  -0.023    2.106   -0.224   1.000  V(H4,C1)      47882  22609   87.11  2
7  -3.468   -1.263   -0.896   1.000  V(H5,C2)      47948  22708   87.49  2
8  -1.045   -3.695   -1.118   1.000  V(H6,C2)      47884  22608   87.11  2
9  -0.534   -0.794   -0.671   0.961  V(C1,C2)       4635   4580   17.65  3
```

You can see that the output is using the common ELF notation (Sect. 4.3.3.1): C(C) for the carbon cores, and V(C–C) and V(C,H) for the C–C and C–H bonds, respectively.

Geometrical parameters are also printed out as follows:

```
 distances from nuclei (A)

V(H1,C2)                C2
                       1.128
V(H2,C1)                C1
                       1.128
V(H3,C1)                C1
                       1.128
V(H4,C1)                C1
                       1.128
V(H5,C2)                C2
                       1.128
V(H6,C2)                C2
                       1.128
V(C1,C2)                C1        C2
                       0.766     0.766

 angles around core attractors

V(H2,C1)          C(C1)          V(H3,C1)          107.608
V(H2,C1)          C(C1)          V(H4,C1)          107.618
V(H2,C1)          C(C1)          V(C1,C2)          111.276
V(H3,C1)          C(C1)          V(H4,C1)          107.604
V(H3,C1)          C(C1)          V(C1,C2)          111.278
V(H4,C1)          C(C1)          V(C1,C2)          111.267
V(H1,C2)          C(C2)          V(H5,C2)          107.628
V(H1,C2)          C(C2)          V(H6,C2)          107.631
V(H1,C2)          C(C2)          V(C1,C2)          111.257
V(H5,C2)          C(C2)          V(H6,C2)          107.629
V(H5,C2)          C(C2)          V(C1,C2)          111.261
V(H6,C2)          C(C2)          V(C1,C2)          111.248

 angles around polysynaptic attractors

C1  V(C1,C2)           C2           179.995
```

This allows you to see for example that the bond lies on the internuclear line (the angle C1–V(C1,C2)–C2 is 179.995$\simeq$ 180°) and as expected for a homonuclear molecule, it is in the middle: the distance V(C1, C2)–C1 is the same as the distance V(C1, C2)–C2 (0.766 a.u.).

In the output, you will also obtain the file `title_ebas.sbf` which identifies the attractors. It can be used in the visualization to color the basins. It suffices to transform them with the `bas_to_syn` utility by running

```
name_ebas.sbf
n
```

#### 9.4.1.3 pop09

The input for `pop09` looks as follows:

1. Input the `.wfn` file.
2. Integration threshold (recommended values are 7–9).
3. Number of the ELF and the AIM basins to integrate (2 numbers—one for ELF and one for AIM, respectively). If we input 0, all the basins are taken into account in the calculation.
4. Skip if the number of ELF basins in 3) is 0. Otherwise, label(s) of the considered ELF basins in the order of the `bas09` output.
5. Skip if the number of QTAIM basins in 3) is 0. Otherwise, label(s) of the considered ELF basins in the order of the `bas09` output.

If we want to integrate all ELF basins, the input will thus look as follows:

```
ethane.wfn
9
0 0
```

This will lead to the following population integrals:

```
basin              vol.    pop.    pab     paa     pbb     sigma2  std. dev.


1 C(C1)            0.81    2.08    2.17    0.17    0.17    0.26    0.51
2 C(C2)            0.81    2.08    2.17    0.17    0.17    0.26    0.51
3 V(H1,C2)        87.32    2.00    2.00    0.32    0.32    0.64    0.80
4 V(H2,C1)        87.34    2.00    2.00    0.32    0.32    0.64    0.80
5 V(H3,C1)        87.48    2.00    2.01    0.32    0.32    0.64    0.80
6 V(H4,C1)        87.11    2.00    2.00    0.32    0.32    0.64    0.80
7 V(H5,C2)        87.49    2.00    2.01    0.32    0.32    0.64    0.80
8 V(H6,C2)        87.11    2.00    2.00    0.32    0.32    0.64    0.80
9 V(C1,C2)        17.65    1.81    1.64    0.40    0.40    0.97    0.98



sum of populations    17.989337
```

The columns `vol` and `pop` are where you find the volumes and populations, respectively. They are represented in Fig. 9.5. We have the following:

- 2.08 electrons for the carbon cores, C(C).
- 2.0 electrons for the C–H bond, V(C,H).
- 1.81 electrons for the C–C bond, V(C,C).

You can verify whether the grid is good enough by checking the total number of electrons found upon integration. In our case, see that the `sum of populations` is 17.99 electrons. The number of electrons expected is 18, so for our purposes, the grid is good enough. Nonetheless, notice that the quality of QTAIM integrations in ToPMoD is considerably lower than those in other codes.

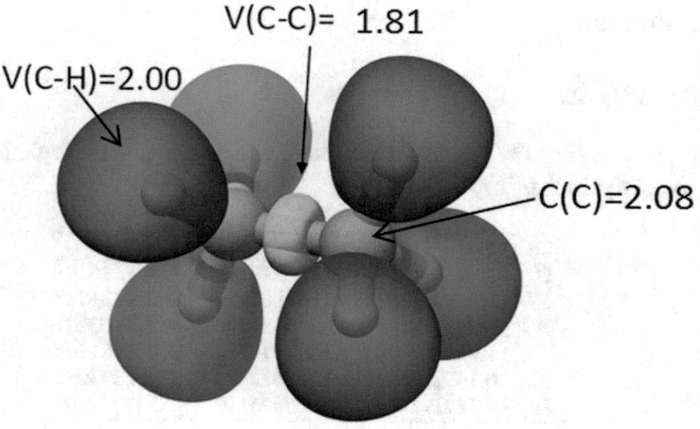


**Fig. 9.5** Representation of the ELF basins: C–H in blue, C–C in green and their respective populations

#### 9.4.1.4 sbf_to_cube

The input expected for `sbf_to_cube` is as follows:

1. Input `.wfn` file name.
2. Input `.sbf` file to be converted.
3. Step size in each direction for the transformation. If we are just keeping the original size, it suffices to introduce `1 1 1`.

This leads in our example to

```
ethane.wfn
ethane_elf.sbf
1 1 1
```

With this input, we obtain a `.cube` file that we can visualize with most visualization packages, yielding isosurfaces as the ones in Fig. 9.5. You can repeat the same procedure to obtain the `.cube` file to color the basins from the `esyn.sbf` file.

## 9.4.2 *Exercises*

### 9.4.2.1 Bond Order

1. Obtain the wfn files for ethane (coordinates used above) and ethylene (coordinates are given in Å):

| Ethylene | | | |
|---|---|---|---|
| C | 0.191801072 | 0.150379199 | −0.447336288 |
| C | −0.814379158 | −0.636861454 | −0.832221523 |
| H | −0.768186149 | −1.214524499 | −1.751803007 |
| H | 1.099507078 | 0.247678083 | −1.036781561 |
| H | 0.145193357 | 0.727672678 | 0.472343052 |
| H | −1.722191344 | −0.733995118 | −0.242469968 |

2. Check that the integration of electrons (total number of electrons) in the output from `pop09` corresponds to the total number of electrons in the system. This should be done each time to check the accuracy of the integration.

3. Draw the Lewis structure of both compounds and identify all the pairs predicted by Lewis theory. Add also core electrons, since they will also appear in the ELF results.

| Compound | Expected Lewis pop. | | | | |
|---|---|---|---|---|---|
| | **TOTAL** | Core C | Core H | C–C Bond | C–H Bond |
| Ethane | | | | | |
| Ethylene | | | | | |

4. Look at the output from `bas09`:
   (i) Identify the attractors with your Lewis picture and explain the ELF notation for each of them.
   (ii) Look at the bonding attractors: How many are there in each case? Why? Where and why are they located? (check xyz-position and angles)

| Compound | ELF Nomenclature | | | |
|---|---|---|---|---|
| | Core C | Core H | C–C Bond | C–H Bond |
| Ethane | | | | |
| Ethylene | | | | |

   (iii) Look at the H–C–H and H–C–C angles and check how they deform from the ideal value according (or not) to VSEPR.

5. Open the `.cube` files. How do both bonds look like? How many maxima are there in ethylene? Coloring the basins thanks to the `title_esyn.cube` file can help you identify the bonds.

6. Look at the output from `pop09`:
   (i) What is the charge for the single and for the double bond according to the output? Does it agree with what you expected? Tip: sum up the various bonding basins for the same bond.
   (ii) Look at the volumes. Do they agree with VSEPR? Excluding hydrogenated bonds (as we have seen, hydrogen having only one electron, the ELF volumes are non-meaningful), comment on cores, non-hydrogenated bonds and lone pairs.

| Integrals | TOTAL | C(C) | | V(C,H) | | ∑V(C,C) | |
|---|---|---|---|---|---|---|---|
| | pop | vol | pop | vol | pop | vol | pop |
| Ethane | | | | | | | |
| Ethylene | | | | | | | |

#### 9.4.2.2 Lone Pairs

1. Obtain the `.wfn` files for ethanol and chloroethane (coordinates in Å):

| Ethanol | | | | Chloroethane | | | |
|---|---|---|---|---|---|---|---|
| C | 0.199529314 | 0.081192401 | –0.086800008 | C | 0.202241146 | 0.041881702 | –0.085155007 |
| C | –0.754224432 | –0.858132285 | –0.805280908 | C | –0.765091796 | –0.870712770 | –0.817915586 |
| H | 0.624531587 | 0.483266302 | 1.797121848 | H | –0.657956575 | –0.716130846 | –1.898526173 |
| H | –0.653082705 | –0.747723791 | –1.890388771 | H | –0.560182522 | –1.922356221 | –0.599214048 |
| H | 1.238214696 | –0.138864143 | –0.376070036 | H | 1.240883014 | –0.173415401 | –0.340771812 |
| H | –0.014214145 | 1.124148291 | –0.365854248 | H | –0.002072126 | 1.095782496 | –0.280066816 |
| H | –1.789975214 | –0.638415584 | –0.528184068 | H | –1.800625356 | –0.654675666 | –0.540899806 |
| H | –0.541160399 | –1.898387725 | –0.540112152 | Cl | 0.081309042 | –0.163204545 | 1.723021036 |
| O | 0.023980175 | –0.104311673 | 1.321707229 | | | | |

2. Check that the integration of electrons (total number of electrons) corresponds to the expected number.

3. Draw the Lewis structure of both compounds and identify all the pairs predicted by Lewis theory.

| Compound | Expected pop. | | | | | |
|---|---|---|---|---|---|---|
| | TOTAL | Core C | Core H | C–C Bond | C–H Bond | Lone Pairs |
| Ethanol | | | | | | |
| Chloroethane | | | | | | |

4. Look at the output from `bas09`:
   (i) Identify the attractors with your Lewis picture and explain the ELF notation for each of them. Find the lone pair attractor(s). What is the ELF notation for them?

(ii) Look at the lone pair attractors: How many are there in each case? Why? (check xyz-position and angles). Compare the distance from the lone pair and the C–O bond to the oxygen. Why are they different?[1]
(iii) Look at the angles around the oxygen and check how they deform from the ideal value according (or not) to VSEPR.

| Compound | ELF Nomenclature | | | | |
|---|---|---|---|---|---|
| | Core C | Core H | C–X Bond | C–H Bond | Lone pairs |
| Ethanol | | | | | |
| Chloroethane | | | | | |

5. Visualize the `.cube` files
   (i) How do the lone pairs look like? Coloring the basins thanks to the `title_esyn.cube` file can help you identify the lone pairs.
   (ii) Check what halogen bonds are. How does this bond relate to what you see in chloroethane?
6. Look at the output from `pop09`:
   (i) Hydrogens were all equivalent in ethane and ethylene, but not in these molecules. Why?
   (ii) Look at the electron population of the C–O and C–Cl bond basin. How is it? Why?
   (iii) Calculate the total charge in the O and Cl lone pairs. How is it with respect to your prediction from the Lewis structure?
   (iv) Look at the volumes of the lone pairs. Are they bigger or smaller than non-hydrogenated bonds? Does this agree with VSEPR?
   (v) Why is the oxygen core smaller than the carbon one?

| Integrals | **TOTAL** | C(C) | | V(C,H) | | V(C,X) | | ∑V(X) | |
|---|---|---|---|---|---|---|---|---|---|
| | pop | vol | pop | vol | pop | vol | pop | vol | pop |
| Ethanol | | | | | | | | | |
| Chloroethane | | | | | | | | | |

[1] You might see that one of the lone pairs of $C_2H_5Cl$ appears as shared. If so, just ignore it in your analysis and take it as a mere lone pair since the distances are very different to C and to Cl (the program uses the gradient to find the connections).

## 9.5 NCIPLOT

We will focus on `NCIPLOT` for the analysis of non covalent interactions in molecules. At the time of writing this book, the latest release of NCIPLOT, NCIPLOT4.2. It is, is publicly available at
https://github.com/juliacontrerasgarcia/nciplot.
There is a manual available in the same repository that covers the details of the current release.

### *9.5.1 Basic Instructions*

The main input of NCIPLOT is formed by two pieces of information:

- Line 1 contains a natural number with the number of molecular files to analyze.
- The following lines contain the name of the files to be analyzed, as many as indicated in line 1.

The extension of the files also determines the type of calculation to be run:

- Promolecular if extension is of the type `.xyz`. This is recommended for big systems.
- Wavefunction if the extension is `.wfn` or `.wfx`.

Note that this also means that all files should be of the same nature.

For example, if we are interested in the non-covalent interactions of one system, the water dimer, the input will look as follows:

```
1
waterdimer.xyz
```

If instead, we have the two molecules in separate files, the input will look as follows:

```
2
water1.xyz
water2.xyz
```

Note that in the second case, the two molecules must be within the same reference system so that they are correctly oriented with respect to one another.

In order to run the program, it suffices to invoke it as follows:

```
nciplot.x < inputfile > outputfile
```

This will lead to four main output files:

- `name.dat` is a data file with two columns, sign($\lambda_2$)$\rho$ and RDG. It allows making the 2D plots.
- Two `.cube` files, `name-grad.cube` and `name-dens.cube`. These are 3D data files for, respectively, plotting and coloring the isosurfaces.

- `name.vmd` is a script for visualizing the `.cube` files with VMD. [1]

One of the interesting advantages of introducing the molecules separately is that atoms are identified as belonging to different molecules, so that intermolecular interactions can be identified from intramolecular ones. This is done by adding the extra keyword `intermolecular`:

```
2
ligand.xyz
protein.xyz
INTERMOLECULAR
```

Additionally, the user can define the cutoffs for $\rho$ and RDG, which will define the boundaries for the search of NCIs. For example, if a rather strong interaction is expected, it is possible to change the default values so that the interaction is not discarded. In this case, the input looks as follows:

```
1
aceticacid1.wfn
aceticacid2.wfn
INTERMOLECULAR
CUTPLOT 0.16 0.5
```

The value 0.16 has been chosen so as to include an NCI peak from the 2D plot which lies below $\rho = 0.16$ a.u. but above the default density cutoff.

This change of default values is also needed when interactions with a hole inside the surface appear, revealing that the default value coincides with a peak, leaving a part of it outside of the study range. An example is shown in Fig. 9.6: the cutoff needs to be increased from $\rho = 0.05$ to for example 0.08 in order to obtain full disks for the hydrogen interactions.

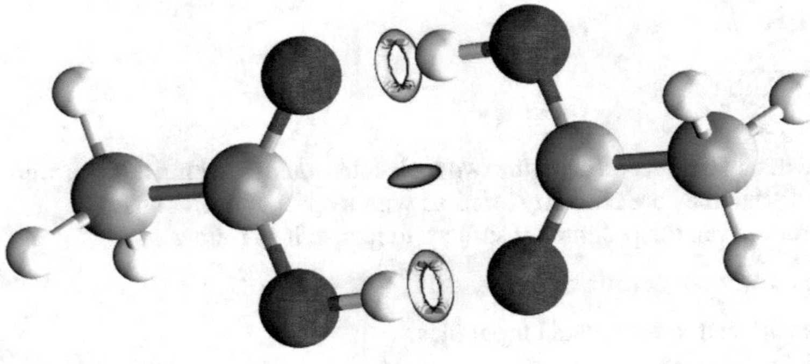

**Fig. 9.6** $s(\mathbf{r}) = 0.5$ isosurfaces in the acetic acid dimer with a $\rho(\mathbf{r}) = 0.05$ cutoff. Reproduced with permission from [3]

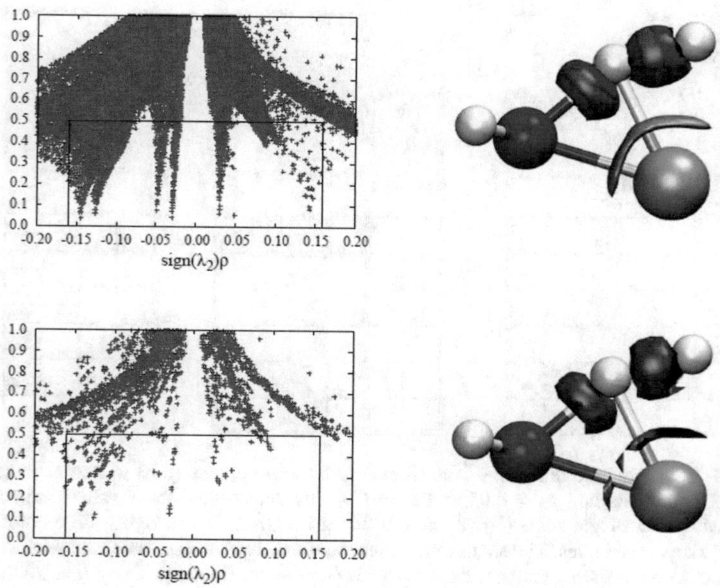


**Fig. 9.7** 2D (left) and 3D (right) NCI plots with default (top) and bigger `INCREMENTS` (bottom)

The default value for the grid step size is 0.1 (Fig. 9.7-top). It is possible to increase this value, which allows for a faster calculation but less accurate (more pixelated) picture (Fig. 9.7-bottom):

```
1
linh4.wfn
CUTPLOT 0.16 0.5
INCREMENTS 0.2 0.2 0.2
```

Alternatively, it is also possible to speed up the calculation with a small accuracy loss using an adaptive grid. A very coarse grid is used in all the space around the molecule in order to identify the regions prone to show non-covalent interactions. Finer user-defined grids are then successively built in these regions. This can allow a $\sim$ 10-fold acceleration (see Sect.8.2.3 for more details). This option requires the use of the keyword `CG2FG` (Coarse Grid To Fine Grid) followed by the number of grids and their relative size. The last step should always be equal to "1" (i.e. the basic `increment`). By default, the adaptative grid is switched off, which is equivalent to `CG2FG 1 1`. In practice, 3- or 4-levels are recommended (`CG2FG 3 4 2 1` or `CG2FG 4 8 4 2 1`) (Fig. 9.8):

```
1
bigmolecule.xyz
CG2FG 4 8 4 2 1
```

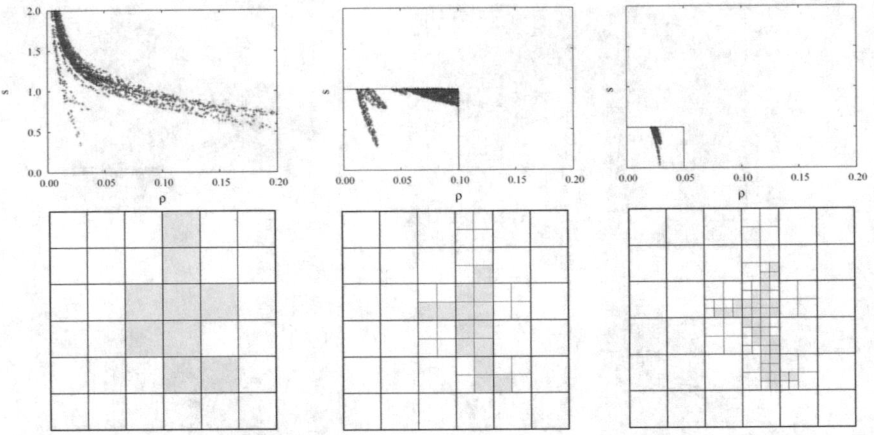


**Fig. 9.8** 2D schematic diagram of multi-level grids, from coarse (left) to fine grids (right) with $4:2:1$ increments and $\rho_c = 0.05$ and $s_c = 1.0$. The density and the $s$ values are computed at the active points of the coarse, medium and fine grids (from left to right). Top: cutoff on ($\rho$,$s$). Bottom: size of the boxes. The small blue boxes are the active boxes of the finest grid, which really contribute to form the isosurface. Adapted with permission from [2]. Copyright 2020 American Chemical Society

It is possible to focus on the visualization of a given interaction type following various criteria:

1. **Strength.** It is possible to select different electron density cutoffs which will enable to isolate interactions by their position in the 2D diagram. This can be done directly for weak interactions or by using the script `range` for isolating stronger interactions such as hydrogen bonds.
2. **Geometry.** In some cases, we might be interested in a given interaction. It is possible to isolate it from a definition of its position in space. Three different options are available in this case:

   - `ATCUBE`. A cubic box is built around a selected list of atoms.
   - `RADIUS`. The user defines an origin and a length.
   - `CUBE`. The cubic box is defined by two opposite points of the cube (similar to the construction of the initial grid in Topmod—Sect. 9.4).

3. **Centered around a given molecule.** In some complicated cases such as enzyme–substrate complexes, it is fundamental to analyze all the interactions (intra and intermolecular) of the substrate and the active site. This is automatically done by the keyword `LIGAND`, which builds a cubic box not around a single point, but around a whole molecule.

Quantification with NCI integrals can be done also following the above Strength criterion. In order to define integration ranges, the keyword RANGE is invoked, and the $sign(\lambda_2)\rho$ intervals subsequently defined:

```
1
bigsystem.xyz
RANGE 3
-0.1 -0.02
-0.02 0.02
 0.02 0.1
```

This provides both the total integrals in the NCI region as well as by ranges. The total integrals look as follows:

```
------------------------------------------------------------------
 Integration  over the volumes of rho^n
------------------------------------------------------------------
 n=1.0            :          3.71920693
 n=1.5            :          0.80392978
 n=2.0            :          0.17992719
 n=2.5            :          0.04164749
 n=3.0            :          0.00995114
 n=4/3            :          1.33570670
 n=5/3            :          0.48721346
 Volume           :         88.37619958
 rho-sum_i rho_i  :          0.00178346

------------------------------------------------------------------
 Integration  over the volumes of sign(lambda2)(rho)^n
------------------------------------------------------------------
 n=1.0            :         -3.71920693
 n=1.5            :         -0.80392978
 n=2.0            :         -0.17992719
 n=2.5            :         -0.04164749
 n=3.0            :         -0.00995114
 n=4/3            :         -1.33570670
 n=5/3            :         -0.48721346
------------------------------------------------------------------
```

The first set of results corresponds to the integrals $\int \rho^n d\mathbf{r}$, and the second one to $\int \text{sign}(\lambda_2)\rho^n d\mathbf{r}$.

On top of this, three ranges are defined in the input file: [–0.1, –0.02], [–0.0, 0.02] and [0.02, 0.1 ], to differentiate hydrogen bonds, van der Waals and steric crowding, respectively. For example, if we are interested in strongly attractive non-covalent interactions, we will look at the output of the [–0.1, 0.02] range:

```
-----------------------------------------------------------------------
Interval            :        -0.10000000         -0.02000000
-----------------------------------------------------------------------
Integration  over the volumes of rho^n
-----------------------------------------------------------------------
n=1.0               :         3.71920693
n=1.5               :         0.80392978
n=2.0               :         0.17992719
n=2.5               :         0.04164749
n=3.0               :         0.00995114
n=4/3               :         1.33570670
n=5/3               :         0.48721346
Volume              :        88.37619958
rho-sum_i rho_i :             0.00178346
```

By comparing the total and the range integration, we see that in this example all the non-covalent interactions are in the strongly attractive regime.

## 9.5.2 *Exercises*

### 9.5.2.1 Model System

We will study the $H_2$ molecule at several distances.

1. Calculate NCI at $d = 2.0$ and 2.5 Å. Visualize the isosurface $s = 0.3$ and a range in between –3 to 3. What kind of interaction do you see for each distance?
2. As we have seen in Sect. 5.3.2.3, we can also use NCI to visualize covalent interactions. It suffices to look at higher densities with the keyword `CUTPLOT`. The difference from non-covalent to covalent is easy to visualize in the shape. What changes do you observe?
3. Remember that the color of the surface is given by the density. What is the density at the point $s = 0$? You can use Gnuplot to plot $s(\rho)$ in the file `.dat`.

| d | Color | Shape | $\rho_{BCP}$ | Interaction type (non-covalent/covalent) |
|---|---|---|---|---|
| d = 0.7 Å | | | | |
| d = 2.0 Å | | | | |
| d = 2.5 Å | | | | |

4. Instead of using wavefunction files, you can use xyz files with promolecular densities. This is easily done for example for 0.7 Å as follows:[2]

```
2

H 0.0 0.0 0.0
H 0.0 0.0 0.7
```

[2] Note there is a blank line between the number and the atomic positions. The "2" makes reference to the number of atoms.

Do the same for $d = 2.0$ and $d = 2.5$ Å. Evaluate the agreement between the promolecular approximation and the real wavefunction. Do they agree better at long or short distances? Why?

#### 9.5.2.2 Non-covalent Interaction Types

This exercise will enable you to see the difference between localized and delocalized interactions. Whereas localized interactions are mainly between two atoms (like the ones we saw for $H_2$), delocalized ones are between several atoms (so the surface spreads out).

1. Calculate $CH_4$ dimer, $CH_4 \cdots H_2O$, $CH_4 \cdots HF$ and $CH_4 \cdots NH_3$ and visualize NCI. Which interactions are local (mainly between two atoms) or delocalized?

| | | $CH_4 \cdots CH_4$ | | | | $CH_4 \cdots NH_3$ | | |
|---|---|---|---|---|---|---|---|---|
| Monomer A | C | 0.000000000 | 0.000000000 | 1.859161821 | C | 0.000000000 | 0.000000000 | −1.990874139 |
| | H | −0.888621462 | −0.513045841 | 1.494593368 | H | 0.000000000 | −1.023189940 | −2.357536774 |
| | H | 0.000000000 | 1.026091683 | 1.494593368 | H | 0.886108478 | 0.511594970 | −2.357536774 |
| | H | 0.888621462 | −0.513045841 | 1.494593368 | H | −0.886108478 | 0.511594970 | −2.357536774 |
| | H | 0.000000000 | 0.000000000 | 2.948285161 | H | 0.000000000 | 0.000000000 | −0.903801408 |
| Monomer B | C | −0.000000000 | −0.000000000 | −1.859161821 | N | 0.000000000 | 0.000000000 | 1.877826056 |
| | H | −0.000000000 | −0.000000000 | −2.948285161 | H | 0.000000000 | 0.937978099 | 2.258958053 |
| | H | −0.000000000 | −1.026091683 | −1.494593368 | H | 0.812312864 | −0.468989052 | 2.258958053 |
| | H | 0.888621462 | 0.513045841 | −1.494593368 | H | −0.812312864 | −0.468989052 | 2.258958053 |
| | H | −0.888621462 | 0.513045841 | −1.494593368 | | | | |

#### 9.5.2.3 Toward Bigger Systems

We will analyze NCI for benzene, naphthalene and anthracene. These systems have one, two and three fused rings. All coordinates can be found in Tables 9.6, 9.7 and 9.8 at the end of this section.

| | | $CH_4 \cdots H_2O$ | | | | $CH_4 \cdots NH$ | | |
|---|---|---|---|---|---|---|---|---|
| Monomer A | C | 0.034380593 | 1.901663427 | 0.000000000 | C | 0.000000000 | 0.000000000 | −1.563190287 |
| | H | 0.542217494 | 2.273137971 | 0.885991138 | H | 0.000000000 | 0.000000000 | −2.651172456 |
| | H | −0.993020634 | 2.255906023 | −0.000000000 | H | 0.000000000 | 1.025759206 | −1.200529232 |
| | H | 0.542217494 | 2.273137971 | −0.885991138 | H | −0.888333531 | −0.512879600 | −1.200529232 |
| | H | 0.047430638 | 0.815178301 | 0.000000000 | H | 0.888333531 | −0.512879600 | −1.200529232 |
| Monomer B | O | 0.034380593 | −1.806674259 | −0.000000000 | H | 0.000000000 | 0.000000000 | 0.736810066 |
| | H | −0.310086665 | −2.286973385 | −0.758506116 | F | 0.000000000 | 0.000000000 | 1.655010202 |
| | H | −0.310086665 | −2.286973385 | 0.758506116 | | | | |

| Dimer | Color | Shape | Interaction type (localized/delocalized) |
|---|---|---|---|
| $CH_4 \cdots CH_4$ | | | |
| $CH_4 \cdots H_2O$ | | | |
| $CH_4 \cdots HF$ | | | |
| $CH_4 \cdots NH_3$ | | | |

1. Calculate NCI. Since in this case, we are using rather big systems, we will choose promolecular densities, faster to calculate. We will further accelerate the calculation with an adaptive grid:

```
2
MonomerA.xyz
MonomerB.xyz
INTERMOLECULAR
CG2FG 3 4 2 1
```

2. Visualize the non-covalent interactions in the parallel conformation (P) and the T-shape conformation (T) for benzene and classify them as delocalized or localized.
3. Visualize now naphthalene and anthracene P and T conformations. What happens as the size increases for each interaction type?
4. You can quantify the size effect with NCIPLOT4.2. Use the following input type to obtain the volume of the non-covalent interactions of the parallel and T-shape series with respect to the number of units.

```
2
BenzeneA.xyz
BenzeneB.xyz
INTERMOLECULAR
RANGE 3
-0.07 -0.01
-0.01 0.01
0.01 0.07
```

Plot the charge ($n = 1$) with respect to the benzene units (1 to 3). What interaction becomes favored with size? Check the fit to a straight line of the local interaction, what does this mean?

| Benzene units | Parallel conformation volume | T-shape conformation volume |
|---|---|---|
| 1 | | |
| 2 | | |
| 3 | | |

**Table 9.6** Benzene coordinates in Å

| | Parallel Benzene | | | | T-shape Benzene | | | |
|---|---|---|---|---|---|---|---|---|
| Monomer A | C | −2.236157102 | 1.303338143 | 0.088680973 | C | −1.398746107 | 0.000000000 | −2.464460286 |
| | C | −2.756218188 | 0.213683080 | 0.794038024 | C | 1.398746107 | 0.000000000 | −2.464460286 |
| | C | −2.455383224 | −1.090191029 | 0.386572992 | C | −0.699341052 | 1.211105093 | −2.464461286 |
| | C | −1.633411171 | −1.302760085 | −0.724282091 | C | −0.699341052 | −1.211105093 | −2.464461286 |
| | C | −1.111556084 | −0.213447025 | −1.427490142 | C | 0.699341052 | −1.211105093 | −2.464461286 |
| | C | −1.413057044 | 1.088874087 | −1.021244112 | C | 0.699341052 | 1.211105093 | −2.464461286 |
| | H | −2.467952073 | 2.317216229 | 0.406231996 | H | −2.485737191 | 0.000000000 | −2.458152281 |
| | H | −3.395036232 | 0.379796126 | 1.658303092 | H | 2.485737191 | 0.000000000 | −2.458152281 |
| | H | −2.859355294 | −1.937880075 | 0.935019035 | H | −1.243224094 | 2.152423163 | −2.460153280 |
| | H | −1.392394202 | −2.315917173 | −1.036178114 | H | −1.243224094 | −2.152423163 | −2.460153280 |
| | H | −0.458282039 | −0.378919068 | −2.279689212 | H | 1.243224094 | −2.152423163 | −2.460153280 |
| | H | −0.997320977 | 1.935180133 | −1.561799154 | H | 1.243224094 | 2.152423163 | −2.460153280 |
| Monomer B | C | 1.633361196 | 1.302760964 | 0.724243022 | C | 0.000000000 | 0.000000000 | 1.063621987 |
| | C | 1.111545109 | 0.213448903 | 1.427481072 | C | 0.000000000 | −1.210102091 | 1.764477041 |
| | C | 1.413098071 | −1.088872208 | 1.021276043 | C | 0.000000000 | −1.211353092 | 3.163402147 |
| | C | 2.236205125 | −1.303339259 | −0.088644042 | C | 0.000000000 | 0.000000000 | 3.863261200 |
| | C | 2.756223215 | −0.213685202 | −0.794034098 | C | 0.000000000 | 1.211353092 | 3.163402147 |
| | C | 2.455342250 | 1.090189912 | −0.386604066 | C | 0.000000000 | 1.210102091 | 1.764477041 |
| | H | 1.392315222 | 2.315920052 | 1.036117041 | H | 0.000000000 | 0.000000000 | −0.020904099 |
| | H | 0.458271064 | 0.378919946 | 2.279681137 | H | 0.000000000 | −2.150558163 | 1.218346995 |
| | H | 0.997409001 | −1.935178254 | 1.561869085 | H | 0.000000000 | −2.152939164 | 3.707588186 |
| | H | 2.468048097 | −2.317218351 | −0.406156064 | H | 0.000000000 | 0.000000000 | 4.950696284 |
| | H | 3.395052256 | −0.379798243 | −1.658289164 | H | 0.000000000 | 2.152939164 | 3.707588186 |
| | H | 2.859292317 | 1.937876958 | −0.935071106 | H | 0.000000000 | 2.150558163 | 1.218346995 |

**Table 9.7** Naphtalene coordinates in Å

| | Parallel Naphtalene | | | | T-shape Naphtalene | | | |
|---|---|---|---|---|---|---|---|---|
| Monomer A | C | 0.632524047 | −2.648397202 | 1.722275134 | C | 1.405298110 | 1.243865097 | −2.464638142 |
| | C | 2.012193156 | −2.645086203 | 1.708831132 | C | 0.708795052 | 2.434569185 | −2.462744143 |
| | C | 2.723970206 | −1.419807108 | 1.693536131 | C | −0.708795052 | 2.434569185 | −2.462744143 |
| | C | 2.043491159 | −0.220234019 | 1.686825127 | C | −1.405298110 | 1.243865097 | −2.464638142 |
| | C | 0.624845050 | −0.187791014 | 1.697887127 | C | −0.719370056 | 0.000000000 | −2.469384143 |
| | C | −0.097732005 | −1.430555110 | 1.718679131 | C | 0.719370056 | 0.000000000 | −2.469384143 |
| | C | −0.104776010 | 1.029894081 | 1.678929130 | C | −1.405298110 | −1.243865097 | −2.464638142 |
| | C | −1.517531117 | −1.396941105 | 1.729047132 | C | 1.405298110 | −1.243865097 | −2.464638142 |
| | C | −2.194899168 | −0.195873017 | 1.715809133 | C | 0.708795052 | −2.434569185 | −2.462744143 |
| | C | −1.482912114 | 1.028322080 | 1.688910127 | C | −0.708795052 | −2.434569185 | −2.462744143 |
| | H | 0.082163007 | −3.587218274 | 1.736167132 | H | 2.493049193 | 1.239870094 | −2.456671146 |
| | H | 2.559403193 | −3.584766273 | 1.712507130 | H | 1.246137093 | 3.379407257 | −2.455090144 |
| | H | 3.811167293 | −1.429903111 | 1.682300127 | H | −1.246137093 | 3.379407257 | −2.455090144 |
| | H | 2.585498199 | 0.722342053 | 1.662873125 | H | −2.493049193 | 1.239870094 | −2.456671146 |
| | H | 0.446058034 | 1.966826150 | 1.644131127 | H | −2.493049193 | −1.239870094 | −2.456671146 |
| | H | −2.062380156 | −2.338977178 | 1.745065131 | H | 2.493049193 | −1.239870094 | −2.456671146 |
| | H | −3.281907249 | −0.183475013 | 1.715603130 | H | 1.246137093 | −3.379407257 | −2.455090144 |
| | H | −2.028912156 | 1.966800151 | 1.656219129 | H | −1.246137093 | −3.379407257 | −2.455090144 |

(continued)

**Table 9.7** (continued)

| | Parallel Naphtalene | | | | T-shape Naphtalene | | | |
|---|---|---|---|---|---|---|---|---|
| Monomer B | C | 0.104776010 | −1.029894081 | −1.678929130 | C | 0.000000000 | 1.242227092 | 1.057699123 |
| | C | 1.482912114 | −1.028322080 | −1.688910127 | C | 0.000000000 | 2.433128188 | 1.753788179 |
| | C | 2.194899168 | 0.195873017 | −1.715809133 | C | 0.000000000 | 2.435217184 | 3.171644284 |
| | C | 1.517531117 | 1.396941105 | −1.729047132 | C | 0.000000000 | 1.244994097 | 3.870146341 |
| | C | 0.097732005 | 1.430555110 | −1.718679131 | C | 0.000000000 | 0.000000000 | 3.185381286 |
| | C | −0.624845050 | 0.187791014 | −1.697887127 | C | 0.000000000 | 0.000000000 | 1.746652176 |
| | C | −0.632524047 | 2.648397202 | −1.722275134 | C | 0.000000000 | −1.244994097 | 3.870146341 |
| | C | −2.043491159 | 0.220234019 | −1.686825127 | C | 0.000000000 | −1.242227092 | 1.057699123 |
| | C | −2.723970206 | 1.419807108 | −1.693536131 | C | 0.000000000 | −2.433128188 | 1.753788179 |
| | C | −2.012193156 | 2.645086203 | −1.708831132 | C | 0.000000000 | −2.435217184 | 3.171644284 |
| | H | −0.446058034 | −1.966826150 | −1.644131127 | H | 0.000000000 | 1.238313096 | −0.027599959 |
| | H | 2.028912156 | −1.966800151 | −1.656219129 | H | 0.000000000 | 3.376752259 | 1.213218135 |
| | H | 3.281907249 | 0.183475013 | −1.715603130 | H | 0.000000000 | 3.380971256 | 3.708507330 |
| | H | 2.062380156 | 2.338977178 | −1.745065131 | H | 0.000000000 | 1.242717095 | 4.958521424 |
| | H | −0.082163007 | 3.587218274 | −1.736167132 | H | 0.000000000 | −1.242717095 | 4.958521424 |
| | H | −2.585498199 | −0.722342053 | −1.662873125 | H | 0.000000000 | −1.238313096 | −0.027599959 |
| | H | −3.811167293 | 1.429903111 | −1.682300127 | H | 0.000000000 | −3.376752259 | 1.213218135 |
| | H | −2.559403193 | 3.584766273 | −1.712507130 | H | 0.000000000 | −3.380971256 | 3.708507330 |

**Table 9.8** Anthracene coordinates in Å

| | Parallel Anthracene | | | | T-shape Anthracene | | | |
|---|---|---|---|---|---|---|---|---|
| Monomer A | C | 1.297490099 | −2.199290170 | −0.783280062 | C | 2.477400188 | −2.464100186 | 1.409310110 |
| | C | 2.395520185 | −1.495050112 | −0.272610020 | C | 3.661250277 | −2.462820186 | 0.712930053 |
| | C | 2.235910172 | −0.725840057 | 0.944810073 | C | 3.661250277 | −2.462820186 | −0.712930053 |
| | C | 0.985330075 | −0.696540053 | 1.574030118 | C | 2.477400188 | −2.464100186 | −1.409310110 |
| | C | −0.112130008 | −1.401460109 | 1.064990079 | C | 1.223920094 | −2.469830186 | −0.724940054 |
| | C | 0.049120005 | −2.177670167 | −0.147550011 | C | 1.223920094 | −2.469830186 | 0.724940054 |
| | C | −1.391440106 | −1.376740107 | 1.696860132 | C | 0.000160002 | −2.466390190 | −1.407630109 |
| | C | −1.077680083 | −2.892580222 | −0.655470051 | C | 0.000160002 | −2.466390190 | 1.407630109 |
| | C | −2.289940174 | −2.846190216 | −0.012170002 | C | −1.223590094 | −2.469970190 | 0.724940054 |
| | C | −2.449430186 | −2.079630160 | 1.177820088 | C | −1.223590094 | −2.469970190 | −0.724940054 |
| | H | 1.417330108 | −2.778750211 | −1.697610129 | H | 2.474170191 | −2.455640187 | 2.496930189 |
| | H | 0.862030067 | −0.106300009 | 2.479780189 | H | 4.607620350 | −2.455010186 | 1.247460094 |
| | H | −1.512410115 | −0.778940057 | 2.596880198 | H | 4.607620350 | −2.455010186 | −1.247460094 |
| | H | −0.955120071 | −3.474240265 | −1.567090122 | H | 2.474170191 | −2.455640187 | −2.496930189 |
| | H | −3.141530241 | −3.388850261 | −0.414470031 | H | 0.000160002 | −2.457340190 | −2.496050188 |
| | H | −3.421150262 | −2.038440157 | 1.661080124 | H | 0.000160002 | −2.457340190 | 2.496050188 |
| | C | 3.362150258 | −0.016199999 | 1.458270113 | C | −2.477080190 | −2.464380190 | 1.409310110 |
| | C | 3.671570282 | −1.503930113 | −0.913290070 | C | −2.477080190 | −2.464380190 | −1.409310110 |
| | C | 4.730150363 | −0.803860062 | −0.387680031 | C | −3.660930279 | −2.463240188 | −0.712930053 |
| | H | 5.696440434 | −0.820640061 | −0.886070065 | H | −4.607300351 | −2.455540188 | −1.247460094 |
| | C | 4.574610347 | −0.053890003 | 0.814730061 | C | −3.660930279 | −2.463240188 | 0.712930053 |
| | H | 5.423380412 | 0.493750039 | 1.217010092 | H | −4.607300351 | −2.455540188 | 1.247460094 |
| | H | 3.236100248 | 0.563900045 | 2.369440182 | H | −2.473850187 | −2.455900188 | 2.496930189 |
| | H | 3.787780288 | −2.079240156 | −1.829630137 | H | −2.473850187 | −2.455900188 | −2.496930189 |

(continued)

**Table 9.8** (continued)

| | Parallel Anthracene | | | | T-shape Anthracene | | | |
|---|---|---|---|---|---|---|---|---|
| Monomer B | C | −0.985290074 | 0.696500052 | −1.574000119 | C | 2.474330188 | 1.054460082 | 0.000000000 |
| | C | 0.112130008 | 1.401490108 | −1.064970081 | C | 3.659230281 | 1.748970131 | 0.000000000 |
| | C | −0.049160006 | 2.177750168 | 0.147520012 | C | 3.662230282 | 3.175160243 | 0.000000000 |
| | C | −1.297550097 | 2.199350168 | 0.783230060 | C | 2.479370188 | 3.874390295 | 0.000000000 |
| | C | −2.395530181 | 1.495020112 | 0.272590023 | C | 1.223810094 | 3.192700241 | 0.000000000 |
| | C | −2.235870171 | 0.725760056 | −0.944790071 | C | 1.222930094 | 1.742610135 | 0.000000000 |
| | C | −3.671590279 | 1.503880117 | 0.913260071 | C | −0.000270002 | 3.876000295 | 0.000000000 |
| | C | −3.362060256 | 0.016030001 | −1.458220111 | C | −0.000050002 | 1.058280080 | 0.000000000 |
| | C | −4.574530351 | 0.053680005 | −0.814690061 | C | −1.223140092 | 1.742420134 | 0.000000000 |
| | C | −4.730130360 | 0.803720063 | 0.387670030 | C | −1.224250094 | 3.192500244 | 0.000000000 |
| | H | −0.861950066 | 0.106240006 | −2.479740189 | H | 2.470500188 | −0.030670002 | 0.000000000 |
| | H | −1.417430107 | 2.778880214 | 1.697520132 | H | 4.603980351 | 1.210430091 | 0.000000000 |
| | H | −3.787850288 | 2.079260159 | 1.829550142 | H | 4.609890351 | 3.708610281 | 0.000000000 |
| | H | −3.235970245 | −0.564130041 | −2.369350179 | H | 2.478930188 | 4.962660379 | 0.000000000 |
| | H | −5.423260415 | −0.494030037 | −1.216940093 | H | −0.000359999 | 4.965110379 | 0.000000000 |
| | H | −5.696430438 | 0.820490060 | 0.886040066 | H | 0.000029999 | −0.027810000 | 0.000000000 |
| | C | 1.391460104 | 1.376780108 | −1.696820131 | C | −2.479930190 | 3.873990295 | 0.000000000 |
| | C | 1.077600082 | 2.892740219 | 0.655430050 | C | −2.474420190 | 1.054060082 | 0.000000000 |
| | C | 2.289870175 | 2.846370216 | 0.012140003 | C | −3.659430278 | 1.748360133 | 0.000000000 |
| | H | 3.141430242 | 3.389100260 | 0.414420030 | C | −3.662670282 | 3.174550245 | 0.000000000 |
| | C | 2.449410189 | 2.079750156 | −1.177800090 | H | −4.610410352 | 3.707840281 | 0.000000000 |
| | H | 3.421130259 | 2.038580156 | −1.661060127 | H | −4.604090351 | 1.209660090 | 0.000000000 |
| | H | 1.512460117 | 0.778950059 | −2.596820200 | H | −2.479670189 | 4.962260379 | 0.000000000 |
| | H | 0.955000075 | 3.474470267 | 1.566990118 | H | -2.470380186 | -0.031070002 | 0.000000000 |

## 9.6 CRITIC2

### 9.6.1 Basic Instructions

The input of CRITIC2 is a single file containing keyword sections to describe the crystal structure, the source of the scalar field to be analyzed and the tasks to be performed.

As the crystal structure is regarded, the code accepts formats inherited from other codes, like WIEN2k's `struct` format, crystallographic `cif` files, VASP's `POSCAR`, Quantum ESPRESSO outputs, abinit density `den` and ELK's `geometry.out` formats. A full description of the crystal structure can alternatively be given by the user with the environment:

```
crystal
...
endcrystal
```

The fields are read from WIEN2k's `clsum` files, VASP's `CHGCAR`, Quantum ESPRESSO cube outputs, abinit `den` and ELK' `state.out`, among others (see Sect. 9.2.1 for some examples of how to obtain these files). This is done with a `load` keyword. If an analytical representation of the field is not available (e.g. after reading a `.cube` file), then the field and its derivatives are obtained by a 3D cubic spline interpolation.

In the case that pseudopotentials are used, core densities can be superimposed with the `zpsp` keyword. You can find them within VASP `POTCAR` file under the keyword `ZVAL`. Please note that (i) unlike in the POTCAR, they should be input as natural numbers, and (ii) they should be introduced in the same order as the different atoms have been introduced in the `POTCAR`.

We will work with the diamond cell as our basic example:

| Structure | Diamond |
|---|---|
| Symmetry | Fd-3m (#227) |
| Cell parameters (Å) | a = 3.556478 |
| Atomic positions | C (8a) 0.125 0.125 0.125 |

The input would look like

```
crystal Diamond.density.cube
load Diamond.density.cube
zpsp C 4
yt
```

The meaning of the keywords is

| Keyword | Meaning |
|---|---|
| crystal | Information to be read on the crystal in between these lines |
| zpsp | Atomic symbols and their core charge (see above) in the same order as in the solid calculation |
| load Diamond.density.cube | Density data to be read from the file `Diamond.density.cube` In case you are using vasp files, you load the `CHGCAR` |
| yt | Find critical points and integrate basins with Yu Trinkle algorithm |

You can run critic as follows:

```
critic2g < Diamond.critin > Diamond.critout
```

If you now open `Diamond.critout`, you can find the electron density maxima, which are at the atoms:

```
# Id  cp  ncp  Name  Z  mult    Volume          Pop             Lap
  1   1   1    C_    6  --    3.94220678E+01  4.06810382E+00   1.55745279E-04
  2   2   1    C_    6  --    3.87861006E+01  3.99998518E+00  -2.66899515E-06
  3   3   1    C_    6  --    3.87861006E+01  3.99998518E+00  -2.66899515E-06
  4   4   1    C_    6  --    3.87861003E+01  3.99998518E+00  -2.66010002E-06
  5   5   1    C_    6  --    3.81490555E+01  3.93175568E+00  -1.47155141E-04
  6   6   1    C_    6  --    3.87860693E+01  3.99998191E+00  -1.94384189E-07
  7   7   1    C_    6  --    3.87860693E+01  3.99998191E+00  -1.94384188E-07
  8   8   1    C_    6  --    3.87860696E+01  3.99998191E+00  -2.03279329E-07
```

You see all atoms hold ca. 4 valence electrons (column `Pop`) and have a volume of 38 a.u.

If we want to see the bonds, we have to look at the topology of ELF with an input like this:

```
crystal Diamond.den.cube
load Diamond.den.cube id rho
load Diamond.ELF.cube id elf
zpsp C 4
reference elf
integrable rho
yt nnm
cpreport Diamondcps.xyz cell
```

Since now we are using the ELF field to partition space and we integrate the density, we have added the keywords:

```
reference elf
integrable rho
```

We will now have maxima out of the nuclei, so we add the keyword nnm (non-nuclear maxima) to the integration line:

```
yt nnm
```

Indeed, the cell shows maxima at the carbon cores and at the C–C bonds:

```
# Id   cp   ncp   Name   Z   mult      Volume              Pop          Lap            $rho
  1    1    1     C_     6   --   0.00000000E+00  0.00000000E+00  0.00000000E+00  0.00000000E+00
  2    2    1     C_     6   --   0.00000000E+00  0.00000000E+00  0.00000000E+00  0.00000000E+00
  3    3    1     C_     6   --   0.00000000E+00  0.00000000E+00  0.00000000E+00  0.00000000E+00
  4    4    1     C_     6   --   0.00000000E+00  0.00000000E+00  0.00000000E+00  0.00000000E+00
  5    5    1     C_     6   --   0.00000000E+00  0.00000000E+00  0.00000000E+00  0.00000000E+00
  6    6    1     C_     6   --   0.00000000E+00  0.00000000E+00  0.00000000E+00  0.00000000E+00
  7    7    1     C_     6   --   0.00000000E+00  0.00000000E+00  0.00000000E+00  0.00000000E+00
  8    8    1     C_     6   --   0.00000000E+00  0.00000000E+00  0.00000000E+00  0.00000000E+00
  9    --   --    ??    --   --   1.93929563E+01  8.67913250E+00  6.09898962E-06  1.99998184E+00
  10   --   --    ??    --   --   1.93929563E+01  8.67913250E+00  6.09898961E-06  1.99998184E+00
  11   --   --    ??    --   --   1.93929659E+01  8.67914141E+00  5.92496041E-06  1.99998533E+00
  12   --   --    ??    --   --   1.93929976E+01  8.67915843E+00 -7.28836495E-06  1.99998825E+00
  13   --   --    ??    --   --   1.93929659E+01  8.67914141E+00  5.92496042E-06  1.99998533E+00
  14   --   --    ??    --   --   1.93929563E+01  8.67913250E+00  6.09898962E-06  1.99998184E+00
  15   --   --    ??    --   --   1.93929976E+01  8.67915843E+00 -7.28836495E-06  1.99998825E+00
  16   --   --    ??    --   --   1.93929563E+01  8.67913250E+00  6.09898961E-06  1.99998184E+00
  17   --   --    ??    --   --   1.93929659E+01  8.67914141E+00  5.92496042E-06  1.99998533E+00
  18   --   --    ??    --   --   1.93929563E+01  8.67913250E+00  6.09898961E-06  1.99998184E+00
  19   --   --    ??    --   --   1.93929976E+01  8.67915843E+00 -7.28836497E-06  1.99998825E+00
  20   --   --    ??    --   --   1.93929976E+01  8.67915843E+00 -7.28836496E-06  1.99998825E+00
  21   --   --    ??    --   --   1.93929976E+01  8.67915843E+00 -7.28836496E-06  1.99998825E+00
  22   --   --    ??    --   --   1.93930121E+01  8.67915827E+00 -1.06386292E-05  1.99998425E+00
  23   --   --    ??    --   --   1.93929563E+01  8.67913250E+00  6.09898961E-06  1.99998184E+00
  24   --   --    ??    --   --   1.93929976E+01  8.67915843E+00 -7.28836496E-06  1.99998825E+00
```

Maxima 1–8 are the carbon atoms, and 9–24 are the bonds. Carbon atoms are bare cores (recall we are using pseudopotentials), and all electrons are now in the bonds, which hold ca 1.9999 electrons each (column `$rho`).

Finally, we will probably want to know where the non-nuclear maxima (in this case bonds) are, so we print the list of critical points as an `xyz` file that we can open with a visualization software:

```
cpreport Diamondcps.xyz cell
```

We will visualize their position in the next section. In the case of molecular crystals, very low densities appear throughout the crystal, so you might need to discard spurious maxima. This can be done by adding a cutoff on the ELF value:

```
yt nnm discard "$elf < 0.2"
```

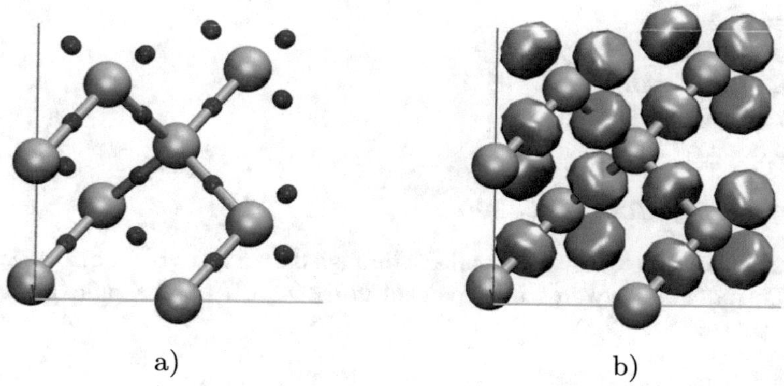


**Fig. 9.9** ELF for diamond. **a** ELF maxima. Bond maxima in small dark blue spheres. **b** ELF = 0.8 isosurface (pink)

In case you have several non-nuclear maxima, you might need to make sure they are considered separately with the keyword `ratom`:

```
yt nnm ratom 1.5 discard "$elf < 0.2"
```

Finally, you might want to print all the critical points within a molecule, even if outside the unit cell, it suffices to make a run with the keyword `molmotif`:

```
crystal urea.scf.in
write ureamolecs.xyz molmotif
```

The output file gives access to the position of the ELF maxima (see Fig. 9.9a). The cell shows maxima at the carbon cores (*n* 1 and *n* 2) and at the C–C bonds (*n* 3). The maxima and ELF isosurfaces (in the `ELFCAR`) can be visualized, as shown in Fig. 9.9b. Other graphical outputs are available, like tracing gradient paths with the `fluxprint` keyword, atomic basins with `basinplot` and isocontours.

**Important!** You might download a version of `CRITIC` that uses slightly different inputs. Two versions of the above input working for other versions are as follows:

```
crystal ELFCAR
load ELFCAR
elf ELFCAR
zpsp O 6 N 5 H 1 C 4
scalar elf
auto
integ gauleg 200 200
```

and

```
crystal
typat O N H C
```

```
zcore 6 5 1 4
struct ELFCAR vasp
elf ELFCAR vasp
endcrystal
scalar elf
auto
integrals gauleg 200 200
```

Finally, plots of NCI are also available through the `nciplot...endnciplot` environment. The keywords implemented in CRITIC2 are the same ones as in NCIPLOT (see Sect. 9.5).

```
crystal tmp1.1.CHGCAR
load tmp.1.1.CHGCAR
zpsp O 6 N 5 H 1 C 4
endcrystal
nciplot
cutoffs 1.0 2.0
endnciplot
```

### 9.6.2 *Exercises*

#### 9.6.2.1 Identifying Interactions

1. Calculate the structures (diamond, Al and NaCl) with the following crystalline structures. Do not forget to create the ELFCAR (VASP) or cube (QE) files. Note that it is easier to visualize conventional cells (`ibrav=1` in QE). For this, you will have to introduce all the atomic positions in the conventional unit cell as provided in Table 1.

| | NaCl | Al | Diamond |
|---|---|---|---|
| SG | Fm-3m (#225) | Fm-3m (#225) | Fd-3m (#227) |
| Cell param (a.u.) | $a = 11.08$ | $a = 7.64$ | $a = 6.77$ |
| All atomic positions | Na 0 0 0<br>Na 0 1/2 1/2<br>Na 1/2 0 1/2<br>Na 1/2 1/2 0<br>Cl 1/2 1/2 1/2<br>Cl 1/2 0 0<br>Cl 0 1/2 0<br>Cl 0 0 1/2 | Al 0 0 0<br>Al 0 1/2 1/2<br>Al 1/2 0 1/2<br>Al 1/2 1/2 0 | C 1/8 1/8 1/8<br>C 1/8 5/8 5/8<br>C 5/8 1/8 5/8<br>C 5/8 5/8 1/8<br>C 7/8 7/8 7/8<br>C 7/8 7/8 3/8<br>C 3/8 7/8 3/8<br>C 3/8 3/8 7/8 |

2. Look at Diamond isosurface ELF = 0.8. Where do you obtain the basins? What is their chemical meaning? How many electrons do you expect in each basin? Identify the bonding type.
3. Look at NaCl isosurface ELF = 0.7. Where do you obtain the basins? What is their chemical meaning? How many electrons do you expect in each basin? Identify the bonding type.
4. Look at Al. What does ELF = 0.5 mean? Play with the ELF value around ELF = 0.5 (0.5,0.55,0.6). Where do you obtain ELF basins? What happens at ELF = 0.6? Is the profile steep or flat? What does this mean? How are these electrons? What model does it remind you of? Identify the bonding type.

#### 9.6.2.2 Quantification

| SG | Cell param | All positions |
|---|---|---|
| $P6_3mmc$ (#194) | $a$ = 7.9608 a.u. | K 0.0 0.0 0.0 |
| | c = 10.882 a.u. | K 0.0 0.0 0.5 |
| | | K 1/3 0.25 2/3 |
| | | K 2/3 1/3 0.75 |

1. Integrate the density basins in NaCl from Exercise 1. What charges do you obtain? Do they agree with what you expected from the bonding type?

2. Integrate the ELF basins in Diamond. What charges do you obtain? Do they agree with what you expected from the bonding type? What would you have obtained if you had integrated the density basins?

3. Now let's try to analyze new materials. Obtain the ELF grid data for the same potassium cell. Look at its ELF surface ELF = 0.8. Compare what you obtain with Na and Diamond from Exercise 1. Which one does it look more alike? How many ELF valence maxima do you find?

   In order to localize these valence maxima, you will need to tweak the parameters:

   ```
   yt nnm ratom 1.5 discard "$elf < 0.4"
   ```

   Check that the number of maxima is the one you expect from the ELF isosurface you visualized. You can obtain the xyz-positions with

   ```
   cpreport Kcps.xyz cell
   ```

   What is the charge of these maxima?

This is a high-pressure structure called an electride. They are insulating metals. Try to explain the observed conductivity from the localization of electrons you have observed.

#### 9.6.2.3 Molecular Crystals

1. Calculate the orthorhombic urea crystal from the following data:

| Cell parameters (a.u.) | $a = 10.516311$, $c = 8.851464$ a.u. |
|---|---|
| Atomic positions | C 1.00000000 0.50000071 0.32600054 |
| | C 0.50000071 0.00000143 0.67400105 |
| | O 1.00000000 0.50000071 0.59530085 |
| | O 0.50000071 0.00000143 0.40470074 |
| | N 0.14590021 0.64590092 0.17660028 |
| | N 0.64590092 0.85410121 0.82340130 |
| | N 0.35410050 0.14590164 0.82340130 |
| | N 0.85409979 0.35410050 0.17660028 |
| | H 0.25750037 0.75750109 0.28270047 |
| | H 0.75750109 0.74250105 0.71730112 |
| | H 0.24250034 0.25750180 0.71730112 |
| | H 0.74249963 0.24250034 0.28270047 |
| | H 0.14410025 0.64410096 0.96200143 |
| | H 0.64410096 0.85590118 0.03800015 |
| | H 0.35590047 0.14410167 0.03800015 |
| | H 0.85589975 0.35590047 0.96200143 |

2. Obtain the atoms that complete the molecules in the cell (keyword `MOLMOTIF`).
3. Calculate the ELF critical points and obtain the list (keyword `CPREPORT`). Visualize them. Are they where you expected?
4. Check the charges. Do you obtain the correct total number of valence electrons? Are the oxygen lone pairs where you expected? Comment on the unusual distribution of nitrogen lone pairs.
5. Plot NCI (`plot_num=19` in QE). How many hydrogen bonds do you obtain for each N? (don't forget periodicity!).

## References

1. VMD Software. https://www.ks.uiuc.edu/Research/vmd/
2. Boto, R.A., Peccati, F., Laplaza, R., Quan, C., Carbone, A., Piquemal, J.P., Maday, Y., Contreras-García, J.: Nciplot4: fast, robust, and quantitative analysis of noncovalent interactions. J. Chem. Theory Comput. **16**, 4150–4158 (2020)

3. Laplaza, R., Peccati, F., A. Boto, R., Quan, C., Carbone, A., Piquemal, J.P., Maday, Y., Contreras-García, J.: Nciplot and the analysis of noncovalent interactions using the reduced density gradient. WIREs Computational Molecular Science **11**(2), e1497 (2021).
4. Schaftenaar, G.: The molden program. The code is available at http://www3.cmbi.umcn.nl/molden/
5. Zou, W.: The molden2aim code (2020). The code is available at http://www.github.com/zorkzou/Molden2AIM

# Chapter 10
# Solutions

**Abstract** This chapter contains the solutions to the exercises of Chap. 9.

## 10.1 Topology

### *10.1.1 Exercise 9.1.1*

1. *Obtain the position of the critical points.*
   Given that V(x,y) is a continuous and differentiable function, the critical points of $V(x, y)$ can be calculated as the points $(x_i, y_i)$ satisfying $\nabla V(x_i, y_i) = 0$.

$$\nabla V = \big(V_x(x, y), V_y(x, y)\big) = \big(2x(x-1)(2x-1), 2y\big).$$

   There are three critical points, all of them lying on the same $y = 0$ line: (0,0), (1/2,0) and (1,0).
2. *Construct the Hessian to characterize their nature.*
   In order to determine the nature of the CPs, the Hessian matrix—the matrix of the second derivatives—needs to be calculated.

$$\mathbb{H}V(x, y) = \begin{pmatrix} V_{xx}(x, y) & V_{xy}(x, y) \\ V_{yx}(x, y) & V_{yy}(x, y) \end{pmatrix} = \begin{pmatrix} 2(6x^2 - 6x + 1) & 0 \\ 0 & 2 \end{pmatrix}.$$

   It turns out to be a diagonal matrix, so its eigenvalues can be read directly to classify the three critical points. On the one hand, $V_{yy}(x_i, y_i)$ equals 2 and therefore is always positive (which prevents any of the critical points from being a maximum). Either minimum or saddle point nature will therefore depend on the signature of $V_{xx}(x, y)$:



Á. Martín Pendás and J. Contreras-García, *Topological Approaches to the Chemical Bond*, Theoretical Chemistry and Computational Modelling,
https://doi.org/10.1007/978-3-031-13666-5_10

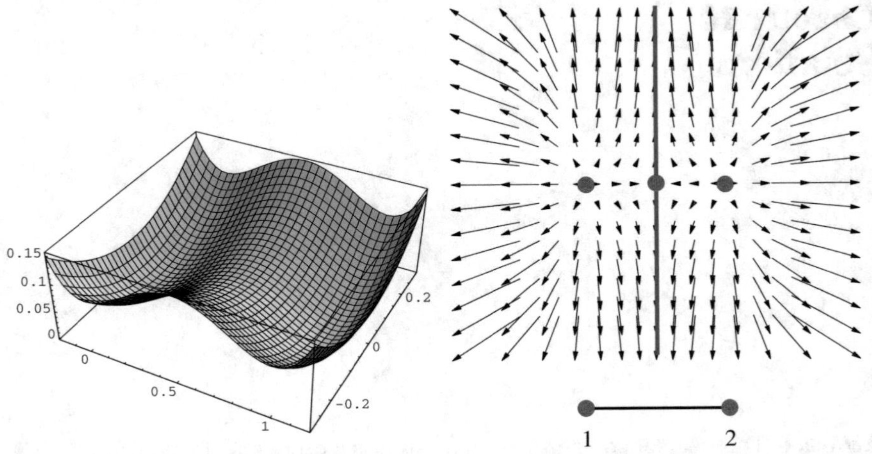


**Fig. 10.1** Representation of the scalar field $V(x, y) = x^2(x-1)^2 + y^2$. The left figure shows the $V(x, y)$ surface, and the right one its associated gradient field. The thick line corresponds to the separatrix between the two basins, and the red dots signal the position of the critical points. Under the gradient field, the adjacency graph of the field is also shown

- $V_{xx}(0, 0) = 2 > 0$. Both eigenvalues are positive, thus corresponding to a minimum.
- $V_{xx}(1/2, 0) = -1 < 0$. It presents a maximum in x, but a minimum in y, which means it is a saddle point.
- $V_{xx}(1, 0) = 2 > 0$. Again, both eigenvalues are positive and thus the critical point is a second minimum.

A representation is found in Fig. 10.1.

3. *Find the repellors, the $\alpha$ limit sets and the separatrix.*
We can now interpret these data in the light of our previous arguments. We see two 2D repellors, with repulsion basins (RBs) that coincide with the $x < \frac{1}{2}$ and $x > \frac{1}{2}$ regions, respectively, and a saddle point. This last CP shows a one-dimensional RB formed by two gradient lines, $x = \frac{1}{2}, y > 0$ and $x = \frac{1}{2}, y < 0$ together with a 1D attraction basin (AB) formed by the $0 < x < \frac{1}{2}, y = 0$ and $\frac{1}{2} < x < 1, y = 0$ lines. The latter connect the two repellors of the dynamical system. *Whenever a separatrix appears between two repellors, we will necessarily find a pair of gradient lines linking them.* It is in this way that binary relations appear among repellors, whose basins are either separated by a common separatrix or completely disconnected. In our simple case, the separatrix of the system is the straight line $x = \frac{1}{2}$.

## 10.2 AIMAll

### 10.2.1 *Exercise 9.3.4.1*

2. *Locate the cage critical point. Order the CPs by increasing density values and discuss the rationale behind the numbers.*
   There are four different types of CPs, C–C BCPs, C–H BCPs, RCPs at the center of the triangular C–C–C faces and a CCP at the center of the tetrahedron.

| CP | Position | $\rho$ | $\nabla^2\rho$ |
|---|---|---|---|
| (3, +3) | (0, 0, 0) | 0.1834 | 0.6024 |
| C1–C2 (3, –1) | (0, 0, 1.1438) | 0.2533 | –0.4560 |
| C1–H5 (3, –1) | (1.7256, 1.7256, 1.7256) | 0.2910 | –1.0772 |
| C1–C2–C3 (3, +1) | (0.3955, –0.3955, 0.3955) | 0.2008 | +0.3196 |

   Notice that the density at the C–H BCP is larger than that at the C–C BCP. Since both atoms are different, this does not necessarily mean that the C–H link is stronger than the C–C one! One can nevertheless compare the densities in regions where the atomic constituents are equivalent. This allows us to compare the C–C BCP density with those at the RCP and CCP. They are seen to decrease from BCP to RCP and to CCP, since we can follow descending gradient lines connecting the BCP to the RCP then to the CCP. Also, notice that the Laplacian at the BCPs is negative, indicating shared interactions.
3. *Do a similar table for the BCPs* containing the difference between bond path lengths and geometric lengths ($\Delta d$). *Relate the behavior to bond strain.*
   In `AIMStudio`, one can load the molecular graphs (File > Open in New Window > file.sumviz) with different values of properties at specific points such as bond critical points (BCP) (they can be also found in the output files). In this way, a given molecule like tetrahedrane can show either the value of the electron density or the bond path length (BPL) as depicted in Fig. 10.2. Recall that all AIMAll quantities are provided in a.u.
   In order to compare the values of the BPLs and the geometric lengths, the latter can be found by clicking on Tools > Selection Tables and selecting the atoms whose distances, angles, etc. are sought. In comparing the previous two distances, there are two kinds: the C–C and the C–H lengths.

| CP | $\rho$ | $\nabla^2\rho$ | $\Delta d$ |
|---|---|---|---|
| C1–C2 | 0.2533 | -0.4560 | 0.0329 |
| C1–H5 | 0.2910 | -1.0772 | -0.0329 |

   In the case of C–C bonds, the BPL is 2.801 a.u., and the corresponding geometrical distance between the nuclei, GPL_II, 2.768, is considerably smaller, with a difference between them of 0.0329 a.u. This can be clearly seen in the noticeably outward curvature of the C–C bond paths in Fig. 10.2.

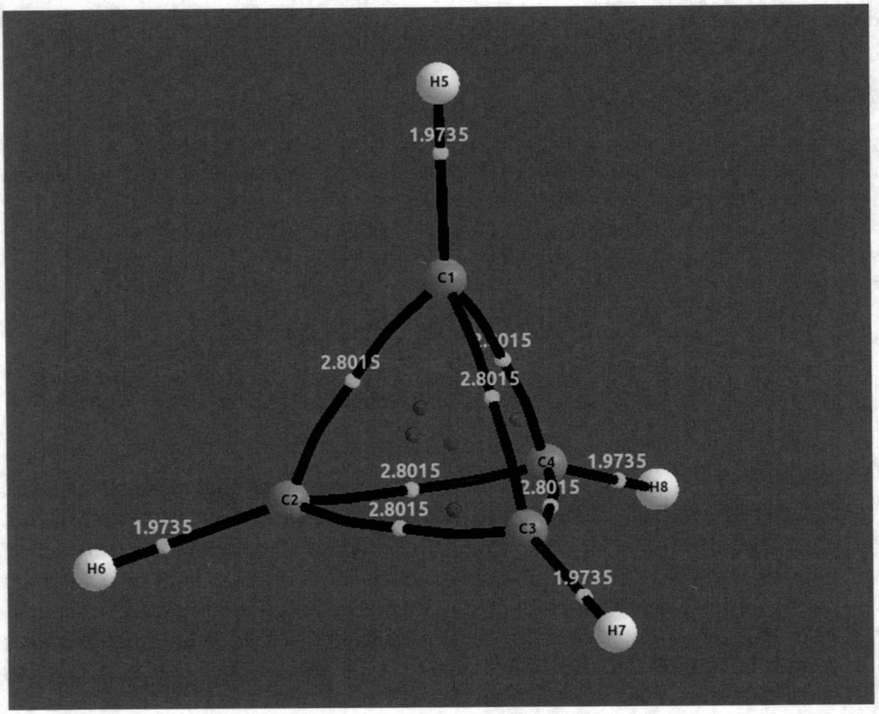


**Fig. 10.2** Molecular graph of tetrahedrane showing the critical points of the electron density field and the values of the BPL on each BCP

An interesting point regards the C–H bonds with a bond path length of 1.9735 a.u. and a GBL_II distance of 2.0064 a.u. A quick look at the graph (Fig. 10.2) shows that these bond paths are remarkably straight. What is going on? The solution to this problem is found when we realize that the (3, –3) maximum of the H atom does not coincide with the H nucleus due to the use of a Gaussian basis set, but is displaced considerably. If we use the distance between the density maxima (BPL_I=1.9735), we solve this puzzle. This should be taken always into account when H atoms are involved.

4. *Compare the densities at the BCP for the C–H bonds (0.2868 a.u. in cyclopropanone, 0.2910 a.u. in cubane). Do the same with the C–C bonds. Why are there two sets of C–C densities in cyclopropanone (0.2757 a.u., 0.2212 a.u.) and only one in cubane (0.2522 a.u.)?*
The densities at BCPs can serve to compare the strength of links when the same kind of atoms and type of bonds are involved. Taking now cyclopropanone, we find two different C–C bonds, the two that share a C atom bonded to the oxygen, with a density of 0.2757 a.u., and a final one with a smaller density, 0.2212. This implies that the proximity to the heteroatom slightly reinforces the C–C

strength, something that follows Pauling's reasonings. In cubane, all C–C links are equivalent, with a BCP density of 0.2522 a.u., slightly larger than that for the more strained three-atom ring equivalent.
Similarly, the C–H bonds in cyclopropanone are characterized by a BCP density of 0.2868 a.u., which increases to 0.2910 in cubane. This shows that ring strain transmits to the C–H links, being smaller in cubane.

### *10.2.2 Exercise 9.3.4.2*

2. *Tick the* `Write Interatomic Surfaces` *option in the* `AIMQB properties` *window. Go to* `File: Open in current window`, *then move to* `atomicfiles` *folder and select the* `.iasviz` *files sequentially. Play with the thickness of the paths and with the presence/absence of strips. Change color. Draw contours, vector maps and Relief surfaces: Select* `New 2D grid` → `Rho`. *Define the plane (to use the molecular plane, get the coordinates of the 3 atoms from the clipboard, click on one atom, right mouse-click and copy, paste). Load in current window the new* `.g2dviz` *file; select* `options`. *Now add a relief map (load the* `.g2dviz`, *select* `Relief map` *and load in the same window). Superimpose a map of the Laplacian (add a new 2D grid and load it in the same window). Draw isosurfaces: Make a 3D grid in the surface dialog. Select the scalar field* `rho`. *Load the* `.3dgviz` *file. Select the Isosurface. Map another field on the surface (the* `ESP`). *Locate nucleophilic and electrophilic regions.*
   In this exercise, you should just play around with the code. For instance, an image of the electrostatic potential mapped onto a 0.001 a.u. density isosurface is shown in Fig. 10.3. It is rather clear that the oxygen's lone pairs are nucleophilic regions, while the H atoms become the electrophilic areas of this molecule.
3. *Examine the CPs of the Laplacian in the $H_2O$ molecule: run* `AIMQB` *with the option to compute the critical points of turned on. Then discover the four types of CPs of the Laplacian. Use the* `Table` *feature to locate the lone pairs. Measure their distance to the O atom. Measure dihedral angles (select* `CPs + Tools + Selection Tables`) *between pairs of CPs. Compare bonding and non-bonding charge concentration regions, and test VSEPR ideas.*
   *Figure 10.4 depicts the CPs of* nabla squared rho$\nabla^2\rho$ *for the water molecule. It has been obtained by selecting each of the files* `.apgviz` *files produced by the code when the option to compute those critical points is activated.*
   The (3, +3) points—in yellow—correspond to the maximal concentration points in each direction. There are seven in total. Three of them are nuclei, and two lie in the bond paths and correspond to the bonded valence shell concentrations (BCCs) of the oxygen atom, and the remaining two lie in the perpendicular plane and can be clearly mapped to the lone pairs. The (3, +1) points—in purple—are those representing concentration in two directions and depletion in one; what becomes the reverse for (3, –1) points—in pink. Finally, the charge depletion

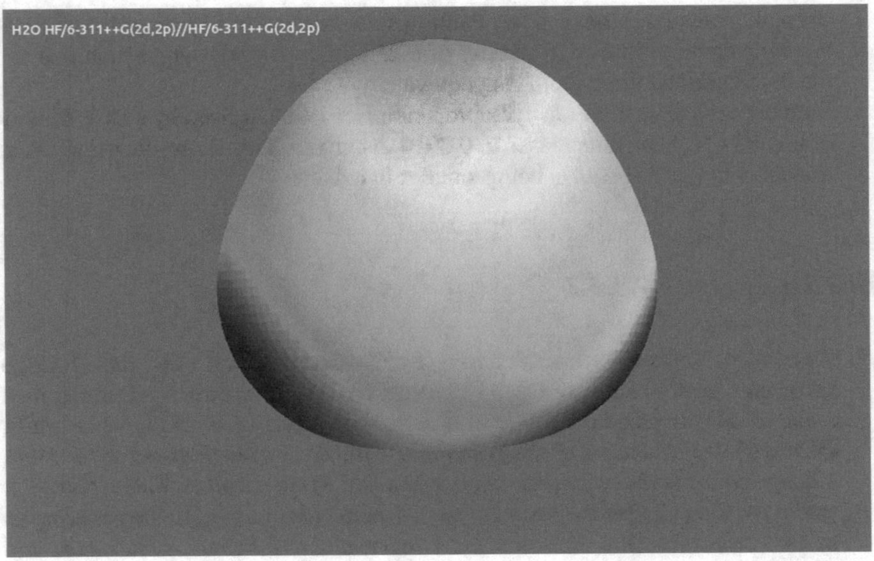


**Fig. 10.3** 0.001 a.u. $\rho$ isosurface of the water molecule on which the electrostatic potential (ESP) has been color-mapped in a scale ranging from red, ESP = –0.08 a.u. to blue, ESP = +0.08 a.u

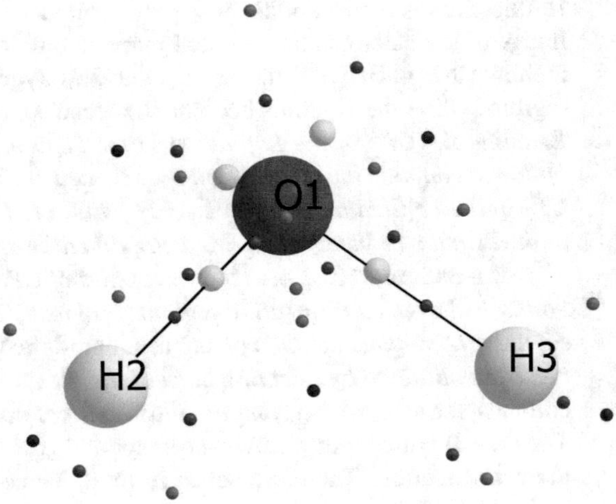


**Fig. 10.4** Water molecule showing the CPs of the Laplacian of the electron density

points in three directions are those in dark green and correspond to (3, –3) points. Notice the complexity of the diagram in such a simple system, and how most of the relevant information is simply stored in the BCCs and lone pairs.

By clicking on `LapCPs > Table`, the list of CPs of $\nabla^2\rho$ appears with information such as their distance to the nucleus. The distance from the O atom to the out-of-plane (3, +3) lone pairs is 0.65 a.u. (a lower value than in the BCC counterparts, 0.72 a.u.). This is what we should expect from the BCCs being attracted or shared between two nuclei. The angles between these two kinds of pairs of minima of $\nabla^2\rho$ can also be selected by clicking trios of CPs in the GUI and turn out to be 122.32° and 106.67°, respectively. This is in very good agreement with the VSEPR rules that state that repulsion between lone pairs is larger than that between bonded pairs (see Table 10.1).

4. *Examine atomic properties in the water molecule: open the* `Atoms` *dialog. Run over Properties: QTAIM charges, integrated Laplacians L(A) and integrated Kinetic Energy. Compare the charge of the H atoms with Mulliken's charge ($Q$(O) $= -0.4954$ a.u.). Explain the difference. Show the atomic dipole moments. Which atom dominates the total molecular dipole? Examine the localization and delocalization indices: get the* `LIs` *and* `DIs`. *Obtain the partition of* upper N Subscript upper A$N_A$ *into localization and delocalization contributions. Why is the localization index smaller than the atomic population? What does it mean that the O–H delocalization index is considerably smaller than one, although we have a* sigma$\sigma$ *bond?*

Table 10.2 collects Mulliken and Bader's atomic charges along with the integrated Laplacian, kinetic energy and atomic dipole moment. Notice that although Mulliken's charges do display the expected polarity in this case, they are far from what is expected given the large electronegativity difference between the two O and H atoms. The integrated Laplacian is close to zero in each case, as QTAIM ensures equal $t$ and $K$ kinetic energies, a criterion that is met when $L$ equals 0. The $t$ kinetic energy shown in each atom is four to six orders of magnitude higher than $L$. It serves to notice how close the two kinetic energies are, since $K = t + L$. Finally, the atomic dipoles are of the same order of magnitude, being the O atom the one that possesses the higher dipole, that will therefore contribute the most to the global molecular dipole. Take into account, however, that dipoles are vector quantities, and that they should be properly added to get the total molecular dipole moment, including the nuclear parts.

The electron population in a molecule can be classified into localized and delocalized parts using localization and delocalization indices. Accordingly, for an atom A $N(A) = \mathrm{LI}(A) + 1/2\sum_{B\neq A}\mathrm{DI}(A, B)$. Taking the water molecule, the atomic populations as well as the LIs and DIs shown in Table 10.3 correspond to strongly polarized bonds. There is a very localized charge on O (0.6 more $e^-$ than the isolated neutral atom) and only 0.06 on hydrogen atoms. Almost all the electrons in H are used in the O–H delocalization. The DIs between O and Hydrogens are close to 0.6, far from 1, the ideal fully delocalized electron pair

**Table 10.1** List of non-equivalent CPs of $\nabla^2\rho$ in the water molecule

| Atomic basin | Multiplicity | Type | Distance from atom | $\rho$ | $x$ | $y$ | $z$ |
|---|---|---|---|---|---|---|---|
| O1 | 1 | (3 , +3) | 0.000001 | 294.858699 | +0.000000 | –0.000000 | +0.213011 |
| O1 | 2 | (3, +1) | 1.087697 | 0.470989 | +0.000000 | +0.866322 | –0.444690 |
| O1 | 2 | (3, –1) | 0.173139 | 20.641435 | –0.000000 | –0.000000 | +0.386151 |
| O1 | 2 | (3, –3) | 0.173335 | 20.576623 | –0.000000 | +0.117307 | +0.340620 |
| O1 | 2 | (3, +1) | 0.174782 | 20.538989 | +0.133600 | +0.000000 | +0.100318 |
| O1 | 4 | (3, +1) | 0.693564 | 0.762723 | +0.387730 | +0.476590 | –0.108792 |
| O1 | 2 | (3, –1) | 0.702403 | 0.682855 | –0.000000 | +0.675855 | +0.404300 |
| O1 | 2 | (3, +3) | 0.646220 | 0.948494 | +0.581213 | –0.000000 | +0.495486 |
| O1 | 1 | (3, –1) | 0.715501 | 0.666415 | +0.000000 | +0.000000 | –0.502489 |
| O1 | 1 | (3, +1) | 0.653200 | 0.891486 | –0.000000 | +0.000000 | +0.866211 |
| O1 | 2 | (3, +3) | 0.715429 | 0.746377 | +0.000000 | +0.571155 | –0.217826 |
| O1 | 2 | (3, +1) | 1.228760 | 0.239069 | +1.219913 | +0.000000 | +0.360195 |
| O1 | 1 | (3, –1) | 1.211086 | 0.234444 | –0.000000 | +0.000000 | +1.424098 |
| O1 | 4 | (3,–1) | 1.200550 | 0.253102 | +0.969202 | +0.552883 | –0.230031 |
| O1 | 2 | (3, –3) | 1.140609 | 0.252954 | –0.000000 | +1.085001 | +0.564810 |
| O1 | 1 | (3,–3) | 1.143145 | 0.276322 | +0.000000 | +0.000000 | –0.930133 |
| H3 | 1 | (3, +3) | 0.008663 | 0.430280 | –0.000000 | –1.414246 | –0.847052 |
| H3 | 2 | (3, +1) | 0.775574 | 0.091209 | +0.719050 | –1.319826 | –1.124407 |
| H3 | 2 | (3,–1) | 0.720801 | 0.083974 | +0.000000 | –1.988765 | –0.407563 |
| H3 | 1 | (3, –3) | 0.641375 | 0.066658 | –0.000000 | –1.965792 | –1.191042 |
| H2 | 1 | (3, +3) | 0.008663 | 0.430280 | +0.000000 | +1.414246 | –0.847052 |
| H2 | 2 | (3, +1) | 0.775574 | 0.091209 | +0.719050 | +1.319826 | –1.124407 |
| H2 | 2 | (3, –1) | 0.766633 | 0.091547 | –0.000000 | +1.041390 | –1.517911 |
| H2 | 1 | (3, –3) | 0.641375 | 0.066658 | –0.000000 | +1.965792 | –1.191042 |

**Table 10.2** Some atomic properties in the water molecule

| Atom | $Q_{Mulliken}$(A) | Q(A) | L(A) | $t$(A) | $\mu_{intra}$(A) |
|---|---|---|---|---|---|
| O | -0.4954 | -1.2540 | 2.5119E-05 | 75.3640 | 0.3510 |
| H | 0.2477 | 0.6270 | 1.3010E-05 | 0.3441 | 0.1532 |
| H | 0.2477 | 0.6270 | 1.3010E-05 | 0.3441 | 0.1532 |

(see Sect. 3.5.3). This is in line with the strongly polarized character of their bonds. The H,H DI is negligible, pointing to no chemical bond between them.

**Table 10.3** Atomic populations and localization and delocalization indices for the water molecule

| Atom | N(A) | LI(A) | DI(A, B) |
|---|---|---|---|
| O | 9.2540 | 8.6374 | (O1, H2) 0.6166 |
| H | 0.3730 | 0.0613 | (O1, H3) 0.6166 |
| H | 0.3730 | 0.0613 | (H2, H3) 0.0068 |

**Table 10.4** Density at the C–C BCPs and delocalization indices for the C–C bond in ethane, ethene, ethyne and benzene. To avoid non-nuclear attractors, the distance between the C atoms in ethyne has been elongated slightly over its optimized value. All data in a.u

| Property | C–C (ethane) | C–C (ethene) | C–C (ethyne) | C–C (benzene) |
|---|---|---|---|---|
| $\rho(\mathbf{r}_{BCP})$ | 0.254 | 0.376 | 0.404 | 0.336 |
| DI(C,C) | 0.992 | 1.885 | 2.861 | 1.389 |

## 10.2.3 Exercise 9.3.4.3

2. *Open also one spreadsheet and annotate the values of the densities at the C–C BCPs.*

   The representative ethane, ethene and ethyne molecules display increasing densities at their C–C BCPs that, in principle, reflect their single, double or triple bonds (see Table 10.4).

3. *Accepting that the standard bond orders of the C–C bond are 1, 2 and 3 for ethane, ethylene and acetylene, fit the densities to a straight line and compute the expected bond order for benzene from its density. Use the C–C delocalization indices to make the correlation.*

By fitting $\rho(\mathbf{r}_{BCP})$ against the nominal bond order (BO = 1, 2, 3), we obtain a regression line that yields an interpolated BO of 1.88 for benzene, in between the single C–C and the double C=C, but closer to double bond and relatively far from the textbook value of 1.5. If the DIs are considered instead, notice that the BO in benzene becomes much closer to the textbook value of 1.5.

## 10.2.4 Exercise 9.3.4.4

2. *What is the topological charge of Li? Comment on its difference with respect to that of Mulliken's analysis (0.1 a.u.). Compute the self-energy of each atom and compare it to that of the isolated atoms. Why is the energy of the Li atom so different from the reference?*

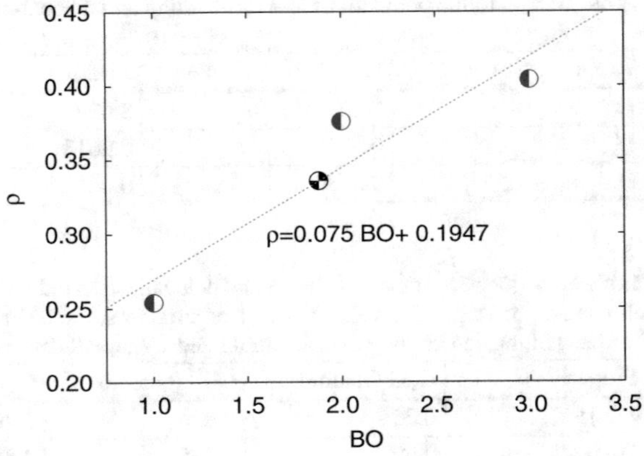


**Fig. 10.5** Correlation between the nominal bond order (BO) in ethane, ethene and ethyne and $\rho$ at the BCP. A linear fit is used to obtain a benzene C–C BO = 1.88 from the density at its BCP

The optimized distance of LiH at the M06-2x/cc-pVDZ level is 3.028633 a.u. Running `AIMQB` with full IQA-enabled options provides a complete energetic decomposition of the total molecular energy into intra- and interatomic contributions. AIMAll also provides IQA descriptors such as the interaction energies and their components between topological atoms, which are related to the bond strength and ionic or covalent character of the bonds. These can be found in the `.sumviz` files. Take into account that some interaction energies as coming from AIMAll are only the A,B integrals,and must be multiplied by two. For instance, `E_IQA_Inter(A,B)/2` is half the interaction energy between the $A$ and $B$ atoms. Also notice that the classical interaction energy, $E_{class}$, has to be found by subtracting the exchange-correlation energy from the total interaction energy in the `.sumviz` file.
As can be seen in Table 10.5, the self-energies of Li and H in LiH ($E_{self}$) are about 117.3 and 7.5 kcal/mol above the neutral atoms. This means that the total self-energy reference is 124.8 kcal/mol above the dissociation limit. This is a considerably large energy which can be rationalized if we notice that the energy of the $Li^+$ cation and the hydride anion at the same level of theory is -7.283 and -0.519 a.u., respectively. This means that if we consider the ionic $Li^+$-$H^-$ reference instead of the neutral one, the total deformation energy decreases to 13.8 kcal/mol. This clearly points out the fact that our system is ionized in the

**Table 10.5** Atomic charges, localization and delocalization indices and IQA energetic descriptors of the LiH molecule. All quantities in a.u.

| Atom | Q | LI | DI | $E_{isolated}$ | $E_{self}$ | $E_{int}$ | $E_{xc}$ | $E_{class}$ |
|---|---|---|---|---|---|---|---|---|
| Li | 0.907 | 1.989 | 0.208 | –7.481 | –7.294 | –0.287 | –0.036 | –0.251 |
| H | –0.907 | 1.804 | | –0.498 | –0.486 | | | |

molecule and that the large deformation energy of the Li is due to the ionization cost.

3. *Compute the total interaction energy and decompose it into classical and covalent. Compute also the localization indices and the delocalization index. What does this tell us about the LiH bond?*
As shown in Table 10.5, the interaction is in fact very ionic, since the LI localization indices are close to 2 in both atoms and their DI of around 0.2 only. This is followed by the interaction energies, too. Whereas the $E_{class}$ accounts for almost all the $E_{int}$, $E_{xc}$ is one order of magnitude lower. The total interaction is thus basically due to electrostatics, not to covalency, as expected from the ionic picture.

### 10.2.5 Exercise 9.3.4.5

2. *Open* `h2oh.sumviz` *and build the fragments' self-energies from the IQA components. Open* `h2oh2oh.sumviz` *and get the deformation energy of the proton donor (PD) and proton acceptor (PA) molecules. Get the PD-PA charge transfer. Study the changes in interactions: covalent and classical, and relate them to the polarization induced by the dipolar field.*
This exercise is intended to show how IQA allows the grouping of atomic energies into chemically intuitive fragments. In this case, we consider the water dimer, at the HF/6-311+G(d,p) level. To do that, you also need to compare with the isolated water molecule at the same level. The header of the `.wfn` file containing its geometry is

```
GAUSSIAN              5 MOL ORBITALS      52 PRIMITIVES        3 NUCLEI
  O    1    (CENTER  1)    0.00000000  0.00000000  0.20066456  CHARGE =  8.0
  H    2    (CENTER  2)   -1.43263141  0.00000000 -0.85622268  CHARGE =  1.0
  H    3    (CENTER  3)    1.43263141  0.00000000 -0.85622268  CHARGE =  1.0
```

Its energy is -76.0523 a.u. Now take the `.sumviz` output file of the IQA-enabled AIMAll calculation of the dimer and compare the atomic charges of all atoms (see Fig. 10.6). As you can see, there is a considerable polarization of the molecules in the dimer. The proton donor molecule (PD) becomes negatively charged by 8 m$e^-$ (add all the charges), while the proton acceptor (PA) loses an equivalent amount of electrons. Since a positively charged part of the PD points toward the PA, and an oxygen's lone pair of the PA points toward the PD, the electric field of this configuration makes electrons of the H atoms of the PA flow toward the oxygen (so that these H atoms become more positively charged with respect to the isolated molecule). Similarly, electrons flow in the PD molecule toward the non-bridging H atom. You can similarly compare the intramolecular interaction energies and rationalize them in terms of these charge flows.

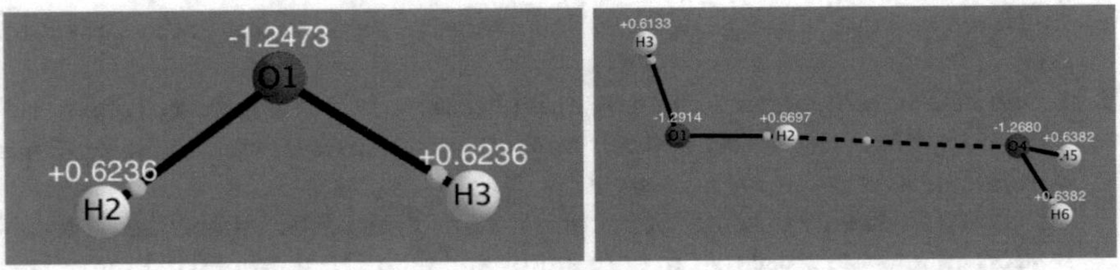


**Fig. 10.6** QTAIM charges in the isolated water molecule and in the optimized water dimer. HF/6-311+G(d,p) calculations. All data in a.u

Now build the fragments quantities. To that end, you need to properly add quantities to obtain the self-energies of the PD and PA molecules (which we call generically $G$ groups). These are obtained as $E^G_{self} = \sum_{A\in G} E^A_{self} + \sum_{A\in G>B\in G} E^{AB}_{int}$. Finally, the interaction energy between the two monomers is $E^{GH}_{int} = \sum_{A\in G, B\in H} E^{AB}_{int}$.

For instance, from these sections of the `.sumviz` file

```
------------------------------------------
Atom A        E_IQA_Intra(A)          T(A)
------------------------------------------
O1       -7.4502054705E+01  7.5349237476E+01
H2       -2.4826330479E-01  3.1650718821E-01
H3       -2.7578824079E-01  3.5094245043E-01
O4       -7.4502071433E+01  7.5351290048E+01
H5       -2.6493709953E-01  3.3485042130E-01
H6       -2.6493672949E-01  3.3485229776E-01
```

as well as,

```
------------------------------------------
Atom A     Atom B     E_IQA_Inter(A,B)/2
------------------------------------------
O1         H2         -3.0062612146E-01
O1         H3         -2.8858472918E-01
H2         H3          8.0285095702E-02
O4         H5         -2.9245826060E-01
O4         H6         -2.9245799620E-01
H5         H6          7.9805046140E-02
```

we can get easily $E^{PD}_{self}$ =−7.4502054705E+01 −2.4826330479E-01 −2.757882407 9E-01 −3.0062612146E-01×2 −2.8858472918E-01×2 +8.0285095702E-02×2, which is equal to −76.043957 a.u. Similarly, we get −76.042168 a.u. for the PA. This leads to deformation energies of about 5.2 and 6.4 kcal/mol, respectively, making a total deformation of 11.6 kcal/mol. Notice how large quantities cancel out overall small deformations.

2. *Compute the covalent and ionic contributions of the hydrogen bond. Why is the O–O delocalization large?*
   In a similar way, we can obtain the total interaction energy between the two monomers, which is equal to −16.8 kcal/mol, partitioned into −10.3 kcal/mol

due to exchange-correlation and –6.5 kcal/mol due to electrostatics. Notice how the covalent contribution to the interaction is far from negligible, although in these relatively weak interactions, it is canceled out by the deformation: $E_{def} + E_{xc} = +1.3$ kcal/mol.
This is the rationale behind considering these hydrogen-bonded systems as electrostatically binded. The electrostatic interaction is close to the binding energy (–5.1 kcal/mol). But this should not be confused with the absence of covalency. It is also interesting to examine the H2–O4 pair, containing the atoms participating in the hydrogen bridge. The total interaction is large, –130.9 kcal/mol, and is divided into –124.8 and –6.1 kcal/mol for the exchange-correlation and classical parts, respectively. Notice how the bridge covalent interaction is again close to the total dimer's binding energy. The large negative electrostatic attraction between H2 and O4 is largely balanced out by the electrostatic repulsion between the two oxygens. The O1–O4 interaction energy is 173.3 kcal/mol, with a non-negligible –4.1 kcal/mol exchange-correlation component. This points toward a considerable three-center character of the full hydrogen bond, in which the three atoms O1–H2–O4 act as an entity. Actually, the delocalization indices of the O1–H2, H2–O4 and O1–O4 pairs in this set are 0.517, 0.045 and 0.047, respectively.

## 10.3 TopMod

### *10.3.1 Exercise 9.4.2.1*

2. *Check that the integration of electrons (total number of electrons) in the output from* `pop09` *corresponds to the expected number. This should be done each time to check the accuracy of the integration.*

The total number of electrons can be found looking for `sum of populations` in the output.

**Ethane**: The expected number of electrons for $CH_3CH_3$ is given by 6+3+6+3=18 electrons. The integration provides `sum of populations 17.989337`.

This means a 0.06% error in the integration.

**Ethylene**: The expected number of electrons for $CH_2 = CH_2$ is given by 6+2+6+2=16 electrons. The integration provides `sum of populations 15.987283`.

This means a 0.08% error in the integration.

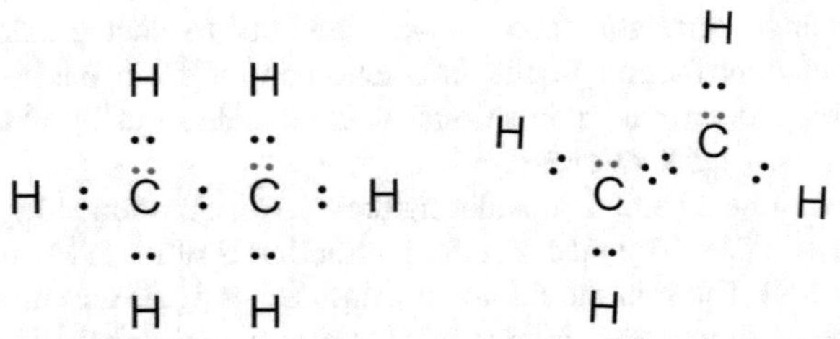


**Fig. 10.7** Lewis structure for ethane (left) and ethylene (right) with core electrons added in red

**Table 10.6** Formal charges in a Lewis picture of ethane and ethylene

| Compound | Expected Lewis pop. | | | | |
|---|---|---|---|---|---|
| | **Total** | Core C | Core H | C–C Bond | C–H Bond |
| Ethane | 18 | 2 | 0 | 2 | 2 |
| Ethylene | 16 | 2 | 0 | 4 | 2 |

In both cases, errors are acceptable, so we can continue our analysis.

3. *Draw the Lewis structure of both compounds and identify all the pairs predicted by Lewis theory. Add also core electrons, since they will also appear in the ELF results.*

   Valence electrons distributed according to Lewis theory (in black) +core electrons (in red) are shown in Fig. 10.7. The expected number of electrons for each chemical unit is shown in Table 10.6.

4. Look at the output from `bas09`:

   (i) *Identify the attractors with your Lewis picture and explain the ELF notation for each of them.*

   The identification of basins with Lewis entities is summarized in Table 10.7. Cores are characterized by a "C", and valence basins by a "V". The atom(s) involved are given in parenthesis, so that carbon core is called "C(C)", and shared valences such as C–C and C–H bonds are named "V(C,C)" and "V(C,H)", respectively. Note that all electrons of hydrogen are involved in bonding, so there are no hydrogen core electrons.

   (ii) *Look at the bonding attractors: How many are there in each case? Why? Where and why are they located? (check xyz-position and angles)*
   The number of maxima and their positions are easily visualized if we represent them in 3D (see Fig. 10.8. In order to do that, it suffices to recover their xyz-position in the output. We can see that one maximum is present for the single C–C bond in ethane. Instead, as we saw in Sect. 4.3.3.1, two maxima appear for ethylene, above and below the bonding line.

**Table 10.7** Nomenclature of ELF basins in ethane and ethylene

| Compound | ELF nomenclature | | | |
|---|---|---|---|---|
| | Core C | Core H | C–C bond | C–H bond |
| Ethane | C(C) | – | V(C,C) | V(C,H) |
| Ethylene | C(C) | – | V(C,C) (appears twice) | V(C,H) |

(iii) *Look at the H–C–H and H–C–C angles and check how they deform from the ideal value according (or not) to VSEPR.*

**Ethane:** The following data is obtained in the output for the distances:

```
V(C1,C2)                C1        C2
                        0.766     0.766
```

and for the angles:

```
C1  V(C1,C2)            C2              179.995
```

Hence, the bonding maximum (as seen in Fig. 10.8, is along the internuclear line (it forms an angle of $\simeq 180°$ with the carbon cores). As expected for a homonuclear molecule, it is midway of both atoms (same distance to each of them, that is, 0.766Å from each).

**Ethylene:** The following data is obtained in the output for the distances:

```
V(C1,C2)                C1        C2
                        0.745     0.745
V(C1,C2)                C1        C2
                        0.745     0.745
```

and for the angles:

```
C1  V(C1,C2)            C2              127.016
C1  V(C1,C2)            C2              127.014
```

In this case, the maxima are symmetrically located upward and downward the internuclear line: since it is still an homonuclear molecule, they have the same distances to each carbon atom. However, we can see from the angles that they do not lie on the internuclear line (Fig. 10.8).

5. *Open the cube files. How do both bonds look like? How many maxima are there in ethylene? Coloring the basins thanks to the* `title_esyn.cube` *file can help you identify the bonds.*

We can see in Fig. 10.9 that the coloring of surfaces allows us to quickly identify their nature. C–H volumes are always very big due to the absence of the Pauli

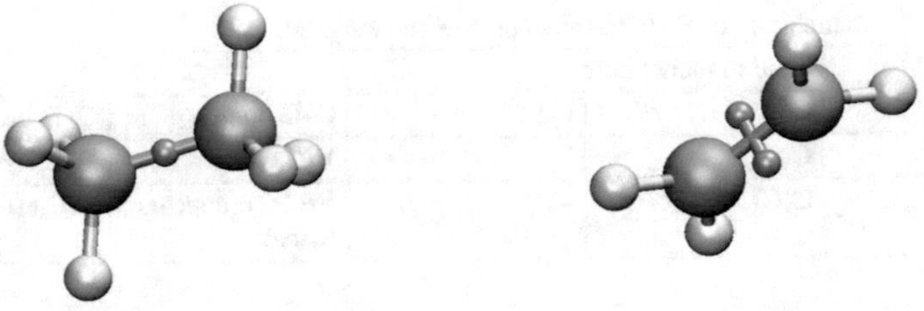

**Fig. 10.8** ELF C–C maxima (in pink) in ethane and ethylene

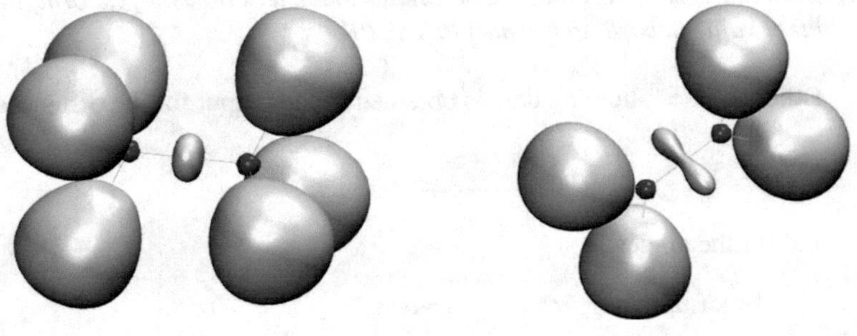

**Fig. 10.9** Ethane (left) and ethylene (right) 3D isosurface ELF = 0.9 with basins colored according to the synapticity: cores in dark blue, hydrogen basins in light pink and bonds in light blue (C–C and C=C)

repulsion in the hydrogen atom (no other same spin electrons). The single bond in ethane has axial symmetry. Instead, it is easy to see now that the two maxima we found for the double C=C bond are reminiscent of the banana picture of a double bond. Moreover, they are connected at high ELF values (the isosurface in the picture is ELF = 0.9!). Thus, they form a unique chemical unit.

6. Look at the output from `pop09`:

   (i) *What is the charge for the single and for the double bond according to the output? Does it agree with what you expected? Tip: sum up the various bonding basins for the same bond.*

   Results for the volumes and populations are collected in Table 10.8. Since we have seen that ethylene bonding maxima should be taken as one chemical unit, their volumes and charges have been added. We can see that cores hold close to two electrons (2.08 $e^-$) and so do the C–H bonds (2.00 $e^-$). The C–C single bond has a slightly lower population (1.81 $e^-$), but still the general picture is very similar to the one we drafted from the Lewis theory in Fig. 10.7. The case of ethylene deviates more: after summing up the contributions from both maxima we are still well below the 4 $e^-$ expected from Lewis, it only has 3.42 $e^-$. This is generally the case in multiple bonds: volumes in

**Table 10.8** Populations (in electrons) and volumes (in a.u.) for ELF basins in ethane and ethylene

| Integrals | **Total** | C(C) | | V(C,H) | | $\sum$V(C,C) | |
|---|---|---|---|---|---|---|---|
| | pop | vol | pop | vol | pop | vol | pop |
| Ethane | 17.99 | 0.81 | 2.08 | $\simeq$87.4 | 2.00 | 17.65 | 1.81 |
| Ethylene | 15.99 | 0.81 | 2.08 | $\simeq$92.8 | 2.1 | $62.8\times2=$ 125.6 | $1.71\times2=3.42$ |

multiple bonds are small and correlation tends to decrease the population from nominal Lewis' values (see Sect. 4.3.3.3).

(ii) *Look at the volumes. Do they agree with VSEPR? Excluding hydrogenated bonds (as we have seen, hydrogen having only one electron, the ELF volumes are non-meaningful), comment on cores, non-hydrogenated bonds and lone pairs.*

If we look now to the volumes in Table 10.8, we can see that the relative volume of cores (0.81 bohr$^3$) is several orders of magnitude smaller than all the other basins. If we now compare the C–C bonds, we can see that the single bond itself (17.65 bohr$^3$) is also one order of magnitude smaller than the double bond (125.6 bohr$^3$). This ordering completely corresponds to the one expected from VSEPR:

$$\text{Double bonds} > \text{Single bonds} > \text{Cores.}$$

As indicated in the exercise, hydrogenated basins are not taken into account. You can see that they occupy very big spaces ($\simeq$ 90 bohr$^3$) compared with the C–C bond. This is always the case, as we already mentioned, due to the absence of core electrons in hydrogen. Hence, we generally do not take them into consideration in these reasonings.

## 10.3.2 Exercise 9.4.2.2

2. *Check that the integration of electrons (total number of electrons) corresponds to the expected number.*

Checking again in the `sum of populations`, we have the following:

**Ethanol**: The expected number of electrons for $CH_3$–$CH_2OH$ is given by $6+3+6+2+8+1=26$ electrons. The integration provides `sum of populations 25.988082`

This means a 0.05% error in the integration.

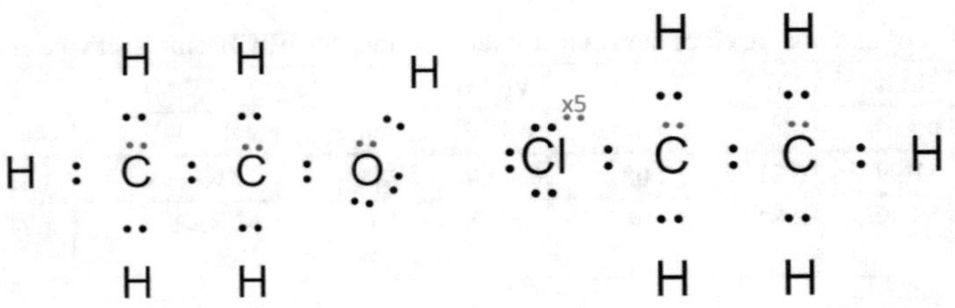

**Fig. 10.10** Lewis structure for ethanol (left) and chloroethane (right) with core electrons added in red

**Table 10.9** Formal charges in a Lewis picture of ethanol and chloroethane

| Compound | Expected Lewis pop. | | | | | |
|---|---|---|---|---|---|---|
| | TOTAL | Core C | Core H | C–X Bond | C–H Bond | Lone Pairs |
| Ethanol | 26 | 2 | 0 | 2 | 2 | 4 |
| Chloroethane | 34 | 2 | 0 | 2 | 2 | 6 |

**Chloroethane**: The expected number of electrons for $CH_3–CH_2Cl$ is given by $6 + 3 + 6 + 2 + 17 = 34$ electrons. The integration provides

```
sum of populations 33.987372
```

This means a 0.04% error in the integration.

In both cases, the errors are acceptable and we can proceed with the analysis.

3. *Draw the Lewis structure of both compounds and identify all the pairs predicted by Lewis theory.*

   Valence and core electrons are shown in Fig. 10.10 in black and red, respectively. The expected number of electrons per chemical unit is shown in Table 10.9.

4. Look at the output from `bas09`:

   (i) *Identify the attractors with your Lewis picture and explain the ELF notation for each of them. Find the lone pair attractor(s). What is the ELF notation for them?*

   The identification of basins is summarized in Table 10.10. Similarly to Exercise 9.4.2.1, we have carbon cores, C(C), C–C bonds, V(C–C) and C–H bonds, V(C–H). On top of these, in this case, we also find lone pairs: V(O) for the lone pairs in ethanol and V(Cl) for the lone pairs in chloroethane.

   (ii) *Look at the lone pair attractors: How many are there in each case? Why? (check xyz-position and angles). Compare the distance from the lone pair and the C–O bond to the oxygen. Why are they different?*

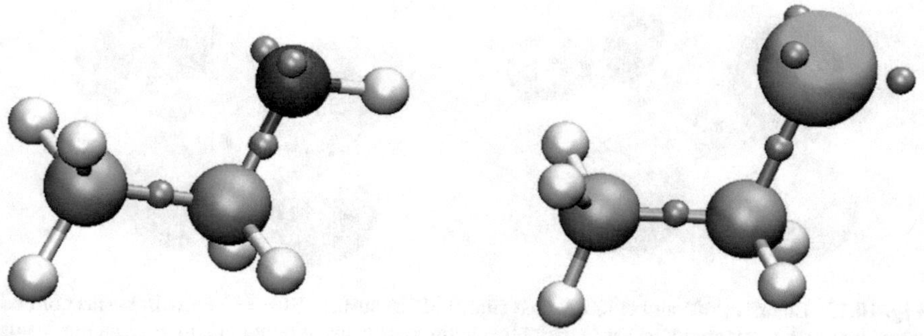

**Fig. 10.11** ELF valence maxima (in pink) in ethanol (C–C, C–O and V(O)) and chloroethane (C–C, C–Cl and V(Cl))

As expected from the Lewis picture (Fig. 10.10), we have two lone pairs in the oxygen:

```
12 V(O)          0.333  -1.234   2.706   0.922   0.000    1   52.28  1
13 V(O)         -0.986   0.104   2.718   0.922   0.000    1   53.33  1
```

and three in the chlorine atom:

```
10 V(Cl)         0.665  -1.844   3.568   0.919   0.000    1   88.15  1
12 V(Cl)         1.251   0.735   3.904   0.919   0.000    1   86.86  1
13 V(Cl)        -1.353   0.220   3.667   0.919   0.000    1   88.06  1
```

They are easily visualized in Fig. 10.11.

If we check the distances of the bonds and the lone pairs in the case of ethanol, we have the following:

```
V(C1,O)           C1       O
                  0.706    0.726
V(O)              O
                  0.580
V(O)              O
                  0.580
```

So lone pairs are at a distance of 0.58 Å from the oxygen, whereas the bond is 0.726 Å away. Note also that in this case we have an heteronuclear bond so it does not lie anymore in the middle of the internuclear line: it is 0.706 Å from the carbon atom and 0.726Å from the oxygen. Note that the fact that the alcohol unit is present and also slightly distorts the C–C bond itself:

```
V(C1,C2)          C1       C2
                  0.740    0.780
```

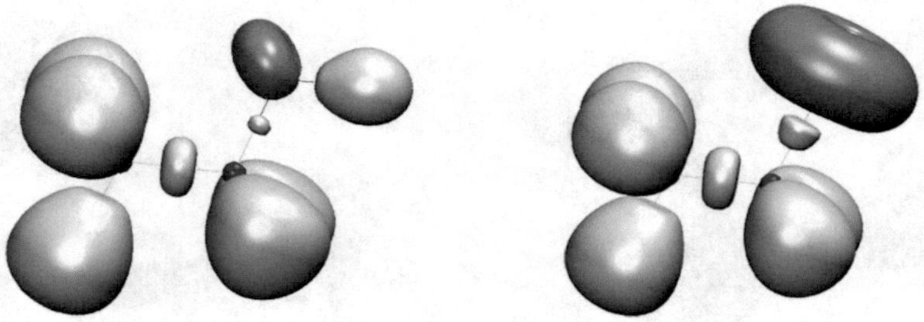

**Fig. 10.12** Ethanol (left) and chloroethane (right) 3D isosurface ELF = 0.85 with basins colored according to the synapticity: cores in dark blue, hydrogen basins in light pink, monosynaptic basins in dark pink-red (lone pairs) and bonds in light blue (C–C, C–O and C–Cl)

A similar analysis can be carried out for chloroethane. In this case, we find the following distances for the lone pairs:

```
V(Cl)           Cl
                3.957
V(Cl)           Cl
                4.349
V(Cl)           Cl
                3.959
```

Note that since the chlorine atom belongs to the third row, they are much further away than in oxygen. Also, the lone pair pointing opposite to the molecule is much further away than the other two since it feels a smaller nuclear attraction. This is clearly visible in Fig. 10.11.

(iii) *Look at the angles around the oxygen and check how they deform from the ideal value according (or not) to VSEPR.*

If we now analyze the angles:

```
V(H6,O)     C(O)     V(C1,O)     106.746
V(H6,O)     C(O)     V(O)        108.682
V(H6,O)     C(O)     V(O)        108.702
V(C1,O)     C(O)     V(O)        107.153
V(C1,O)     C(O)     V(O)        107.164
V(O)        C(O)     V(O)        117.860
```

we see that the angle between the lone pairs (117°) is much bigger than the rest, as predicted by the VSEPR theory.

5. Visualize the cube files

(i) *How do the lone pairs look like? Coloring the basins thanks to the* `title_esyn.cube` *file can help you identify the lone pairs.*

ELF isosurfaces are shown in Fig. 10.12. It can be seen that lone pair maxima appear united at high ELF values, showing that they constitute a unique chemical entity. Hence, we will take all the populations from these basins together when looking at the output of `pop09`.

(ii) *Check what halogen bonds are. How does this bond relate to what you see in chloroethane?*

Halogen bonding was reviewed in Sect. 6.1.1.2. It can be understood as an attractive interaction between an electrophilic region associated with a halogen atom in a molecular entity and a nucleophilic region in another, or the same, molecular entity. Its high directionality stems from an anisotropic distribution of electron density around the halogen nucleus, known as the $\sigma$ hole. The picture of chloroethane enables us to easily see this "hole" in the localization of its lone pairs (opposite to the C–Cl bond).

6. Look at the output from `pop09`:

Results for integrations are summarized in Table 10.11.

(i) *Hydrogens were all equivalent in ethane and ethylene, but not in these molecules. Why?*

In the case of ethanol, we obtain the following results for the hydrogenated basins:

```
basin              vol.    pop.    pab     paa     pbb     sigma2  std. dev.
4 V(H1,C2)         87.97   2.00    2.00    0.32    0.32    0.65    0.80
5 V(H2,C1)         80.59   2.05    2.10    0.34    0.34    0.64    0.80
6 V(H3,C1)         81.62   2.10    2.20    0.37    0.37    0.64    0.80
7 V(H4,C2)         83.34   1.98    1.96    0.31    0.31    0.64    0.80
8 V(H5,C2)         84.15   2.02    2.03    0.33    0.33    0.65    0.81
9 V(H6,O)          48.75   1.72    1.48    0.28    0.28    0.80    0.89
```

The most apparent difference appears for the hydrogen at the alcohol position. Its volume is much smaller (48.75 bohr$^3$, whereas the others are all ca. 80 bohr$^3$). The reason for this is easily found in the population. Due to the electronegativity of oxygen, the population of the basin is much smaller, only 1.72$e^-$, whereas it is close to 2$e^-$ in all the other cases.

**Table 10.10** Nomenclature of ELF basins in ethanol and chloroethane

| Compound | ELF Nomenclature | | | | |
|---|---|---|---|---|---|
| | Core C | Core H | C–C Bond | C–H Bond | Lone pairs |
| Ethanol | C(C) | – | V(C,C) | V(C,H) | V(O) |
| Chloroethane | C(C) | – | V(C,C) | V(C,H) | V(Cl) |

In the case of chloroethanol, it is interesting to check the effect of the voluminous atom on the hydrogen volume distribution:

| basin | vol. | pop. | pab | paa | pbb | sigma2 | std. dev. |
|---|---|---|---|---|---|---|---|
| 4 V(H1,C2) | 86.75 | 1.98 | 1.97 | 0.32 | 0.32 | 0.65 | 0.81 |
| 5 V(H2,C1) | 76.15 | 2.06 | 2.13 | 0.36 | 0.36 | 0.65 | 0.81 |
| 6 V(H3,C1) | 76.05 | 2.08 | 2.16 | 0.37 | 0.37 | 0.65 | 0.81 |
| 7 V(H4,C2) | 81.80 | 1.99 | 1.98 | 0.32 | 0.32 | 0.65 | 0.80 |
| 8 V(H5,C2) | 81.79 | 2.01 | 2.02 | 0.33 | 0.33 | 0.65 | 0.81 |

We can see that the electron population stays very close to Lewis'. However, volumes of the hydrogen basins on C1, where the Cl is attached, are much smaller, obliged by the Cl lone pairs.

(ii) *Look at the electron population of the C–O and C–Cl bond basin. How is it? Why?*
We obtain the following results for ethanol:

| basin | vol. | pop. | pab | paa | pbb | sigma2 | std. dev. |
|---|---|---|---|---|---|---|---|
| 10 V(C1,O) | 7.02 | 1.22 | 0.75 | 0.16 | 0.16 | 0.80 | 0.90 |

and chloroethanol:

| basin | vol. | pop. | pab | paa | pbb | sigma2 | std. dev. |
|---|---|---|---|---|---|---|---|
| 11 V(C1,Cl) | 12.00 | 1.27 | 0.80 | 0.19 | 0.19 | 0.84 | 0.91 |

In both cases, the bonding basin is depleted, with ca. $1.2e^-$, due to the electronegativity of the heteroatoms, oxygen and chlorine.

(iii) *Calculate the total charge in the O and Cl lone pairs. How is it with respect to your prediction from the Lewis structure?*

The total sum of lone pair populations is collected in Table 10.11. We can see that we obtain $4.7e^-$ and $6.66e^-$ for V(O) and V(Cl), respectively. In both cases, the population is bigger than the one predicted by Lewis. This is due to the electronegativity of oxygen and chlorine, which stabilize ionic resonant forms, increasing the population of the lone pairs.

(iv) *Look at the volumes of the lone pairs. Are they bigger or smaller than non-hydrogenated bonds? Does this agree with VSEPR?*

Table 10.11 shows that in spite of the too big volume predicted for the hydrogenated bonds, the lone pair volumes are bigger. This agrees with what is expected from VSEPR:

$$\text{Lone pairs} > \text{Bonds}.$$

(v) *Why is oxygen core smaller than the carbon one?*

Checking back again in the output, we have that

**Table 10.11** Populations (in electrons) and volumes (in a.u.) for ELF basins in ethanol and chloroethane

| Integrals | **Total** | C(C) | | V(C,H) | | V(C,X) | | $\sum$V(X) | |
|---|---|---|---|---|---|---|---|---|---|
| | pop | vol | pop | vol | pop | vol | pop | vol | pop |
| Ethanol | 25.99 | 0.80 | 2.08 | $\simeq$83 | $\simeq$2 | 7.02 | 1.22 | $52.3 + 53.3 = 105.6$ | $2.3 + 2.4 = 4.7$ |
| Chloroethane | 33.98 | 0.82 | 2.08 | $\simeq$80 | $\simeq$2 | 11.99 | 1.26 | $88.15 + 88.01 + 86.9 = 263.06$ | $2.25 + 2.28 + 2.13 = 6.66$ |

```
basin          vol.    pop.    pab     paa     pbb     sigma2  std. dev.
1 C(C1)         0.3    2.08    1.67    0.40    0.26    0.35    0.36    0.805
2 C(O)          0.1    2.11    1.64    0.47    0.35    0.41    0.42    0.776
3 C(C2)         0.3    2.08    1.69    0.40    0.27    0.34    0.35    0.810
```

This is expected from general atomic size arguments. If the oxygen core holds more protons, the atomic radii (and also the shell radii) are smaller.

## 10.4 NCIPLOT

### *10.4.1 Exercise 9.5.2.1*

1. *Calculate NCI at* dd $= 2.0$ *and* 2.5 Å. *Visualize the isosurface* $s = 0.3$ *and a range between –3 and 3. What kind of interaction do you see for each distance?*

   Results for CCSD/aug-cc-pvtz are shown in Fig. 10.13. You can see that vdW interactions, green and rather flat surface, appear at 2.5 Å. As we start to compress, the surface becomes thicker, and the color becomes bluer. These results are summarized in Table 10.12.

2. *As we have seen in Sect.* 5.3.2.3, *we can also use NCI to visualize covalent interactions. It suffices to look at higher densities with the keyword* CUTPLOT. *The difference from non-covalent to covalent is easy to visualize in the shape. What changes do you observe?*

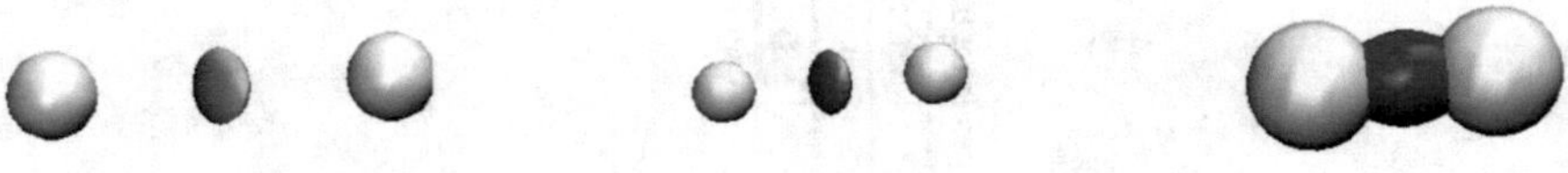

**Fig. 10.13** NCI surfaces at 2.5 Å (right), 2.0 Å (center) and 0.7Å (right). Wavefunction obtained at CCSD/aug-cc-pvtz level, isosurface $s = 0.3$ and color range between –3 and 3

**Table 10.12** NCI characteristics (color, shape and electron density) and the identification of the bonding type

| d | Color | Shape | $\rho_{BCP}$ (a.u.) | Interaction type (non-covalent/covalent) |
|---|---|---|---|---|
| d = 0.7 Å | Deep blue | Oblong | 0.2698 | Covalent |
| d = 2.0 Å | Blue-green | Oblate sphere | 0.0203 | Non-covalent (strong) |
| d = 2.5 Å | Green | Oblate sphere | 0.0071 | Non-covalent (weak) |

Using an input like this one:

```
1
h2_d.0.7xyz
rhoplot 3.0 2.0
cutplot 3.0 0.3
```

NCI is also able to reveal covalent bonds like the one expected for the hydrogen bond distance, 0.7 Å. In this case, the thickness of the surface becomes so important that it becomes oblong. Simultaneously, for the color range we were using, this means we are achieving the extreme of the scale, so the color becomes deep blue. These results are summarized in Table 10.12.

3. *Remember that the color of the surface is given by the density. What is the density at the point $s = 0$? You can use* `Gnuplot` to plot $s(\rho)$ *in the file* `.dat`.

The density at $s = 0$ corresponds to the BCP for QTAIM critical points (Sect. 5.3.4.3). Hence, if we plot $s$ vs $\rho$ (or $s$ vs sign$(\lambda_2)\rho$), we can qualitatively determine the density at the BCP. We can see in red the results of the CCSD calculation. At 2.5 Å , the density of the peak is very close to zero (ca. 0.07 a.u.). There is negligible density reconstruction at this distance, so the peaks' surface is very small, they are nearly two lines. As the distance decreases, the density increases to ca. 0.02 a.u. and the surface of the peak increases. Finally, in the covalent case, the density is much bigger (ca. 0.26 a.u.) and the surface of the peak is huge, as expected for a covalent bond, where the density reconstruction is very important. Note that we can use AIMall as introduced in Sect. 9.3 to accurately determine the density at $s = 0$. Quantitative results are collected in Table 10.12. The evolution is shown in Fig. 10.15. We can see how the density increases exponentially as we diminish the distance, as expected from the promolecular model.

4. *Evaluate the agreement between the promolecular approximation and the real wavefunction. Do they agree better at long or short distances? Why?*

After re-running the three distances with the promolecular approach, we obtain their `.dat` files and we can compare them with the ones we already had. This is shown in Fig. 10.14. It is interesting to note that very small differences are found for $d = 2.5$Å and $d = 2.0$Å, whereas a big shift appears at $d = 0.7$Å. This allows to visualize the validity of the promolecular model for non-covalent interactions, but not for covalent ones. The shift in the covalent case allows us to visualize the need to shift the $s$ value and coloring ranging. When we use the promolecular approximation, we need to use higher $s$ values and bigger coloring ranges to obtain pictures equivalent to their wavefunction analogues.

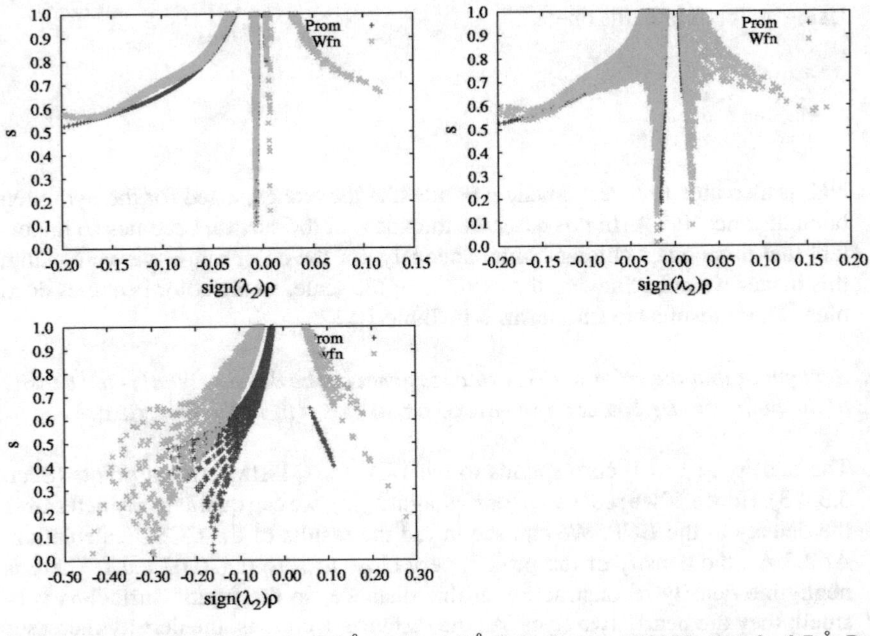


**Fig. 10.14** Top: NCI 2D plots for 2.5 Å (right), 2.0 Å (left). Bottom: NCI 2D plots for 0.7 Å. Red for Wavefunction data at CCSD/aug-cc-pvtz level and green for promolecular

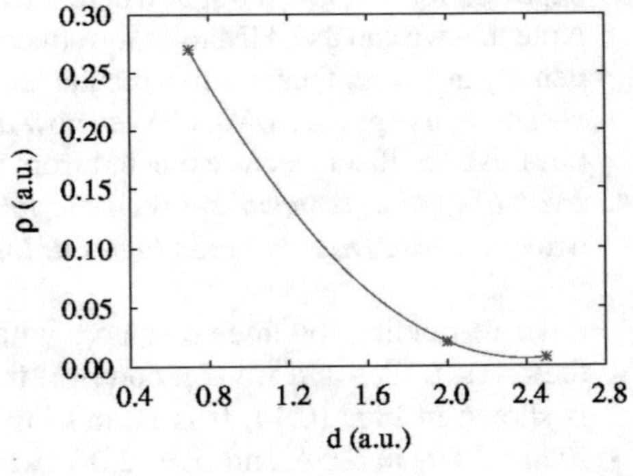


**Fig. 10.15** Density evolution with the compression of the H–H distance

## 10.4.2 Exercise 9.5.2.2

1. *Calculate $CH_4$ dimer, $CH_4 \cdots H_2O$, $CH_4 \cdots HF$ and $CH_4 \cdots NH_3$ and visualize NCI. Which interactions are local (mainly between two atoms) or delocalized?*

   We have seen that one of the main advantages of NCI is a quick identification of localized vs. delocalized interactions. This is illustrated by this set of examples. We can see that methane establishes localized C–H$\cdots$O and C–H$\cdots$N hydrogen

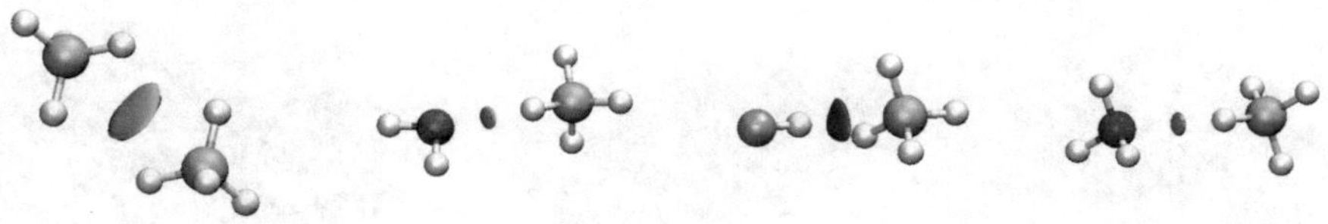

a) b) c) d)

**Fig. 10.16** NCI 3D plots for $CH_4$ dimer (**a**), $CH_4 \cdots H_2O$ (**b**), $CH_4 \cdots HF$ (**c**) and $CH_4 \cdots NH_3$ (**d**). Wavefunction obtained at B3LYP/6-31G level

**Table 10.13** NCI shape and the identification of the bonding localization features

| Dimer | Color | Shape | Interaction type (Localized/delocalized) |
|---|---|---|---|
| $CH_4 \cdots CH_4$ | Green | Diffuse and flat | Delocalized |
| $CH_4 \cdots H_2O$ | Green | Oblate sphere | Localized |
| $CH_4 \cdots HF$ | Blue | Slightly diffuse and flat | Delocalized |
| $CH_4 \cdots NH_3$ | Green | Oblate sphere | Localized |

interactions with water (Fig. 10.16b) and ammonia (Fig. 10.16d), respectively. Instead, a delocalized van der Waals interaction is stabilizing methane dimer (Fig. 10.16a). The interaction with HF showcases an interesting case, where the interaction is strong and delocalized. Results are summarized in Table 10.13.

### *10.4.3 Exercise 9.5.2.3*

1. Calculate NCI. Since in this case we are using rather big systems, we will choose promolecular densities, faster to calculate.

2. Visualize the non-covalent interactions in the parallel conformation (P) and the T-shape conformation (T) for benzene and classify them as delocalized or localized. NCI surfaces for promolecular calculations are shown in Fig. 10.17. We can see that the stacking conformation leads to a delocalized interaction that extends all throughout the benzene system. Instead, the T-shape conformation is stabilized by a localized C–H$\cdots\pi$ interaction.
3. *Visualize now naphthalene and anthracene P and T conformations. What happens as the size increases for each interaction type?*
   As the system grows, the delocalized interactions grow with the system, extending all throughout the $\pi$ system. Instead, the T-shape, being localized, is mainly an addition of new C–H$\cdots\pi$ interactions.

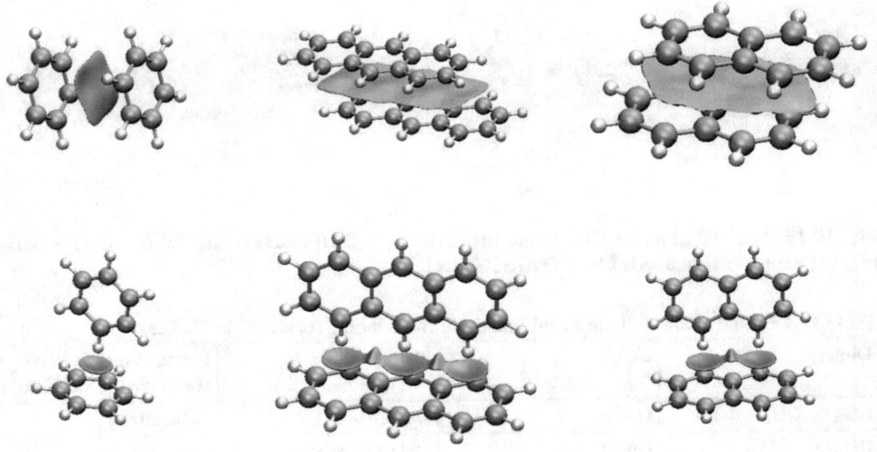

**Fig. 10.17** NCI surface along the Benzene–Anthracene–Naphtalene series in the parallel (top) and T-shape (bottom) conformations

4. *Plot the charge (n = 1) with respect to the benzene units (1–3). What interaction becomes favored with size? Check the fit to a straight line of the local interaction, what does this mean?*
   The integration for $n = 1$ is found in the output of NCIPLOT-4.0 at the line `n=1.0` (note that since it is a density integration, it corresponds to a charge). For example, for the benzene parallel conformation, we find a charge of `0.70727487`:

```
--------------------------------------------------------------
 Integration  over the volumes of rho^n
--------------------------------------------------------------
 n=1.0              :          0.70727487
 n=1.5              :          0.06867501
 n=2.0              :          0.00700973
 n=2.5              :          0.00074746
 n=3.0              :          0.00008279
 n=4/3              :          0.14877154
 n=5/3              :          0.03202473
 Volume             :         89.38800000
 rho-sum_i rho_i :             0.00006956
```

   If we do the same for $n = 1$–3 in both the parallel and the T-shape conformation, we can complete Table 10.14.

   If we plot these results (Fig. 10.18), we see that T-shape interaction charges evolve in a linear way. A regression of $R^2 = 0.999998$ is found! Instead, the parallel conformation charge increases more rapidly, as expected from the 3D pictures.

**Table 10.14** Integral of the electron density over an increasing number of benzene units for the Parallel and T-shape conformations

| Benzene units | Parallel conformation volume | T-shape conformation volume |
|---|---|---|
| 1 | 0.70727487 | 0.64513427 |
| 2 | 1.74240469 | 1.32932673 |
| 3 | 2.98020102 | 2.00981552 |

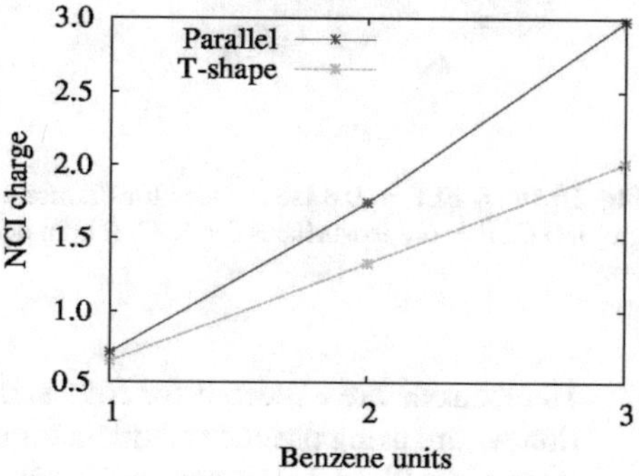


**Fig. 10.18** Evolution of NCI charge with the number of Benzene units. $R^2 = 0.999998$ for the T-shape conformations

# 10.5 CRITIC2

## *10.5.1 Exercise 9.6.2.1*

2. *Look at Diamond isosurface ELF = 0.8. Where do you obtain the basins? What is their chemical meaning? How many electrons do you expect in each basin? Identify the bonding type.*

   Isosurfaces are shown in Fig. 10.19a. Diamond shows a typical covalent pattern with bonding basins in between neighboring carbon atoms leading to a fourfold coordination. Since the bonding is in between the same atoms, the bonding basin is totally symmetric along the internuclear line and appears in the middle of this line. In this case, carbon cores are constituted by the K shell ($2\ e^-$) and the four electrons left are distributed uniformly between the four bonds. Since each carbon contributes equally to each bond, this means each bond should hold ca. $2\ e^-$.

3. *Look at NaCl isosurface ELF = 0.7. Where do you obtain the basins? What is their chemical meaning? How many electrons do you expect in each basin? Identify the bonding type.*

   Isosurfaces are shown in Fig. 10.19b. We can identify an ionic pattern: shells surround the atoms, no bonding basins are present and there has been a complete charge transfer, so that no shells are seen around $Na^+$. Hence, ca. 10 and

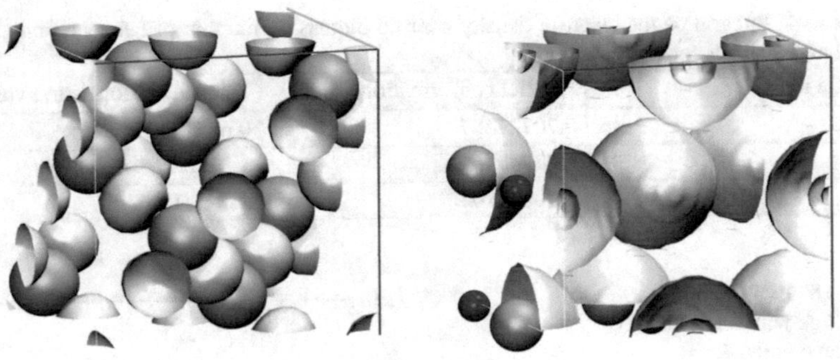

a) b)

**Fig. 10.19** **a** ELF = 0.8 isosurfaces for diamond (C atoms in light blue and ELF isosurfaces in grey). **b** ELF = 0.7 isosurfaces for NaCl (Na in deep blue, Cl in light blue and ELF isosurfaces in grey)

18 electrons are expected for $Na^+$ and $Cl^-$, respectively. Since in our calculation we are using pseudopotentials for the K+L shells of both atoms (10 $e^-$), this means that 0 and 8 electrons will be found around Na and Cl atoms, respectively.

4. *Look at Al. What does ELF = 0.5 mean? Play with the ELF value around ELF = 0.5 (0.5, 0.55, 0.6). Where do you obtain ELF basins? What happens at ELF = 0.6 is the profile steep or flat? What does this mean? How are these electrons? What model does it remind you of? Identify the bonding type.*

We saw in Sect. 4.3.1.2, ELF = 0.5 for the homogeneous electron gas ($D^\alpha = D_h^\alpha$). Figure 10.20 shows the different representations we are asked for. We can see that the isosurfaces change a lot (which was not the case in Fig. 10.19) upon small changes of the surface. This means that the topology is very rich around this area, with maxima and saddle points that lead to different pictures when we go beyond some of the values at these points. This means that ELF oscillates around these values in the valence, i.e. it is very flat and wavy in this region. This means that electrons are delocalized all throughout the valence, as described in a homogeneous electron gas. This is the typical picture of a metallic solid. Note, nevertheless, that many metals do not show such a clear homogeneous electron gas behavior.

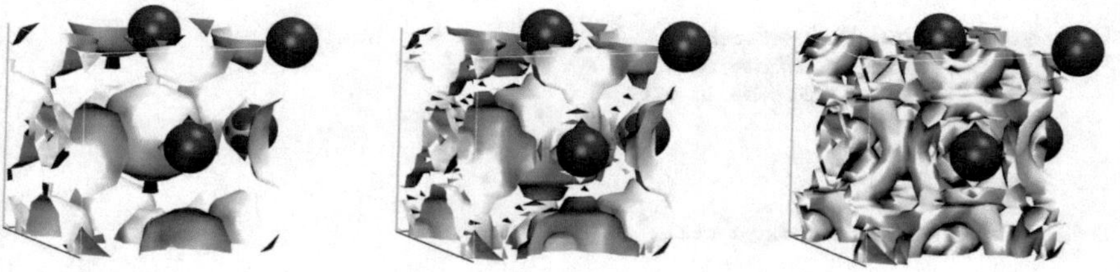


**Fig. 10.20** Isosurfaces around ELF = 0.5 for Aluminum. Isosurfaces in grey and Al atoms in pink

### *10.5.2 Exercise 9.6.2.2*

1. *Integrate the density basins in NaCl from Exercise 1. What charges do you obtain? Do they agree with what you expected from the bonding type?*

   In order to integrate the basins QTAIM basins in NaCl calculated with QE, we can use the following input:

```
crystal ./NaCl.den.cube
load ./NaCl.den.cube
auto
yt
```

   This leads to the following information in the output:

```
* Integrated atomic properties
# Integrable properties 1 to 3
# Id   cp   ncp   Name  Z    mult      Volume           Pop              Lap
  1    1    1     Na    11   --   8.32820865E-01  1.49069080E-03 -2.38662049E-03
  2    2    2     Cl    17   --   3.39229186E+02  7.99850904E+00  2.38662049E-03
```

   It suffices to read the column Pop to obtain the charge on each atom. By doing so, we can see that we have ca. 0.001 $e^-$ on Na and 7.999 $e^-$ on Cl. This is in very good agreement with our estimate from Exercise 9.6.2.1.

2. *Integrate the ELF basins in Diamond. What charges do you obtain? Do they agree with what you expected from the bonding type? What would you have obtained if you had integrated the density basins?*

   Since now we are interested in bonding charges, it is more advisable to use the ELF partition, rather than the QTAIM one. In order to obtain charges within the ELF partition, we have to load both the ELF cube file (for the partition) and the density file (for the integration). This is done as follows:

```
crystal ./Diamond.den.cube
load ./Diamond.den.cube id rho
load ./Diamond.ELF.cube id elf
zpsp C 4
reference elf
integrable rho
yt nnm
cpreport Diamondcps.xyz cell
```

Results look as follows:

```
* Integrated atomic properties
# Integrable properties 1 to 4
# Id  cp  ncp  Name  Z  mult      Volume           Pop              Lap             $rho
  1   1   1    C_    6   --   0.00000000E+00  0.00000000E+00  0.00000000E+00  0.00000000E+00
  2   2   1    C_    6   --   0.00000000E+00  0.00000000E+00  0.00000000E+00  0.00000000E+00
  3   3   1    C_    6   --   0.00000000E+00  0.00000000E+00  0.00000000E+00  0.00000000E+00
  4   4   1    C_    6   --   0.00000000E+00  0.00000000E+00  0.00000000E+00  0.00000000E+00
  5   5   1    C_    6   --   0.00000000E+00  0.00000000E+00  0.00000000E+00  0.00000000E+00
  6   6   1    C_    6   --   0.00000000E+00  0.00000000E+00  0.00000000E+00  0.00000000E+00
  7   7   1    C_    6   --   0.00000000E+00  0.00000000E+00  0.00000000E+00  0.00000000E+00
  8   8   1    C_    6   --   0.00000000E+00  0.00000000E+00  0.00000000E+00  0.00000000E+00
  9    --  --  ??   --   --   1.93929563E+01  8.67913250E+00  6.09898962E-06  1.99998184E+00
  10   --  --  ??   --   --   1.93929563E+01  8.67913250E+00  6.09898961E-06  1.99998184E+00
  11   --  --  ??   --   --   1.93929659E+01  8.67914141E+00  5.92496041E-06  1.99998533E+00
  12   --  --  ??   --   --   1.93929976E+01  8.67915843E+00 -7.28836495E-06  1.99998825E+00
  13   --  --  ??   --   --   1.93929659E+01  8.67914141E+00  5.92496042E-06  1.99998533E+00
  14   --  --  ??   --   --   1.93929563E+01  8.67913250E+00  6.09898962E-06  1.99998184E+00
  15   --  --  ??   --   --   1.93929976E+01  8.67915843E+00 -7.28836495E-06  1.99998825E+00
  16   --  --  ??   --   --   1.93929563E+01  8.67913250E+00  6.09898961E-06  1.99998184E+00
  17   --  --  ??   --   --   1.93929659E+01  8.67914141E+00  5.92496042E-06  1.99998533E+00
  18   --  --  ??   --   --   1.93929563E+01  8.67913250E+00  6.09898961E-06  1.99998184E+00
  19   --  --  ??   --   --   1.93929976E+01  8.67915843E+00 -7.28836497E-06  1.99998825E+00
  20   --  --  ??   --   --   1.93929976E+01  8.67915843E+00 -7.28836496E-06  1.99998825E+00
  21   --  --  ??   --   --   1.93929976E+01  8.67915843E+00 -7.28836496E-06  1.99998825E+00
  22   --  --  ??   --   --   1.93930121E+01  8.67915827E+00 -1.06386292E-05  1.99998425E+00
  23   --  --  ??   --   --   1.93929563E+01  8.67913250E+00  6.09898961E-06  1.99998184E+00
  24   --  --  ??   --   --   1.93929976E+01  8.67915843E+00 -7.28836496E-06  1.99998825E+00
------------------------------------------------------------------------------------------------
```

Since we have loaded two cube files, now the integral is under the column `rho`. Given that we are using a pseudopotential for the carbon core, no charge is found on the carbons. The agreement with the bond charges that we did in Exercise 9.6.2.1 is very good: we find ca. 1.99998 $e^-$ on the C–C bonds, V(C–C). If we had integrated the density, we should have obtained the atomic charge (for the pseudopotential), i.e. 4 electrons per carbon.

3. *Now let's try to analyze new materials. Obtain the ELF grid data for the potassium in Table 9.6.2.2. Look at its ELF surface ELF = 0.8. Compare what you obtain with Na and Diamond from Exercise 1. Which one does it look more alike? How many ELF valence maxima do you find?*

   ELF = 0.8 isosurface is shown in Fig. 10.21. We find isolated ELF maxima around the core and also two maxima in the voids, but not along the internuclear lines. Thus, it looks like the NaCl picture in Exercise 9.6.2.1, but without cores at the valence maxima.

   Check that the number of maxima is the one you expect from the ELF isosurface you visualized. What is the charge of these maxima?

With the tips provided in the exercise, we are able to write the following input:

```
crystal ./K.den.cube
load ./K.den.cube id rho
load ./K.ELF.cube id elf
reference elf
integrable rho
yt nnm ratom 1.5 discard "$elf < 0.4"
cpreport Kcps.xyz cell
```

which allows to obtain the following output:

```
* Integrated atomic properties
# Integrable properties 1 to 4
# Id   cp   ncp   Name  Z    mult      Volume            Pop              Lap              $rho
  1    1    1     K_    19   --    9.02211755E+01  3.22475816E+01 -5.79347815E-03  8.27179514E+00
  2    2    1     K_    19   --    9.02192304E+01  3.22470660E+01 -6.06397524E-03  8.27183395E+00
  3    3    2     K_    19   --    8.63447320E+01  3.16977964E+01  2.09566967E-01  8.21638130E+00
  4    4    2     K_    19   --    8.63447320E+01  3.16977964E+01  2.09566967E-01  8.21638130E+00
  5    --   --    ??    --   --    1.21765003E+02  4.21011888E+01 -2.03638240E-01  1.51183180E+00
  6    --   --    ??    --   --    1.21765003E+02  4.21011888E+01 -2.03638240E-01  1.51183180E+00
```

We see that the list of maxima agrees with Fig. 10.21: two valence maxima are obtained. When we check their charge, we see that they are holding ca. 1.51 $e^-$ each.

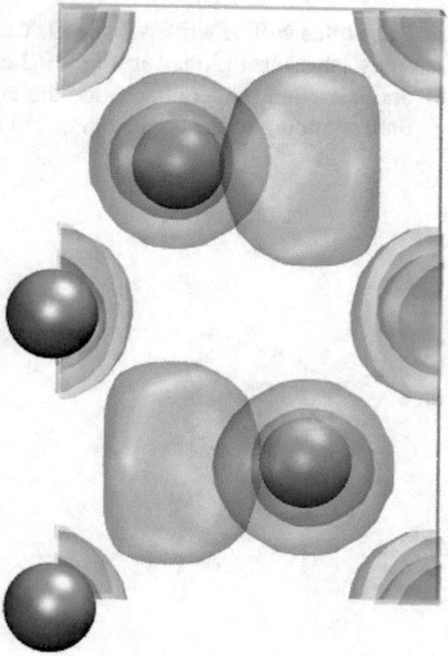

**Fig. 10.21** ELF = 0.8 isosurfaces (in pink) for high-pressure potassium (K atoms also in pink)

This is a high-pressure structure called an electride. They are insulating metals. Try to explain the observed conductivity from the localization of electrons you have observed.
We have seen a very localized picture that resembles that of NaCl, very different from the one in metallic aluminum. This is related to the insulating properties of the material. Under very high pressures, delocalized electrons are destabilized because they occupy a big volume. Thus, localizing them in the voids becomes favorable, leading to insulating phases for elements that typically behave like metals at normal pressure.

### 10.5.3 Exercise 9.6.2.3

2. *Obtain the atoms that complete the molecules in the cell (keyword* MOLMOTIF*).*

In order to obtain the list of atoms that complete the molecules, we can write the following input:

```
crystal urea.scf.in
write ureamolecs.xyz molmotif
```

The atoms will be written in the file `ureamolecs.xyz`. It contains 64 atoms. They have been plotted in Fig. 10.22a: the atoms in the unit cell are in ball-and-stick and the atoms outside the unit cell are in lines. This allows a better understanding of the organization and the interactions in the molecular crystal.

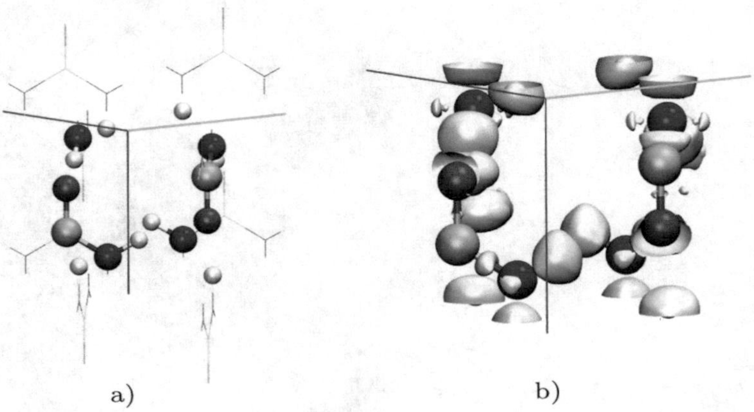


**Fig. 10.22** **a** Results from `molmotif` keyword in lines. Unit cell atoms in ball-and-sticks. **b** ELF = 0.85 isosurface within the unit cell

3. *Calculate the ELF critical points and obtain the list (keyword* `CPREPORT`*).* Visualize them. Are they where you expected them?

In order to obtain the critical points and being able to plot them, we can use the following input:

```
crystal ./urea.den.cube
load ./urea.den.cube id rho
load ./urea.ELF.cube id elf
zpsp C 4 O 6 N 5 H 1
nocore
reference elf
integrable rho
yt nnm discard "$elf < 0.2"
cpreport ureaelfcps.xyz molmotif cell
```

We can visualize the ELF of urea in Fig. 10.22b. If we look at the valence electrons (the ones with a `??`), we can find the following valence maxima according to ELF notation: V(C,O), V(C,N), V(O) and V(N). We can identify them by comparing their position with the isosurfaces. This has been added to the output table in the last column (`MEANING`):

```
# Id   cp   ncp   Name  Z   mult     Volume            Pop              Lap              $rho   MEANING
17     --   --    ??    --  --   2.01254713E+01  8.23676057E+00  2.54840386E-02  2.14223464E+00 V(C,N)
18     --   --    ??    --  --   2.01197822E+01  8.23506634E+00  2.54809587E-02  2.14205466E+00 V(C,N)
19     --   --    ??    --  --   2.01230300E+01  8.23600743E+00  2.54749018E-02  2.14218852E+00 V(C,N)
20     --   --    ??    --  --   2.01237321E+01  8.23610553E+00  2.54871125E-02  2.14214056E+00 V(C,N)
21     --   --    ??    --  --   6.18146987E-01  1.56578824E-01 -1.04956674E-03  1.22131638E-02 V(O)
22     --   --    ??    --  --   6.13834309E-01  1.55687882E-01 -1.04472934E-03  1.21499266E-02 V(O)
23     --   --    ??    --  --   6.17820330E-01  1.56502799E-01 -1.04825753E-03  1.22077260E-02 V(O)
24     --   --    ??    --  --   6.14934926E-01  1.55881800E-01 -1.04557707E-03  1.21635990E-02 V(O)
25     --   --    ??    --  --   1.41208256E+01  5.24635373E+00 -2.60210274E-02  1.54636940E+00 V(C,O)
26     --   --    ??    --  --   1.41200378E+01  5.24629558E+00 -2.60332754E-02  1.54635512E+00 V(C,O)
27     --   --    ??    --  --   3.20634780E+01  7.15013707E+00  6.38601454E-03  6.52920432E-01 V(N)
28     --   --    ??    --  --   3.20517006E+01  7.14823972E+00  6.40699537E-03  6.52892146E-01 V(N)
29     --   --    ??    --  --   3.20611420E+01  7.14869588E+00  6.35859082E-03  6.52933039E-01 V(N)
30     --   --    ??    --  --   3.20467091E+01  7.14704200E+00  6.42225129E-03  6.52807993E-01 V(N)
31     --   --    ??    --  --   3.20508776E+01  7.14603220E+00  6.40956283E-03  6.52799612E-01 V(N)
32     --   --    ??    --  --   3.20447314E+01  7.14708384E+00  6.43721829E-03  6.52935955E-01 V(N)
33     --   --    ??    --  --   3.20462357E+01  7.14712790E+00  6.40033833E-03  6.52891836E-01 V(N)
34     --   --    ??    --  --   3.20591884E+01  7.14805549E+00  6.38408948E-03  6.52848864E-01 V(N)
------------------------------------------------------------------------------------------------------
Sum                              9.78866313E+02  2.80504973E+02  3.91244088E-04  4.80002482E+01
```

4. *Check the charges. Do you obtain the correct total number of valence electrons? Are the oxygen lone pairs where you expected them? Comment on the unusual distribution of nitrogen lone pairs.*

The urea molecule has a formula unit of $CH_4N_2O$. This makes a total of $6 + 4 + 14 + 8 = 32\, e^-$. Since we are using pseudopotentials for the K shell of C, N and O, we have a total of $32–2\times4 = 24\, e^-$. Since we have two formula units per unit cell, we should have 48 electrons. The sum of populations (at the end of column `rho`) makes `4.80002482E+01`. Hence, the integration is trustworthy.
If we start with the atomic positions, we have the following:

```
* Integrated atomic properties
# Integrable properties 1 to 4
```

```
# Id   cp   ncp   Name  Z   mult     Volume             Pop              Lap             $rho
  1    1    1     C_    6   --    0.00000000E+00  0.00000000E+00  0.00000000E+00  0.00000000E+00
  2    2    1     C_    6   --    0.00000000E+00  0.00000000E+00  0.00000000E+00  0.00000000E+00
  3    3    2     O_    8   --    1.14671019E+02  3.07726625E+01  5.60796151E-03  6.26341610E+00
  4    4    2     O_    8   --    1.14664990E+02  3.07738846E+01  5.61497833E-03  6.26352327E+00
  5    5    3     N_    7   --    1.67427397E+00  1.16739577E+00 -3.40958804E-02  6.62443191E-01
  6    6    3     N_    7   --    1.67425584E+00  1.16737603E+00 -3.41024059E-02  6.62421319E-01
  7    7    3     N_    7   --    1.67398817E+00  1.16718630E+00 -3.40962042E-02  6.62343075E-01
  8    8    3     N_    7   --    1.67411903E+00  1.16728325E+00 -3.40988489E-02  6.62396115E-01
  9    9    4     H_    1   --    4.90466629E+01  1.42071104E+01  3.58681315E-02  1.99610846E+00
  10   10   4     H_    1   --    4.90646567E+01  1.42086997E+01  3.58094073E-02  1.99620571E+00
  11   11   4     H_    1   --    4.90700474E+01  1.42102390E+01  3.57929651E-02  1.99631461E+00
  12   12   4     H_    1   --    4.90603100E+01  1.42092436E+01  3.58392611E-02  1.99624814E+00
  13   13   5     H_    1   --    4.47430517E+01  1.40526859E+01 -2.86594182E-02  1.97644481E+00
  14   14   5     H_    1   --    4.47412000E+01  1.40523465E+01 -2.86513517E-02  1.97641718E+00
  15   15   5     H_    1   --    4.47424350E+01  1.40524917E+01 -2.86576234E-02  1.97641406E+00
  16   16   5     H_    1   --    4.47436251E+01  1.40527130E+01 -2.86693672E-02  1.97644491E+00
```

We see that carbon and nitrogen cores, represented by the pseudopotentials, hold a negligible amount of electrons. Instead, hydrogen and oxygen behave like very ionic units, with 6.62 and 1.98-199 $e^-$, respectively.

If we now look at the valence charges in the previous section, we can see that all the charge of oxygen appears in the core, so that the valence integration is negligible. Hence, we can summarize the system as follows:

| Entity | Charge ($e^-$) |
|---|---|
| V(C,O) | 1.55 |
| V(C,N) | 2.14 |
| V(C,H) | 1.98 |
| V(O) | 6.27 |
| V(N) | 0.65×2 |

Oxygen lone pairs appear opposite to the C=O bond, which is where they are expected from a Lewis structure. The oxygen being very ionic, the C–O "double" bond only holds 1.55 $e^-$. If we look at Sect. 9.6.2.1, we can see that the ELF isosurface is very small, confirming this interpretation. C–N bond is itself polar-covalent: shape pointing toward the carbon atom, but holding a charge close to 2 $e^-$. The nitrogen atom shows a lone pair which is formed of two small basins, as expected for a planar NH2 structure.

5. *Plot NCI* (`plot_num=19` *in QE). How many hydrogen bonds do you obtain for each N? (don't forget periodicity!).*

The QE input looks as follows:

```
&INPUTPP
   prefix      = 'urea'
   outdir      = './'
   plot_num    = 19
   filplot       = 'urea.rdg.dat'
/

&PLOT
   iflag          = 3
```

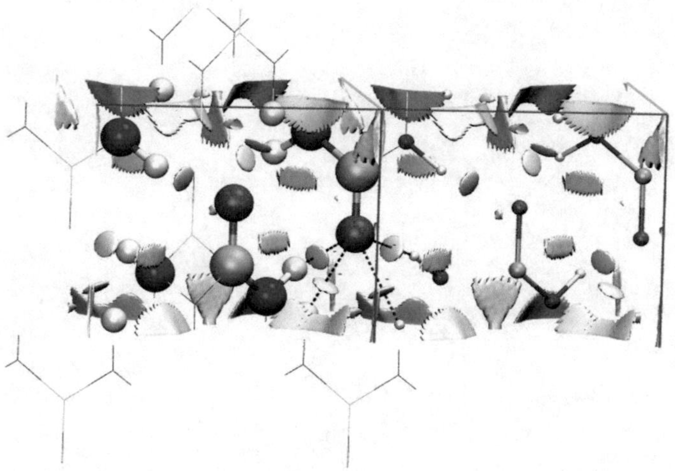

**Fig. 10.23** NCI plot corresponding to question 5

```
    output_format = 6
    fileout       = 'urea.rdg.cube'
    nx = 500
    ny = 500
    nz = 500
  /
```

We can then obtain the ċube files for plotting NCI with the following input:

```
crystal ./urea.den.cube
load ./urea.den.cube id rho
load ./urea.ELF.cube id elf
zpsp C 4 O 6 N 5 H 1
nciplot
cutoffs 1.0 2.0
endnciplot
```

The result is shown in Fig. 10.23. We can see that once the neighboring molecules are taken into account, each oxygen establishes four hydrogen bonds.

# Epilogue

You certainly enjoyed reading this book and you are eager to apply these exciting methods to your own research. You have already downloaded various dedicated software and maybe left the book on your desk. If you have decided to skip this afterword, I will not be upset, I know that this kind of supplementary text is written to be printed rather than to be read. However, the expected role of a postface is to draw perspectives on a topic and it is what I will try to do.

In my opinion, the topological methods described in this book constitute an epistemologically rather safe alternative to the mainstream interpretative methods which, according to C. A. Coulson, "give a physical significance to quantities, such as molecular orbitals or valence bond structures, appearing as intermediates during the course of solution of the many-body Schrödinger equation". They participate in a scientific program aiming to build a mathematical bridge between chemical ideas and quantum mechanics. "Chemical ideas" have to be understood as the set of concepts and empirical rules developed before the formulation of quantum mechanics or independently during the XIXth and XXth centuries in order to explain the structure and the reactivity of molecules and solids. At the atomic and subatomic levels, matter is made of interacting particles which are ruled by the laws of quantum mechanics. As quantum objects, they are given to an observer by the footprints they leave in the perceptual world. These footprints, for example, definite proportion law and stoichiometry, have been identified by chemists and interpreted in terms of a representation of the matter in terms of atoms linked by bonds. It is interesting to note that the hypothesis of electrons and of electronic shells has been expressed by Laming in 1845 a few months after the introduction of the electrochemical equivalent by Faraday. However, most chemical concepts related to the structure of the matter such as those of bond, covalence, aromaticity, have epistemologically weak definitions and do not correspond to physical observables or pseudo observables. Although the predictive power of quantum mechanics is indisputable, its ability to provide explanation has been questioned by many scientists and philosophers of science among them I can mention René Thom, Karl Popper and Niels Bohr.



Á. Martín Pendás and J. Contreras-García*Topological Approaches to the Chemical Bond*, Theoretical Chemistry and Computational Modelling,
https://doi.org/10.1007/978-3-031-13666-5

Chemistry has its own concepts, its own level of understanding and its own representation of the microscopic matter which explains the properties of the matter from its composition in terms of elements and their position in the periodic table rather than from physical laws mastered by the Schrödinger equation. Though we are conscious of the dangers of the temptation of reductionism, which at the end leads to throw away without further ado essential "prequantum" concepts of Chemistry, it is one of our tasks as theoretical chemists to build (when it is possible) rigorous bridges between Chemistry and Quantum Mechanics. To do that, we can take advantage of the ability of quantum chemistry to provide rather reliable data to design numerical experiments in order to question the content of the chemical concepts and of the chemical laws. We can expect that such experiments would be able to point the pros and the cons of the current definitions and rules as well as suggest improvements. In this respect, topological methods are among the most convenient instruments for such experiments because they are based on mathematical theories independent of the actual quantum chemical calculation schemes. These theories, the differential geometry and differential topology, give access to a rigorous qualitative thinking.

The topological methods rely on the analysis of potential functions which carry the physical and/or chemical information. These functions are either on density and density of property functions which can be evaluated as the expectation value of an operator and have, therefore, a clear signification or more or less complicated functions of these latter densities and for which an interpretation is necessary. As an example, the Becke and Edgecombe electron localization function ELF has received many ad hoc interpretations based on a term by term identification of its formula. Its success is not due to the physical/chemical meaning of its values but rather to its ability to perform a partition of the position space in basins of attractors matching the VSEPR model electronic domains and therefore the Lewis's picture of the bonding. I think it is fundamental to improve our tools and to explore new local functions as it has been done by M. Kohout with Shannon's information entropy as well as new properties.

In addition to methodological improvements, the critical evaluation of the concepts is another challenging task. In general, chemists have a naïve tendency to consider the empirical rules as inalienable laws, in this way, the octet rule may be violated, atomic shell populations must be as close as possible to integral numbers. In many papers published on the shell structure of atoms, this latter idealization is considered an absolute criterion of the reliability of the partition method. If the temperature indicated by the thermometer does not correspond to our expectation, is the thermometer necessarily wrong? The outputs of the topological analysis provided they have an unambiguous physical meaning can be used to corroborate or invalidate the expectations of the current bonding models. They can also be used to revisit the content of the underlying concepts and, if necessary, introduce new concepts. During the development of the "elfology", we encountered a problem with the Lewis's valence model which considers that an electron pair can be shared by two atomic valence shells. This restriction is certainly due to the representation of bonds by lines drawn in the plane of the sheet of paper. We found in some crystals and molecules that a bonding valence basin can be shared by several atomic valence shell which led us to introduce the synaptic order concepts. The introduction of the synaptic order should be considered a generalization of Lewis's ideas rather than a refutation. Removing a constraint satisfies, moreover, the parsimony principle. It is worth noting that, for the same reason of writing the formula, the arrangement of substituents around a carbon center was believed to be planar before the hypothesis of Le Bel and Van't Hoff.

For many applications, the output of topological methods inspires and supports explanations. I recommend chemical and physical explanations as your first choice. By chemical explanations, I mean those based on chemical concepts and rules such as VSEPR, whereas physical explanations rely on molecular mechanics or on intermolecular system theory. When both apply, these explanations are often complementary. The other level of explanation, the quantum chemical one, tells how the approximate wave functions are built. It has the advantage to be predictive but has an inherent epistemological weakness. It is insecure to try connections between chemical explanations supported by topological methods and quantum chemical explanations, particularly to characterize bonding. When topological methods invalidate an expected and often accepted explanation be careful, examine the assumptions and all the possible sources of discrepancy. Topological tools are instruments, not theories.

Topological methods have their *terræ incognitæ* among which is the realm of excited states, an El Dorado for photochemistry. Only a few preliminary explorations have been undertaken in this direction although the issues are very challenging. The understanding of excited states lacks chemical explanations and one can expect the heuristic skills of topological methods to enable the emergence of dedicated concepts and dedicated rules.

I hope that these considerations have been an incentive for the use and development of topological methods. I have the last argument, maybe the most convincing one: during the quarter of a century, I have played with these toys, I always had fun and I never got bored. The game is not over.

Douarnenez and Bures sur Yvette, August–November 2020

Bernard Silvi
Emeritus Professor, Sorbonne Université, France

# Index






Á. Martín Pendás and J. Contreras-García*Topological Approaches to the Chemical Bond*, Theoretical Chemistry and Computational Modelling,
https://doi.org/10.1007/978-3-031-13666-5

Printed in the United States
by Baker & Taylor Publisher Services